Tumor-Associated Antigens

Identification, Characterization, and Clinical Applications

Edited by
Olivier Gires and Barbara Seliger

WILEY-VCH Verlag GmbH & Co. KGaA

The Editors

Dr. Olivier Gires
HNO Forschung
Klinikum der Univ.München
Marchioninistr. 15
81377 München

Prof. Dr. Barbara Seliger
MLU Halle-Wittenberg
Institut für Med. Immunologie
Magdeburger Str. 2
06097 Halle

All books published by Wiley-VCH are carefully produced. Nevertheless, authors, editors, and publisher do not warrant the information contained in these books, including this book, to be free of errors. Readers are advised to keep in mind that statements, data, illustrations, procedural details or other items may inadvertently be inaccurate.

Library of Congress Card No.: applied for

British Library Cataloguing-in-Publication Data
A catalogue record for this book is available from the British Library.

Bibliographic information published by the Deutsche Nationalbibliothek
The Deutsche Nationalbibliothek lists this publication in the Deutsche Nationalbibliografie; detailed bibliographic data are available on the Internet at http://dnb.d-nb.de.

© 2009 WILEY-VCH Verlag GmbH & Co. KGaA, Weinheim

All rights reserved (including those of translation into other languages). No part of this book may be reproduced in any form – by photoprinting, microfilm, or any other means – nor transmitted or translated into a machine language without written permission from the publishers. Registered names, trademarks, etc. used in this book, even when not specifically marked as such, are not to be considered unprotected by law.

Typesetting Thomson Digital, Noida, India
Printing betz-druck GmbH, Darmstadt
Binding Litges & Dopf GmbH, Heppenheim
Cover Design Adam-Design, Weinheim

Printed in the Federal Republic of Germany
Printed on acid-free paper

ISBN: 978-3-527-32084-4

Contents

List of Contributors XV

Part One Tumor-associated Antigens (TAAs): Subclasses of TAAs 1

1 T Cell Antigens in Cancer 3
Annette Paschen
1.1 Introduction 3
1.2 Generation of T-cell Epitopes 4
1.2.1 Subclasses of Tumor-associated T-cell Antigens 5
1.2.1.1 Unique Tumor Antigens 6
1.2.1.2 Cancer Testis Antigens 6
1.2.1.3 Differentiation Antigens 7
1.2.1.4 Overexpressed Antigens 7
1.3 Identification of T-cell Antigens and their Epitopes 8
1.3.1 T-cell Antigens for Cancer Immunotherapy – How are Candidates Selected? 10
1.4 Conclusions 12
References 12

2 Human Tumor Antigens as Targets of Immunosurveillance and Candidates for Cancer Vaccines 23
Olivera J. Finn, Robert J. Binder, Anthony G. Brickner, Lisa H. Butterfield, Robert L. Ferris, Pawel Kalinski, Hideho Okada, Walter J. Storkus, Theresa L. Whiteside, and Hassane M. Zarour
2.1 Introduction 23
2.2 Tumor Antigen Classes 24
2.2.1 Oncofetal Antigens 24
2.2.2 Oncogenes as Tumor Antigens 25
2.2.3 Overexpressed Normal Molecules as Tumor Antigens 26
2.2.4 Cancer-Testis (CT) Antigens 26

Tumor-Associated Antigens. Edited by Olivier Gires and Barbara Seliger
Copyright © 2009 WILEY-VCH Verlag GmbH & Co. KGaA, Weinheim
ISBN: 978-3-527-32084-4

2.2.5	Minor Histocompatibility Antigens (mHAgs) as Tumor Antigens 27
2.2.6	Human Melanoma Antigens 28
2.2.7	Human Glioma Antigens and Immunosurveillance in the CNS 29
2.2.8	Heat Shock Proteins, Efficient Carriers of Tumor Antigens 30
2.2.9	Dendritic Cells, Efficient Cross-presenters of Tumor Antigens 31
2.2.10	Spontaneous and Vaccine-induced Immunity and the Tumor Microenvironment 33
2.3	Summary 34
	References 34

Part Two Methods to Detect TAAs 45

3 Humoral Immune Responses against Cancer Antigens: Serological Identification Methods. Part I: SEREX 47
Carsten Zwick, Klaus-Dieter Preuss, Frank Neumann, and Michael Pfreundschuh

3.1	Introduction 47
3.2	The SEREX Approach 48
3.2.1	Identification of Human Tumor Antigens by SEREX 49
3.2.2	Specificity of Human Tumor Antigens 50
3.2.2.1	Shared Tumor Antigens 50
3.2.2.2	Differentiation Antigens 51
3.2.2.3	Antigens Encoded by Mutated Genes 51
3.2.2.4	Viral Genes 52
3.2.2.5	Overexpressed Genes 52
3.2.2.6	Amplified Genes 52
3.2.2.7	Splice Variants of Known Genes 52
3.2.2.8	Cancer-related Autoantigens 52
3.2.2.9	Non-cancer-related Autoantigens 52
3.2.2.10	Products of Underexpressed Genes 53
3.2.3	Significance of Antibodies against SEREX Antigens 53
3.2.4	Reverse T Cell Immunology 53
3.2.5	Functional Significance of Human Tumor Antigens 54
3.2.6	The Human Cancer Immunome 54
3.2.7	Perspectives for Vaccine Development 56
3.3	Conclusions 57
	References 58

4 Humoral Immune Responses against Cancer Antigens: Serological Identification Methods. Part II: Proteomex and AMIDA 63
Barbara Seliger and Olivier Gires

4.1	Introduction 63
4.1.1	A Humoral Response against Self-antigens: The Notion of Tumor-associated Antigens 64

4.2		Implementation of Serum Antibodies: Serological Screening Technologies 65
4.2.1		PROTEOMEX, alias SERPA and SPEARS 66
4.2.1.1		PROTEOMEX Technology and its 'Pros' and 'Cons' 67
4.2.1.2		Candidate Biomarkers Identified by PROTEOMEX 67
4.2.1.3		Implementation of Candidate Biomarkers in the Clinic 68
4.2.2		AMIDA 68
4.2.2.1		Autologous AMIDA 69
4.2.2.2		Allo-AMIDA 69
4.2.3		Advantages and Disadvantages of AMIDA 71
4.3		AMIDA Antigens and Clinical Application 71
4.3.1		Diagnostic TAAs Detected with AMIDA Screening 71
4.3.2		Therapeutic Markers 72
4.4		Conclusions 72
		References 73
5		**cDNA and Microarray-based Technologies** 79
		Ena Wang, Ping Jin, Hui Lui Liu, David F. Stroncek, and Francesco M. Marincola
5.1		Introduction 79
5.2		Technical Aspects 80
5.2.1		Handling of Samples and the Need for Consistent Messenger RNA Amplification 81
5.2.1.1		Collection of Source Material and RNA Isolation 82
5.2.1.2		Single Strand cDNA Synthesis 83
5.2.1.3		Double-stranded cDNA (ds-cDNA) Synthesis 84
5.2.2		RNA Amplifications 85
5.2.2.1		Linear Amplification 85
5.2.2.2		PCR-based Exponential Amplification 86
5.2.2.3		Target Labeling for cDNA Microarray using Amplified RNA 88
5.2.2.4		Bioinformatics Tools 89
5.2.2.5		Limitations of Transcriptional Profiling 90
5.2.2.6		Usefulness of Transcriptional Profiling for Antigen Discovery and the Understanding of Tumor–Host Interactions 90
5.3		Summary 93
		References 93
6		**Detection and Identification of TAA by SELDI-TOF** 103
		Ferdinand von Eggeling and Christian Melle
6.1		Introduction 103
6.2		SELDI (ProteinChip) Technology 104
6.2.1		The Procedural Method 104
6.2.2		SELDI-MS in TAA Identification 106
6.2.2.1		Tumor Samples 106
6.2.2.2		Body Fluids 106

6.2.2.3	SELDI-MS TAAs and Clinical Potential	*107*
6.3	Conclusions and Future Perspectives	*108*
	References	*108*

Part Three TAAs and Their Usefulness *113*

7	**Tumor-associated Antigens in Childhood Cancer** *115*	
	Uta Behrends and Josef Mautner	
7.1	Introduction to Childhood Cancer	*115*
7.1.1	Incidence, Etiology and Types of Cancer in Children	*115*
7.1.2	Cure Rates, Treatment Failure and Toxicity	*115*
7.2	TAA for Pediatric Cancer Therapy	*116*
7.2.1	Potential Clinical Impact of Childhood TAA and Strategies for their Identification	*116*
7.2.2	TAA in Childhood Leukemia	*117*
7.2.3	TAA in Childhood Brain Tumors	*119*
7.2.4	TAA in Childhood Lymphoma	*121*
7.2.5	TAA in Pediatric Neuroblastoma	*122*
7.2.6	TAA in Rhabdomyosarcoma and other Soft Tissue Sarcomas	*124*
7.2.7	TAA in Osteosarcoma	*125*
7.2.8	TAA in Tumors of the Ewing Family	*126*
7.2.9	TAA in Wilms' Tumor	*127*
7.3	Conclusion	*128*
	References	*129*

8	**Epigenetically-regulated Therapeutic Tumor-associated Antigens** *143*	
	Hugues J. M. Nicolay, Luca Sigalotti, Sandra Coral, Elisabetta Fratta, Alessia Covre, Ester Fonsatti, and Michele Maio	
8.1	Introduction	*143*
8.2	Epigenetics	*144*
8.2.1	DNA Methylation	*144*
8.2.2	Histone Post-translational Modifications	*145*
8.3	TAA	*146*
8.3.1	Classification of TAA	*146*
8.3.2	Epigenetically-regulated TAA	*146*
8.3.2.1	HMW-MAA	*146*
8.3.2.2	Mucins	*147*
8.3.2.3	CTA	*148*
8.4	Perspectives and Conclusion	*153*
	References	*153*

9	**Cancer Testis Antigens** *161*	
	Jonathan Cebon, Otavia Caballero, Thomas John, and Oliver Klein	
9.1	Introduction	*161*

9.2	Definitions and Classification 162
9.3	Tissue Distribution 163
9.3.1	Normal tissues 163
9.3.2	Tumors 163
9.4	Function 164
9.5	Immunology 166
9.6	Clinical Trials 167
9.7	Conclusions 172
	References 173

10	**Rationale for Treatment of Colorectal Cancer with EpCAM Targeting Therapeutics** 179
	Patrick A. Baeuerle and Gert Riethmueller
10.1	Introduction 179
10.2	EpCAM Expression in Colorectal Cancer 181
10.2.1	Frequent and High Level Expression of EpCAM in Primary Tumors of Colon Cancer 181
10.2.2	EpCAM Expression on Colon Cancer Metastases 183
10.2.3	EpCAM Expression on Circulating Colon Cancer Cells 184
10.2.4	EpCAM Expression on Colon Cancer Stem Cells 184
10.2.5	EpCAM Expression on Human Colon Cancer Cell Lines 185
10.3	EpCAM-directed Therapeutic Approaches 185
10.3.1	Clinical Results with Anti-EpCAM Murine Antibody Edrecolomab in Stage II and III CRC 185
10.3.2	Clinical Results with Edrecolomab as a Vaccine for Induction of Anti-idiotypic Response 188
10.3.3	EpCAM Protein as a Vaccine in Colorectal Cancer 189
10.3.4	Adecatumumab, a Novel Fully Human Anti-EpCAM Antibody 190
10.4	Therapeutic Window of EpCAM-directed Therapies 192
10.5	Conclusions 196
	References 198

11	**Carcinoembryonic Antigen** 201
	Wolfgang Zimmermann and Robert Kammerer
11.1	CEA Biology 201
11.1.1	CEA Gene Family, Genomic Localization, Protein Structure 201
11.1.2	Evolution 203
11.1.3	Expression 203
11.1.4	Biological Functions 204
11.2	Clinical Relevance of CEA 206
11.2.1	CEA as a Tumor Marker for Prognosis and Post-surgery Follow-up 206
11.2.2	Targeting CEA for Tumor Localization and Therapy 207
11.2.2.1	Animal Models for CEA 207
11.2.2.2	Tumor Localization and Therapy with Anti-CEA Antibodies 207

11.2.2.3	CEA-based Vaccines	209
11.2.3	Immune Monitoring	211
11.3	Conclusion	213
	References	213

12 HER-2 as a Tumor Antigen 219
Barbara Seliger
- 12.1 Introduction 219
- 12.2 Biology of HER2 219
- 12.2.1 Features of HER2 219
- 12.2.2 The Self-antigen HER2 as a Potential Therapeutic Target 220
- 12.2.3 Determination of HER2 Status in Tumors 222
- 12.3 General Approaches for Targeting HER2 using Anti-cancer Agents 222
- 12.4 Active HER2-based Cancer Vaccines 222
- 12.4.1 HER2-specific Peptides 223
- 12.4.2 T Cell-mediated Immunity to HER2 in Cancer Patients 224
- 12.5 Passive Immunotherapy Targeting HER2 224
- 12.6 HER2 Effects on Immunogenicity: Both Sides of the Coin 225
- 12.7 Conclusions 226
- References 227

13 Epstein-Barr Virus-associated Antigens 231
Christoph Mancao and Wolfgang Hammerschmidt
- 13.1 Introduction 231
- 13.2 Functions of EBV Antigens in Latently Infected Cells 232
- 13.2.1 EBV Nuclear Antigen 1 (EBNA1) 233
- 13.2.2 EBV Nuclear Antigen 2 and Leader Protein (EBNA2 and EBNA-LP) 233
- 13.2.3 EBV Nuclear Antigen Family 3 (EBNA3) 234
- 13.2.4 Latent Membrane Protein 1 (LMP1) 234
- 13.2.5 Latent Membrane Protein 2 (LMP2) 235
- 13.3 T-cell Responses to EBV Antigens 236
- 13.3.1 $CD8^+$ T-cell Responses 236
- 13.3.2 $CD4^+$ T-cell Response 237
- 13.4 EBV-associated Malignancies 237
- 13.4.1 Post-Transplant Lymphoproliferative Disease (PTLD) 237
- 13.4.2 Adoptive T-cell Therapy 238
- 13.4.3 Burkitt's Lymphoma and Hodgkin's Lymphoma 240
- References 241

14 Human Papillomavirus (HPV) Tumor-associated Antigens 249
Andreas M. Kaufmann
- 14.1 Introduction 249
- 14.2 HPV-encoded TAA 250
- 14.2.1 The Proteins of HPV 250
- 14.2.2 Vaccine Development 253

14.2.3	Therapeutic Vaccine Strategies 253
14.2.4	HPV-induced Cellular TAA 256
14.3	Conclusions and Future Perspectives 256
	References 257

15 Circulating TAAs: Biomarkers for Cancer Diagnosis, CA125 261
Angela Coliva, Ettore Seregni, and Emilio Bombardieri

15.1	Introduction 261
15.2	Definition and Classification of Mucins 261
15.3	Structure of MUC 16 263
15.3.1	Gene and Protein Structure 263
15.3.1.1	Domain I 263
15.3.1.2	Domain II 263
15.3.1.3	Domain III 265
15.3.2	Evolutionary Considerations 266
15.4	Distribution of MUC16 266
15.5	MUC16 Function in Normal Tissues 266
15.6	Release of MUC16: CA125 Tumor Marker 267
15.6.1	History 267
15.6.2	Assay Methods 268
15.6.3	Serum CA125 Levels in Healthy Subjects 269
15.6.3.1	Age 269
15.6.3.2	Menstrual cycle 270
15.6.3.3	Pregnancy 270
15.6.3.4	Race 270
15.6.3.5	Other Factors that may affect CA125 Levels 270
15.6.4	Serum CA125 Levels in Patients with Benign Diseases 270
15.6.5	Serum CA125 Levels in Non-ovarian Cancer 270
15.6.6	Serum CA125 Levels in Ovarian Cancers 270
15.7	Clinical Applications 271
15.7.1	Screening 271
15.7.2	Evaluating Pelvic Masses 272
15.7.3	Assessing Prognosis 272
15.7.4	Assessing Response to Therapy 273
15.7.5	Follow-up after Completion of Initial Therapy 274
15.8	Novel Markers 274
15.9	Conclusions 276
	References 276

Part Four Clinical Applications of TAAs 281

16 Overview of Cancer Vaccines 283
John Copier and Angus Dalgleish

| 16.1 | Introduction 283 |

16.2	Cellular Immune Responses in Tumor Rejection	283
16.3	Selection of TAAs for the Development of Cancer Vaccines	284
16.4	Types of Cancer Vaccine	285
16.4.1	Defined Antigen Approaches	285
16.4.1.1	Peptides	285
16.4.1.2	Recombinant Protein	286
16.4.1.3	DNA	287
16.4.1.4	Viral Vectors	288
16.4.2	Undefined Antigens	289
16.4.2.1	Autologous Tumor Cell Vaccines	289
16.4.2.2	Allogeneic Tumor Cell Vaccines	290
16.4.2.3	Loading DCs with Tumor Preparations	290
16.5	Immune Biomarkers	292
16.6	Problems with the Assessment of Clinical Trials for Cancer Vaccines	292
16.6.1	Biomarkers	294
16.6.2	Schedule and Protocol	294
16.7	Combined Therapies; Synergy with Conventional Approaches	294
16.8	Conclusion	295
	References	296

17 Tumor-associated Antigenic Peptides as Vaccine Candidates 303
Dhiraj Hans, Paul R. Young, and David P. Fairlie

17.1	Introduction	303
17.2	Different Types of Antigens	304
17.2.1	Mutation Antigens	304
17.2.2	Shared Tumor-specific Antigens	306
17.2.3	Differentiation Antigens	308
17.2.4	Over-Expressed Antigens	309
17.2.5	Viral Antigens	311
17.3	Conclusions	312
	References	313

18 DNA Vaccines for the Human Papilloma Virus 317
Barbara Ma, Chien-Fu Hung, and T.-C. Wu

18.1	Introduction	317
18.1.1	HPV and Cervical Cancer	317
18.1.2	Events in the Progression from HPV Infection to Cervical Cancer	318
18.1.3	HPV Antigens for Vaccine Development	318
18.2	DNA Vaccines for HPV	319
18.2.1	Preventive HPV DNA Vaccines	319
18.2.2	Therapeutic HPV DNA Vaccines	321
18.2.2.1	Importance of DCs in Enhancing DNA Vaccine Potency	322
18.2.2.2	Clinical Progress in HPV DNA Vaccine Development	329

18.2.3	Combined Approaches	*331*
18.2.3.1	Combined Preventive and Therapeutic HPV DNA Vaccines	*331*
18.2.3.2	HPV DNA Vaccines in Combination with Other Therapies	*331*
18.3	Conclusion	*332*
	References	*332*

19 Adoptive T-cell Transfer in Cancer Treatment *339*
Andreas Moosmann and Angela Krackhardt

19.1	Introduction	*339*
19.2	T Cell-based Therapies	*339*
19.2.1	Therapy of Virus-associated Cancers with Antigen-specific T Cells	*339*
19.2.2	T Cells with Specificity for Self-antigens	*342*
19.2.3	T Cells with Specificity for Alloantigens	*343*
19.2.4	Engineered T Cells	*344*
19.2.5	Safety Concerns and Suicide Gene Transfer	*345*
19.3	Conclusion	*346*
	References	*346*

Index *355*

List of Contributors

Patrick A. Baeuerle
Micromet Inc.
Staffelseestr. 2
81477 München
Germany

Uta Behrends
Munich University of Technology
Department of Pediatrics
Kölner Platz 1
80804 München
Germany

and

Institute of Clinical Molecular Biology
and Tumour Genetics
Helmholtz Centre
Marchioninistr. 25
81377 Munich
Germany

Robert J. Binder
University of Pittsburgh Cancer
Institute
Cancer Immunology, Immunotherapy
and Immunoprevention Program
Pittsburgh
PA 15261
USA

Emilio Bombardieri
Fondazione IRCCS
'Istituto Nazionale dei Tumori'
Nuclear Medicine Department
Via Venezian, 1
20133-Milano (I)
Italy

Anthony G. Brickner
University of Pittsburgh Cancer
Institute
Cancer Immunology, Immunotherapy
and Immunoprevention Program
Pittsburgh
PA 15261
USA

Lisa H. Butterfield
University of Pittsburgh Cancer
Institute
Cancer Immunology, Immunotherapy
and Immunoprevention Program
Pittsburgh
PA 15261
USA

Otavia Caballero
Ludwig Institute for Cancer Research
New York Branch of Human Cancer Immunology
Memorial Sloan-Kettering Cancer Center
1275 York Avenue, BOX 32
New York, NY 10021–6007 USA

Jonathan Cebon
Ludwig Institute for Cancer Research
Melbourne Centre for Clinical Sciences
Austin Hospital
Studley Road, Heidelberg VIC 3084
Australia

Angela Coliva
Fondazione IRCCS
Istituto Nazionale dei Tumori
Nuclear Medicine Department
Via Venezian, 1
20133-Milano (I)
Italy

John Copier
St George's University of London
Department of Cellular and Molecular Medicine
Cranmer Terrace
London SW17 0RE
UK

Sandra Coral
Istituto di Ricovero e Cura a Carattere Scientifico
Centro di Riferimento Oncologico
Cancer Bioimmunotherapy Unit
Department of Medical Oncology
33081 Aviano
Italy

Alessia Covre
Istituto di Ricovero e Cura a Carattere Scientifico
Centro di Riferimento Oncologico
Cancer Bioimmunotherapy Unit
Department of Medical Oncology
33081 Aviano
Italy

Angus Dalgleish
St George's University of London
Department of Cellular and Molecular Medicine
Cranmer Terrace
London SW17 0RE
UK

Ferdinand von Eggeling
Universitätsklinikum Jena
Institut für Humangenetik und Anthropologie
Core Unit Chip Application (CUCA)
Leutragraben 3
07743 Jena
Germany

David P Fairlie
University of Queensland
Centre for Drug Design and Development
Institute for Molecular Bioscience
Brisbane
Queensland 4072
Australia

Robert L. Ferris
University of Pittsburgh Cancer Institute
Cancer Immunology, Immunotherapy and Immunoprevention Program
Pittsburgh
PA 15261
USA

Olivera J. Finn
University of Pittsburgh Cancer
Institute
Cancer Immunology, Immunotherapy
and Immunoprevention Program
Pittsburgh
PA 15261
USA

Ester Fonsatti
University Hospital of Siena
Istituto Toscano Tumori
Division of medical Oncology and
Immunotherapy
Department of Oncology
53100 Siena
Italy

Elisabetta Fratta
Istituto di Ricovero e Cura a Carattere
Scientifico
Centro di Riferimento Oncologico
Cancer Bioimmunotherapy Unit
Department of Medical Oncology
33081 Aviano
Italy

Olivier Gires
LMU Munich
CCG Molecular Oncology
Head and Neck Research Department
Marchioninistr. 15
81377 Munich

and

GSF-National Research Center for
Environment and Health
Marchioninistr. 25
81377 Munich
Germany

Wolfgang Hammerschmidt
German Research Center for
Environmental Health
Helmholtz Center Munich
Department of Gene Vectors
Marchioninistrasse 25
81377 Munich
Germany

Dhiraj Hans
University of Queensland
Centre for Drug Design and
Development
Institute for Molecular Bioscience
Brisbane
Queensland 4072
Australia

Chien-Fu Hung
The John Hopkins Medical Institutions
Departments of Pathology, Obstetrics
and Gynecology, and Oncology
Baltimore
MD 21231
USA

Ping Jin
National Institutes of Health
Infectious Disease and
Immunogenetics Section
Cell Processing Laboratory
Department of Transfusion Medicine
Clinical Center
R-1C711, Bldg 10
9000 Rockville Pike
Bethesda
MD 20892
USA

Thomas John
Ludwig Institute for Cancer Research
Melbourne Centre for Clinical Sciences
Austin Hospital
Studley Road
Heidelberg
VIC 3084
Australia

Pawel Kalinski
University of Pittsburgh Cancer Institute
Cancer Immunology, Immunotherapy and Immunoprevention Program
Pittsburgh
PA 15261
USA

Robert Kammerer
Friedrich-Loeffler-Institute (FLI)
Institute of Immunology
Paul-Ehrlich- Str. 28
72076 Tübingen
Germany

Andreas M. Kaufmann
Charité-Universitätsmedizin Berlin
Gynäkologie
Gynäkologische Tumorimmunologie
Campus Benjamin Franklin und Campus Mitte
Hindenburgdamm 30
12200 Berlin
Germany

Oliver Klein
Ludwig Institute for Cancer Research
Melbourne Centre for Clinical Sciences
Austin Hospital
Studley Road, Heidelberg VIC 3084
Australia

Angela Krackhardt
Institute of Molecular Immunology
Helmholtz-Zentrum München
German Research Center for Environmental Health
Marchioninistr. 25
81377 München
Germany

Hui Lui Liu
National Institutes of Health
Infectious Disease and Immunogenetics Section (IDIS)
Department of Transfusion Medicine
Clinical Center
R-1C711, Bldg 10
9000 Rockville Pike
Bethesda
MD 20892
USA

Barbara Ma
The John Hopkins Medical Institutions
Department of Pathology
Baltimore
MD 21231
USA

Michele Maio
University Hospital of Siena
Division of Medical Oncology and Immunotherapy
Department of Oncology
Strada delle Scotte 14
53100 Siena
Italy

and

Istituto di Ricovero e Cura a Carattere Scientifico
Centro di Riferimento Oncologico
Cancer Bioimmunotherapy Unit
Department of Medical Oncology
33081 Aviano
Italy

Christoph Mancao
German Research Center for
Environmental Health
Helmholtz Center Munich
Department of Gene Vectors
Marchioninistrasse 25
81377 Munich
Germany

Francesco M. Marincola
National Institutes of Health
Infections Disease and
Immunogenetics Section (IDIS)
Department of Transfusion Medicine
Clinical Center
R-1C711, Bldg 10
9000 Rockville Pike
Bethesda
MD 20892
USA

Josef Mautner
Munich University of Technology
Department of Pediatrics
Kölner Platz 1
80804 Munich
Germany

and

Institute of Clinical Molecular Biology
and Tumour Genetics
Helmholtz Centre Munich
Marchioninistr. 25
81377 Munich
Germany

Christian Melle
Universitätsklinikum Jena
Institut für Humangenetik und
Anthropologie
Core Unit Chip Application (CUCA)
Leutragraben 3
07743 Jena
Germany

Andreas Moosmann
Ludwig-Maximilians-Universität,
Munich
Department of Otorhinolaryngology
Clinical Cooperation Group Molecular
Oncology
Marchioninistr. 25
81377 München
Germany

Frank Neumann
Saarland University Medical School
Med. Klinik und Poliklinik, Innere
Medizin I
66421 Homburg
Germany

Hugues J. M. Nicolay
University Hospital of Siena
Istituto Toscano Tumori
Division of Medical Oncology and
Immunotherapy
Department of Oncology
53100 Siena
Italy

and

Istituto di Ricovero e Cura a Carattere
Scientifico
Centro di Riferimento Oncologico
Cancer Bioimmunotherapy Unit
Department of Medical Oncology
33801 Aviano
Italy

Hideho Okada
University of Pittsburgh Cancer
Institute
Cancer Immunology, Immunotherapy
and Immunoprevention Program
Pittsburgh
PA 15261
USA

Annette Paschen
University Medical School Mannheim
Clinical Cooperation Unit Dermato-
Oncology of the German Cancer
Research Center
Theodor Kutzer Ufer 1
68135 Mannheim
Germany

Michael Pfreundschuh
Saarland University Medical School
Med. Klinik und Poliklinik, Innere
Medizin I
66421 Homburg
Germany

Klaus-Dieter Preuss
Saarland University Medical School
Med. Klinik und Poliklinik, Innere
Medizin I
66421 Homburg
Germany

Gert Riethmueller
Ludwig-Maximilians-University
Institute for Immunology
Goethestr. 31
83333 Munich
Germany

Barbara Seliger
Martin-Luther-Universität Halle-
Wittenberg
Klinikum der Medizinischen Fakultät
Institut für Medizinische Immunologie
Magdeburgerstr. 2
06097 Halle (Saale)
Germany

Ettore Seregni
Fondazione IRCCS
'Istituto Nazionale dei Tumori'
Nuclear Medicine Department
Via Venezian, 1
20133-Milano (I)
Italy

Luca Sigalotti
Istituto di Ricovero e Cura a Carattere
Scientifico
Centro di Riferimento Oncologico
Cancer Bioimmunotherapy Unit
Department of Medical Oncology
33081 Aviano
Italy

Walter J. Storkus
University of Pittsburgh Cancer
Institute
Cancer Immunology, Immunotherapy
and Immunoprevention Program
Pittsburgh
PA 15261
USA

Dourid F. Stroncek
National Institutes of Health
Cell Processing Laboratory
Department of Transfusion Medicine
Clinical Center
R-1C711, Bldg 10
9000 Rockville Pike
Bethesda
MD 20892

Ena Wang
National Institutes of Health
Infectious Disease and
Immunogenetics Section (DIS)
Department of Transfusion Medicine
Clinical Center
R-1C711, Bldg 10
9000 Rockville Pike
Bethesda
MD 20892
USA

Theresa L. Whiteside
University of Pittsburgh Cancer
Institute
Cancer Immunology, Immunotherapy
and Immunoprevention Program
Pittsburgh
PA 15261
USA

T-C. Wu
The John Hopkins Medical Institutions
Departments of Pathology, Obstetrics
and Gynecology, Oncology, and
Molecular
Microbiology and Immunology
Cancer Research Building II,
Room 309
1550 Orleans Street
Baltimore
MD 21231
USA

Paul R. Young
University of Queensland
School of Molecular and Microbial
Sciences
Brisbane
Queensland 4072
Australia

Hassane M. Zarour
University of Pittsburgh Cancer
Institute
Cancer Immunology, Immunotherapy
and Immunoprevention Program
Pittsburgh
PA 15261
USA

Wolfgang Zimmermann
Ludwig-Maximilian University
Tumour Immunology Laboratory
LIFE Center
University Clinic Grosshadern
81377 Munich
Germany

Carsten Zwick
Saarland University Medical School
Med. Klinik und Poliklinik, Innere
Medizin I
66421 Homburg
Germany

Part One
Tumor-associated Antigens (TAAs): Subclasses of TAAs

1
T Cell Antigens in Cancer
Annette Paschen

1.1
Introduction

The immune system has the ability to discriminate between normal and malignant cells. Studies in different animal tumor model systems demonstrate that innate and adaptive immune cells cooperate to eliminate cancer cells [1]. Clinical observations such as the increased rate of tumor formation in immune-compromised individuals and the spontaneous, though rare, regressions of tumors also indicate the presence of anti-tumor activity in the human immune system [2, 3]. Of the different cellular effectors involved in anti-tumor immunity, cytotoxic $CD8^+$ T lymphocytes (CTLs) are of particular interest due to their ability to specifically and effectively kill autologous tumor cells [4–7], leading to increased efforts by scientists and clinicians to also exploit this anti-tumor potential for cancer immunotherapy.

How do cytotoxic $CD8^+$ T lymphocytes distinguish between normal and malignant cells? Initially, T lymphocytes screen target cells for their protein composition. During the continual turnover of cellular proteins (antigens) small peptide fragments are generated, which are sampled and exposed on the cell surface by major histocompatibility complex (MHC) class I molecules. CTLs monitor these degradation products (epitopes) using their T cell receptors (TCRs). Due to mutational events and epigenetic alterations, tumor cells differ from normal cells in their protein composition and degradation products. The emerging aberrant antigen epitope repertoire presented on MHC class I molecules can be recognized by autologous CTLs leading to the specific killing of the tumor cells [4–7].

Before acquiring cytotoxic effector function, naïve $CD8^+$ T lymphocytes must be primarily activated (T cell priming). A few studies in mouse tumor models demonstrated that malignant cells upon migration to peripheral lymphoid organs have the ability to directly prime antigen-specific $CD8^+$ T cells [8]. More frequently, T cell priming has been described to be dependent on the activity of a specific cellular mediator, the dendritic cell (DC). Briefly, DCs internalize tumor antigens in the periphery, migrate to the draining lymph nodes and present the processed antigens to the resident naïve T cell repertoire. In addition to this antigen stimulus, DCs provide accessory signals (e.g. cytokines, co-stimulatory molecules) known to be of importance for the effective priming of $CD8^+$ T cells [9].

Fragments of the sampled and processed antigens are not only exposed on MHC class I, but also on MHC class II molecules of the DCs, the latter being recognized by $CD4^+$ T cells. Upon binding to specific peptide–MHC class II complexes $CD4^+$ T cells become activated and express the surface molecule CD40 ligand (CD40L). Interaction of CD40L with CD40 on DCs strongly enhances their $CD8^+$ T cell priming capacity [10–12]. In addition to primary T cell activation, helper $CD4^+$ T cells are also required for the maintenance of antigen-specific $CD8^+$ T cell responses as well as for the activation of antigen-specific B cells and subsequent antibody production [13–15]. Consequently, therapeutically effective anti-tumor immunity may be dependent on the activation of both antigen-specific cytotoxic $CD8^+$ T cells and helper $CD4^+$ T cells.

1.2
Generation of T-cell Epitopes

MHC class I molecules are composed of a β_2-microglobulin subunit that non-covalently associates with the polymorphic heavy chain containing the peptide binding groove. Several hundreds of human heavy chain alleles are known (listed at http//www.anthonynolan.com/HIG/index.html), albeit that specific alleles are preferentially expressed in certain population groups (e.g. the MHC class I molecule HLA-A*0201 is present in about 50% of Caucasians). Each individual can express a maximum of six different alleles, and the peptide binding motifs have been identified in some cases. The peptide ligands in general consist of 9 to 10 amino acids (aa), though exceptionally long variants have also been described [16, 17]. Binding of the peptide ligands is a prerequisite for stable surface expression of MHC class I molecules.

The major source of MHC class I peptides are endogenous proteins degraded by the proteasome in the cytosol. The proteasome is a multi-catalytic complex composed of several subunits. The three subunits $\beta1$, $\beta2$, and $\beta5$, mediate its standard catalytic activity. In response to IFN-γ, this subunit composition is changed, leading to the formation of the so-called immunoproteasome containing the catalytic subunits $\beta1i$ (LMP2), $\beta2i$ (MECL-1), and $\beta5i$ (LMP7). Standard and immunoproteasome are characterized by different cleavage preferences [18]. It has been demonstrated that some tumor antigen epitopes are dependent on the activity of the standard proteasome in order to be efficiently generated whereas others require the immunoproteasome [19, 20]. The proteasome generates the correct C-terminus of the majority of MHC class I ligands, albeit that a few exceptions have been described [21, 22]. Recent evidence has established a peptide splicing property of the proteasome which leads to the generation of a $CD8^+$ T cell epitope from two discontinuous fragments of a long precursor peptide. By splicing peptides, the proteasome generates additional diversity to the pool of antigenic peptides *in vivo* [23].

Peptides generated in the cytosol are translocated via the transporter associated with antigen processing (TAP) into the endoplasmic reticulum (ER) where loading onto MHC class I molecules takes place. In principle, the proteasome can generate peptides of the correct size that directly fit into the groove of MHC class I molecules. However, the majority of peptides are produced as N-terminal-extended precursors that require additional processing by cytosolic and/or ER-localized peptidases before binding to the MHC class I groove can occur [24].

MHC class I peptides are not exclusively generated from endogenous proteins, but can also be derived from endocytosed exogenous antigens via different intracellular pathways, collectively known as cross-presentation pathways [9]. Until recently, cross-presentation was described as a specific feature of professional antigen presenting cells (pAPCs), such as macrophages and especially DCs. Interestingly, Godefroy et al. demonstrated that CTLs from a melanoma patient recognized a cross-presented epitope on autologous tumor cells generated from an exogenous protein after receptor-mediated internalization [25]. However, the relative contribution of this pathway to the overall generation of MHC class I peptides by tumor cells remains to be elucidated.

Unlike MHC class I molecules, MHC class II $\alpha\beta$ heterodimers present ligands of 12 to 26 amino acids in length. Constitutive MHC class II presentation is restricted to a few cell types, mainly pAPCs as B cells, macrophages and DCs characterized by high endocytic activity. Exogenous proteins internalized via endocytosis (phagocytosis, receptor-mediated endocytosis, pinocytosis) are a major source of the peptides presented by MHC class II molecules. Unfolding and proteolysis of the internalized antigens as well as loading of the fragments onto MHC class II molecules occur in the late endosomes and lysosomes of the cell [26]. However, MHC class II molecules can also present peptides derived from endogenous proteins originating from different cellular compartments including the cytosol. These antigens access the endosome/lysosome by different pathways. For instance cytoplasmic proteins and organelles can be engulfed by autophagosomes, which then fuse with lysosomes for protein degradation and MHC class II loading [26]. Although constitutive surface exposure of MHC class II molecules is normally restricted to pAPCs, some tumors show *in situ* MHC class II expression. In the case of melanoma this expression is associated with prolonged patient survival [27].

1.2.1
Subclasses of Tumor-associated T-cell Antigens

A report characterizing the first human tumor antigen recognized by CTLs from a melanoma patient and designated MAGE-1 was published in 1991, while the first $CD8^+$ T cell epitope was described in 1992 [4, 28]. In the following years data concerning human tumor-associated T cell antigens and their epitopes continuously increased (detailed information available at http://www.cancerimmunity.org/links/databases.htm). Although the antigens are heterogenous in nature, they can be categorized by their expression pattern as unique antigens, cancer testis antigens, differentiation antigens, and overexpressed antigens.

1.2.1.1 Unique Tumor Antigens

Altered proteins originating from gene mutations and fusion proteins arising from chromosomal aberrations in tumor cells are the source of neo-antigens which are recognized by $CD4^+$ and $CD8^+$ T cells [29–42]. Since the accumulation of genetic alterations is a hallmark of cancer, it can be assumed that each tumor expresses multiple of these unique antigens, which are truly tumor-specific and accordingly not present in any normal tissue. Consequently, tolerance mechanisms acting on all self-reactive T cells do not affect the unique antigen-specific T-cell repertoire, i.e. high affinity TCRs specifically recognizing unique tumor antigens are neither deleted in the thymus during lymphocyte maturation nor deleted or anergized in peripheral lymphoid and non-lymphoid tissues [43–47].

For the majority of unique antigens described so far, expression is restricted to the tumor cells of the patients from whom they have been isolated [33, 34, 37, 38], whereas only a few of these antigens are shared by different tumors of the same histology. The shared expression of these antigens, such as the mutated BRAF, K-RAS and CDKN2A, is attributable to their biological function, which is known to be of importance for tumor formation and maintenance indicating that high antigen stability is advantageous for the tumor [31, 35, 36]. In addition to mutations, the activity of cryptic promoters or the partial and thus incomplete splicing of RNA in tumor cells can provide another source of 'shared unique' antigens [32, 48].

With respect to the tumor-specific expression and the recognition by high affinity T cells, 'shared unique' T cell antigens fulfill the criteria of ideal target structures for cancer immunotherapy. However, their therapeutic targeting would still be limited to a very small subset of patients due to the peptide–MHC restriction. For instance, a mutated antigen differing from the wild-type protein by a single amino acid might only give rise to a single neo-epitope presented by only one MHC allele. Researchers might overcome this limitation within the next years by screening cancer cells for shared genetic alterations [49]. This will lead to the prediction of new potential antigens, whose presentation by the tumors and recognition by T cells has then to be validated.

In the case of some malignancies, an association with viral infections has been described. Antigens expressed by the virus provide another source of unique tumor antigens. For instance, infections of B cells with Epstein Barr Virus (EBV) can result in the formation of B-cell malignancies. Such tumor cells can be recognized by T cells directed against EBV antigens [50, 51]. Similarly, infection of cervix epithelial cells with human papilloma virus (HPV) can induce the outgrowth of cervical carcinoma. The viral oncoproteins, E6 and E7 expressed in tumor cells, function as T cell antigens [52, 53].

1.2.1.2 Cancer Testis Antigens

In addition to mutations, cancer cells are characterized by epigenetic alterations that induce expression of otherwise silenced genes. For instance, demethylation which occurs in many tumor cells, elicits the transcription of a specific set of genes whose products belong to the group of cancer testis antigens (CTAs) [54]. These antigens are detectable in many tumors of different histology, but not in normal tissues except for MHC-negative testicular germ cells and placental trophoblasts. Due to this highly restricted expression pattern CTAs might also be classified as tumor-specific, comparable to unique tumor antigens. However, medullar thymic epithelium cells have been demonstrated to express the RNA of different CTA. Thus, the possibility that the CTA-specific TCR repertoire is affected by thymic deletion cannot be completely excluded [43, 44, 55].

The CTA group encompasses several antigen families including the well-known family of MAGE antigens. So far, this family has been found to consist of nine antigens, which are known to be recognized by specific $CD4^+$ and $CD8^+$ T cells [4, 20, 28, 56–60]. Also, members of the SSX antigen family have been demonstrated to be targets of T helper cells and CTL responses [61–63]. Another well-characterized CTA is NY-ESO-1. This antigen was initially defined as an antibody target structure recognized by sera from cancer patients suffering from a variety of malignancies and was subsequently verified as an antigen that could be detected by $CD4^+$ and $CD8^+$ T cells [64–67]. Based on their highly specific and broad expression profile, CTAs seem to be well suited to cancer therapy and thus have been targeted in many different clinical trials on cancer immunotherapy [68].

1.2.1.3 Differentiation Antigens

These antigens are expressed in malignant and normal cells of the same lineage. Thus, the expression of melanoma differentiation antigens (MDAs) is also detectable in normal melanocytes. Due to this expression pattern, central as well as peripheral tolerance mechanisms act vigorously on antigen-specific T cells [43–47, 55, 69] suggesting that the remaining specific T-cell repertoire is of low affinity. However, there is evidence that even for these antigens self-tolerance is incomplete. MDA-specific T cells of high affinity have been isolated from the blood and tumors of melanoma patients and the adoptive transfer of *ex vivo* expanded, activated autologous T cells into patients mediated regression of metastatic tumors [70–73]. Furthermore, in *vitiligo* (= hypopigmentary skin disorder) MDA-specific CTLs are involved in the auto-immune destruction of normal skin melanocytes indicating a breakdown of natural tolerance. T cell-mediated local elimination of melanocytes can also sometimes be observed in melanoma patients [74, 75]. Several MDAs such as Melan-A/MART-1, gp100, tyrosinase, tyrosinase-related protein 1 (TRP1) and TRP2 have been described as specific targets of T helper cells and CTLs [5–7, 76–79]. In addition to melanoma, T cell differentiation antigens are known to be present in other malignancies, for example CEA in cases of gut carcinoma and PSA in prostate carcinoma [80, 81]. Targeting differentiation antigens in clinical trials, as has frequently been the case, is associated with the risk of inducing autoimmunity. In cases where the normal cells (e.g. melanocytes) are dispensable autoimmune toxicity is tolerable.

1.2.1.4 Overexpressed Antigens

In comparison to normal cells, tumor cells downregulate the expression of some gene products, whereas others are strongly overexpressed. Peptides derived from several overexpressed proteins are recognized by specific $CD4^+$ and $CD8^+$ T lymphocytes [16, 25, 82–89]. The distribution of these antigens in normal tissues is heterogenous, with some antigens being expressed in a few normal tissues, while others can be detected ubiquitously. T cells responding to these antigens ignore normal cells *in vitro*, most probably due to the very low level of antigen expression. Again it can be assumed that central and peripheral mechanisms act vigorously on the antigen-specific T cell repertoire in order to maintain self-tolerance [43–47]. This tolerance may be circumvented in the case of several hundred-fold overexpression and the subsequent increase in the presentation of a given antigen [90].

Overexpressed antigens such as Her2/neu, MUC1, PRAME, survivin, and telomerase not only exhibit a heterogenous distribution pattern in normal tissues, but also show a heterogenous pattern in terms of their biological function [82–89]. In some cases this function is known to be important for the survival of cancer cells (e.g. the anti-apoptotic activity of survivin), suggesting highly stable expression of the antigen in tumor cells.

Within the group of overexpressed antigens the cell surface antigen, mucin MUC1 has some specific features. The glycosylation pattern of the MUC1 protein expressed on tumor cells and normal cells is different. Interestingly, helper $CD4^+$ T cells are capable of recognizing glycosylated MUC1 peptide epitopes and of discriminating different glycosylation patterns, thereby distinguishing between modified peptides originating from cancer cells and normal cells [69, 91]. There is accumulating evidence that peptide epitopes carrying different types of post-translational modifications can be recognized specifically by $CD4^+$ and $CD8^+$ T cells [92]. Very recently, phosphopeptides have been eluted from MHC class I molecules of tumor cells. These modified peptides, the majority of which are derived from the aberrant phosphorylation of signaling proteins, can be recognized by $CD8^+$ T cells and might represent new targets for cancer immunotherapy [93].

Do patients' T cells preferentially recognize antigens from any of the four subclasses described above? Several studies have demonstrated that T lymphocytes isolated from one cancer patient respond to multiple tumor antigens of different subclasses [37, 94, 95]. Lennerz *et al.* carried out a comprehensive analysis on the specificity of T cells obtained from the peripheral blood of a melanoma patient at different time points during disease progression and demonstrated that in this individual the predominant anti-tumor immune response was due to T cells responding to unique antigens [37]. In contrast, by analyzing the T-cell infiltrate of a regressing melanoma metastasis Coulie and colleagues demonstrated that the different established T cell clones specifically recognized CTAs and MDAs [95].

1.3
Identification of T-cell Antigens and their Epitopes

In order to identify tumor-associated T-cell antigens/epitopes, researchers followed very different strategies. The major approaches, of which there are several variants, are presented in subsequent chapters in this book (for detailed information see Chapters 3 and 4 on SEREX, and Proteomex and AMIDA respectively).

1.3 Identification of T-cell Antigens and their Epitopes

The classical strategy employed to define antigens/epitopes is based on the use of tumor cell lines and autologous T cells as screening tools, thus restricting its application to only a few malignancies such as melanoma. Initially, T cells isolated from the tumor or the peripheral blood of the cancer patient are co-cultured with autologous tumor cells for epitope sensitization and induction of proliferation. Subsequently, stimulated T cells are used to screen HLA-matched target cells transfected with the tumor cDNA expression library for the presence of the antigen. Specific killing of the transfectants indicates synthesis and processing of the tumor antigen. The antigen-specific T cells are then co-incubated with target cells expressing only truncated antigen fragments in order to determine the epitope-containing region. Subsequently, overlapping synthetic peptides covering the antigen fragment of interest are loaded onto the target cell in order to define the minimal T cell epitope. Several members of all four subclasses of antigens were characterized using this strategy, which is still the major approach for defining patient-specific unique $CD4^+$ and $CD8^+$ T cell antigens/epitopes [4–7, 16, 25, 28–30, 32–35, 37–39].

Alternatively, peptides eluted from MHC molecules and subsequently characterized by liquid chromatography in combination with mass spectroscopy, can be employed to define potential T-cell epitopes and their corresponding antigen sources. In several studies peptides were eluted from MHC class I molecules of tumor cells [58, 93, 96], and in another approach ligands were isolated from MHC class II molecules of dendritic cells pulsed with tumor lysate [97]. In any case, the immunogenicity of the identified MHC ligands has to be proven by sensitizing (priming) T cells against the corresponding synthetic peptides *in vitro* and the expression profile of the corresponding antigen source in normal and malignant cells has to be determined.

More recently, researchers exploited antibodies from the sera of cancer patients as a tool for the identification of potential B and T cell antigens. According to the SEREX technology antibodies are used to screen tumor-derived cDNA expression libraries for target proteins. These proteins should also be recognized by helper T cells, since antibody production by plasma cells is dependent on the helper function of antigen-specific $CD4^+$ T cells [64, 98]. In another more rational approach potential T-cell targets are selected based on their gene, RNA and/or protein expression profile in tumor cells assuming that specifically expressed and overexpressed tumor proteins should function as T-cell antigens [49, 99]. In any case, T-cell recognition of the candidate proteins either after selection by serological techniques or based on their expression profile, must be demonstrated.

In order to prove the recognition of potential antigens by T cells, the reverse immunology approach can be applied. First, the protein of interest is screened by predictive computer algorithms for peptides that might bind to a specific MHC molecule. Predicted candidate sequences are synthesized and analyzed for their MHC binding ability. High affinity binders are then loaded onto DCs for *in vitro* priming of autologous T cells. Finally, peptide-reactive T cells are employed to demonstrate the generation and presentation of the corresponding epitope either by tumor cells expressing the target antigen endogenously (in the case of $CD8^+$ T cells) or by antigen-loaded pAPCs (in the case of $CD4^+$ T cells) [99].

Although many tumor-associated antigens and epitopes were defined using the reverse immunology strategy, a major disadvantage of this approach is that, following extensive T cell culture, only a few predicted peptides can be verified as true epitopes. To partially overcome this limitation, different strategies for peptide pre-selection prior to *in vitro* T-cell sensitization were developed. *In vitro* digestion of polypeptides encompassing the potential epitope using purified proteasomes can be undertaken to narrow down the spectrum of predicted candidate peptides [86, 100]. Furthermore, mice transgenic for HLA class I and II molecules can be employed for pre-selection of candidate sequences. Peptides recognized by murine T cells in response to vaccination are chosen for subsequent *in vitro* sensitization of human T lymphocytes, this procedure has been applied to peptides derived from the MAGE-A4 and SAGE ($CD8^+$ T cell epitopes) as well as TRP-2 and gp100 ($CD4^+$ T cell epitopes) antigens [60, 79, 101, 102].

Of the different approaches described above, combinations are currently applied to define patients' individual tumor-associated T-cell antigen repertoire [103]. Although techniques and experimental strategies are continuously improving, definition of tumor antigens and epitopes, especially comprehensive personalized analyses, is still a challenging, however mandatory task.

1.3.1
T-cell Antigens for Cancer Immunotherapy – How are Candidates Selected?

The identification of the first tumor-associated antigens and their epitopes in the early 1990s initiated a new area in clinical cancer immunotherapy. In principle two treatment concepts can be distinguished: (1) antigen-specific vaccination (active immunotherapy) in order to induce and boost the anti-tumor activity of specific $CD4^+$ and $CD8^+$ T cells within the tumor host and (2) adoptive T-cell therapy (passive immunotherapy) for elimination of cancer cells by antigen-specific, *ex vivo* expanded and adoptively transferred autologous CTLs.

Over the past 15 years specific immunotherapy (in the case of solid tumors) has mainly been applied to melanoma patients receiving different types of vaccines [68], consisting of synthetic peptide epitopes and recombinant tumor antigens combined with adjuvants or of recombinant viral vectors as well as antigen-loaded DCs. Unfortunately, therapeutic effects were only detectable in very few individuals, and similar observations were also reported in adoptive T-cell transfer studies with specificity for the gp100 antigen [68, 104]. However, it has recently become very clear that in order to be effective T cell-based therapy should be used with strategies that overcome immune suppression induced by the tumor. Via an inhibitory network of different mechanisms the tumor counteracts an effective anti-tumor immune response: malignant cells release immune inhibitory factors such as PGE-2, IL-10, and TGF-β. Furthermore, the tumor microenvironment is enriched with immune-suppressive cells, e.g. regulatory $CD4^+CD25^+$ T cells and myeloid suppressor cells [1, 105]. Thus immune suppression must be reversed, as was demonstrated impressively in recent clinical trials on adoptive T cell therapy: 50% of melanoma patients showed objective therapeutic responses when treated with autologous *in vitro* expanded tumor-reactive T cells after non-myeloablative chemotherapy, which eliminates immune-suppressive cells and in addition stimulates the supportive activity of innate immune cells [71, 73, 106]. However the removal of immune barriers strongly increases the risk of inducing fatal autoimmunity. Consequently, T-cell antigens for cancer immunotherapy need to be selected very carefully by taking the following criteria into consideration:

1. *Specificity of antigen expression*: antigens/epitopes should only be targeted if induction of severe autoimmune responses by antigen-specific T cells can be excluded.

2. *T cell affinity*: antigens/epitopes should only be targeted if high affinity T cells are known to exist in the patient. Preclinical studies have demonstrated that CTLs with high affinity for their specific peptide–MHC complexes are superior in eradicating tumor cells [107]. Interestingly, the affinity of a TCR for its peptide–MHC complex can be enhanced by amino acid substitutions in the peptide sequence. Such altered MHC ligands in contrast to the native epitope, strongly activate specific T cells that can cross-react with the wild-type sequence exposed on tumor cells [108], though a certain risk of inducing non-cross-reactive T cells cannot be excluded. On the other hand, even the low affinity tumor-specific T-cell repertoire of cancer patients might be exploited for adoptive immunotherapy, in cases where the T cells have been genetically engineered *in vitro* to express high affinity tumor antigen-specific TCRs [109].

3. *Stability of the peptide–MHC complex*: a recent preclinical *in vivo* study clearly demonstrated that the poor immunogenicity of a self-tumor antigen is not necessarily the consequence of central and peripheral tolerance mechanisms acting on the specific T-cell repertoire, but can be due to the low affinity of the antigen epitope for its specific MHC class I molecule [110]. Again amino acid substitutions can improve the binding of the peptide ligand to the MHC molecule. Such peptide analogs are currently under intense clinical investigation in vaccination therapies [111], but again the risk of inducing suboptimal T cells (see above) has to be considered [112].

4. *Antigen stability, processing, and expression level*: based on the cancer immuno-editing theory defined by Schreiber and colleagues, the immune system will select tumor cells that by different mechanism resist T-cell attack [1, 113]. In accordance with this model, loss of antigen expression in human tumors has been described [105, 113]. In order to impede the generation of antigen loss variants it has been suggested that antigens which are important for the oncogenic process should be targeted; alternatively, multiple antigens should be selected for therapy.

The generation of antigen epitopes is controlled by the activity of the proteasome in addition to cytosolic and ER-localized peptidases [18, 21, 24]. Tumor cells differ in terms of the expression of specific catalytic and regulatory proteasome subunits and peptidases, resulting in the presentation of a divergent epitope repertoire [114, 115]. Thus, specific tumor antigen epitopes should only be targeted if cancer cells are known to present the epitopes effectively, otherwise even T cells of high therapeutic potential cannot act on their target cells [116, 117].

Ideally the antigen of interest should be expressed at high level in the tumor cells. As indicated by recent preclinical *in vivo* studies, high expression of the tumor antigen will allow T cells to kill cancer cells and in addition, cells in the tumor stroma that internalize and cross-present the tumor antigen, leading to an effective elimination of well-established tumors [118].

1.4
Conclusions

So far, more than 100 tumor-associated T-cell antigens have been characterized, the majority of which was identified by employing T cells and tumor cells from melanoma patients as screening tools. Several of these antigens are shared by tumors of different histology. However, since each tumor entity is characterized by a specific signature of genetic and epigenetic alterations, it has to be assumed that its unique antigen pattern still needs to be defined. In future clinical trials antigen-specific immunotherapy should be combined with strategies to overcome tumor-induced immune suppression. Whether under conditions of reduced immune inhibition the specificity of T cells directed against self-antigens (CTA, differentiation antigens, overexpressed antigens) is sufficiently selective to prevent fatal autoimmune disease, still needs to be determined in well-designed clinical studies. On the other hand, complete remission of solid tumors in advanced patients, though rare, can be achieved by 'simple' vaccination [68]. Understanding the molecular and cellular mechanisms behind this process might pave the way for inducing effective tumor immunity without creating severe autoimmunity.

References

1 Dunn, G.P., Old, L.J. and Schreiber, R.D. (2004) The three Es of cancer immunoediting. *Annual Review of Immunology*, **22**, 329–360.

2 Buell, J.F., Gross, T.G. and Woodle, E.S. (2005) Malignancy after transplantation. *Transplantation*, **80**, S254–S264.

3 Baldo, M., Schiavon, M., Cicogna, P.A., Boccato, P. and Mazzoleni, F. (1992) Spontaneous regression of subcutaneous metastasis of cutaneous melanoma. *Plastic and Reconstructive Surgery*, **90**, 1073–1076.

4 van der Bruggen, P., Traversari, C., Chomez, P., Lurquin, C., De Plaen, E., Van den Eynde, B., Knuth, A. and Boon, T. (1991) A gene encoding an antigen recognized by cytolytic T lymphocytes on a human melanoma. *Science*, **254**, 1643–1647.

5 Brichard, V., Van Pel, A., Wolfel, T., Wolfel, C., De Plaen, E., Lethe, B., Coulie, P. and Boon, T. (1993) The tyrosinase gene codes for an antigen recognized by autologous cytolytic T lymphocytes on HLA-A2 melanomas.

The Journal of Experimental Medicine, **178**, 489–495.

6 Kawakami, Y., Eliyahu, S., Delgado, C.H., Robbins, P.F., Rivoltini, L., Topalian, S.L., Miki, T. and Rosenberg, S.A. (1994) Cloning of the gene coding for a shared human melanoma antigen recognized by autologous T cells infiltrating into tumor. *Proceedings of the National Academy of Sciences of the United States of America*, **91**, 3515–3519.

7 Bakker, A.B., Schreurs, M.W., de Boer, A.J., Kawakami, Y., Rosenberg, S.A., Adema, G.J. and Figdor, C.G. (1994) Melanocyte lineage-specific antigen gp100 is recognized by melanoma-derived tumor-infiltrating lymphocytes. *The Journal of Experimental Medicine*, **179**, 1005–1009.

8 Ochsenbein, A.F., Sierro, S., Odermatt, B., Pericin, M., Karrer, U., Hermans, J., Hemmi, S., Hengartner, H. and Zinkernagel, R.M. (2001) Roles of tumour localization, second signals and cross priming in cytotoxic T-cell induction. *Nature*, **411**, 1058–1064.

9 Shen, L. and Rock, K.L. (2006) Priming of T cells by exogenous antigen cross-presented on MHC class I molecules. *Current Opinion in Immunology*, **18**, 85–91.

10 Schoenberger, S.P., Toes, R.E., van der Voort, E.I., Offringa, R. and Melief, C.J. (1998) T-cell help for cytotoxic T lymphocytes is mediated by CD40-CD40L interactions. *Nature*, **393**, 480–483.

11 Ridge, J.P., Di Rosa, F. and Matzinger, P. (1998) A conditioned dendritic cell can be a temporal bridge between a CD4+ T-helper and a T-killer cell. *Nature*, **393**, 474–478.

12 Bennett, S.R., Carbone, F.R., Karamalis, F., Flavell, R.A., Miller, J.F. and Heath, W.R. (1998) Help for cytotoxic-T-cell responses is mediated by CD40 signalling. *Nature*, **393**, 478–480.

13 Janssen, E.M., Lemmens, E.E., Wolfe, T., Christen, U., von Herrath, M.G. and Schoenberger, S.P. (2003) CD4+ T cells are required for secondary expansion and memory in CD8+ T lymphocytes. *Nature*, **421**, 852–856.

14 Hamilton, S.E., Wolkers, M.C., Schoenberger, S.P. and Jameson, S.C. (2006) The generation of protective memory-like CD8+ T cells during homeostatic proliferation requires CD4+ T cells. *Nature Immunology*, **7**, 475–481.

15 Mills, D.M. and Cambier, J.C. (2003) B lymphocyte activation during cognate interactions with CD4+ T lymphocytes: molecular dynamics and immunologic consequences. *Seminars in Immunology*, **15**, 325–329.

16 Probst-Kepper, M., Stroobant, V., Kridel, R., Gaugler, B., Landry, C., Brasseur, F., Cosyns, J.P., Weynand, B., Boon, T. and Van Den Eynde, B.J. (2001) An alternative open reading frame of the human macrophage colony-stimulating factor gene is independently translated and codes for an antigenic peptide of 14 amino acids recognized by tumor-infiltrating CD8 T lymphocytes. *The Journal of Experimental Medicine*, **193**, 1189–1198.

17 Burrows, S.R., Rossjohn, J. and McCluskey, J. (2006) Have we cut ourselves too short in mapping CTL epitopes? *Trends in Immunology*, **27**, 11–16.

18 Strehl, B., Seifert, U., Krüger, E., Heink, S., Kuckelkorn, U. and Kloetzel, P.M. (2005) Interferon-gamma, the functional plasticity of the ubiquitin–proteasome system, and MHC class I antigen processing. *Immunological Reviews*, **207**, 19–30.

19 Morel, S., Levy, F., Burlet-Schiltz, O., Brasseur, F., Probst-Kepper, M., Peitrequin, A.L., Monsarrat, B., Van Velthoven, R., Cerottini, J.C., Boon, T., Gairin, J.E. and Van den Eynde, B.J. (2000) Processing of some antigens by the standard proteasome but not by the immunoproteasome results in poor presentation by dendritic cells. *Immunity*, **12**, 107–117.

20 Schultz, E.S., Chapiro, J., Lurquin, C., Claverol, S., Burlet-Schiltz, O., Warnier, G., Russo, V., Morel, S., Levy, F., Boon, T., Van den Eynde, B.J. and van der Bruggen, P. (2002) The production of a new MAGE-3 peptide presented to cytolytic T lymphocytes by HLA-B40 requires the immunoproteasome. *The Journal of Experimental Medicine*, **195**, 391–399.

21 Seifert, U., Maranon, C., Shmueli, A., Desoutter, J.F., Wesoloski, L., Janek, K., Henklein, P., Diescher, S., Andrieu, M., de la Salle, H., Weinschenk, T., Schild, H., Laderach, D., Galy, A., Haas, G., Kloetzel, P.M., Reiss, Y. and Hosmalin, A. (2003) An essential role for tripeptidyl peptidase in the generation of an MHC class I epitope. *Nature Immunology*, **4**, 375–379.

22 Guil, S., Rodríguez-Castro, M., Aguilar, F., Villasevil, E.M., Antón, L.C. and Del Val, M. (2006) Need for tripeptidyl-peptidase II in major histocompatibility complex class I viral antigen processing when proteasomes are detrimental. *The Journal of Biological Chemistry*, **281**, 39925–39934.

23 Vigneron, N., Stroobant, V., Chapiro, J., Ooms, A., Degiovanni, G., Morel, S., van der Bruggen, P., Boon, T. and Van den Eynde, B.J. (2004) An antigenic peptide produced by peptide splicing in the proteasome. *Science*, **304**, 587–590.

24 Saveanu, L., Carroll, O., Hassainya, Y. and van Endert, P. (2005) Complexity, contradictions, and conundrums: studying post-proteasomal proteolysis in HLA class I antigen presentation. *Immunological Reviews*, **207**, 42–59.

25 Godefroy, E., Moreau-Aubry, A., Diez, E., Dreno, B., Jotereau, F. and Guilloux, Y. (2005) alpha v beta3-dependent cross-presentation of matrix metalloproteinase-2 by melanoma cells gives rise to a new tumor antigen. *The Journal of Experimental Medicine*, **202**, 61–72.

26 Li, P., Gregg, J.L., Wang, N., Zhou, D., O'Donnell, P., Blum, J.S. and Crotzer, V.L. (2005) Compartmentalization of class II antigen presentation: contribution of cytoplasmic and endosomal processing. *Immunological Reviews*, **207**, 206–217.

27 Anichini, A., Mortarini, R., Nonaka, D., Molla, A., Vegetti, C., Montaldi, E., Wang, X. and Ferrone, S. (2006) Association of antigen-processing machinery and HLA antigen phenotype of melanoma cells with survival in American Joint Committee on Cancer stage III and IV melanoma patients. *Cancer Research*, **66**, 6405–6411.

28 Traversari, C., van der Bruggen, P., Luescher, I.F., Lurquin, C., Chomez, P., Van Pel, A., De Plaen, E., Amar-Costesec, A. and Boon, T. (1992) A nonapeptide encoded by human gene MAGE-1 is recognized on HLA-A1 by cytolytic T lymphocytes directed against tumor antigen MZ2-E. *The Journal of Experimental Medicine*, **176**, 1453–1457.

29 Wolfel, T., Hauer, M., Schneider, J., Serrano, M., Wolfel, C., Klehmann-Hieb, E., De Plaen, E., Hankeln, T., Meyer zum Buschenfelde, K.H. and Beach, D. (1995) A p16INK4a-insensitive CDK4 mutant targeted by cytolytic T lymphocytes in a human melanoma. *Science*, **269**, 1281–1284.

30 Robbins, P.F., El-Gamil, M., Li, Y.F., Kawakami, Y., Loftus, D., Appella, E. and Rosenberg, S.A. (1996) A mutated beta-catenin gene encodes a melanoma-specific antigen recognized by tumor infiltrating lymphocytes. *The Journal of Experimental Medicine*, **183**, 1185–1192.

31 Gjertsen, M.K., Bjorheim, J., Saeterdal, I., Myklebust, J. and Gaudernack, G. (1997) Cytotoxic CD4+ and CD8+ T lymphocytes, generated by mutant p21-ras (12Val) peptide vaccination of a patient, recognize 12Val-dependent nested epitopes present within the vaccine peptide and kill autologous tumour cells carrying this mutation. *International Journal of Cancer*, **72**, 784–790.

32 Lupetti, R., Pisarra, P., Verrecchia, A., Farina, C., Nicolini, G., Anichini, A.,

Bordignon, C., Sensi, M., Parmiani, G. and Traversari, C. (1998) Translation of a retained intron in tyrosinase-related protein (TRP) 2 mRNA generates a new cytotoxic T lymphocyte (CTL)-defined and shared human melanoma antigen not expressed in normal cells of the melanocytic lineage. *The Journal of Experimental Medicine*, **188**, 1005–1016.

33 Wang, R.F., Wang, X., Atwood, A.C., Topalian, S.L. and Rosenberg, S.A. (1999) Cloning genes encoding MHC class II-restricted antigens: mutated CDC27 as a tumor antigen. *Science*, **284**, 1351–1354.

34 Novellino, L., Renkvist, N., Rini, F., Mazzocchi, A., Rivoltini, L., Greco, A., Deho, P., Squarcina, P., Robbins, P.F., Parmiani, G. and Castelli, C. (2003) Identification of a mutated receptor-like protein tyrosine phosphatase kappa as a novel, class II HLA-restricted melanoma antigen. *Journal of Immunology (Baltimore, Md: 1950)*, **170**, 6363–6370.

35 Huang, J., El-Gamil, M., Dudley, M.E., Li, Y.F., Rosenberg, S.A. and Robbins, P.F. (2004) T cells associated with tumor regression recognize frameshifted products of the CDKN2A tumor suppressor gene locus and a mutated HLA class I gene product. *Journal of Immunology (Baltimore, Md: 1950)*, **172**, 6057–6064.

36 Sharkey, M.S., Lizée, G., Gonzales, M.I., Patel, S. and Topalian, S.L. (2004) CD4(+) T-cell recognition of mutated B-RAF in melanoma patients harboring the V599E mutation. *Cancer Research*, **64**, 1595–1599.

37 Lennerz, V., Fatho, M., Gentilini, C., Frye, R.A., Lifke, A., Ferel, D., Wölfel, C., Huber, C. and Wölfel, T. (2005) The response of autologous T cells to a human melanoma is dominated by mutated neoantigens. *Proceedings of the National Academy of Sciences of the United States of America*, **102**, 16013–16018.

38 Takenoyama, M., Baurain, J.F., Yasuda, M., So, T., Sugaya, M., Hanagiri, T., Sugio, K., Yasumoto, K., Boon, T. and Coulie, P.G. (2006) A point mutation in the NFYC gene generates an antigenic peptide recognized by autologous cytolytic T lymphocytes on a human squamous cell lung carcinoma. *International Journal of Cancer*, **118**, 1992–1997.

39 Graf, C., Heidel, F., Tenzer, S., Radsak, M.P., Solem, F.K., Britten, C.M., Huber, C., Fischer, T. and Wölfel, T. (2007) A neoepitope generated by an FLT3 internal tandem duplication (FLT3-ITD) is recognized by leukemia-reactive autologous CD8+ T cells. *Blood*, **109**, 2985–2988.

40 Bocchia, M., Korontsvit, T., Xu, Q., Mackinnon, S., Yang, S.Y., Sette, A. and Scheinberg, D.A. (1996) Specific human cellular immunity to bcr-abl oncogene-derived peptides. *Blood*, **87**, 3587–3592.

41 Wang, R.F., Wang, X. and Rosenberg, S.A. (1999) Identification of a novel major histocompatibility complex class II-restricted tumor antigen resulting from a chromosomal rearrangement recognized by CD4(+) T cells. *The Journal of Experimental Medicine*, **189**, 1659–1668.

42 van den Broeke, L.T., Pendleton, C.D., Mackall, C., Helman, L.J. and Berzofsky, J.A. (2006) Identification and epitope enhancement of a PAX-FKHR fusion protein breakpoint epitope in alveolar rhabdomyosarcoma cells created by a tumorigenic chromosomal translocation inducing CTL capable of lysing human tumors. *Cancer Research*, **66**, 1818–1823.

43 Kyewski, B. and Klein, L. (2006) A central role for central tolerance. *Annual Review of Immunol*, **24**, 571–606.

44 Gallegos, A.M. and Bevan, M.J. (2006) Central tolerance: good but imperfect. *Immunological Reviews*, **209**, 290–296.

45 Redmond, W.L. and Sherman, L.A. (2005) Peripheral tolerance of CD8 T lymphocytes. *Immunity*, **22**, 275–284.

46 Nichols, L.A., Chen, Y., Colella, T.A., Bennett, C.L., Clausen, B.E. and Engelhard, V.H. (2007) Deletional

self-tolerance to a melanocyte/melanoma antigen derived from tyrosinase is mediated by a radio-resistant cell in peripheral and mesenteric lymph nodes. *Journal of Immunology (Baltimore, Md: 1950)*, **179**, 993–1003.

47 Lee, J.W., Epardaud, M., Sun, J., Becker, J.E., Cheng, A.C., Yonekura, A.R., Heath, J.K. and Turley, S.J. (2007) Peripheral antigen display by lymph node stroma promotes T cell tolerance to intestinal self. *Nature Immunology*, **8**, 181–190.

48 Guilloux, Y., Lucas, S., Brichard, V.G., Van Pel, A., Viret, C., De Plaen, E., Brasseur, F., Lethé, B., Jotereau, F. and Boon, T. (1996) A peptide recognized by human cytolytic T lymphocytes on HLA-A2 melanomas is encoded by an intron sequence of the N-acetylglucosaminyltransferase V gene. *The Journal of Experimental Medicine*, **183**, 1173–1183.

49 Strausberg, R.L., Simpson, A.J., Old, L.J. and Riggins, G.J. (2004) Oncogenomics and the development of new cancer therapies. *Nature*, **429**, 469–474.

50 Lee, S.P., Brooks, J.M., Al-Jarrah, H., Thomas, W.A., Haigh, T.A., Taylor, G.S., Humme, S., Schepers, A., Hammerschmidt, W., Yates, J.L., Rickinson, A.B. and Blake, N.W. (2004) CD8 T cell recognition of endogenously expressed Epstein-Barr virus nuclear antigen 1. *The Journal of Experimental Medicine*, **199**, 1409–1420.

51 Adhikary, D., Behrends, U., Moosmann, A., Witter, K., Bornkamm, G.W. and Mautner, J. (2006) Control of Epstein-Barr virus infection in vitro by T helper cells specific for virion glycoproteins. *The Journal of Experimental Medicine*, **203**, 995–1006.

52 Evans, E.M., Man, S., Evans, A.S. and Borysiewicz, L.K. (1997) Infiltration of cervical cancer tissue with human papillomavirus-specific cytotoxic T-lymphocytes. *Cancer Research*, **57**, 2943–2950.

53 Höhn, H., Pilch, H., Günzel, S., Neukirch, C., Hilmes, C., Kaufmann, A., Seliger, B. and Maeurer, M.J. (1999) CD4+ tumor-infiltrating lymphocytes in cervical cancer recognize HLA-DR-restricted peptides provided by human papillomavirus-E7. *Journal of Immunology (Baltimore, Md: 1950)*, **163**, 5715–5722.

54 De Smet, C., Lurquin, C., Lethé, B., Martelange, V. and Boon, T. (1999) DNA methylation is the primary silencing mechanism for a set of germ line- and tumor-specific genes with a CpG-rich promoter. *Molecular and Cellular Biology*, **19**, 7327–7335.

55 Gotter, J., Brors, B., Hergenhahn, M. and Kyewski, B. (2004) Medullary epithelial cells of the human thymus express a highly diverse selection of tissue-specific genes colocalized in chromosomal clusters. *The Journal of Experimental Medicine*, **199**, 155–166.

56 Chaux, P., Luiten, R., Demotte, N., Vantomme, V., Stroobant, V., Traversari, C., Russo, V., Schultz, E., Cornelis, G.R., Boon, T. and van der Bruggen, P. (1999) Identification of five MAGE-A1 epitopes recognized by cytolytic T lymphocytes obtained by in vitro stimulation with dendritic cells transduced with MAGE-A1. *Journal of Immunology (Baltimore, Md: 1950)*, **163**, 2928–2936.

57 Chaux, P., Vantomme, V., Stroobant, V., Thielemans, K., Corthals, J., Luiten, R., Eggermont, A.M., Boon, T. and van der Bruggen, P. (1999) Identification of MAGE-3 epitopes presented by HLA-DR molecules to CD4(+) T lymphocytes. *The Journal of Experimental Medicine*, **189**, 767–778.

58 Pascolo, S., Schirle, M., Guckel, B., Dumrese, T., Stumm, S., Kayser, S., Moris, A., Wallwiener, D., Rammensee, H.G. and Stevanovic, S. (2001) A MAGE-A1 HLA-A*0201 epitope identified by mass spectrometry. *Cancer Research*, **61**, 4072–4077.

59 Consogno, G., Manici, S., Facchinetti, V., Bachi, A., Hammer, J., Conti-Fine, B.M., Rugarli, C., Traversari, C. and Protti, M.P. (2003) Identification of immunodominant regions among promiscuous HLA-DR-restricted CD4+ T-cell epitopes on the tumor antigen MAGE-3. *Blood*, **101**, 1038–1344.

60 Miyahara, Y., Naota, H., Wang, L., Hiasa, A., Goto, M., Watanabe, M., Kitano, S., Okumura, S., Takemitsu, T., Yuta, A., Majima, Y., Lemonnier, F.A., Boon, T. and Shiku, H. (2005) Determination of cellularly processed HLA-A2402-restricted novel CTL epitopes derived from two cancer germ line genes, MAGE-A4 and SAGE. *Clinical Cancer Research*, **11**, 5581–5589.

61 Ayyoub, M., Stevanovic, S., Sahin, U., Guillaume, P., Servis, C., Rimoldi, D., Valmori, D., Romero, P., Cerottini, J.C., Rammensee, H.G., Pfreundschuh, M., Speiser, D. and Levy, F. (2002) Proteasome-assisted identification of a SSX-2-derived epitope recognized by tumor-reactive CTL infiltrating metastatic melanoma. *Journal of Immunology (Baltimore, Md: 1950)*, **168**, 1717–1722.

62 Neumann, F., Wagner, C., Stevanovic, S., Kubuschok, B., Schormann, C., Mischo, A., Ertan, K., Schmidt, W. and Pfreundschuh, M. (2004) Identification of an HLA-DR-restricted peptide epitope with a promiscuous binding pattern derived from the cancer testis antigen HOM-MEL-40/SSX2. *International Journal of Cancer*, **112**, 661–668.

63 Ayyoub, M., Merlo, A., Hesdorffer, C.S., Rimoldi, D., Speiser, D., Cerottini, J.C., Chen, Y.T., Old, L.J., Stevanovic, S. and Valmori, D. (2005) CD4+ T cell responses to SSX-4 in melanoma patients. *Journal of Immunology (Baltimore, Md: 1950)*, **174**, 5092–5099.

64 Chen, Y.T., Scanlan, M.J., Sahin, U., Tureci, O., Gure, A.O., Tsang, S., Williamson, B., Stockert, E., Pfreundschuh, M. and Old, L.J. (1997) A testicular antigen aberrantly expressed in human cancers detected by autologous antibody screening. *Proceedings of the National Academy of Sciences of the United States of America*, **94**, 1914–1918.

65 Jager, E., Chen, Y.T., Drijfhout, J.W., Karbach, J., Ringhoffer, M., Jager, D., Arand, M., Wada, H., Noguchi, Y., Stockert, E., Old, L.J. and Knuth, A. (1998) Simultaneous humoral and cellular immune response against cancer-testis antigen NY-ESO-1: definition of human histocompatibility leukocyte antigen (HLA)-A2-binding peptide epitopes. *The Journal of Experimental Medicine*, **187**, 265–270.

66 Jager, E., Jager, D., Karbach, J., Chen, Y.T., Ritter, G., Nagata, Y., Gnjatic, S., Stockert, E., Arand, M., Old, L.J. and Knuth, A. (2000) Identification of NY-ESO-1 epitopes presented by human histocompatibility antigen (HLA)-DRB4*0101-0103 and recognized by CD4(+) T lymphocytes of patients with NY-ESO-1-expressing melanoma. *The Journal of Experimental Medicine*, **191**, 625–630.

67 Zeng, G., Touloukian, C.E., Wang, X., Restifo, N.P., Rosenberg, S.A. and Wang, R.F. (2000) Identification of CD4+ T cell epitopes from NY-ESO-1 presented by HLA-DR molecules. *Journal of Immunology (Baltimore, Md: 1950)*, **165**, 1153–1159.

68 Boon, T., Coulie, P.G., Van den Eynde, B.J. and van der Bruggen, P. (2006) Human T cell responses against melanoma. *Annual Review of Immunology*, **24**, 175–208.

69 Cloosen, S., Arnold, J., Thio, M., Bos, G.M., Kyewski, B. and Germeraad, W.T. (2007) Expression of tumor-associated differentiation antigens, MUC1 glycoforms and CEA, in human thymic epithelial cells: implications for self-tolerance and tumor therapy. *Cancer Research*, **67**, 3919–3926.

70 Yee, C., Savage, P.A., Lee, P.P., Davis, M.M. and Greenberg, P.D. (1999) Isolation of high avidity

70 melanoma-reactive CTL from heterogeneous populations using peptide-MHC tetramers. *Journal of Immunology (Baltimore, Md: 1950)*, **162**, 2227–2234.

71 Dudley, M.E., Wunderlich, J.R., Robbins, P.F., Yang, J.C., Hwu, P., Schwartzentruber, D.J., Topalian, S.L., Sherry, R., Restifo, N.P., Hubicki, A.M., Robinson, M.R., Raffeld, M., Duray, P., Seipp, C.A., Rogers-Freezer, L., Morton, K.E., Mavroukakis, S.A., White, D.E. and Rosenberg, S.A. (2002) Cancer regression and autoimmunity in patients after clonal repopulation with antitumor lymphocytes. *Science*, **298**, 850–854.

72 Yee, C., Thompson, J.A., Byrd, D., Riddell, S.R., Roche, P., Celis, E. and Greenberg, P.D. (2002) Adoptive T cell therapy using antigen-specific CD8+ T cell clones for the treatment of patients with metastatic melanoma: in vivo persistence, migration, and antitumor effect of transferred T cells. *Proceedings of the National Academy of Sciences of the United States of America*, **99**, 16168–16173.

73 Dudley, M.E., Wunderlich, J.R., Yang, J.C., Sherry, R.M., Topalian, S.L., Restifo, N.P., Royal, R.E., Kammula, U., White, D.E., Mavroukakis, S.A., Rogers, L.J., Gracia, G.J., Jones, S.A., Mangiameli, D.P., Pelletier, M.M., Gea-Banacloche, J., Robinson, M.R., Berman, D.M., Filie, A.C., Abati, A. and Rosenberg, S.A. (2005) Adoptive cell transfer therapy following non-myeloablative but lymphodepleting chemotherapy for the treatment of patients with refractory metastatic melanoma. *Journal of Clinical Oncology*, **23**, 2346–2357.

74 Yee, C., Thompson, J.A., Roche, P., Byrd, D.R., Lee, P.P., Piepkorn, M., Kenyon, K., Davis, M.M., Riddell, S.R. and Greenberg, P.D. (2000) Melanocyte destruction after antigen-specific immunotherapy of melanoma: direct evidence of t cell-mediated vitiligo. *The Journal of Experimental Medicine*, **192**, 1637–1644.

75 Uchi, H., Stan, R., Turk, M.J., Engelhorn, M.E., Rizzuto, G.A., Goldberg, S.M., Wolchok, J.D. and Houghton, A.N. (2006) Unraveling the complex relationship between cancer immunity and autoimmunity: lessons from melanoma and vitiligo. *Advances in Immunology*, **90**, 215–241.

76 Sun, Y., Song, M., Stevanovic, S., Jankowiak, C., Paschen, A., Rammensee, H.G. and Schadendorf, D. (2000) Identification of a new HLA-A*0201-restricted T-cell epitope from the tyrosinase-related protein 2 (TRP2) melanoma antigen. *International Journal of Cancer*, **87**, 399–404.

77 Robbins, P.F., El-Gamil, M., Li, Y.F., Zeng, G., Dudley, M. and Rosenberg, S.A. (2002) Multiple HLA class II-restricted melanocyte differentiation antigens are recognized by tumor-infiltrating lymphocytes from a patient with melanoma. *Journal of Immunology (Baltimore, Md: 1950)*, **169**, 6036–6047.

78 Paschen, A., Jing, W., Drexler, I., Klemm, M., Song, M., Müller-Berghaus, J., Nguyen, X.D., Osen, W., Stevanovic, S., Sutter, G. and Schadendorf, D. (2005) Melanoma patients respond to a new HLA-A*01-presented antigenic ligand derived from a multi-epitope region of melanoma antigen TRP-2. *International Journal of Cancer*, **116**, 944–948.

79 Paschen, A., Song, M., Osen, W., Nguyen, X.D., Mueller-Berghaus, J., Fink, D., Daniel, N., Donzeau, M., Nagel, W., Kropshofer, H. and Schadendorf, D. (2005) Detection of spontaneous CD4+ T-cell responses in melanoma patients against a tyrosinase-related protein-2-derived epitope identified in HLA-DRB1*0301 transgenic mice. *Clinical Cancer Research*, **11**, 5241–5247.

80 Tsang, K.Y., Zaremba, S., Nieroda, C.A., Zhu, M.Z., Hamilton, J.M. and Schlom, J. (1995) Generation of human cytotoxic T cells specific for human

carcinoembryonic antigen epitopes from patients immunized with recombinant vaccinia-CEA vaccine. *Journal of the National Cancer Institute*, **87**, 982–990.

81 Correale, P., Walmsley, K., Nieroda, C., Zaremba, S., Zhu, M., Schlom, J. and Tsang, K.Y. (1997) In vitro generation of human cytotoxic T lymphocytes specific for peptides derived from prostate-specific antigen. *Journal of the National Cancer Institute*, **89**, 293–300.

82 Fisk, B., Blevins, T.L., Wharton, J.T. and Ioannides, C.G. (1995) Identification of an immunodominant peptide of HER-2/neu protooncogene recognized by ovarian tumor-specific cytotoxic T lymphocyte lines. *The Journal of Experimental Medicine*, **181**, 2109–2117.

83 Rongcun, Y., Salazar-Onfray, F., Charo, J., Malmberg, K.J., Evrin, K., Maes, H., Kono, K., Hising, C., Petersson, M., Larsson, O., Lan, L., Appella, E., Sette, A., Celis, E. and Kiessling, R. (1999) Identification of new HER2/neu-derived peptide epitopes that can elicit specific CTL against autologous and allogeneic carcinomas and melanomas. *Journal of Immunology (Baltimore, Md: 1950)*, **163**, 1037–1044.

84 Hiltbold, E.M., Ciborowski, P. and Finn, O.J. (1998) Naturally processed class II epitope from the tumor antigen MUC1 primes human CD4+ T cells. *Cancer Research*, **58**, 5066–5070.

85 Brossart, P., Heinrich, K.S., Stuhler, G., Behnke, L., Reichardt, V.L., Stevanovic, S., Muhm, A., Rammensee, H.G., Kanz, L. and Brugger, W. (1999) Identification of HLA-A2-restricted T-cell epitopes derived from the MUC1 tumor antigen for broadly applicable vaccine therapies. *Blood*, **93**, 4309–4317.

86 Kessler, J.H., Beekman, N.J., Bres-Vloemans, S.A., Verdijk, P., van Veelen, P.A., Kloosterman-Joosten, A.M., Vissers, D.C., ten Bosch, G.J., Kester, M.G., Sijts, A., Wouter Drijfhout, J., Ossendorp, F., Offringa, R. and Melief, C.J. (2001) Efficient identification of novel HLA-A(*)0201-presented cytotoxic T lymphocyte epitopes in the widely expressed tumor antigen PRAME by proteasome-mediated digestion analysis. *The Journal of Experimental Medicine*, **193**, 73–88.

87 Schmitz, M., Diestelkoetter, P., Weigle, B., Schmachtenberg, F., Stevanovic, S., Ockert, D., Rammensee, H.G. and Rieber, E.P. (2000) Generation of survivin-specific CD8+ T effector cells by dendritic cells pulsed with protein or selected peptides. *Cancer Research*, **60**, 4845–4849.

88 Vonderheide, R.H., Hahn, W.C., Schultze, J.L. and Nadler, L.M. (1999) The telomerase catalytic subunit is a widely expressed tumor-associated antigen recognized by cytotoxic T lymphocytes. *Immunity*, **10**, 673–679.

89 Schroers, R., Huang, X.F., Hammer, J., Zhang, J. and Chen, S.Y. (2002) Identification of HLA DR7-restricted epitopes from human telomerase reverse transcriptase recognized by CD4+ T-helper cells. *Cancer Research*, **62**, 2600–2605.

90 Zinkernagel, R.M. and Hengartner, H. (2001) Regulation of the immune response by antigen. *Science*, **293**, 251–253.

91 Vlad, A.M., Muller, S., Cudic, M., Paulsen, H., Otvos, L., Jr, Hanisch, F.G. and Finn, O.J. (2002) Complex carbohydrates are not removed during processing of glycoproteins by dendritic cells: processing of tumor antigen MUC1 glycopeptides for presentation to major histocompatibility complex class II-restricted T cells. *The Journal of Experimental Medicine*, **196**, 1435–1446.

92 Engelhard, V.H., Altrich-Vanlith, M., Ostankovitch, M. and Zarling, A.L. (2006) Post-translational modifications of naturally processed MHC-binding epitopes. *Current Opinion in Immunology*, **18**, 92–97.

93 Zarling, A.L., Polefrone, J.M., Evans, A.M., Mikesh, L.M., Shabanowitz, J.,

Lewis, S.T., Engelhard, V.H. and Hunt, D.F. (2006) Identification of class I MHC-associated phosphopeptides as targets for cancer immunotherapy. *Proceedings of the National Academy of Sciences of the United States of America*, **103**, 14889–14894.

94 Sensi, M., Nicolini, G., Zanon, M., Colombo, C., Molla, A., Bersani, I., Lupetti, R., Parmiani, G. and Anichini, A. (2005) Immunogenicity without immunoselection: a mutant but functional antioxidant enzyme retained in a human metastatic melanoma and targeted by CD8(+) T cells with a memory phenotype. *Cancer Research*, **65**, 632–640.

95 Germeau, C., Ma, W., Schiavetti, F., Lurquin, C., Henry, E., Vigneron, N., Brasseur, F., Lethé, B., De Plaen, E., Velu, T., Boon, T. and Coulie, P.G. (2005) High frequency of antitumor T cells in the blood of melanoma patients before and after vaccination with tumor antigens. *The Journal of Experimental Medicine*, **201**, 241–248.

96 Skipper, J.C., Gulden, P.H., Hendrickson, R.C., Harthun, N., Caldwell, J.A., Shabanowitz, J., Engelhard, V.H., Hunt, D.F. and Slingluff, C.L. Jr, (1999) Mass-spectrometric evaluation of HLA-A*0201-associated peptides identifies dominant naturally processed forms of CTL epitopes from MART-1 and gp100. *International Journal of Cancer*, **82**, 669–677.

97 Röhn, T.A., Reitz, A., Paschen, A., Nguyen, X.D., Schadendorf, D., Vogt, A.B. and Kropshofer, H. (2005) A novel strategy for the discovery of MHC class II-restricted tumor antigens: identification of a melanotransferrin helper T-cell epitope. *Cancer Research*, **65**, 10068–10078.

98 Sahin, U., Tureci, O., Schmitt, H., Cochlovius, B., Johannes, T., Schmits, R., Stenner, F., Luo, G., Schobert, I. and Pfreundschuh, M. (1995) Human neoplasms elicit multiple specific immune responses in the autologous host. *Proceedings of the National Academy of Sciences of the United States of America*, **92**, 11810–11813.

99 Viatte, S., Alves, P.M. and Romero, P. (2006) Reverse immunology approach for the identification of CD8 T-cell-defined antigens: advantages and hurdles. *Immunology and Cell Biology*, **84**, 318–330.

100 Kessler, J.H., Bres-Vloemans, S.A., van Veelen, P.A., de Ru, A., Huijbers, I.J., Camps, M., Mulder, A., Offringa, R., Drijfhout, J.W., Leeksma, O.C., Ossendorp, F. and Melief, C.J. (2006) BCR-ABL fusion regions as a source of multiple leukemia-specific CD8+ T-cell epitopes. *Leukemia*, **20**, 1738–1750.

101 Touloukian, C.E., Leitner, W.W., Topalian, S.L., Li, Y.F., Robbins, P.F., Rosenberg, S.A. and Restifo, N.P. (2000) Identification of a MHC class II-restricted human gp100 epitope using DR4-IE transgenic mice. *Journal of Immunology (Baltimore, Md: 1950)*, **164**, 3535–3542.

102 Santomasso, B.D., Roberts, W.K., Thomas, A., Williams, T., Blachère, N.E., Dudley, M.E., Houghton, A.N., Posner, J.B. and Darnell, R.B. (2007) A T cell receptor associated with naturally occurring human tumor immunity. *Proceedings of the National Academy of Sciences of the United States of America*, **104**, 19073–19078.

103 Weinschenk, T., Gouttefangeas, C., Schirle, M., Obermayr, F., Walter, S., Schoor, O., Kurek, R., Loeser, W., Bichler, K.H., Wernet, D., Stevanovic, S. and Rammensee, H.G. (2002) Integrated functional genomics approach for the design of patient-individual antitumor vaccines. *Cancer Research*, **62**, 5818–5827.

104 Dudley, M.E., Wunderlich, J., Nishimura, M.I., Yu, D., Yang, J.C., Topalian, S.L., Schwartzentruber, D.J., Hwu, P., Marincola, F.M., Sherry, R., Leitman, S.F. and Rosenberg, S.A. (2001) Adoptive transfer of cloned melanoma-reactive T lymphocytes for the treatment of patients with metastatic

105 Drake, C.G., Jaffee, E. and Pardoll, D.M. (2006) Mechanisms of immune evasion by tumors. *Advances in Immunology*, **90**, 51–81.

106 Paulos, C.M., Wrzesinski, C., Kaiser, A., Hinrichs, C.S., Chieppa, M., Cassard, L., Palmer, D.C., Boni, A., Muranski, P., Yu, Z., Gattinoni, L., Antony, P.A., Rosenberg, S.A. and Restifo, N.P. (2007) Microbial translocation augments the function of adoptively transferred self/tumor-specific CD8+ T cells via TLR4 signaling. *The Journal of Clinical Investigation*, **117**, 2197–2204.

107 Zeh, H.J., 3rd, Perry-Lalley, D., Dudley, M.E., Rosenberg, S.A. and Yang, J.C. (1999) High avidity CTLs for two self-antigens demonstrate superior *in vitro* and *in vivo* antitumor efficacy. *Journal of Immunology (Baltimore, Md: 1950)*, **162**, 989–994.

108 Slansky, J.E., Rattis, F.M., Boyd, L.F., Fahmy, T., Jaffee, E.M., Schneck, J.P., Margulies, D.H. and Pardoll, D.M. (2000) Enhanced antigen-specific antitumor immunity with altered peptide ligands that stabilize the MHC-peptide-TCR complex. *Immunity*, **13**, 529–538.

109 Morgan, R.A., Dudley, M.E., Wunderlich, J.R., Hughes, M.S., Yang, J.C., Sherry, R.M., Royal, R.E., Topalian, S.L., Kammula, U.S., Restifo, N.P., Zheng, Z., Nahvi, A., de Vries, C.R., Rogers-Freezer, L.J., Mavroukakis, S.A. and Rosenberg, S.A. (2006) Cancer regression in patients after transfer of genetically engineered lymphocytes. *Science*, **314**, 126–129.

110 Yu, Z., Theoret, M.R., Touloukian, C.E., Surman, D.R., Garman, S.C., Feigenbaum, L., Baxter, T.K., Baker, B.M. and Restifo, N.P. (2004) Poor immunogenicity of a self/tumor antigen derives from peptide-MHC-I instability and is independent of tolerance. *The Journal of Clinical Investigation*, **114**, 551–559.

111 Speiser, D.E., Liénard, D., Rufer, N., Rubio-Godoy, V., Rimoldi, D., Lejeune, F., Krieg, A.M., Cerottini, J.C. and Romero, P. (2005) Rapid and strong human CD8+ T cell responses to vaccination with peptide, IFA, and CpG oligodeoxynucleotide 7909. *The Journal of Clinical Investigation*, **115**, 739–746.

112 Carrabba, M.G., Castelli, C., Maeurer, M.J., Squarcina, P., Cova, A., Pilla, L., Renkvist, N., Parmiani, G. and Rivoltini, L. (2003) Suboptimal activation of CD8(+) T cells by melanoma-derived altered peptide ligands: role of Melan-A/MART-1 optimized analogues. *Cancer Research*, **63**, 1560–1567.

113 Marincola, F.M., Wang, E., Herlyn, M., Seliger, B. and Ferrone, S. (2003) Tumors as elusive targets of T-cell-based active immunotherapy. *Trends in Immunology*, **24**, 335–342.

114 Sun, Y., Sijts, A., Song, M., Janke, K., Nussbaum, A., Kral, S., Schirle, M., Stevanovic, S., Paschen, A., Schild, H., Kloetzel, P.M. and Schadendorf, D. (2002) Expression of the proteasome activator PA28 rescues the presentation of a cytotoxic T lymphocyte epitope on human melanoma cells. *Cancer Research*, **62**, 2875–2882.

115 Fruci, D., Ferracuti, S., Limongi, M.Z., Cunsolo, V., Giorda, E., Fraioli, R., Sibilio, L., Carroll, O., Hattori, A., van Endert, P.M. and Giacomini, P. (2006) Expression of endoplasmic reticulum aminopeptidases in EBV-B cell lines from healthy donors and in leukemia/lymphoma, carcinoma, and melanoma cell lines. *Journal of Immunology (Baltimore, Md: 1950)*, **176**, 4869–4879.

116 Dutoit, V., Taub, R.N., Papadopoulos, K.P., Talbot, S., Keohan, M.L., Brehm, M., Gnjatic, S., Harris, P.E., Bisikirska, B., Guillaume, P., Cerottini, J.C., Hesdorffer, C.S., Old, L.J. and Valmori, D. (2002) Multiepitope CD8(+) T cell response to a NY-ESO-1 peptide vaccine results in imprecise tumor targeting. *The Journal of Clinical Investigation*, **110**, 1813–1822.

117 So, T., Hanagiri, T., Chapiro, J., Colau, D., Brasseur, F., Yasumoto, K., Boon, T. and Coulie, P.G. (2007) Lack of tumor recognition by cytolytic T lymphocyte clones recognizing peptide 195–203 encoded by gene MAGE-A3 and presented by HLA-A24 molecules. *Cancer Immunology, Immunotherapy*, **56**, 259–269.

118 Zhang, B., Bowerman, N.A., Salama, J.K., Schmidt, H., Spiotto, M.T., Schietinger, A., Yu, P., Fu, Y.X., Weichselbaum, R.R., Rowley, D.A., Kranz, D.M. and Schreiber, H. (2007) Induced sensitization of tumor stroma leads to eradication of established cancer by T cells. *The Journal of Experimental Medicine*, **204**, 49–55.

2
Human Tumor Antigens as Targets of Immunosurveillance and Candidates for Cancer Vaccines

Olivera J. Finn, Robert J. Binder, Anthony G. Brickner, Lisa H. Butterfield, Robert L. Ferris, Pawel Kalinski, Hideho Okada, Walter J. Storkus, Theresa L. Whiteside, and Hassane M. Zarour[*]

2.1
Introduction

Cancer remains a leading cause of death in the developed world and is slowly rising to the top in the developing world. Surgery, radiation, and chemotherapy can lead to a temporary remission, but are not effective in preventing cancer recurrence. The newly acquired knowledge in cancer biology yields potential new targets for more specific and thus potentially more effective targeted therapies of cancer. The prototype of a targeted therapy is immunotherapy. Sophisticated animal models and improved understanding of the various immune effector mechanisms have revealed that the immune system can effectively control cancer growth [1]. However, under certain circumstances tumors escape immune control and become clinical disease. Learning about the successes and the failures of immune surveillance of cancer can reveal important effector mechanisms and guide approaches towards more effective immunotherapies. The goal would be to prepare the patient's immune system for the encounter with primary cancer using prophylactic vaccines, or recurrent cancer using therapeutic vaccines.

Development of cancer vaccines at the bench and in the clinic has followed two major approaches: the use of tumor cells or cell lysates as immunogens or the use of well-characterized tumor antigens. In this chapter, we will focus on the second approach. A large number of tumor antigens have been characterized over the last two decades [2]. Most are derived from self-proteins that are either mutated or otherwise differentially expressed between normal and tumor cells. We have selected several major categories of tumor antigens and one or more representative molecules from each category, to review what has been learned about them as targets

---

[*] Co-authors listed in alphabetical order.

for tumor immunosurveillance and as candidate antigens for prophylactic and therapeutic cancer vaccines.

2.2
Tumor Antigen Classes

2.2.1
Oncofetal Antigens

Molecules that are highly expressed during fetal development, but transcriptionally silenced after birth were among the earliest candidates for tumor-specific antigens. Tumors were found to re-express these molecules, while the corresponding normal cells remained negative. Some of the most studied are the Oncofetal Antigen/Immature Laminin Receptor (OFA/iLRP) (reviewed in [3]), Glypican3, a cell-surface heparan sulfate proteoglycan [4], and the Alpha Fetoprotein (AFP), a member of the albumin family [5]. The carcinoembryonic antigen (CEA), originally considered to be an oncofetal antigen strictly expressed in the fetal colon and colon cancer, exhibits low expression in the healthy adult colon and is overexpressed rather than re-expressed in colon cancer [6] (see also chapter 11).

AFP is the major fetal serum protein (3 mg/ml at 10–13 weeks of development) that is reduced to 30–100 µg/ml at birth, and 1–3 ng/ml in adult life [5, 7]. Fifty to 80% of hepatocellular carcinomas (HCC) reactivate AFP, secreting up to 1 mg/ml in serum. In the 1970s and 1980s targeting AFP with antibody-based therapies was largely ineffective. The rekindled interest in this target during recent years is based on the availability of more suitable animal models, as well as on the novel knowledge that molecules like AFP provide targets for specific T cells. In a murine model, AFP can serve as a tumor rejection antigen [8]. Human and murine dendritic cells (DC) can process and present multiple AFP peptide epitopes to $CD8^+$ T cells [8–12].

The adult immune system is exposed to AFP during some non-malignant events, such as pregnancy or hepatic injury, which leads to a low, but detectable frequency of AFP-specific, IFN-γ producing $CD4^+$ T cells [13]. AFP-specific $CD4^+$ T cells are detected in HCC patients [14, 15], but are reduced or eliminated in advanced stages of disease [13]. AFP-specific CTL have also been found in HCC patients [11, 16] with frequencies varying with stage and severity of disease [15, 17, 18] or type of therapy [18]. Clearly, re-expression of this molecule in adults is under the surveillance of the immune system, although further studies will be required to determine whether AFP plays a role in immunosurveillance of tumors.

In a published phase I clinical vaccine trial, stage IV HCC patients received four HLA-A2.1-restricted AFP peptides (100–500 mg) emulsified in Montanide ISA-51 adjuvant [11]. The vaccine stimulated antigen-specific T cells but no clinical responses were detected. In the second trial, stage III and IV patients were immunized with the same peptides loaded on autologous DC. IFN-γ-producing AFP-specific T cells were elicited [10], again demonstrating immunogenicity of the AFP-based vaccine and supporting its further development and testing.

2.2.2
Oncogenes as Tumor Antigens

Mutations in cellular proto-oncogenes can generate two types of antigenic epitopes, those that are tumor-specific (peptides that encompass the mutated residues) and those that are tumor-associated (wild-type peptides over-represented owing to abnormally high expression of the mutated oncogenes in tumor cells). Good examples of the first type are the family of RAS oncogenes which, as a result of mutations, drive carcinogenesis through aberrant growth signals without changes in protein expression [19, 20]. The second type is represented by p53 and HER-2, which can be either mutated and/or overexpressed in tumor cells [20].

Mutations in the RAS genes are frequently found in human malignancies. *K-RAS* mutations are particularly characteristic of pancreatic and colorectal carcinoma [21]. Most RAS mutations are confined to codons 12, 13 and 61 [22] and T-cell responses to mutated peptides and proteins can either occur spontaneously in patients or can be elicited through vaccination of healthy individuals [23–29] and cancer patients [30, 31]. Both $CD4^+$ and $CD8^+$ T cells can be elicited that recognize HLA-matched cancer cell lines expressing the corresponding mutation [32, 33] demonstrating that relevant peptide epitopes are generated by the endogenous processing of the mutant molecule in tumor cells. Thus, the peripheral blood repertoire in healthy individuals and cancer patients contains T cells capable of recognizing mutant RAS and these T cells can be selectively expanded by vaccination.

Missense mutations in the *p53* tumor suppressor gene are the most common genetic event associated with human cancer [34]. Numerous HLA class I-restricted cytotoxic T lymphocyte (CTL)-defined tumor epitopes could theoretically be generated that incorporate these mutations [35, 36] providing that the mutation does not abrogate its processing or presentation [37]. However, because of the constraints imposed by HLA class I antigen processing and presentation, p53 missense mutations generate very few epitopes, and vaccines targeting these mutations are generally considered to have limited applicability. In contrast, epitopes derived from the wild-type sequence of overexpressed p53 can be numerous and are considered good candidates for developing broadly applicable cancer vaccines [38–42].

Work in preclinical models [36, 43] demonstrated efficient rejection of p53-expressing tumors in mice immunized with DCs pulsed with wild-type p53 peptides. This led to the development of p53 vaccines in patients with colon, ovarian, and head and neck cancer. A phase I trial at the University of Pittsburgh tested a multi-epitope vaccine based on wild-type p53 peptides as targets of cytotoxic and T helper cells, while other trials have used p53-transfected DC as vaccines [44, 45]. p53 vaccines have induced immune responses [46] and in some instances improved clinical responses to subsequent chemotherapy [45].

2.2.3
Overexpressed Normal Molecules as Tumor Antigens

This is a large category of molecules whose tumor-specific expression is characterized by gross overexpression compared to the levels in normal cells. This overexpression often leads to multiple posttranslational modifications, such as aberrant glycosylation, phosphorylation, or inappropriate localization in the cell. Certain oncogenes, like p53 and HER-2, in addition to being mutated in some tumors, can also be overexpressed thereby further modulating their function. Other well known tumor antigens in this category are tumor mucins [47], cell cyclins [48], CEA [6], and PSA [49].

The carcinoembryonic antigen (CEA) is an oncofetal glycoprotein that is expressed in many cancer types, including colorectal, gastric and pancreatic carcinomas [50]. Circulating CEA can be detected in the majority of patients with CEA^+ tumors and has been used as a marker to monitor patients' responses to therapy. Preclinical studies in transgenic animal models have shown that induction of immunity to CEA through various vaccination protocols can protect animals from tumor challenge [51–53]. Over the years, several groups have developed and tested CEA-based vaccines in the clinic [54–57] showing that these vaccines are safe, immunogenic and in some patients appear to have beneficial outcome.

Mucins are a large family of epithelial cell surface or secreted glycoproteins whose expression drastically changes in malignant cells. Thus they have been explored both as markers of malignancy as well as targets for therapy [47]. MUC1 is the first mucin to be characterized as a target of human cellular and humoral immune responses due to changes in its expression and glycosylation in tumor cells [58]. Patients with $MUC1^+$ tumors, which include most human adenocarcinomas, have antibodies and T cells specific for various MUC1 epitopes and these immune responses are involved in the control of tumor growth [59, 60]. Abnormal expression and glycosylation of MUC1 has been associated with several non-malignant conditions including infections and inflammation of ductal epithelia. These conditions can also lead to the generation of antibodies and T cells specific for abnormal MUC1, which appear to be involved in immunosurveillance and protection against $MUC1^+$ tumors [61, 62]. MUC1-based vaccines employing various adjuvants and delivery systems are being tested in phase I and phase II trials in patients with breast, colon, lung and pancreatic cancer [63–67]. The vaccines are safe, immunogenic and in most of the trials beneficial effects have been observed in the form of disease stabilization or longer survival of patients.

2.2.4
Cancer-Testis (CT) Antigens

These molecules are expressed by tumors of many histological types [68], but not by normal tissues except testis (spermatogonia and primary spermatocytes) [69]. Since male germline cells do not express HLA molecules, they cannot present CT-derived epitopes to T cells and thus should not serve as targets of CT-antigen-based anti-cancer immune responses [70]. This was confirmed in mice vaccinated with the CT antigen P1A since no autoimmune damage of the testes was found [68]. The human MAGE-1 antigen was the first CT antigen to be discovered [71]. The list now includes MAGE, BAGE [72], GAGE [73], NY-ESO-1/LAGE [74, 75], SCP-1/HOM-Tes-14, SSX-2/HOMEL-40 [76], CT9 [77], CT10 [78], SAGE [79], and CAGE [80]. Twenty-two CT antigens map to chromosome X (CT-X), and 12 of them cluster to the telomeric end of the q arm. Twenty CT antigens are distributed throughout the genome (non-X CTs). The restricted expression of the CT antigens, as well as the availability of multiple CT epitopes (see http://www.cancerimmunity.org/peptidedatabase/Tcellepitopes.htm) serves as a basis for their inclusion in several cancer vaccines. The recent discovery of CT-derived class II epitopes that can either be broadly presented by multiple MHC class II molecules [81, 82] or by MHC molecules commonly expressed by a large number of cancer patients [83, 84] has rekindled interest in CT-based cancer vaccines.

NY-ESO-1 appears to be the best at inducing spontaneous T cell and antibody responses in patients with NY-ESO-1-expressing cancer [85]. In those who have no preexisting immunity, intradermal injection of HLA-A2-restricted NY-ESO-1 peptides and GM-CSF vaccine induced CTL responses [85]. Evidence of tumor regression was also seen in these patients. A placebo-controlled clinical trial evaluated the safety and immunogenicity of intramuscular injections of NY-ESO-1 in Iscomatrix adjuvant in patients with NY-ESO-1$^+$ tumors. The vaccine was well tolerated and able to induce high-titer antibody responses, strong DTH reactions, and circulating CD8$^+$ and CD4$^+$ T cells specific for a broad range of NY-ESO-1 epitopes [86]. Most recently, a small trial of NY-ESO-1 protein and CpG in Montanide in patients with NY-ESO-1 positive or negative tumors resulted in the early induction of specific CD4$^+$ T cells and antibody responses in most vaccinated patients followed by the later development of CD8$^+$ T cell responses in a fraction of patients [87].

2.2.5
Minor Histocompatibility Antigens (mHAgs) as Tumor Antigens

The so-called graft versus leukemia (GVL) effect in hematologic malignancies is the most extensively studied and clinically validated example of immunological antitumor responses after allogeneic hematopoietic cell transplant (alloHCT). Hematopoietic malignancies exhibit different sensitivities to the GVL effect. Chronic myelogenous leukemia (CML) is particularly sensitive to GVL as exemplified by the 5.4-fold increased risk of relapse in recipients of T-cell depleted alloHCT [88] and the ~60–80% complete molecular remission rate observed in patients who receive donor lymphocyte infusion (DLI) for post-alloHCT relapse [89, 90]. There is also compelling clinical evidence for a graft versus tumor (GVT) effect contributing to the efficacy of alloHCT as treatment for mature B cell malignancies, particularly chronic lymphocytic leukemia (CLL), follicular lymphoma, and multiple myeloma [91–93]. Both potent graft versus tumor (GVT) and graft versus host disease (GVHD) are mediated by high-avidity donor T cells recognizing minor histocompatibility antigens (mHAgs) on recipient cells [94]. mHAgs originate from allelic polymorphisms for which MHC-matched alloHCT donor and recipient are disparate. mHAg allelic variations range from a deletion of an entire gene [95] to single nucleotide polymorphisms resulting in differences in mRNA splicing [96], protein translation [97, 98], proteasomal cleavage [99], protein splicing [100], peptide transport [101], MHC binding [102], and T cell receptor contact [103]. Recent studies have shown that antibodies are also capable of recognizing a subset of human mHAgs [104].

Approaches to using mHAg vaccines to boost GVL or GVT effects include vaccinating the alloHCT donor or the recipient post-alloHCT or post-DLI. Vaccination of a healthy donor, unless performed *ex vivo* using donor dendritic cells, is still under ethical constraints. Therefore administration of mHAg vaccine to the patient to boost mHAg-specific donor T cells primed by mHAg on recipient's cells currently represents the most feasible approach [105]. A clinical trial is currently ongoing testing a vaccine based on the well-known mHAgs HA-1/HA-2 in HLA-A2-positive patients with advanced hematologic malignancies receiving appropriately mHAg-mismatched alloHCT (Dr. Koen van Besien, University of Chicago, personal communication). Another ongoing trial is a phase I trial of a vaccine based on mHAgs ACC-1, ACC-2 and HA-1 peptides in patients with recurring hematological malignancies post-alloHCT (reviewed in [106]). As these trials show feasibility and safety, and as additional minor antigens are identified, more clinical testing of mHAg-based vaccines can be expected.

2.2.6
Human Melanoma Antigens

Using antibodies or T cells isolated from melanoma patients or melanoma mouse models, modern genomic and proteomic approaches have yielded a library of gene products that are differentially (over)-expressed in melanomas compared to normal melanocytes [107–109]. Given that some of these antigens are also expressed by other cancers and discussed elsewhere in this chapter, the focus here will be on melanoma-associated antigens (MAA) shared with melanocytes, the specialized skin cell type from which melanomas evolve.

MAA of the melanocytic lineage, such as tyrosinase, gp100/pmel17, MART-1, tyrosinase-related proteins-1/-2, and melanocortin receptor, are involved in melanogenesis and cutaneous pigmentation [107, 110, 111]. Although these MAA represent normal 'selfproteins', antigen-specific, spontaneous or inducible $CD4^+$ and $CD8^+$ T-cell responses specific for epitopes derived from these molecules have been reported in melanoma patients [112–115]. This suggests an active surveillance against neoplastic/malignant melanocytes that would be expected to result in the selection of antigen-/epitope-deficient tumor variants over extended periods of progressive tumor growth [112, 116]. Melanoma variants may evolve by the loss of MAA gene transcription, loss of HLA, or via alterations in the antigen-processing machinery of melanoma cells that may impair their ability to generate antigenic peptides [117]. This is best exemplified by the careful inspection of recurrent melanoma lesions in a patient who had previously demonstrated a dramatic response to a vaccine consisting of MAA peptide epitopes admixed with Incomplete Freund's Adjuvant (IFA). In this individual, one recurrent lesion displayed the coordinate loss of multiple MAA, while a second distal lesion exhibited loss of a relevant HLA allele [118]. Several other examples suggest that the frequency and intensity of expression of MAA may decrease as a consequence of immunotherapy, particularly in patients who showed clinical responses [119–122]. Such modulation in MAA expression are both melanoma intrinsic and extrinsic, since IFN-γ produced by type-1 tumor-infiltrating T cells has been reported to loco-regionally suppress melanoma expression of MART-1, gp100, and TRP-1 in a coordinate fashion *in situ* [123].

These observations prove that the adaptive immune system monitors and regulates melanoma *in vivo* and suggest that if a sufficiently large and diverse response could be prompted by vaccination and sustained in patients, regression (or prevention of recurrence) of antigenically heterogeneous disease might be achieved. Over the past 15 years, a substantial number of phase I/II clinical trials investigating the safety and efficacy of vaccines based on MAA lineage antigen-derived peptides or cDNA have been undertaken. Some of the recently published trials have used a large variety of vaccine delivery methods and antigens [124–131]. Data from these trials show that vaccines can be used safely, and that in the majority of patients, they enhance circulating frequencies of MAA-specific T cells. However, objective clinical responses are still only seen in a minority of patients. This may be due to the action of many tumor-associated suppression mechanisms (as discussed elsewhere in this chapter), melanoma heterogeneity (particularly in the advanced-stage disease), inability of activated T cells to traffic into tumor sites, or the failure to expand sufficient numbers of high avidity T cells.

2.2.7
Human Glioma Antigens and Immunosurveillance in the CNS

Although immune mechanisms maintaining the 'immunologically privileged' status of the central nervous system (CNS) and CNS tumors have been well characterized during the past decade, it has also become clear that this 'privileged' status is not absolute. Systemic immunization with CNS-specific antigens in mice induces auto-immune responses that are manifested in the CNS [132]. Also, in paraneoplastic cerebellar degeneration [133, 134], the essential pathogenesis has been revealed to be a specific T-cell response against an antigen shared by breast and ovarian cancers and normal cerebellum. These findings show that a systemic immune response can access the CNS. Thus vaccines based on carefully selected brain tumor antigens, administered in the periphery should be able to induce an immune response that would act in the CNS, the site of the tumor.

Glioblastoma multiforme (GBM) is by far the most common and most malignant glial tumor which also includes anaplastic astrocytomas (AA). Patients with GBM have a median survival of approximately 12 months, whereas those with AA have a median survival of 24 to 36 months. In addition, low-grade gliomas often progress to more malignant gliomas when they recur [135]. Despite extensive research, no significant advances in the treatment of GBM have been achieved in the past 25 years. The primary reason is that the tumor is beyond the reach of local control when it is first detected clinically or radiologically. As antigen-specific CD8$^+$ cytotoxic T-lymphocytes (CTL) have been shown in animal models to be able to survey the CNS parenchyma and selectively kill astrocytes that express a model antigen hemagglutinin (HA) without collaterally damaging neurons and oligodentrocytes or myelin [136], T cell targeting of glioma-associated antigens (GAAs) may be a promising therapeutic approach. This has prompted efforts to analyze the antigenic profile of glioblastoma cells.

The following antigens are examples of known targets on other tumor types that are also expressed on brain tumors. IL-13Rα2 is a membrane glycoprotein that is overexpressed by >80% of malignant gliomas, but not by normal brain tissues or other normal organs except the testis. IL-13Rα2 has attracted significant attention as a target for glioma therapy [137]. EphA2 is a tyrosine kinase receptor known to play a role in carcinogenesis [138, 139]. In normal cells, EphA2 localizes to sites of cell–cell contact [140], where it may negatively regulate cell growth. In cancers, EphA2 is frequently overexpressed and often functionally dysregulated, which contributes to their malignant phenotype [141]. Peptide EphA2$_{883-891}$ can elicit an HLA-A2-restricted CTL response against glioma cell lines [142]. Recent studies have revealed that a majority of malignant gliomas express high levels of EphA2 [142]. EphA2 mRNA overexpression was found to correlate inversely with survival [143]. These findings support the proposal that targeting of EphA2 by immunotherapy may have a major impact in controlling tumor growth and prolonging patients' survival. Survivin is an apoptosis inhibitory protein that is overexpressed in most human cancers and inhibition of its function results in increased apoptosis [144]. Survivin-based vaccines have shown immunogenicity and efficacy in pancreatic cancer [145] and advanced melanoma [146]. Astrocytomas (grades II–IV), but not normal brain tissues were shown to express high levels of survivin which directly correlated with poor prognosis [147]. Encouraging results from vaccine trials in other tumors support using survivin peptide-based vaccines for brain cancer as well. WT1 an oncogenic transcription factor and a product of the Wilm's tumor oncogene, overexpressed in leukemias and solid tumors [148, 149]. Analysis of primary GBM showed expression of WT1 protein in the majority of tumor samples [150], while no expression was detected in normal glial cells [151]. Epidermal Growth Factor Receptor (EGFR) vIII, the type III variant of the epidermal growth factor receptor (EGFRvIII), is present in 20–25% of GBM. A peptide derived from EGFRvIII was shown to induce modest IFN-γ and CTL responses against HLA-A2$^+$ EGFRvIII$^+$ glioma cells [152].

2.2.8
Heat Shock Proteins, Efficient Carriers of Tumor Antigens

Immunization of mice with irradiated carcinogen-induced tumor cells elicits immune protection to a tumor challenge, but only if the challenge tumor is the same as the immunizing tumor. When tumor cell proteins were biochemically fractionated and each fraction tested for its ability to reject a tumor challenge upon immunization, immunogenicity was found in those fractions containing heat shock proteins (HSPs). To date the immunogenic HSPs that have been tested include gp96, hsp90, hsp70, calreticulin, hsp110, and grp170. HSPs are associated with numerous peptides, among, which are unknown tumor-specific peptides responsible for eliciting tumor rejection responses [153]. The HSP molecule *per se* is therefore not a tumor antigen but rather a carrier of the entire antigenic peptide repertoire of the tumor cell.

In the prevention setting, mice immunized with tumor-derived HSPs reject a subsequent challenge of live tumor cells. Therapeutic vaccination of tumor-bearing mice with HSPs retards the growth of the primary tumor in a majority of mice [154] and significantly impairs metastasis. Depletion of $CD8^+$ or $CD4^+$ T cells or Natural Killer (NK) cells abrogates these effects.

Clinical testing of HSP-based vaccines began in 1995 in patients with renal cell carcinoma and melanoma [155–157]. These trials demonstrated that autologous HSP vaccines elicited robust and safe T-cell responses. They also demonstrated the feasibility of the manufacture of autologous HSP vaccines from each patient's tumor. Subsequent phase I and II trials were carried out in patients with pancreatic cancer and colon cancer with similar outcomes [158–160]. In the first completed phase III trial, approximately 300 patients with stage IV melanoma were accrued for a randomized, open label, multi-center study. Patients were vaccinated weekly for the first 4 weeks and every other week after that for as long as the vaccine lasted. The patients in the control arm received the physician's choice of treatment that consisted of some combination of dacarbazine, temozolomide, Interleukin-2, and surgery. The intent-to-treat analysis showed no significant difference between the two groups. However, the prospective subset analysis showed a significant correlation between numbers of immunizations and survival [161].

2.2.9
Dendritic Cells, Efficient Cross-presenters of Tumor Antigens

Difficulties in eliciting a strong immune response to various tumor antigens in the therapeutic setting revealed multiple deficiencies in the patients' immune system, one of which was an overall reduction in the stimulatory capacity of their DC [162, 163]. This suggested the use of *ex-vivo*-generated DCs as carriers of cancer vaccines [164].

DCs provide T cells with the Ag-specific 'signal 1' and a co-stimulatory 'signal 2' [165–167] required for the proper activation and expansion of antigen-specific T cells. DCs also provide an additional 'signal 3' (polarization), driving the development of immune responses towards type-1 or type-2 immunity [168] associated with differential involvement of particular effector mechanisms and different abilities to induce cancer rejection [25–35]. An additional signal (tentatively termed 'Signal 4') may regulate organ-specific trafficking of immune cells [169–175]. In addition to their role as the initiators of antigen-specific $CD4^+$ and $CD8^+$ T cells, DCs have also been shown to support the tumoricidal activity of NK cells [176].

Following their original application as vaccines in patients with B cell lymphoma [177] and melanoma [178], DC-based vaccines have been tested in over 10 000 patients as therapeutic agents against advanced cancer. However, despite frequently observed massive expansion of tumor-specific T cells, particularly in patients treated with mature DCs, and prolonged survival in some patients [179–181], the frequency of objective clinical responses induced by the currently used DC-based vaccines is still below expectations [179, 180, 182, 183]. As most of these trials have been phase I or phase II, the results are not unexpected and the trials have served their intended role by lightning barriers to better success of DC-based cancer vaccines.

Several features of DCs, including their maturation status, migratory potential, and cytokine production were shown to be important. Effective induction of anti-tumor CTL responses required mature DCs that express high levels of co-stimulatory molecules and could migrate in response to CCL19 or CCL21, the lymph node-produced CCR7 ligands [184–186]. In addition, high IL-12p70 secretion was shown (1) to dramatically enhance the ability of DCs to induce tumor-specific Th1 cells and CTLs and (2) to promote tumor rejection in therapeutic mouse models [54–64]. Unfortunately, since the maturation stage of DCs obtained in the early protocols inversely correlated with their ability to produce IL-12p70 [168, 187], the desirable combination of all three main features: high immunostimulatory function, high migratory activity, and high capacity to produce IL-12p70, were not all simultaneously attained or even reciprocally exclusive.

The 'first-generation' DC-based vaccines utilized relatively immature or only partially mature DCs that were immunogenic [177, 178], but suboptimal with regard to their lymph-node homing ability and T cell-stimulating potential [184, 185]. 'Second-generation' DC-based vaccines, matured in the presence of an IL-1β/TNF-α/IL-6/PGE_2-containing cytokine cocktail [188], showed a desirable fully mature status, but a reduced ability to produce bioactive IL-12p70 [168, 187, 189]. Thus, although such 'second generation' DC-based vaccines are clearly superior to immature DCs with respect to their immunostimulatory capacity [184, 190] and migratory responses to LN-associated chemokines [185, 191, 192], the desirable combination of the main three features was not attained.

With this knowledge, improvements are being made to the latest DC-based vaccines resulting in a dramatic increase in their capacity to induce long-lived tumor-specific T cells with strongly pronounced anti-tumor effector functions in human *in vitro* and mouse *in vivo* models. For example, combination of IFN-γ with LPS or DC maturation-inducing cytokines TNF-α and IL-1β overcomes the DC maturation-associated 'exhaustion' yielding stable type-1 polarized DCs that produce strongly elevated levels of IL-12p70 upon interaction with CD40L-expressing $CD4^+$ Th cells and induce much stronger Th1 type and CTL responses [193, 194]. DC1 with similar properties can be effectively induced by the 'two-signal-activated' autologous NK cells or memory-type $CD8^+$ T cells [195, 196].

2.2.10
Spontaneous and Vaccine-induced Immunity and the Tumor Microenvironment

The tumor microenvironment is comprised of proliferating tumor cells, the tumor stroma, blood vessels, infiltrating inflammatory cells and a variety of associated tissue cells. It is a unique environment that emerges in the course of tumor progression as a result of its interactions with the host. It is at all times shaped and dominated by the tumor, which orchestrates the molecular and cellular events taking place in surrounding tissues.

Inflammatory infiltrates are usually an abundant component of the tumor microenvironment [197], although the origin, cellular composition, and functional attributes of the cells accumulating in pre-malignant as well as malignant lesions are variable. Inflammatory cells either contribute to tumor progression or actively interfere with its development (reviewed in [198]). The presence of inflammatory cells in pre-cancerous lesions has been taken as evidence for *'immune surveillance'*, that is the ability of the host immune system to detect tumor cells and eliminate them [198]. However, in a growing tumor, immune-mediated demise of susceptible tumor cells may lead to *'immune selection'* or replacement of immuno-sensitive by immuno-resistant tumor cells thus favoring tumor progression [199]. As tumors continue to grow, they establish mechanisms responsible for *'immune evasion'* which plays a key role in down-regulating anti-tumor host responses [200]. Ultimately, this process results in *'immune deviation'* or re-shaping of the initially competent host immune system in such a way that anti-tumor immune responses are suppressed, thus allowing tumor dissemination [200, 201].

Immune cells present in the tumor include those mediating adaptive immunity, T lymphocytes, dendritic cells (DC) and occasional B cells, as well as effectors of innate immunity, macrophages, polymorphonuclear leukocytes, and rarely NK cells [198]. In some tumors, e.g. medullary breast carcinomas, infiltrating lymphocytes form lymph node-like structures suggesting that the immune response is operating *in situ* [202]. Indeed, tumor-infiltrating lymphocytes (TIL) with specificity for tumor-associated antigens (TAA) are often a major component of the tumor microenvironment, as indicated by tetramer analysis of $CD8^+$ T cells isolated from human tumors [203]. Also, TIL are a source of tumor-specific lymphocytes used for adoptive transfers after expansion in IL-2-containing culture media [204]. TIL clones with the specificity for a broad variety of the TAA discussed above can be grown from human tumors confirming that immune responses directed not only against 'unique' antigens expressed by the tumor, but also against a range of differentiation or tissue-specific antigens, are generated by the host [205]. While accumulations of these effector T cells in the tumor might be considered as evidence of immunosurveillance by the host, they are largely ineffective in arresting tumor growth. This is due to immunosuppressive mechanisms mediated by the tumor leading to a partial or complete inhibition of local and systemic innate as well as adaptive anti-tumor immunity. Thus, tumors are not passive targets for host immunity; they actively down-regulate all phases of anti-tumor immune responses and the tumor microenvironment, once established, represents a consistently effective barrier to immune cell functions.

2.3
Summary

The failure of immunosurveillance in tumor-bearing hosts has been one of the major incentives for the development of cancer immunotherapy, including anti-tumor vaccines. Numerous animal tumor models have provided strong evidence that in the presence of effective anti-tumor immunity, tumors fail to progress and established tumors regress [199]. Hence, recovery of immune surveillance and protection of immune cells from tumor-induced suppression are becoming well-rationalized objectives of anti-tumor therapies. As the molecular mechanisms involved in tumor-induced immune suppression are uncovered and defined, these objectives are becoming more realistic. While prophylactic tumor-specific cancer vaccines would obviously represent a preferable means of arresting tumor development or recurrence by *a priori* strengthening immune surveillance, in humans, this approach is currently limited to anti-viral vaccines targeting carcinogenic viruses such as HPV, EBV, HBV, or HCV. The remarkably successful development and wide-scale application of anti-HPV vaccines is likely to make a tremendous impact on the incidence of cervical carcinoma in the future [205]. Therapeutic cancer vaccines currently in clinical trials worldwide aim at generating or strengthening tumor-specific immune responses in patients with established and often advanced disease. To date, these trials have yielded a significant amount of scientific information, but they have been a major disappointment in terms of clinical results and correlations between immunological responses and clinical outcome [183]. This limited success of cancer vaccines despite the improved vaccine designs and the use of well-defined TAA as immunogens, suggests that different and novel approaches to therapeutic cancer vaccines are necessary. With an improved understanding of mechanisms underlying tumor-induced immune suppression, future vaccines will likely combine approaches designed to restore anti-tumor immune responses, eliminate tumor escape, and correct tumor-induced immune deviation to enable the host's immune system to more effectively deal with the tumor.

References

1 Dunn, G.P., Koebel, C.M. and Schreiber, R.D. (2006) *Nature Reviews. Immunology*, **6**, 836–848.

2 Graziano, D.F. and Finn, O.J. (2005) *Cancer Treatment and Research*, **123**, 89–111.

3 Coggin, J.H., Jr. Barsoum, A.L. and Rohrer, J.W. (1998) *Immunology Today*, **19**, 405–408.

4 Lander, A.D. and Selleck, S.B. (2000) *The Journal of Cell Biology*, **148**, 227–232.

5 Mizejewski, G.J. (1995) *Critical Reviews in Eukaryotic Gene Expression*, **5**, 281–316.

6 Hammarstrom, S. (1999) *Seminars in Cancer Biology*, **9**, 67–81.

7 Mizejewski, G.J. and MacColl, R. (2003) *Molecular Cancer Therapeutics*, **2**, 1243–1255.

8 Butterfield, L.H., Koh, A., Meng, W., Vollmer, C.M., Ribas, A., Dissette, V., Lee, E., Glaspy, J.A., McBride, W.H. and

Economou, J.S. (1999) *Cancer Research*, **59**, 3134–3142.

9 Butterfield, L.H., Meng, W.S., Koh, A., Vollmer, C.M., Ribas, A., Dissette, V.B., Faull, K., Glaspy, J.A., McBride, W.H. and Economou, J.S. (2001) *Journal of Immunology (Baltimore, Md: 1950)*, **166**, 5300–5308.

10 Butterfield, L.H., Ribas, A., Dissette, V.B., Lee, Y., Yang, J.Q., De la Rocha, P., Duran, S.D., Hernandez, J., Seja, E., Potter, D.M., McBride, W.H., Finn, R., Glaspy, J.A. and Economou, J.S. (2006) *Clinical Cancer Research*, **12**, 2817–2825.

11 Butterfield, L.H., Ribas, A., Meng, W.S., Dissette, V.B., Amarnani, S., Vu, H.T., Seja, E., Todd, K., Glaspy, J.A., McBride, W.H. and Economou, J.S. (2003) *Clinical Cancer Research*, **9**, 5902–5908.

12 Butterfield, L.H., Ribas, A., Potter, D.M. and Economou, J.S. (2007) *Cancer Immunology, Immunotherapy*, **56**, 1931–1943.

13 Evdokimova, V.N., Liu, Y., Potter, D.M. and Butterfield, L.H. (2007) *Journal of Immunotherapy*, **30**, 425–437.

14 Alisa, A., Ives, A., Pathan, A.A., Navarrete, C.V., Williams, R., Bertoletti, A. and Behboudi, S. (2005) *Clinical Cancer Research*, **11**, 6686–6694.

15 Hanke, P., Rabe, C., Serwe, M., Bohm, S., Pagenstecher, C., Sauerbruch, T. and Caselmann, W.H. (2002) *Scandinavian Journal of Gastroenterology*, **37**, 949–955.

16 Ritter, M., Ali, M.Y., Grimm, C.F., Weth, R., Mohr, L., Bocher, W.O., Endrulat, K., Wedemeyer, H., Blum, H.E. and Geissler, M. (2004) *Journal of Hepatology*, **41**, 999–1007.

17 Bei, R., Budillon, A., Reale, M.G., Capuano, G., Pomponi, D., Budillon, G., Frati, L. and Muraro, R. (1999) *Cancer Research*, **59**, 5471–5474.

18 Mizukoshi, E., Nakamoto, Y., Tsuji, H., Yamashita, T. and Kaneko, S. (2006) *International Journal of Cancer*, **118**, 1194–1204.

19 Diaz-Flores, E. and Shannon, K. (2007) *Genes and Development*, **21**, 1989–1992.

20 Spandidos, D.A. (2007) *Journal of the Balkan Union of Oncology*, **12** (Suppl 1), S9–S12.

21 Almoguera, C., Shibata, D., Forrester, K., Martin, J., Arnheim, N. and Perucho, M. (1988) *Cell*, **53**, 549–554.

22 Capella, G., Cronauer-Mitra, S., Pienado, M.A. and Perucho, M. (1991) *Environmental Health Perspectives*, **93**, 125–131.

23 Gedde-Dahl, T., 3rd Spurkland, A., Eriksen, J.A., Thorsby, E. and Gaudernack, G. (1992) *International Immunology*, **4**, 1331–1337.

24 Fossum, B., Gedde-Dahl, T., 3rd Breivik, J., Eriksen, J.A., Spurkland, A., Thorsby, E. and Gaudernack, G. (1994) *International Journal of Cancer*, **56**, 40–45.

25 Jung, S. and Schluesener, H.J. (1991) *The Journal of Experimental Medicine*, **173**, 273–276.

26 Fossum, B., Gedde-Dahl, T., 3rd Hansen, T., Eriksen, J.A., Thorsby, E. and Gaudernack, G. (1993) *European Journal of Immunology*, **23**, 2687–2691.

27 Tsang, K., Nieroda, C., DeFilippi, R., Chung, Y., Yamauc, H. Greiner, J. *et al.*, (1994) *Vaccine Research*, **3**, 183–193.

28 Qin, H., Chen, W., Takahashi, M., Disis, M.L., Byrd, D.R., McCahill, L., Bertram, K.A., Fenton, R.G., Peace, D.J. and Cheever, M.A. (1995) *Cancer Research*, **55**, 2984–2987.

29 Van Elsas, A., Nijman, H.W., Van der Minne, C.E., Mourer, J.S., Kast, W.M., Melief, C.J. and Schrier, P.I. (1995) *International Journal of Cancer*, **61**, 389–396.

30 Gjertsen, M.K., Bakka, A., Breivik, J., Saeterdal, I., Solheim, B.G., Soreide, O., Thorsby, E. and Gaudernack, G. (1995) *Lancet*, **346**, 1399–1400.

31 Khleif, S.N., Abrams, S.I., Hamilton, J.M., Bergmann-Leitner, E., Chen, A., Bastian, A., Bernstein, S., Chung, Y., Allegra, C.J. and Schlom, J. (1999) *Journal of Immunotherapy*, **22** (1997), 155–165.

32 Fossum, B., Olsen, A.C., Thorsby, E. and Gaudernack, G. (1995) *Cancer Immunology, Immunotherapy*, **40**, 165–172.

33 Abrams, S.I., Khleif, S.N., Bergmann-Leitner, E.S., Kantor, J.A., Chung, Y., Hamilton, J.M. and Schlom, J. (1997) *Cellular Immunology*, **182**, 137–151.

34 Olivier, M., Eeles, R., Hollstein, M., Khan, M.A., Harris, C.C. and Hainaut, P. (2002) *Human Mutation*, **19**, 607–614.

35 Nijman, H.W., Van der Burg, S.H., Vierboom, M.P., Houbiers, J.G., Kast, W.M. and Melief, C.J. (1994) *Immunology Letters*, **40**, 171–178.

36 Mayordomo, J.I., Loftus, D.J., Sakamoto, H., De Cesare, C.M., Appasamy, P.M., Lotze, M.T., Storkus, W.J., Appella, E. and DeLeo, A.B. (1996) *The Journal of Experimental Medicine*, **183**, 1357–1365.

37 Falk, K., Rotzschke, O., Stevanovic, S., Jung, G. and Rammensee, H.G. (1991) *Nature*, **351**, 290–296.

38 Ropke, M., Hald, J., Guldberg, P., Zeuthen, J., Norgaard, L., Fugger, L., Svejgaard, A., Van der Burg, S., Nijman, H.W., Melief, C.J. and Claesson, M.H. (1996) *Proceedings of the National Academy of Sciences of the United States of America*, **93**, 14704–14707.

39 Theobald, M., Biggs, J., Dittmer, D., Levine, A.J. and Sherman, L.A. (1995) *Proceedings of the National Academy of Sciences of the United States of America*, **92**, 11993–11997.

40 Chikamatsu, K., Nakano, K., Storkus, W.J., Appella, E., Lotze, M.T., Whiteside, T.L. and DeLeo, A.B. (1999) *Clinical Cancer Research*, **5**, 1281–1288.

41 McArdle, S.E., Rees, R.C., Mulcahy, K.A., Saba, J., McIntyre, C.A. and Murray, A.K. (2000) *Cancer Immunology, Immunotherapy*, **49**, 417–425.

42 Barfoed, A.M., Petersen, T.R., Kirkin, A.F., Thor Straten, P., Claesson, M.H. and Zeuthen, J. (2000) *Scandinavian Journal of Immunology*, **51**, 128–133.

43 Mayordomo, J.I., Zorina, T., Storkus, W.J., Zitvogel, L., Celluzzi, C., Falo, L.D., Melief, C.J., Ildstad, S.T., Kast, W.M. Deleo, A.B. et al., (1995) *Nature Medicine*, **1**, 1297–1302.

44 Menon, A.G., Kuppen, P.J., van der Burg, S.H., Offringa, R., Bonnet, M.C., Harinck, B.I., Tollenaar, R.A., Redeker, A., Putter, H., Moingeon, P., Morreau, H., Melief, C.J. and van de Velde, C.J. (2003) *Cancer Gene Therapy*, **10**, 509–517.

45 Antonia, S.J., Mirza, N., Fricke, I., Chiappori, A., Thompson, P., Williams, N., Bepler, G., Simon, G., Janssen, W., Lee, J.H., Menander, K., Chada, S. and Gabrilovich, D.I. (2006) *Clinical Cancer Research*, **12**, 878–887.

46 Herrin, V., Behrens, R.J., Achtar, M., Monahan, B., Bernstein, S., Brent-Steele, T., Whiteside, T., Wieckowski, E., Berzofsky, J. and Khleif, S.N. (2003) *Proceedings of the American Society for clinical Oncology*, p. 169 (abstr 678).

47 Hollingsworth, M.A. and Swanson, B.J. (2004) *Nature Reviews. Cancer*, **4**, 45–60.

48 Egloff, A.M., Vella, L.A. and Finn, O.J. (2006) *Cancer Research*, **66**, 6–9.

49 Freedland, S.J. and Partin, A.W. (2006) *Urology*, **67**, 458–460.

50 Thompson, J.A., Grunert, F. and Zimmermann, W. (1991) *Journal of Clinical Laboratory Analysis*, **5**, 344–366.

51 Clarke, P., Mann, J., Simpson, J.F., Rickard-Dickson, K. and Primus, F.J. (1998) *Cancer Research*, **58**, 1469–1477.

52 Zhou, H., Luo, Y., Mizutani, M., Mizutani, N., Becker, J.C., Primus, F.J., Xiang, R. and Reisfeld, R.A. (2004) *The Journal of Clinical Investigation*, **113**, 1792–1798.

53 Zeytin, H.E., Patel, A.C., Rogers, C.J., Canter, D., Hursting, S.D., Schlom, J. and Greiner, J.W. (2004) *Cancer Research*, **64**, 3668–3678.

54 Marshall, J.L., Hawkins, M.J., Tsang, K.Y., Richmond, E., Pedicano, J.E., Zhu, M.Z. and Schlom, J. (1999) *Journal of Clinical Oncology*, **17**, 332–337.

55 Conry, R.M., Khazaeli, M.B., Saleh, M.N., Allen, K.O., Barlow, D.L., Moore, S.E., Craig, D., Arani, R.B., Schlom, J. and LoBuglio, A.F. (1999) *Clinical Cancer Research*, **5**, 2330–2337.

56 Horig, H., Lee, D.S., Conkright, W., Divito, J., Hasson, H., LaMare, M., Rivera, A., Park, D., Tine, J., Guito, K., Tsang, K.W., Schlom, J. and Kaufman, H.L. (2000) *Cancer Immunology, Immunotherapy*, **49**, 504–514.

57 Marshall, J.L., Gulley, J.L., Arlen, P.M., Beetham, P.K., Tsang, K.Y., Slack, R., Hodge, J.W., Doren, S., Grosenbach, D.W., Hwang, J., Fox, E., Odogwu, L., Park, S., Panicali, D. and Schlom, J. (2005) *Journal of Clinical Oncology*, **23**, 720–731.

58 Vlad, A.M., Kettel, J.C., Alajez, N.M., Carlos, C.A. and Finn, O.J. (2004) *Advances in Immunology*, **82**, 249–293.

59 Kotera, Y., Fontenot, J.D., Pecher, G., Metzgar, R.S. and Finn, O.J. (1994) *Cancer Research*, **54**, 2856–2860.

60 von Mensdorff-Pouilly, S., Verstraeten, A.A., Kenemans, P., Snijdewint, F.G., Kok, A., Van Kamp, G.J., Paul, M.A., Van Diest, P.J., Meijer, S. and Hilgers, J. (2000) *Journal of Clinical Oncology*, **18**, 574–583.

61 Jerome, K.R., Kirk, A.D., Pecher, G., Ferguson, W.W. and Finn, O.J. (1997) *Cancer Immunology Immunotherapy*, **43**, 355–360.

62 Cramer, D.W., Titus-Ernstoff, L., McKolanis, J.R., Welch, W.R., Vitonis, A.F., Berkowitz, R.S. and Finn, O.J. (2005) *Cancer Epidemiology, Biomarkers & Prevention: A Publication of the American Association for Cancer Research, Cosponsored by the American Society of Preventive Oncology*, **14**, 1125–1131.

63 Goydos, J.S., Elder, E., Whiteside, T.L., Finn, O.J. and Lotze, M.T. (1996) *The Journal of Surgical Research*, **63**, 298–304.

64 Ramanathan, R.K., Lee, K.M., McKolanis, J., Hitbold, E., Schraut, W., Moser, A.J., Warnick, E., Whiteside, T., Osborne, J., Kim, H., Day, R., Troetschel, M. and Finn, O.J. (2005) *Cancer Immunology, Immunotherapy*, **54**, 254–264.

65 Scholl, S.M., Balloul, J.M., Le Goc, G., Bizouarne, N., Schatz, C., Kieny, M.P., von Mensdorff-Pouilly, S., Vincent-Salomon, A., Deneux, L., Tartour, E., Fridman, W., Pouillart, P. and Acres, B. (2000) *Journal of Immunotherapy*, **23** (1997), 570–580.

66 Loveland, B.E., Zhao, A., White, S., Gan, H., Hamilton, K., Xing, P.X., Pietersz, G.A., Apostolopoulos, V., Vaughan, H., Karanikas, V., Kyriakou, P., McKenzie, I.F. and Mitchell, P.L. (2006) *Clinical Cancer Research*, **12**, 869–877.

67 Butts, C., Murray, N., Maksymiuk, A., Goss, G., Marshall, E., Soulieres, D., Cormier, Y., Ellis, P., Price, A., Sawhney, R., Davis, M., Mansi, J., Smith, C., Vergidis, D., Ellis, P., MacNeil, M. and Palmer, M. 2005 *Journal of Clinical Oncology*, **23**, 6674–6681.

68 Uyttenhove, C., Godfraind, C., Lethe, B., Amar-Costesec, A., Renauld, J.C., Gajewski, T.F., Duffour, M.T., Warnier, G., Boon, T. and Van den Eynde, B.J. (1997) *International Journal of Cancer*, **70**, 349–356.

69 Jungbluth, A.A., Busam, K.J., Kolb, D., Iversen, K., Coplan, K., Chen, Y.T., Spagnoli, G.C. and Old, L.J. (2000) *International Journal of Cancer*, **85**, 460–465.

70 Haas, G.G., Jr. D'Cruz, O.J. and De Bault, L.E. (1988) *American Journal of Reproductive Immunology and Microbiology*, **18**, 47–51.

71 van der Bruggen, P., Traversari, C., Chomez, P., Lurquin, C., De Plaen, E., Van den Eynde, B., Knuth, A. and Boon, T. (1991) *Science*, **254**, 1643–1647.

72 Boel, P., Wildmann, C., Sensi, M.L., Brasseur, R., Renauld, J.C., Coulie, P., Boon, T. and van der Bruggen, P. (1995) *Immunity*, **2**, 167–175.

73 Van den Eynde, B., Peeters, O., De Backer, O., Gaugler, B., Lucas, S. and Boon, T. (1995) *The Journal of Experimental Medicine*, **182**, 689–698.

74 Lethe, B., Lucas, S., Michaux, L., De Smet, C., Godelaine, D., Serrano, A., De Plaen, E. and Boon, T. (1998) *International Journal of Cancer*, **76**, 903–908.

75 Chen, Y.T., Scanlan, M.J., Sahin, U., Tureci, O., Gure, A.O., Tsang, S.,

Williamson, B., Stockert, E., Pfreundschuh, M. and Old, L.J. (1997) *Proceedings of the National Academy of Sciences of the United States of America*, **94**, 1914–1918.

76 Tureci, O., Sahin, U., Schobert, I., Koslowski, M., Scmitt, H., Schild, H.J., Stenner, F., Seitz, G., Rammensee, H.G. and Pfreundschuh, M. (1996) *Cancer Research*, **56**, 4766–4772.

77 Scanlan, M.J., Altorki, N.K., Gure, A.O., Williamson, B., Jungbluth, A., Chen, Y.T. and Old, L.J. (2000) *Cancer Letters*, **150**, 155–164.

78 Gure, A.O., Stockert, E., Arden, K.C., Boyer, A.D., Viars, C.S., Scanlan, M.J., Old, L.J. and Chen, Y.T. (2000) *International Journal of Cancer*, **85**, 726–732.

79 Martelange, V., De Smet, C., De Plaen, E., Lurquin, C. and Boon, T. (2000) *Cancer Research*, **60**, 3848–3855.

80 Cho, B., Lim, Y., Lee, D.Y., Park, S.Y., Lee, H., Kim, W.H., Yang, H., Bang, Y.J. and Jeoung, D.I. (2002) *Biochemical and Biophysical Research Communications*, **292**, 715–726.

81 Kobayashi, H., Song, Y., Hoon, D.S., Appella, E. and Celis, E. (2001) *Cancer Research*, **61**, 4773–4778.

82 Zarour, H.M., Maillere, B., Brusic, V., Coval, K., Williams, E., Pouvelle-Moratille, S., Castelli, F., Land, S., Bennouna, J., Logan, T. and Kirkwood, J.M. (2002) *Cancer Research*, **62**, 213–218.

83 Schultz, E.S., Lethe, B., Cambiaso, C.L., Van Snick, J., Chaux, P., Corthals, J., Heirman, C., Thielemans, K., Boon, T. and van der Bruggen, P. (2000) *Cancer Research*, **60**, 6272–6275.

84 Zeng, G., Wang, X., Robbins, P.F., Rosenberg, S.A. and Wang, R.F. (2001) *Proceedings of the National Academy of Sciences of the United States of America*, **98**, 3964–3969.

85 Jager, E., Gnjatic, S., Nagata, Y., Stockert, E., Jager, D., Karbach, J., Neumann, A., Rieckenberg, J., Chen, Y.T., Ritter, G., Hoffman, E., Arand, M., Old, L.J. and Knuth, A. (2000) *Proceedings of the National Academy of Sciences of the United States of America*, **97**, 12198–12203.

86 Davis, I.D., Chen, W., Jackson, H., Parente, P., Shackleton, M., Hopkins, W., Chen, Q., Dimopoulos, N., Luke, T., Murphy, R., Scott, A.M., Maraskovsky, E., McArthur, G., MacGregor, D., Sturrock, S., Tai, T.Y., Green, S., Cuthbertson, A., Maher, D., Miloradovic, L., Mitchell, S.V., Ritter, G., Jungbluth, A.A., Chen, Y.T., Gnjatic, S., Hoffman, E.W., Old, L.J. and Cebon, J.S. (2004) *Proceedings of the National Academy of Sciences of the United States of America*, **101**, 10697–10702.

87 Gajewski, T.F. (2006) *Clinical Cancer Research*, **12**, 2326s–2330.

88 Goldman, J.M., Gale, R.P., Horowitz, M.M., Biggs, J.C., Champlin, R.E., Gluckman, E., Hoffmann, R.G., Jacobsen, S.J., Marmont, A.M. McGlave, P.B. et al., (1988) *Annals of Internal Medicine*, **108**, 806–814.

89 Porter, D.L., Roth, M.S., McGarigle, C., Ferrara, J.L. and Antin, J.H. (1994) *The New England Journal of Medicine*, **330**, 100–106.

90 Kolb, H.J., Mittermuller, J., Clemm, C., Holler, E., Ledderose, G., Brehm, G., Heim, M. and Wilmanns, W. (1990) *Blood*, **76**, 2462–2465.

91 Gribben, J.G. (2007) *Best Practice & Research Clinical Haematology*, **20**, 513–527.

92 Sorror, M.L., Maris, M.B., Sandmaier, B.M., Storer, B.E., Stuart, M.J., Hegenbart, U., Agura, E., Chauncey, T.R., Leis, J., Pulsipher, M., McSweeney, P., Radich, J.P., Bredeson, C., Bruno, B., Langston, A., Loken, M.R., Al-Ali, H., Blume, K.G., Storb, R. and Maloney, D.G. (2005) *Journal of Clinical Oncology*, **23**, 3819–3829.

93 Salama, M., Nevill, T., Marcellus, D., Parker, P., Johnson, M., Kirk, A., Porter, D., Giralt, S., Levine, J.E., Drobyski, W., Barrett, A.J., Horowitz, M. and Collins, R.H. (2000) *Bone Marrow Transplantation*, **26**, 1179–1184.

94 Goulmy, E. (1997) *Immunological Reviews*, **157**, 125–140.

95 Murata, M., Warren, E.H. and Riddell, S.R. (2003) *The Journal of Experimental Medicine*, **197**, 1279–1289.

96 Kawase, T., Akatsuka, Y., Torikai, H., Morishima, S., Oka, A., Tsujimura, A., Miyazaki, M., Tsujimura, K., Miyamura, K., Ogawa, S., Inoko, H., Morishima, Y., Kodera, Y., Kuzushima, K. and Takahashi, T. (2007) *Blood*, **110**, 1055–1063.

97 de Rijke, B., van Horssen-Zoetbrood, A., Beekman, J.M., Otterud, B., Maas, F., Woestenenk, R., Kester, M., Leppert, M., Schattenberg, A.V., de Witte, T., van de Wiel-van Kemenade, E. and Dolstra, H. (2005) *The Journal of Clinical Investigation*, **115**, 3506–3516.

98 Brickner, A.G., Evans, A.M., Mito, J.K., Xuereb, S.M., Feng, X., Nishida, T., Fairfull, L., Ferrell, R.E., Foon, K.A., Hunt, D.F., Shabanowitz, J., Engelhard, V.H., Riddell, S.R. and Warren, E.H. (2006) *Blood*, **107**, 3779–3786.

99 Spierings, E., Brickner, A.G., Caldwell, J.A., Zegveld, S., Tatsis, N., Blokland, E., Pool, J., Pierce, R.A., Mollah, S., Shabanowitz, J., Eisenlohr, L.C., van Veelen, P., Ossendorp, F., Hunt, D.F., Goulmy, E. and Engelhard, V.H. (2003) *Blood*, **102**, 621–629.

100 Warren, E.H., Vigneron, N.J., Gavin, M.A., Coulie, P.G., Stroobant, V., Dalet, A., Tykodi, S.S., Xuereb, S.M., Mito, J.K., Riddell, S.R. and Van den Eynde, B.J. (2006) *Science*, **313**, 1444–1447.

101 Brickner, A.G., Warren, E.H., Caldwell, J.A., Akatsuka, Y., Golovina, T.N., Zarling, A.L., Shabanowitz, J., Eisenlohr, L.C., Hunt, D.F., Engelhard, V.H. and Riddell, S.R. (2001) *The Journal of Experimental Medicine*, **193**, 195–206.

102 den Haan, J.M., Meadows, L.M., Wang, W., Pool, J., Blokland, E., Bishop, T.L., Reinhardus, C., Shabanowitz, J., Offringa, R., Hunt, D.F., Engelhard, V.H. and Goulmy, E. (1998) *Science*, **279**, 1054–1057.

103 Dolstra, H., Fredrix, H., Maas, F., Coulie, P.G., Brasseur, F., Mensink, E., Adema, G.J., de Witte, T.M., Figdor, C.G. and van de Wiel-van Kemenade, E. (1999) *The Journal of Experimental Medicine*, **189**, 301–308.

104 Miklos, D.B., Kim, H.T., Miller, K.H., Guo, L., Zorn, E., Lee, S.J., Hochberg, E.P., Wu, C.J., Alyea, E.P., Cutler, C., Ho, V., Soiffer, R.J., Antin, J.H. and Ritz, J. (2005) *Blood*, **105**, 2973–2978.

105 Goulmy, E. (2006) *Human Immunology*, **67**, 433–438.

106 Akatsuka, Y., Morishima, Y., Kuzushima, K., Kodera, Y. and Takahashi, T. (2007) *Cancer Science*, **98**, 1139–1146.

107 Boon, T., Coulie, P.G., Van den Eynde, B.J. and van der Bruggen, P. (2006) *Annual Review of Immunology*, **24**, 175–208.

108 Jager, D., Jager, E. and Knuth, A. (2001) *Oncology*, **60**, 1–7.

109 Slingluff, C.L., Jr. Chianese-Bullock, K.A., Bullock, T.N., Grosh, W.W., Mullins, D.W., Nichols, L., Olson, W., Petroni, G., Smolkin, M. and Engelhard, V.H. (2006) *Advances in Immunology*, **90**, 243–295.

110 Touloukian, C.E., Leitner, W.W., Schnur, R.E., Robbins, P.F., Li, Y., Southwood, S., Sette, A., Rosenberg, S.A. and Restifo, N.P. (2003) *Journal of Immunology (Baltimore, Md: 1950)*, **170**, 1579–1585.

111 Jager, D., Stockert, E., Jager, E., Gure, A.O., Scanlan, M.J., Knuth, A., Old, L.J. and Chen, Y.T. (2000) *Cancer Research*, **60**, 3584–3591.

112 Riker, A., Cormier, J., Panelli, M., Kammula, U., Wang, E., Abati, A., Fetsch, P., Lee, K.H., Steinberg, S., Rosenberg, S. and Marincola, F. (1999) *Surgery*, **126**, 112–120.

113 Hofbauer, G.F., Burkhart, A., Schuler, G., Dummer, R., Burg, G. and Nestle, F.O. (2004) *Journal of Immunotherapy*, **27** (1997), 73–78.

114 Chi, D.D., Merchant, R.E., Rand, R., Conrad, A.J., Garrison, D., Turner, R., Morton, D.L. and Hoon, D.S. (1997) *The American Journal of Pathology*, **150**, 2143–2152.

115 de Wit, N.J., van Muijen, G.N. and Ruiter, D.J. (2004) *Histopathology*, **44**, 517–541.

116 Jager, D., Jager, E. and Knuth, A. (2001) *Journal of Clinical Pathology*, **54**, 669–674.

117 Seliger, B. (2005) *BioDrugs: Clinical Immunotherapeutics, Biopharmaceuticals and Gene Therapy*, **19**, 347–354.

118 Khong, H.T., Wang, Q.J. and Rosenberg, S.A. (2004) *Journal of Immunotherapy*, **27** (1997), 184–190.

119 Saleh, F.H., Crotty, K.A., Hersey, P. and Menzies, S.W. (2001) *International Journal of Cancer*, **94**, 551–557.

120 Jager, E., Ringhoffer, M., Karbach, J., Arand, M., Oesch, F. and Knuth, A. (1996) *International Journal of Cancer*, **66**, 470–476.

121 Ohnmacht, G.A., Wang, E., Mocellin, S., Abati, A., Filie, A., Fetsch, P., Riker, A.I., Kammula, U.S., Rosenberg, S.A. and Marincola, F.M. (2001) *Journal of Immunology (Baltimore, Md: 1950)*, **167**, 1809–1820.

122 Yamshchikov, G.V., Mullins, D.W., Chang, C.C., Ogino, T., Thompson, L., Presley, J., Galavotti, H., Aquila, W., Deacon, D., Ross, W., Patterson, J.W., Engelhard, V.H., Ferrone, S. and Slingluff, C.L. Jr. (2005) *Journal of Immunology (Baltimore, Md: 1950)*, **174**, 6863–6871.

123 Le Poole, I.C., Riker, A.I., Quevedo, M.E., Stennett, L.S., Wang, E., Marincola, F.M., Kast, W.M., Robinson, J.K. and Nickoloff, B.J. (2002) *The American Journal of Pathology*, **160**, 521–528.

124 Lesimple, T., Neidhard, E.M., Vignard, V., Lefeuvre, C., Adamski, H., Labarriere, N., Carsin, A., Monnier, D., Collet, B., Clapisson, G., Birebent, B., Philip, I., Toujas, L., Chokri, M. and Quillien, V. 2006 *Clinical Cancer Research*, **12**, 7380–7388.

125 Atzpodien, J. and Reitz, M. (2007) *Cancer Biotherapy & Radiopharmaceuticals*, **22**, 551–555.

126 Banchereau, J., Palucka, A.K., Dhodapkar, M., Burkholder, S., Taquet, N., Rolland, A., Taquet, S., Coquery, S., Wittkowski, K.M., Bhardwaj, N., Pineiro, L., Steinman, R. and Fay, J. (2001) *Cancer Research*, **61**, 6451–6458.

127 Slingluff, C.L., Jr. Petroni, G.R., Yamshchikov, G.V., Hibbitts, S., Grosh, W.W., Chianese-Bullock, K.A., Bissonette, E.A., Barnd, D.L., Deacon, D.H., Patterson, J.W., Parekh, J., Neese, P.Y., Woodson, E.M., Wiernasz, C.J. and Merrill, P. (2004) *Journal of Clinical Oncology*, **22**, 4474–4485.

128 Grover, A., Kim, G.J., Lizee, G., Tschoi, M., Wang, G., Wunderlich, J.R., Rosenberg, S.A., Hwang, S.T. and Hwu, P. (2006) *Clinical Cancer Research*, **12**, 5801–5808.

129 Cassaday, R.D., Sondel, P.M., King, D.M., Macklin, M.D., Gan, J., Warner, T.F., Zuleger, C.L., Bridges, A.J., Schalch, H.G., Kim, K.M., Hank, J.A., Mahvi, D.M. and Albertini, M.R. (2007) *Clinical Cancer Research*, **13**, 540–549.

130 Spaner, D.E., Astsaturov, I., Vogel, T., Petrella, T., Elias, I., Burdett-Radoux, S., Verma, S., Iscoe, N., Hamilton, P. and Berinstein, N.L. (2006) *Cancer*, **106**, 890–899.

131 Lindsey, K.R., Gritz, L., Sherry, R., Abati, A., Fetsch, P.A., Goldfeder, L.C., Gonzales, M.I., Zinnack, K.A., Rogers-Freezer, L., Haworth, L., Mavroukakis, S.A., White, D.E., Steinberg, S.M., Restifo, N.P., Panicali, D.L., Rosenberg, S.A. and Topalian, S.L. (2006) *Clinical Cancer Research*, **12**, 2526–2537.

132 Segal, B.M. (2003) *Curr Allergy and Asthma Reports*, **3**, 86–93.

133 Darnell, R.B. (1999) *The New England Journal of Medicine*, **340**, 1831–1833.

134 Albert, M.L., Darnell, J.C., Bender, A., Francisco, L.M., Bhardwaj, N. and Darnell, R.B. (1998) *Nature Medicine*, **4**, 1321–1324.

135 Ashby, L.S. and Shapiro, W.R. (2004) *Curr Neurology and Neuroscience Reports*, **4**, 211–217.

136 Cabarrocas, J., Bauer, J., Piaggio, E., Liblau, R. and Lassmann, H. (2003) *European Journal of Immunology*, **33**, 1174–1182.

137 Kahlon, K.S., Brown, C., Cooper, L.J., Raubitschek, A., Forman, S.J. and Jensen, M.C. (2004) *Cancer Research*, **64**, 9160–9166.
138 Oba, S.M., Wang, Y.J., Song, J.P., Li, Z.Y., Kobayashi, K., Tsugane, S., Hamada, G.S., Tanaka, M. and Sugimura, H. (2001) *Cancer Letters*, **164**, 97–104.
139 Dodelet, V.C. and Pasquale, E.B. (2000) *Oncogene*, **19**, 5614–5619.
140 Miao, H., Burnett, E., Kinch, M., Simon, E. and Wang, B. (2000) *Nature Cell Biology*, **2**, 62–69.
141 Kinch, M.S., Moore, M.B. and Harpole, D.H. Jr. (2003) *Clinical Cancer Research*, **9**, 613–618.
142 Hatano, M., Eguchi, J., Tatsumi, T., Kuwashima, N., Dusak, J.E., Kinch, M.S., Pollack, I.F., Hamilton, R.L., Storkus, W.J. and Okada, H. (2005) *Neoplasia (New York, NY)*, **7**, 717–722.
143 Liu, F., Park, P.J., Lai, W., Maher, E., Chakravarti, A., Durso, L., Jiang, X., Yu, Y., Brosius, A., Thomas, M., Chin, L., Brennan, C., DePinho, R.A., Kohane, I., Carroll, R.S., Black, P.M. and Johnson, M.D. 2006 *Cancer Research*, **66**, 10815–10823.
144 Blanc-Brude, O.P., Yu, J., Simosa, H., Conte, M.S., Sessa, W.C. and Altieri, D.C. (2002) *Nature Medicine*, **8**, 987–994.
145 Wobser, M., Keikavoussi, P., Kunzmann, V., Weininger, M., Andersen, M.H. and Becker, J.C. (2006) *Cancer Immunology, Immunotherapy*, **55**, 1294–1298.
146 Otto, K., Andersen, M.H., Eggert, A., Keikavoussi, P., Pedersen, L.O., Rath, J.C., Bock, M., Brocker, E.B., Straten, P.T., Kampgen, E. and Becker, J.C. (2005) *Vaccine*, **23**, 884–889.
147 Uematsu, M., Ohsawa, I., Aokage, T., Nishimaki, K., Matsumoto, K., Takahashi, H., Asoh, S., Teramoto, A. and Ohta, S. (2005) *Journal of Neuro-Oncology*, **72**, 231–238.
148 Sugiyama, H. (2002) *International Journal of Hematology*, **76**, 127–132.
149 Oka, Y., Tsuboi, A., Elisseeva, O.A., Udaka, K. and Sugiyama, H. (2002) *Current Cancer Drug Targets*, **2**, 45–54.
150 Nakahara, Y., Okamoto, H., Mineta, T. and Tabuchi, K. (2004) *Brain Tumor Pathology*, **21**, 113–116.
151 Oji, Y., Suzuki, T., Nakano, Y., Maruno, M., Nakatsuka, S., Jomgeow, T., Abeno, S., Tatsumi, N., Yokota, A., Aoyagi, S., Nakazawa, T., Ito, K., Kanato, K., Shirakata, T., Nishida, S., Hosen, N., Kawakami, M., Tsuboi, A., Oka, Y., Aozasa, K., Yoshimine, T. and Sugiyama, H. (2004) *Cancer Science*, **95** 822–827.
152 Wu, A.H., Xiao, J., Anker, L., Hall, W.A., Gregerson, D.S., Cavenee, W.K., Chen, W. and Low, W.C. (2006) *Journal of Neuro-Oncology*, **76**, 23–30.
153 Li, Z. and Srivastava, P.K. (1993) *The EMBO Journal*, **12**, 3143–3151.
154 Tamura, Y., Peng, P., Liu, K., Daou, M. and Srivastava, P.K. (1997) *Science*, **278**, 117–120.
155 Janetzki, S., Palla, D., Rosenhauer, V., Lochs, H., Lewis, J.J. and Srivastava, P.K. (2000) *International Journal of Cancer*, **88**, 232–238.
156 Amato, R.J., Murray, L., Wood, L., Savary, C., Tomasovic, S., Srivastava, P.K. and Reitsma, D. (1999) *ASCO meeting abstract*.
157 Eton, O., East, M.J., Ross, M., Savary, C., Tomasovic, S., Reitsma, D., Hawkins, E. and Srivastava, P.K. (2000) *Proceedings of the American Association for Cancer Research*, **41**, 543.
158 Belli, F., Testori, A., Rivoltini, L., Maio, M., Andreola, G., Sertoli, M.R., Gallino, G., Piris, A., Cattelan, A., Lazzari, I., Carrabba, M., Scita, G., Santantonio, C., Pilla, L., Tragni, G., Lombardo, C., Arienti, F., Marchiano, A., Queirolo, P., Bertolini, F., Cova, A., Lamaj, E., Ascani, L., Camerini, R., Corsi, M., Cascinelli, N., Lewis, J.J., Srivastava, P. and Parmiani, G. (2002) *Journal of Clinical Oncology*, **20**, 4169–4180.
159 Mazzaferro, V., Coppa, J., Carrabba, M.G., Rivoltini, L., Schiavo, M., Regalia, E., Mariani, L., Camerini, T., Marchiano, A., Andreola, S., Camerini, R., Corsi, M.,

Lewis, J.J., Srivastava, P.K. and Parmiani, G. (2003) *Clinical Cancer Research*, **9**, 3235–3245.

160 Li, Z., Qiao, Y., Liu, B., Laska, E.J., Chakravarthi, P., Kulko, J.M., Bona, R.D., Fang, M., Hegde, U., Moyo, V., Tannenbaum, S.H., Menoret, A., Gaffney, J., Glynn, L., Runowicz, C.D. and Srivastava, P.K. (2005) *Clinical Cancer Research*, **11**, 4460–4468.

161 Richards, J., Testori, A., Whitman, E., Mann, G.B., Lutzky, J., Camacho, L., Parmiani, G., Hoos, A., Gupta, R. and Srivastava, P. (2006) *Journal of Clinical Oncology*, **24**, 1.

162 Yang, L. and Carbone, D.P. (2004) *Advances in Cancer Research*, **92**, 13–27.

163 Pinzon-Charry, A., Maxwell, T. and Lopez, J.A. (2005) *Immunology and Cell Biology*, **83**, 451–461.

164 Steinman, R.M. and Banchereau, J. (2007) *Nature*, **449**, 419–426.

165 Banchereau, J. and Steinman, R.M. (1998) *Nature*, **392**, 245–252.

166 Schuler, G., Schuler-Thurner, B. and Steinman, R.M. (2003) *Current Opinion in Immunology*, **15**, 138–147.

167 Schuler, G. and Steinman, R.M. (1997) *The Journal of Experimental Medicine*, **186**, 1183–1187.

168 Kalinski, P., Hilkens, C.M., Wierenga, E.A. and Kapsenberg, M.L. (1999) *Immunology Today*, **20**, 561–567.

169 Mora, J.R., Bono, M.R., Manjunath, N., Weninger, W., Cavanagh, L.L., Rosemblatt, M. and Von Andrian, U.H. (2003) *Nature*, **424**, 88–93.

170 Mora, J.R., Cheng, G., Picarella, D., Briskin, M., Buchanan, N. and von Andrian, U.H. (2005) *The Journal of Experimental Medicine*, **201**, 303–316.

171 Mora, J.R. and von Andrian, U.H. (2004) *Immunity*, **21**, 458–460.

172 Schaerli, P., Loetscher, P. and Moser, B. (2001) *Journal of Immunology (Baltimore, Md: 1950)*, **167**, 6082–6086.

173 Stagg, A.J., Kamm, M.A. and Knight, S.C. (2002) *European Journal of Immunology*, **32**, 1445–1454.

174 Weninger, W., Manjunath, N. and von Andrian, U.H. (2002) *Immunological Reviews*, **186**, 221–233.

175 Calzascia, T., Masson, F., Di Berardino-Besson, W., Contassot, E., Wilmotte, R., Aurrand-Lions, M., Rucgg, C., Dietrich, P.Y. and Walker, P.R. (2005) *Immunity*, **22**, 175–184.

176 Fernandez, N.C., Lozier, A., Flament, C., Ricciardi-Castagnoli, P., Bellet, D., Suter, M., Perricaudet, M., Tursz, T., Maraskovsky, E. and Zitvogel, L. (1999) *Nature Medicine*, **5**, 405–411.

177 Hsu, F.J., Benike, C., Fagnoni, F., Liles, T.M., Czerwinski, D., Taidi, B., Engleman, E.G. and Levy, R. (1996) *Nature Medicine*, **2**, 52–58.

178 Nestle, F.O., Alijagic, S., Gilliet, M., Sun, Y., Grabbe, S., Dummer, R., Burg, G. and Schadendorf, D. (1998) *Nature Medicine*, **4**, 328–332.

179 Srivastava, P.K. (2006) *Current Opinion in Immunology*, **18**, 201–205.

180 Nestle, F.O., Farkas, A. and Conrad, C. (2005) *Current Opinion in Immunology*, **17**, 163–169.

181 Banchereau, J., Ueno, H., Dhodapkar, M., Connolly, J., Finholt, J.P., Klechevsky, E., Blanck, J.P., Johnston, D.A., Palucka, A.K. and Fay, J. (2005) *Journal of Immunotherapy*, **28** (1997), 505–516.

182 Banchereau, J. and Palucka, A.K. (2005) *Nature Reviews. Immunology*, **5**, 296–306.

183 Rosenberg, S.A., Yang, J.C. and Restifo, N.P. (2004) *Nature Medicine*, **10**, 909–915.

184 Dhodapkar, M.V., Steinman, R.M., Sapp, M., Desai, H., Fossella, C., Krasovsky, J., Donahoe, S.M., Dunbar, P.R., Cerundolo, V., Nixon, D.F. and Bhardwaj, N. (1999) *The Journal of Clinical Investigation*, **104**, 173–180.

185 De Vries, I.J., Krooshoop, D.J., Scharenborg, N.M., Lesterhuis, W.J., Diepstra, J.H., Van Muijen, G.N., Strijk, S.P., Ruers, T.J., Boerman, O.C., Oyen, W.J., Adema, G.J., Punt, C.J. and Figdor, C.G. (2003) *Cancer Research*, **63**, 12–17.

186 Adema, G.J., de Vries, I.J., Punt, C.J. and Figdor, C.G. (2005) *Current Opinion in Immunology*, **17**, 170–174.

187 Langenkamp, A., Messi, M., Lanzavecchia, A. and Sallusto, F. (2000) *Nature Immunology*, **1**, 311–316.

188 Jonuleit, H., Kuhn, U., Muller, G., Steinbrink, K., Paragnik, L., Schmitt, E., Knop, J. and Enk, A.H. (1997) *European Journal of Immunology*, **27**, 3135–3142.

189 Mailliard, R.B., Wankowicz-Kalinska, A., Cai, Q., Wesa, A., Hilkens, C.M., Kapsenberg, M.L., Kirkwood, J.M., Storkus, W.J. and Kalinski, P. (2004) *Cancer Research*, **64**, 5934–5937.

190 de Vries, I.J., Lesterhuis, W.J., Scharenborg, N.M., Engelen, L.P., Ruiter, D.J., Gerritsen, M.J., Croockewit, S., Britten, C.M., Torensma, R., Adema, G.J., Figdor, C.G. and Punt, C.J. (2003) *Clinical Cancer Research*, **9**, 5091–5100.

191 Luft, T., Jefford, M., Luetjens, P., Toy, T., Hochrein, H., Masterman, K.A., Maliszewski, C., Shortman, K., Cebon, J. and Maraskovsky, E. (2002) *Blood*, **100**, 1362–1372.

192 Scandella, E., Men, Y., Gillessen, S., Forster, R. and Groettrup, M. (2002) *Blood*, **100**, 1354–1361.

193 Vieira, P.L., de Jong, E.C., Wierenga, E.A., Kapsenberg, M.L. and Kalinski, P. (2000) *Journal of Immunology (Baltimore, Md: 1950)*, **164**, 4507–4512.

194 Xu, S., Koski, G.K., Faries, M., Bedrosian, I., Mick, R., Maeurer, M., Cheever, M.A., Cohen, P.A. and Czerniecki, B.J. (2003) *Journal of Immunology (Baltimore, Md: 1950)*, **171**, 2251–2261.

195 Kalinski, P., Nakamura, Y., Watchmaker, P., Giermasz, A., Muthuswamy, R. and Mailliard, R.B. (2006) *Immunologic Research*, **36**, 137–146.

196 Mailliard, R.B., Son, Y.I., Redlinger, R., Coates, P.T., Giermasz, A., Morel, P.A., Storkus, W.J. and Kalinski, P. (2003) *Journal of Immunology (Baltimore, Md: 1950)*, **171**, 2366–2373.

197 Whiteside, T. (2006) In *In The Link Between Inflammation and Cancer*, (eds Dalgleish A.G., and Haefner B.), Springer, New York. pp. 103–124.

198 Whiteside, T.L. (2006) *Cancer Treatment and Research*, **130**, 103–124.

199 Smyth, M.J., Dunn, G.P. and Schreiber, R.D. (2006) *Advances in Immunology*, **90**, 1–50.

200 Whiteside, T.L., Campoli, M. and Ferrone, S. (2005) In *In Analyzing T Cell Responses* (eds Nagorsen E., and Marincola F.), Springer, pp. 43–82.

201 Bui, J.D. and Schreiber, R.D. (2007) *Current Opinion in Immunology*, **19**, 203–208.

202 Coronella, J.A., Spier, C., Welch, M., Trevor, K.T., Stopeck, A.T., Villar, H. and Hersh, E.M. (2002) *Journal of Immunology (Baltimore, Md: 1950)*, **169**, 1829–1836.

203 Pittet, M.J., Valmori, D., Dunbar, P.R., Speiser, D.F., Lienard, D., Lejeune, F., Fleischhauer, K., Cerundolo, V., Cerottini, J.C. and Romero, P. (1999) *The Journal of Experimental Medicine*, **190**, 705–715.

204 Zhou, J., Dudley, M.E., Rosenberg, S.A. and Robbins, P.F. (2004) *Journal of Immunology (Baltimore, Md: 1950)*, **173**, 7622–7629.

205 Cutts, F.T., Franceschi, S., Goldie, S., Castellsague, X., de Sanjose, S., Garnett, G., Edmunds, W.J., Claeys, P., Goldenthal, K.L., Harper, D.M. and Markowitz, L. (2007) *Bulletin of the World Health Organization*, **85**, 719–726.

Part Two
Methods to Detect TAAs

3
Humoral Immune Responses against Cancer Antigens: Serological Identification Methods. Part I: SEREX

Carsten Zwick, Klaus-Dieter Preuss, Frank Neumann, and Michael Pfreundschuh

3.1
Introduction

The identification and molecular characterization of tumor antigens that elicit specific immune responses in the tumor-bearing host is a major task in tumor immunology. In the 1970s and 1980s of the twentieth century monoclonal antibody (mAb) technology was exploited for the identification of molecules on tumor cells that could be used as diagnostic markers or as target structures for immunotherapeutic approaches with mAb. While some of these efforts have yielded new therapeutic tools, such as the anti-CD20 antibody rituximab, which shows considerable activity and has been licensed for the treatment of B cell lymphomas [1], approaches for *active* immunotherapy require the identification of target structures, which are immunogenic in the autologous tumor-bearing host.

The analysis of humoral and cellular immune responses against such antigens in cancer patients had long indicated the presence of cancer-specific antigens which are recognized by the patient's immune system [2]. To define the molecular nature of these antigens, cloning techniques were developed that used established cytotoxic T lymphocyte (CTL) clones [3] or circulating antibodies [4] as probes for screening tumor-derived expression libraries. While the molecular characterization of the first human tumor antigens was accomplished with cloning techniques that used established CTL clones [3], it is now commonly accepted that immune recognition of tumors is a concerted action. Thus, high-titered circulating tumor-associated antibodies of the IgG class may reflect a significant host–tumor interaction and may identify such gene products for which at least cognate T cell help, but also specific cytotoxic T cells should exist. This rationale led us to design a novel strategy using the antibody repertoire of cancer patients for the molecular definition of antigens. Serologically defined antigens could then be subjected to procedures of 'reverse' T cell immunology for the definition of epitopes which are presented by MHC class I or II molecules, respectively, and are recognized by T lymphocytes.

Tumor-Associated Antigens. Edited by Olivier Gires and Barbara Seliger
Copyright © 2009 WILEY-VCH Verlag GmbH & Co. KGaA, Weinheim
ISBN: 978-3-527-32084-4

3.2
The SEREX Approach

To identify tumor antigens recognized by the antibody repertoire of cancer patients we developed a serological cloning approach termed SEREX (serological analysis of tumor antigens by recombinant cDNA expression cloning). SEREX allows for a systematic and unbiased search for antibody responses to proteins expressed by human neoplasms. The respective tumor antigens are identified by their reactivity with antibodies in the autologous and allogeneic sera of cancer patients. For SEREX, cDNA expression libraries are constructed from fresh tumor specimens, cloned into λ phage expression vectors and the phages are then used to transfect *E. coli*. Recombinant proteins expressed during the lytic infection of the bacteria are transferred onto nitrocellulose membranes, which are then incubated with diluted (1 : 500–1 : 1000) and – most importantly – extensively pre-absorbed serum from the autologous patient. Clones reactive with high-titered antibodies are identified using an enzyme conjugated second antibody specific for human immunoglobulin (IgG). Positive clones are subcloned to monoclonality thus allowing direct molecular characterization by DNA sequencing.

The SEREX approach is technically characterized by several features [4, 5]:

(i) There is no need for established tumor cell lines and pre-characterized CTL clones.

(ii) The use of fresh tumor specimens restricts the analysis to genes that are expressed by the tumor cells *in vivo*. Therefore *in vitro* artifacts that are associated with short- and long-term tumor cell culture are circumvented.

(iii) The use of the polyclonal (polyspecific) patient's serum allows for the identification of multiple antigens with one screening course.

(iv) The screening is restricted to clones against which the patient's immune system has raised high-titered IgG antibody responses indicating the presence of a concomitant T-helper lymphocyte response *in vivo*.

(v) As both the expressed antigenic protein and the coding cDNA are present in the same plaque of the phage immunoscreening assay, identified antigens can be sequenced immediately. Sequence information of excised cDNA inserts can be directly used to determine the expression spectrum of identified transcripts by Northern blot and reverse transcription polymerase chain reaction (RT-PCR).

(vi) The release of periplasmatic proteins involved in protein folding during phage-induced bacterial lysis allows at least the partial folding of recombinant proteins and provides the basis for the identification of linear as well as nonlinear epitopes. This has been confirmed by the expression of transcripts which code for enzymatically-active proteins [6]. In contrast, epitopes derived from eucaryotic post-translational modification (e.g. glycosylation) are not detected by the phage immunoscreening assay.

Meanwhile a number of modifications of the original method have been implemented. Immunoglobulins, which are also recombinantly expressed due to the presence of B-lymphocytes and plasma cells in the tumor specimens used for the cDNA library, may represent >90% of all 'positive' clones in libraries derived from certain tissues. They can be identified by a modified initial screening procedure whereby the nitrocellulose membrane is incubated with enzyme-conjugated antihuman IgG followed by visualization with the appropriate enzymatic color reaction prior to the incubation of the autologous patient's serum [7]. Screening of tumor cell lines rather than fresh tumor specimens circumvents this problem and additionally provides a pure RNA source which is not contaminated with normal stroma [8]. Subtractive approaches allow the addition of tumor-specific transcripts to the cDNA library [9]. cDNA libraries may also be prepared from sources of specific interest, such as amplified chromosomal regions obtained by microdissection or from immunoprivileged tissue such as testis [7–11].

3.2.1
Identification of Human Tumor Antigens by SEREX

Our group has constructed and analyzed cDNA expression libraries derived from a variety of different neoplasms using SEREX. Primary libraries with at least 1×10^6 independent clones were established. The screening of at least 1×10^6 clones per library revealed multiple reactive clones in each library. Some transcripts were repeatedly detected indicating that there were multiple representations within the library. In order to bias for the detection of antigens of the cancer testis (CT) class, libraries were constructed from normal testis tissue and screened with sera from allogeneic tumor patients.

From the plenitude of antigens detected in the primary screening assay, those clones for which a more in-depth analysis appeared to be justified were selected. Selection criteria for further functional analysis of an antigen were (i) its sequence analysis and comparison with databases to reveal identity or homologies with known genes and to identify domains or motifs indicative of a putative function or cellular localization; (ii) the analysis of the expression pattern of the respective antigen in normal tissues and tumors by RT-PCR, Northern blot hybridization with specific probes, and by analysis in expressed sequence tag-containing databases and (iii) an initial survey for antibodies in the sera from healthy controls and allogeneic tumor patients to evaluate the incidence of serum antibodies to the respective antigen.

Four different groups of genes coding for antigens were identified. The first group codes for known tumor antigens such as the melanoma antigens MAGE-1, MAGE-4a and tyrosinase. The second group encodes known classical autoantigens, for which immunogenicity is associated with autoimmune diseases, for example antimitochondrial antibodies or antibodies to U1-snRNP. When patients known to have autoimmune or rheumatic disorders are excluded from SEREX analysis, the incidence of such antigens is not higher than 1%. The third group codes for transcripts that are either identical or highly homologous to known genes, but have not been known to elicit immune responses in cancer patients. Examples include restin, which

Table 3.1 Specificity of tumor antigens detected by SEREX.

Specificity	Example	Source
1. Shared tumor antigens	HOM-MEL-40	Melanoma
2. Differentiation antigens	HOM-Mel-55 (tyrosinase)	Melanoma
3. Mutated genes	NY-COL-2 (p53)	Colorectal carcinoma
4. Splice variants	HOM-HD-397 (restin)	Hodgkin's disease
5. Viral antigens	HOM-RCCC-1.14 (HERV-K10)	Renal cell cancer
6. Overexpression	HOM-HD-21 (galectin-9)	Hodgkin's disease
7. Gene amplifications	HOM-NSCLC-11 (eIF-4)	Lung cancer
8. Cancer-related autoantigens	HOM-MEL-2.4 (CEBP)	Melanoma
9. Cancer-independent autoantigens	NY-ESO-2 (U1-snRNP)	Esophageal cancer
10. Underexpressed genes	HOM-HCC-8	Hepatocellular carcinoma

had originally been identified by a murine mAb specific for Hodgkin and Reed-Sternberg cells and lactate dehydrogenase, an enzyme overexpressed in many human tumors. The fourth group of serologically defined antigens represents products of previously unknown genes.

3.2.2
Specificity of Human Tumor Antigens

According to their expression pattern in normal and malignant tissues, several classes of tumor antigens can be distinguished (Table 3.1): (i) shared tumor antigens, (ii) differentiation antigens, products of (iii) mutated, (iv) viral, (v) overexpressed, and (vi) amplified genes as well as (vii) splice variants, (viii) widely expressed, but cancer-associated autoantigens, the immunogenicity of which is restricted to cancer patients, (ix) common autoantigens to which antibodies are found in sera from patients with other than malignant diseases, and finally (x) products of genes which are under-expressed in the autologous tumor compared to normal tissues.

3.2.2.1 Shared Tumor Antigens

Shared tumor antigens are expressed in a variable proportion of human tumors (10 to 70%, depending on the type of tumor). Interestingly, all human shared tumor antigens identified to date are expressed in a variety of human cancers, but not in normal tissues, except for testis; therefore, the term *cancer testis(CT) antigens* has been coined to describe them [11]. Since it is not the whole testis organ, but only the spermatocytes that express these antigens, the term *cancer germline antigens* might be more appropriate. Many of the genes coding for these antigens exist as multimember gene families. The prototypes of this category, MAGE [3], BAGE [12], and GAGE [13], were initially identified as targets for cytotoxic T cells. The HOM-MEL-40 antigen was detected in a melanoma library and is the first cancer/testis antigen identified by SEREX. It is encoded by the SSX-2 gene [7]. Members of the SSX gene family, SSX1 and SSX2, have been shown to be involved in the t(X;18)(p11.2; q11.2) translocation, which is found in the majority of human synovial sarcomas [14] and fuses the respective SSX gene with the SYT gene from chromosome 18. Using homology cloning and PCR, additional members of the SSX family were identified [15] revealing at least nine genes, of which four (SSX-1, 2, 4 and 5) demonstrate a CT antigen-like expression [16]. Using SEREX, Chen et al. [11] identified NY-ESO-1 as a new CT antigen. NY-ESO-1 mRNA expression is detectable in a wide array of human cancers, including melanomas, breast, bladder, and prostate cancer. A homologous gene, known as LAGE-1 was subsequently isolated by a subtractive cloning approach [17] demonstrating that NY-ESO-1 belongs to a gene family with at least two members. NY-ESO-1 as well as its homolog LAGE-1 were discovered by independent groups using tumor-specific CTL or tumor-infiltrating lymphocytes (TIL) derived from melanoma patients as probes, thus disclosing several HLA-A0201 and HLA-A31 restricted epitopes [18] and demonstrating that NY-ESO-1 is a target for both antibody and CTL responses in the same patient [18, 19]. IgG antibody responses, sometimes at very high titers (up to 1: 100,000), directed against NY-ESO-1 are present in up to 50% of antigen-expressing patients demonstrating the extraordinary immunogenicity of NY-ESO-1 and qualifying this antigen as a frequent target for $CD4^+$ T lymphocytes [20]. Another new CT antigen is HOM-TES-14 [9] which is encoded by the gene coding for the synaptonemal complex protein-1 (SCP-1). HOM-TES-14/SCP-1 was discovered independently during the screening of testis-derived libraries using the sera of patients with seminoma [9], cutaneous [21] and nodal lymphomas [22]. It is one of the CT antigens with the broadest expression spectrum and the highest expression rate.

CT antigens can be divided between those that are encoded on the X chromosome (CT-X antigens) and those that are not (non-X CT antigens). Twenty-two of the 44 known CT antigens, including the principal CT antigen cancer-vaccine candidates, are CT-X antigens. In normal testes the CT-X genes are generally expressed in the spermatogonia, which are proliferating germ cells. The genes tend to form recently expanded gene families associated with inverted DNA repeats. Remarkably, it has been estimated that 10% of the genes on the X chromosome belong to CT-X families. The genes for the non-X CTs, on the other hand, are distributed throughout the genome and do not generally form gene families or reside within genomic repeats. In the testis, their expression appears more dominant in later stages of germ-cell differentiation, such as in the spermatocytes.

3.2.2.2 Differentiation Antigens

Differentiation antigens are expressed in tumors in a lineage-associated expression, but are also found in normal cells of the same origin; examples include tyrosinase and GFAP (glial fibrillary acidic protein), which are antigenic in malignant melanoma and glioma [23], but are also expressed in melanocytes or brain cells, respectively.

3.2.2.3 Antigens Encoded by Mutated Genes

Antigens encoded by mutated genes have been demonstrated only rarely by the serological approach, with mutated p53 being one example [24]. However, it is very difficult to exclude the possibility that an antibody response detected by SEREX was not initially induced by a mutated antigen, since such antibody responses may be directed to epitopes shared between the wild-type and mutated form of the antigen. Thus, wild-type alleles may be picked up during immunoscreening, and sequencing of several independent clones from the same library as well as exclusion of polymorphisms is mandatory.

3.2.2.4 Viral Genes

A virus-encoded antigen that elicits an autologous antibody response is the *env* protein of the human endogenous retrovirus HERV-K10, which has been found in a renal cell cancer and in a seminoma.

3.2.2.5 Overexpressed Genes

Overexpressed genes code for many tumor antigens that have been identified by SEREX, which has an inherent methodological bias for the detection of abundant transcripts. The members of this antigen class are expressed at low levels in normal tissues (usually detectable by RT-PCR, but often missed by Northern blot analysis), but are up to 100-fold overexpressed in tumors. Examples include HOM-RCC-3.1.3, a new carbonic anhydrase which is overexpressed in a fraction of renal cell cancers [6] and the Bax inhibitor protein 1, which is overexpressed in gliomas [23].

3.2.2.6 Amplified Genes

Amplified genes may also code for tumor antigens. The overexpression of a transcript resulting from a gene amplification has been demonstrated for the translation initiation factor eIF-4γ in a squamous cell lung cancer [10].

3.2.2.7 Splice Variants of Known Genes

Splice variants of known genes were also found to be immunogenic in cancer patients. Examples include NY-COL-38 and restin, which represents a splice variant of the formerly described cytoplasmic linker protein CLIP-170 [25].

3.2.2.8 Cancer-related Autoantigens

Cancer-related autoantigens are ubiquitously expressed and at a similar level in healthy as well as malignant tissues. The encoding genes are not altered in tumor samples, however, they elicit antibody responses only in cancer patients, but not in healthy individuals. This might result from tumor-associated post-translational modifications or changes in the antigen processing and/or presentation in tumor cells. An example is HOM-MEL-2.4, which represents the CCAAT enhancer binding protein.

3.2.2.9 Non-cancer-related Autoantigens

Non-cancer-related autoantigens are expressed in most human tissues; in contrast to cancer-related autoantigens, antibodies against these antigens are found in non-tumor bearing controls at a similar frequency as in tumor patients. Examples include HOM-RCC-10, which represents a mitochondrial protein and HOM-TES-11, which is identical to pericentriol material-1 (PCM-1).

3.2.2.10 Products of Underexpressed Genes

An antigen was detected during the SEREX analysis of a hepatocellular carcinoma, which was underexpressed in the malignant tissue when compared to normal liver [26]. This antigen – p14.5 – is a strong translational repressor and displays low mRNA and protein expression in a variety of undifferentiated proliferating liver and kidney tumor cells. This led to the assumption that p14.5 prevents hepatocytes from entering the cell cycle. In accordance with this, we found p14.5 to be down-regulated in four of five HCCs studied by Northern blotting.

3.2.3
Significance of Antibodies against SEREX Antigens

The analysis of sera from patients with various malignant diseases and from healthy controls showed that there are different patterns of serological responses against SEREX antigens. The group of antibodies which occurs exclusively in patients with cancer is clinically the most interesting. Such strictly tumor-associated antibody responses are detected with varying frequencies only in the sera of patients with tumors that express the respective antigen. Examples include antibodies against HOM-TES-14/SCP-1 [9], NY-ESO-I [11], and against several antigens cloned from colon cancer [24]. The incidence of tumor-associated antibodies in unselected tumor patients ranges between 5 and 50% depending on the tumor type and the respective antigen. The antibodies against many SEREX antigens are only detected in the patient whose serum was used for the SEREX assay. A third group of antibodies occurs in cancer patients and healthy controls at a similar rate. While most of these antibodies are directed against non-cancer-associated and widely expressed auto-antigens, e.g. poly-adenosyl-ribosyl transferase, antibodies in this category are also found to be directed against antigens with a very restricted expression pattern, e.g. restin. Restin represents a differentiation antigen, since its expression is limited to Hodgkin- and Reed-Sternberg cells and immature dendritic cells.

There is little information on the clinical significance of antibody responses and their correlation with the clinical course of the malignant disease [27]. Well-designed prospective studies are necessary to answer the question, whether and which antibodies are (alone or in combination with others) valuable for the diagnosis and/or the evaluation of the response to therapy of malignant diseases. Why only a minority of the patients with an antigen-positive tumor develop antibodies to the respective antigen remains enigmatic. From anecdotal observations we have the impression that antibodies are only found in a patient's serum if the tumor expresses the respective antigen. Antibody titers drop and often disappear when the tumor is removed or the patient is in remission.

3.2.4
Reverse T Cell Immunology

SEREX may also be useful for the analysis of the $CD4^+$ and $CD8^+$ T cell repertoire against tumor antigens in a strategy which is also known as 'reverse T cell immunology'. Serologically-defined antigens are valuable tools for the identification and determination of peptide epitopes reacting in the context of MHC molecules with the antigen-specific T cell receptor, since isotype switching and development of high-titered IgG *in vivo* requires cognate $CD4^+$ T-cell help. Several $CD4^+$-specific epitopes of the NY-ESO-1 antigen have been identified [28]. With regard to $CD8^+$ T lymphocytes which recognize SEREX antigens, CTL responses have been demonstrated for NY-ESO-1 and HOM-MEL-40 [28–30] and HOM-TES-14/SCP-1 [31].

3.2.5
Functional Significance of Human Tumor Antigens

Antigens with a known function identified by SEREX include HOM-RCC-3.1.3 [6], which was shown to be a novel member of the carbonic anhydrase (CA) family, designated as CA XII. Overexpression of this transcript was observed in 10% of renal cell cancers (RCC) suggesting a potential significance in this tumor type. In fact, the same transcript was cloned shortly thereafter by another group based on its downregulation by the wild-type von-Hippel-Lindau tumor suppressor gene, the loss of function of which is known to be associated with an increased incidence of RCC [32]. Since the invasiveness of RCC cell lines expressing CA XII has been shown to be inhibited by acetazolamide [33], CA XII might be exploited therapeutically. Bax inhibitor protein 1, which was found to be overexpressed in gliomas, is an antiapoptotic molecule [23].

The first cancer testis antigen to which a physiological function could be ascribed is HOM-TES-14, which is encoded by the SCP-1 (synaptonemal complex protein 1) gene [9]. SCP-1 is known to be selectively expressed during the meiotic prophase of spermatocytes and is involved in the pairing of homologous chromosomes, an essential step for the generation of haploid cells in meiosis I. Transfection of diploid cell lines with SCP-1 induces polyploidy, suggesting that the aberrant expression of this meiosis-specific gene product in the somatic cells of human tumor cells is involved in the induction of chromosomal instabilities in cancer cells [34].

3.2.6
The Human Cancer Immunome

SEREX allows for the identification of an entire profile of antigens using the antibody repertoire of a single cancer patient. The analysis of a variety of neoplasms demonstrated that all hitherto investigated neoplasms are immunogenic in the tumor-bearing host and that immunogenicity is conferred by multiple antigens. For the systematic documentation and archiving of sequence data and immunological characteristics of identified antigens, an electronic SEREX database was initiated by the Ludwig Cancer Research Institutes which is accessible to the public (www.licr.org/SEREX.html). By December 2007, more than 2700 entries had been made into the SEREX database, the majority of them representing independent antigens. The SEREX database is not only meant as a computational interface for discovery information management, but also as a tool for mapping the entire panel of gene products which elicit spontaneous immune responses in the tumor-bearing autologous host, for which the name 'cancer immunome' has been coined by L. J. Old (LICR, New York Branch). The cancer immunome, which is defined by using spontaneously-occurring immune effectors from cancer patients as probes, has attracted increasing interest since it has been shown that many antigens may be valuable as new molecular markers of malignant disease. The value of each of these markers or combinations of them for diagnostic or prognostic evaluation of cancer patients needs to be determined by studies which correlate the presence or absence of these markers with clinical data.

The multitude of tumor-specific antigens identified by the SEREX approach demonstrates that there is ample immune recognition of human tumors by the autologous host's immune system. Together with the identification of T lymphocyte epitopes a picture of the immunological profile of cancer is emerging. Information relating to the cancer immunome provides a new basis for understanding tumor biology and for the development of new diagnostic and therapeutic strategies for cancer. The specificities of the antigens expressed by human tumors and detected by SEREX vary widely, ranging from tumor-specific antigens to common autoantigens. This surprising finding together with the observation that some ubiquitously expressed antigens elicit immune responses only in cancer patients suggests that the context in which a protein is presented to the immune system (e.g. the context of 'danger') is more decisive for its immunogenicity (and breakdown of tolerance) than its more or less tumor-restricted expression [35]. Our results obtained with the SEREX analysis of many human tumors also show that in addition to the rare tumor-specific antigens there is a great majority of widely expressed autoantigens which are presented by the tumor and recognized by the immune system. The presentation of common autoantigens by a given tumor (presumably in the context of 'danger') induces a broad range of autoimmunity and it is only if the tumor happens to express and present tumor-specific molecules that tumor-specific autoimmunity can occur: specific tumor immunity is just a small part of broad autoimmunity that is commonly induced by malignant growth.

Despite the fact that SEREX has enlarged the pool of available tumor antigens, the proportion of tumors for which no tumor antigen is known, is still high, particularly in frequent neoplasms such as colon and prostate cancer. A similar situation applies to myeloid leukemias and lymphomas [22]. Our analysis of the B-cell repertoire against human cancers using the conventional SEREX approach with a bacterial expression system shows that the human *cancer immunome,* i.e. the sum of all the proteins expressed by lymphomas and recognized by the patients immune system, is limited. Our recent experience with the SEREX analysis of gliomas [23] and lymphomas [22] also suggests that antigens that have escaped detection to date by the conventional SEREX approach are very likely to have a very limited expression spectrum, thus lacking the clinical importance of having the potential for use in a widely applicable vaccine. One example of such a cancer testis antigen is HOM-MM-1/cylicin-2, which was found to be expressed in only 3/121 human neoplasms [36]. In addition, it is noteworthy that very few of the SEREX antigens detected to date are located on the cell surface. A modified SEREX approach using a eukaryotic expression system should open a whole new dimension of novel antigens, since the spectrum of antigens to be discovered will not be limited to genomically determined proteins, but will include all posttranslational modifications of these molecules. Posttranslational modifications play an important role not only in the function of many proteins, but also in their immunogenicity. Hence, the characterization of such modifications is also very important for the development of recombinant protein-based vaccines. The spectrum of posttranslational modifications covers a wide range of modifications from glycosylation, lipidation, phosphorylation, methylation, acetylation, citrullination, and deimination to posttranslational truncation [37] and modification of peptides [38, 39]. Aberrant glycosylation of glycosphingolipids and glycoproteins in malignant cells is an important mechanism in the pathogenesis of many neoplasias, since it influences the function of many adhesion and signal transduction molecules in these cells. Recent evidence indicates that up to 0.1% of the antigenic peptides presented by MHC class I molecules are glycosylated [40]. As a consequence, a eukaryotic cDNA library expression system in yeast was established to overcome some of the inherent problems discussed above [41]. Yeast has the advantage that recombinant proteins can be expressed on the cell surface as part of the cell wall in a more naturally folded and partially glycosylated manner [42]. Because of the display of antigens on the yeast cell surface, fluorescence-activated cell sorting (FACS) can be used with high statistical confidence for the rapid quantitative isolation of rare clones with defined parameters [43]. However, the reproducibility of the yeast system is not reliable and so far no novel antigens, which may have escaped detection by the classical SEREX approach, have been discovered using the yeast system.

Several research groups have used antibody phage display selection strategies to isolate antibodies that bind to surface markers on tumor cells. This has resulted in the establishment of phage antibodies with specificities for human melanoma cells [44–47], human lung carcinoma [48], and colorectal carcinoma cells [49, 50]. However, the step from the selection of cell-specific phage antibodies to the molecular characterization of the cellular target is still a major hurdle. For the identification of antigens, a genomic approach using cDNA expression cloning as well as various proteomic approaches (see Chapter 4) using for example two-dimensional gel electrophoresis, affinity chromatography and mass spectrometry can be employed.

3.2.7
Perspectives for Vaccine Development

The fact that tumors present a majority of molecules that are also expressed in normal tissue and only a minority of tumor-specific molecules in the context of their MHC molecules also implies that vaccines using whole tumor-cell preparations are rather unlikely to be successful, because the induction of tumor-specific immunity by such vaccines would be a 'quantité négligeable' compared to the majority of immune responses (or tolerance) induced to normal autoantigens by such a vaccine.

The main goal of cancer vaccination is the induction of an effective specific catalytic response against tumor cells which spares the cells of normal tissues. With respect to specificity several classes of antigens may be suitable targets; in addition to the CT antigens, these include differentiation antigens, tumor-associated overexpressed gene products, mutated gene products, and tumor-specific splice variants. Clinically the most interesting class of antigens is that of the shared tumor antigens or cancer-testis antigens. To cope with the rapidly growing number of CT antigens, a new nomenclature has been suggested [51]. According to the order of their initial identification the individual genes are designated numerically. Since individual CT antigens are only expressed in a variable proportion of tumors, only the availability of several CT antigens could significantly enlarge the proportion of patients eligible for vaccination studies. In this regard it is interesting that members of a given gene family tend to be expressed in a co-regulated fashion, whereas different gene families are preferentially expressed in other sets of tumors. It is therefore reasonable to choose antigens from different CT families to cover as many tumors as possible.

Moreover, immunohistological investigations for MAGE antigens and HOM-TES-14/SCP-1 have demonstrated heterogeneity of antigen expression even in the same tumor specimen [52–54]. Thus, vaccinating a patient with a mixture of several antigens would have the potential of reducing or even preventing the *in vivo* selection of antigen-negative clones and would also address the problem of heterogeneous expression of a given antigen in an individual tumor specimen.

Identification of antigen-derived peptide epitopes which are capable of priming or activating specific CTL or T helper cells, is an indispensable prerequisite for the development of molecular vaccine strategies. Because of the diversity of peptides presented by the highly polymorphic HLA alleles, the definition of these epitopes by 'reverse T cell immunology' represents an enormous challenge. However, we and others have demonstrated that by using straightforward strategies, the identification of epitopes from SEREX-defined antigens which bind and activate either $CD8^+$ or $CD4^+$ T lymphocytes is feasible and has been successful for each SEREX antigen for which the definition of such epitopes has been pursued.

3.3
Conclusions

The identification and availability of a large number of human tumor antigens and their MHC-binding epitopes has increased the potential for the development of polyvalent vaccines for a wide spectrum of human cancers using pure preparations of antigenic proteins or peptide fragments. However, more antigens with a restricted expression in malignant tissues need to be discovered if the availability of mono- or even polyvalent vaccines for the majority of cancer patients is the goal. To this end, current screening techniques need to be modified in order to prevent posttranlsationally modified human tumor antigens being overlooked. Moreover, the study and long-term follow-up of large numbers of patients will help to determine the diagnostic and prognostic relevance of tumor-related/specific autoantibodies in patients' sera and of antigen expression in tumors, as well as the correlation with CTL responses and specific T helper cells. Well-designed clinical trials will answer the question of whether our increased knowledge about these antigens can be translated into gene and immunotherapeutic strategies with an improved prognosis for patients with malignant disease that is not curable using current standard therapy. Last, but not least, the determination of the functional role of cancer-associated antigens provides us with a more profound insight into genetic and molecular alterations that play a role in the pathogenesis and growth of cancer.

References

1 Maloney, D.G., Grillo-Lopez, A.J., White, C.A., Bodkin, D., Schilder, R.J. Neidhart, J.A. *et al.* (1997) IDEC-C2B8 (Rituximab) anti-CD20 monoclonal antibody therapy in patients with relapsed low-grade non-Hodgkin's lymphoma. *Blood*, **90** (6), 2188–2195.

2 Old, L.J. (1981) Cancer immunology: the search for specificity – G. H. A. Clowes Memorial lecture. *Cancer Research*, **41** (2), 361–375.

3 van der Bruggen, P., Traversari, C., Chomez, P., Lurquin, C., De Plaen, E. Van den, E.B. *et al.* (1991) A gene encoding an antigen recognized by cytolytic T lymphocytes on a human melanoma. *Science*, **254** (5038), 1643–1647.

4 Sahin, U., Tureci, O., Schmitt, H., Cochlovius, B., Johannes, T. Schmits, R. *et al.* (1995) Human neoplasms elicit multiple specific immune responses in the autologous host. *Proceedings of the National Academy of Sciences of the United States of America*, **92** (25), 11810–11813.

5 Sahin, U., Tureci, O. and Pfreundschuh, M. (1997) Serological identification of human tumor antigens. *Current Opinion in Immunology*, **9** (5), 709–716.

6 Tureci, O., Sahin, U., Vollmar, E., Siemer, S., Gottert, E. and Seitzm, G. *et al.* (1998) Human carbonic anhydrase XII: cDNA cloning, expression, and chromosomal localization of a carbonic anhydrase gene that is overexpressed in some renal cell cancers. *Proceedings of the National Academy of Sciences of the United States of America*, **95** (13), 7608–7613.

7 Tureci, O., Sahin, U., Schobert, I., Koslowski, M., Scmitt, H. Schild, H.J. *et al.* (1996) The SSX-2 gene, which is involved in the t(X;18) translocation of synovial sarcomas, codes for the human tumor antigen HOM-MEL-40. *Cancer Research*, **56** (20), 4766–4772.

8 Gure, A.O., Stockert, E., Arden, K.C., Boyer, A.D., Viars, C.S. and Scanlan, M.J. *et al.* (2000) CT10: a new cancer-testis (CT) antigen homologous to CT7 and the MAGE family, identified by representational-difference analysis. *International Journal of Cancer*, **85** (5), 726–732.

9 Tureci, O., Sahin, U., Zwick, C., Koslowski, M., Seitz, G. and Pfreundschuh, M. (1998) Identification of a meiosis-specific protein as a member of the class of cancer/testis antigens. *Proceedings of the National Academy of Sciences of the United States of America*, **95** (9), 5211–5216.

10 Brass, N., Heckel, D., Sahin, U., Pfreundschuh, M., Sybrecht, G.W. and Meese, E. (1997) Translation initiation factor eIF-4gamma is encoded by an amplified gene and induces an immune response in squamous cell lung carcinoma. *Human Molecular Genetics*, **6** (1), 33–39.

11 Chen, Y.T., Scanlan, M.J., Sahin, U., Tureci, O., Gure, A.O. Tsang, S. *et al.* (1997) A testicular antigen aberrantly expressed in human cancers detected by autologous antibody screening. *Proceedings of the National Academy of Sciences of the United States of America*, **94** (5), 1914–1918.

12 Boel, P., Wildmann, C., Sensi, M.L., Brasseur, R., Renaud, J.C. Coulie, P. et al. (1995) BAGE: a new gene encoding an antigen recognized on human melanomas by cytolytic T lymphocytes. *Immunity*, **2** (2), 167–175.

13 van den Eynde, B., Peeters, O., De Backer, O., Gaugler, B., Lucas, S. and Boon, T. (1995) A new family of genes coding for an antigen recognized by autologous cytolytic T lymphocytes on a human melanoma. *The Journal of Experimental Medicine*, **182** (3), 689–698.

14 Clark, J., Rocques, P.J., Crew, A.J., Gill, S., Shipley, J. Chan, A.M. et al. (1994) dentification of novel genes, SYT and SSX, involved in the t(X;18)(p11.2;q11.2) translocation found in human synovial sarcoma. *Nature Genetics*, **7** (4), 502–508.

15 Gure, A.O., Wei, I.J., Old, L.J. and Chen, Y.T. (2002) The SSX gene family: Characterization of 9 complete genes. *International Journal of Cancer*, **101** (5), 448–453.

16 Tureci, O., Chen, Y.T., Sahin, U., Gure, A.O., Zwick, C. Villena, C. et al. (1998) Expression of SSX genes in human tumors. *International Journal of Cancer*, **77** (1), 19–23.

17 Lethe, B., Lucas, S., Michaux, L., De Smet, C., Godelaine, D. Serrano, A. et al. (1998) LAGE-1, a new gene with tumor specificity. *International Journal of Cancer*, **76** (6), 903–908.

18 Jager, E., Chen, Y.T., Drijfhout, J.W., Karbach, J., Ringhoffer, M. Jager, D. et al. (1998) Simultaneous humoral and cellular immune response against cancer-testis antigen NY-ESO-1: definition of human histocompatibility leukocyte antigen (HLA)-A2-binding peptide epitopes. *The Journal of Experimental Medicine*, **187** (2), 265–270.

19 Jager, E., Stockert, E., Zidianakis, Z., Chen, Y.T., Karbach, J. Jager, D. et al. (1999) Humoral immune responses of cancer patients against "Cancer-Testis" antigen NY-ESO-1: correlation with clinical events. *International Journal of Cancer*, **84** (5), 506–510.

20 Stockert, E., Jager, E., Chen, Y.T., Scanlan, M.J., Gout, I. Karbach, J. et al. (1998) A survey of the humoral immune response of cancer patients to a panel of human tumor antigens. *The Journal of Experimental Medicine*, **187** (8), 1349–1354.

21 Eichmuller, S., Usener, D., Dummer, R., Stein, A., Thiel, D. and Schadendorf, D. (2001) Serological detection of cutaneous T-cell lymphoma-associated antigens. *Proceedings of the National Academy of Sciences of the United States of America*, **98** (2), 629–634.

22 Huang, S., Preuss, K.D., Xie, X., Regitz, E. and Pfreundschuh, M. (2002) Analysis of the antibody repertoire of lymphoma patients. *Cancer Immunology and Immunotherapy*, **51**, 655–662.

23 Schmits, R., Cochlovius, B., Treitz, G., Regitz, E., Ketter, R. Preuss, K.D. et al. (2002) Analysis of the antibody repertoire of astrocytoma patients against antigens expressed by gliomas. *International Journal of Cancer*, **98** (1), 73–77.

24 Scanlan, M.J., Chen, Y.T., Williamson, B., Gure, A.O., Stockert, E. Gordan, J.D. et al. (1998) Characterization of human colon cancer antigens recognized by autologous antibodies. *International Journal of Cancer*, **76** (5), 652–658.

25 Pierre, P., Scheel, J., Rickard, J.E. and Kreis, T.E. (1992) CLIP-170 links endocytic vesicles to microtubules. *Cell*, **70** (6), 887–900.

26 Stenner-Liewen, F., Luo, G., Sahin, U., Tureci, O., Koslovski, M. Kautz, I. et al. (2000) Definition of tumor-associated antigens in hepatocellular carcinoma. *Cancer Epidemiology, Biomarkers & Prevention*, **9** (3), 285–290.

27 Obata, Y., Takahashi, T., Sakamoto, J., Tamaki, H., Tominaga, S. Hamajima, N. et al. (2000) SEREX analysis of gastric cancer antigens. *Cancer Chemotherapy and Pharmacology*, **46**, Suppl. S37–S42.

28 Jager, E., Jager, D., Karbach, J., Chen, Y.T., Ritter, G. Nagata, Y. et al. (2000) Identification of NY-ESO-1 epitopes presented by human histocompatibility antigen (HLA)-DRB4*0101-0103 and recognized by CD4(+) T lymphocytes of patients with NY-ESO-1 expressing melanoma. *The Journal of Experimental Medicine*, **191** (4), 625–630.

29 Ayyoub, M., Stevanovic, S., Sahin, U., Guillaume, P., Servis, C. Rimoldi, D. et al. (2002) Proteasome-assisted identification of a SSX-2-derived epitope recognized by tumor-reactive CTL infiltrating metastatic melanoma. *Journal of Immunology (Baltimore, Md: 1950)*, **168** (4), 1717–1722.

30 Wagner, C., Neumann, F., Kubuschok, B., Regitz, E., Mischo, A. and Stevanovic, S. et al. (2003) Identification of an HLA-A*02 restricted immunogenic peptide derived from the cancer testis antigen HOM-MEL-40/SSX2. *Cancer Immunity*, **3**, 18.

31 Neumann, F., Wagner, C., Preuss, K.D., Kubuschok, B., Schormann, C. Stevanovic, S. et al. (2005) Identification of an epitope derived from the cancer testis antigen HOM-TES-14/SCP1 and presented by dendritic cells to circulating CD4+T cells. *Blood*, **106** (9), 3105–3113.

32 Ivanov, S.V., Kuzmin, I., Wei, M.H., Pack, S., Geil, L. Johnson, B.E. et al. (1998) Down-regulation of transmembrane carbonic anhydrases in renal cell carcinoma cell lines by wild-type von Hippel-Lindau transgenes. *Proceedings of the National Academy of Sciences of the United States of America*, **95** (21), 12596–12601.

33 Parkkila, S., Rajaniemi, H., Parkkila, A.K., Kivela, J., Waheed, A. Pastorekova, S. et al. (2000) Carbonic anhydrase inhibitor suppresses invasion of renal cancer cells in vitro. *Proceedings of the National Academy of Sciences of the United States of America*, **97** (5), 2220–2224.

34 Koslowski, M., Bell, C., Seitz, G., Lehr, H.A., Roemer, K. Muntefering, H. et al. (2004) Frequent nonrandom activation of germ-line genes in human cancer. *Cancer Research*, **64** (17), 5988–5993.

35 Matzinger, P. (2002) The danger model: a renewed sense of self. *Science*, **296** (5566), 301–305.

36 Xie, X., Renner, C., Preuss, K.D.K.B. and Pfreundschuh, M. (2002) Systematic search and molecular characterization of the antigenic targets of myeloma immunoglobulins: A monoclonal IgA from a female patient targeting sperm-specific cyclicin II. *Cancer Immunity*, **1**, 11–19.

37 Doyle, H.A. and Mamula, M.J. (2001) Post-translational protein modifications in antigen recognition and autoimmunity. *Trends in Immunology*, **22** (8), 443–449.

38 Stoltze, L., Schirle, M., Schwarz, G., Schroter, C., Thompson, M.W. Hersh, L.B. et al. (2000) Two new proteases in the MHC class I processing pathway. *Nature Immunology*, **1** (5), 413–418.

39 Valmori, D., Pittet, M.J., Vonarbourg, C., Rimoldi, D., Lienard, D. Speiser, D. et al. (1999) Analysis of the cytolytic T lymphocyte response of melanoma patients to the naturally HLA-A*0201-associated tyrosinase peptide 368–376. *Cancer Research*, **59** (16), 4050–4055.

40 Haurum, J.S., Hoier, I.B., Arsequell, G., Neisig, A., Valencia, G. Zeuthen, J. et al. (1999) Presentation of cytosolic glycosylated peptides by human class I major histocompatibility complex molecules in vivo. *The Journal of Experimental Medicine*, **190** (1), 145–150.

41 Wadle, A., Mischo, A., Imig, J., Wullner, B., Hensel, D. Watzig, K. et al. (2005) Serological identification of breast cancer-related antigens from a *Saccharomyces cerevisiae* surface display library. *International Journal of Cancer*, **117** (1), 104–113.

42 Boder, E.T. and Wittrup, K.D. (1997) Yeast surface display for screening combinatorial polypeptide libraries. *Nature Biotechnology*, **15** (6), 553–557.

43 Daugherty, P.S., Iverson, B.L. and Georgiou, G. (2000) Flow cytometric screening of cell-based libraries. *Journal of Immunological Methods*, **243** (1–2), 211–227.

44 Cai, X. and Garen, A. (1995) Anti-melanoma antibodies from melanoma patients immunized with genetically modified autologous tumor cells: selection of specific antibodies from single-chain Fv fusion phage libraries. *Proceedings of the National Academy of Sciences of the United States of America*, **92** (14), 6537–6541.

45 Kupsch, J.M., Tidman, N., Bishop, J.A., McKay, I., Leigh, I. and Crowe, J.S. (1995) Generation and selection of monoclonal antibodies, single-chain Fv and antibody fusion phage specific for human melanoma-associated antigens. *Melanoma Research*, **5** (6), 403–411.

46 Noronha, E.J., Wang, X., Desai, S.A., Kageshita, T. and Ferrone, S. (1998) Limited diversity of human scFv fragments isolated by panning a synthetic phage-display scFv library with cultured human melanoma cells. *Journal of Immunology (Baltimore, Md: 1950)*, **161** (6), 2968–2976.

47 Pereira, S., Van Belle, P., Elder, D., Maruyama, H., Jacob, L. Sivanandham, M. *et al.* (1997) Combinatorial antibodies against human malignant melanoma. *Hybridoma*, **16** (1), 11–16.

48 Ridgway, J.B., Ng, E., Kern, J.A., Lee, J., Brush, J. Goddard, A. *et al.* (1999) Identification of a human anti-CD55 single-chain Fv by subtractive panning of a phage library using tumor and nontumor cell lines. *Cancer Research*, **59** (11), 2718–2723.

49 Roovers, R.C., van der, L.E., de Bruine, A.P., Arends, J.W. and Hoogenboom, H.R. (2001) Identification of colon tumour-associated antigens by phage antibody selections on primary colorectal carcinoma. *European Journal of Cancer (Oxford, England: 1990)*, **37** (4), 542–549.

50 Roovers, R.C., van der, L.E., Zijlema, H., de Bruine, A., Arends, J.W. and Hoogenboom, H.R. (2001) Evidence for a bias toward intracellular antigens in the local humoral anti-tumor immune response of a colorectal cancer patient revealed by phage display. *International Journal of Cancer*, **93** (6), 832–840.

51 Old, L.J. and Chen, Y.T. (1998) New paths in human cancer serology. *The Journal of Experimental Medicine*, **187** (8), 1163–1167.

52 Hofbauer, G.F., Schaefer, C., Noppen, C., Boni, R., Kamarashev, J. Nestle, F.O. *et al.* (1997) MAGE-3 immunoreactivity in formalin-fixed, paraffin-embedded primary and metastatic melanoma: frequency and distribution. *The American Journal of Pathology*, **151** (6), 1549–1553.

53 Jungbluth, A.A., Chen, Y.T., Stockert, E., Busam, K.J., Kolb, D. Iversen, K. *et al.* (2001) Immunohistochemical analysis of NY-ESO-1 antigen expression in normal and malignant human tissues. *International Journal of Cancer*, **92** (6), 856–860.

54 Xie, X., Wacker, H.H., Huang, S., Regitz, E., Preuss, K.D. Romeike, B. *et al.* (2003) Differential expression of cancer testis genes in histological subtypes of non-Hodgkin's lymphomas. *Clinical Cancer Research*, **9** (1), 167–173.

4
Humoral Immune Responses against Cancer Antigens: Serological Identification Methods. Part II: Proteomex and AMIDA

Barbara Seliger and Olivier Gires

4.1
Introduction

In 1890 the German pathologist von Hansemann and colleagues first described mitotic abnormalities in cancer cells of epithelial origin, eventually leading to the term 'aneuploidy' [1]. More than two decades later, Theodor Boveri hypothesized that tumors originate from a single cell with chromosomal changes, which may support uncontrolled growth and provide survival advantages [2]. This postulate was termed 'clonal theory' and formally proven in the 1960s and 1970s, when the first chromosomal alteration, a translocation occurring in 98% of chronic myeloid leukemia (CML) patients, now referred to as the Philadelphia chromosome, was identified [3, 4]. The fusion resulting from this chromosomal translocation was characterized as the BCR-ABL protein, a constitutively active mutant of the *c-abl* gene that drives cells into proliferation and is therefore an excellent marker and therapeutic target for CML cells [5, 6]. From thereon, biomedical sciences and later molecular biology approaches were dedicated to defining genetic changes, which are concomitant or causal factors of cancer.

Despite the increased knowledge about pathogenesis, risk factors, and the molecular processes that impede controlled cell growth and hence are relevant to tumorigenesis, cancer remains a prime health concern. In past years basic and translational medical research have focussed on the identification of biomarkers and target molecules for the diagnosis, prognosis, therapy monitoring and treatment of cancer. So far only a few markers have been approved for clinical use due to conflicting results from different studies, their low sensitivity and specificity. Thus clinically reliable tumor markers are urgently needed for the early diagnosis, prognosis, and monitoring of the disease to further facilitate therapy decisions in cases of cancer. A plethora of technologies dedicated to the identification of target molecules has been developed including those that rely on the humoral response against tumor-associated antigens (TAAs) in diseased individuals. As is the case for other diseases, cancers elicit immune responses that result in the

Tumor-Associated Antigens. Edited by Olivier Gires and Barbara Seliger
Copyright © 2009 WILEY-VCH Verlag GmbH & Co. KGaA, Weinheim
ISBN: 978-3-527-32084-4

induction of T- and B-lymphocytes specific for tumor-associated proteins, largely self-antigens, but also those comprising viral and bacterial proteins. In this respect, cancer-specific serum antibodies are valuable tools for the isolation and subsequent identification of their cognate antigens. Clearly, the SEREX technology (serological analysis of recombinant cDNA expression libraries) was the first serological screening procedure described [7–10] (see Chapter 3). SEREX allows screening for TAAs after recombinant expression of cDNAs derived from tumor samples in bacteria, which guarantees comparably high sensitivity at the expense of potential post-translational modifications occurring in cancer cells. Later proteomics-based methods such as PROTEOMEX, also known as SERPA (serological and proteomic identification of antibody responses) and SPEARS (serological and proteomic evaluation of antibody responses) [11–14] together with AMIDA [15, 16] were developed to characterize TAAs in their native context, i.e. in lysates from tumor cells, but at the expense of high sensitivity screening.

4.1.1
A Humoral Response against Self-antigens: The Notion of Tumor-associated Antigens

It had long been postulated in the past that self-antigens never elicit an immune response *in vivo*. A simple reason for this former belief was the stringent negative selection of autoreactive B- and T-cells during their development. In the course of their differentiation process in the bone marrow, thymus, and periphery, self-reactive lymphocytes are removed from the pool of viable immune cells via induction of apoptosis or anergy [17–28]. The remaining cells of the adaptive immune system, which have 'passed all tests', would warrant the eradication of foreign organisms upon recognition of specific antigens. However, the occurrence of autoimmune diseases had already changed this dogma as the target for specific B- and T-cells in these autoaggressive disorders were self-antigens such as for example myelin autoantigens in multiple sclerosis [29, 30]. In addition, it has been reliably demonstrated that tumor immune responses can occur in cancer patients based on the identification of antibodies to a number of different intracellular and surface antigens in patients with various types of cancer. These tumor-associated antigens (TAAs), i.e. proteins that elicit immune responses in cancer patients, were often self-antigens. In the early stages of malignancy, the concentration of TAAs in tumor cells or in the circulation is usually very low, but a significant amplification of signals obtained from TAA is generated by the immune response. TAAs represent target structures for adaptive and innate immunity including B-cells, T-cells, professional antigen presenting cells (APC), and NK cells [31–35]. Therefore, tumor antigens and the autoantibodies they elicit are the source of candidate cancer biomarkers. Such antibodies have been discovered in serum using different processes such as screening tumor-derived cDNA expression libraries (SEREX) and proteome-based strategies. In comparison to molecular biological methods the proteomic approach has the advantage of relying on intact proteins and potentially critical post-translation modifications can be recognized by autoantibodies in patients' serum. Using these approaches a number of tumor autoantigens have been discovered such as the carcinoma embryonic antigens (CEA), the carbohydrate antigen 125 (CA-125), the cytokeratin 19 fragment marker (CYFRA 21–1), the ubiquitin carboxyl terminale hydrolase 1, (UCHL1), the superoxide dismutase, and others. The frequency of occurrence of such autoantibodies was significantly higher in serum of tumor patients as compared with sera from healthy individuals. These markers might play important roles in cancer screening, monitoring of cancer progression, treatment response and surveillance, although they do not represent ideal tools for the detection of cancer owing to their low specificity and/or sensitivity.

Based on these findings, there are a number of open issues concerning autoantigens: *What qualifies autoantigenic TAAs compared with self-proteins that do not induce autoimmunity?* One way of inducing immune responses against self-antigens is the several hundred-fold over-expression of the respective self-antigen [36]. Enhanced presentation of antigenic epitopes by APCs may result in the activation of immune cells [37, 38] including $CD4^+$ T-helper cells along with the production of tumor-specific antibodies by plasma B-cells [39–41].

However self-antigens manage to induce immune cells, the main net effect is the presence of serum antibodies with specificity for a given cancer. Such serum antibodies may represent potent indicators (i.e. candidate biomarkers) of disease, of treatment success, and/or of recurrence [42, 43]. Apart from their diagnostic value, serum antibodies can be used to selectively isolate and identify novel tumor markers, namely their cognate antigen protein [44].

4.2
Implementation of Serum Antibodies: Serological Screening Technologies

In the late 1970s and early 1980s several groups used the emerging monoclonal antibody technology [45] to identify potentially new TAAs [46–49]. The method was known as *autologous typing* and was based on the binding of serum antibodies to cultured autologous carcinoma cell lines. Following repeated rounds of antibody absorption to carcinoma cells, a panel of valid and valuable tumor antigens was identified including for example the p53 tumor suppressor [50], the CD20 antigen on B-cell lymphomas [51], and the epithelial antigen EpCAM on colon carcinomas [52, 53]. Eventually, two decades later, due to the availability of modern molecular biology and genetic engineering methods, the initial perception of using serum antibodies for identification purposes has been considerably refined. At present the aim of biomedical science is to determine accurately the tumor-relevant autoantibodies and their cognate antigens. This particular scientific field has been very recently summarized as 'cancer immunomics' by Caron *et al.* [42]. Two proteome-based technologies can nowadays be implemented for the serological screening of TAAs: *PROTEOMEX* and *AMIDA*. This chapter is dedicated to the illustration of the potency as well as the disadvantages of both techniques.

4.2.1
PROTEOMEX, alias SERPA and SPEARS

In general, the methods for tumor antigen identification are often expensive, time consuming, and incompatible with the screening of a large number of tumor samples. Furthermore, the use of cDNA expression libraries precludes the identification of immunogens that are dependent on potentially decisive post-translational modifications. PROTEOMEX, a combination of proteomics and SEREX, which is also known as SERPA or SPEARS, has the potential to overcome some of these limitations [12, 14, 54–56]. PROTEOMEX is based on the separation of proteins derived from tumor and normal tissues by two-dimensional gel electrophoresis (2DE), followed by the subsequent mass spectrometric identification of proteins, and has been implemented in screening for renal cell carcinoma (RCC) [11, 13, 14, 54, 55, 57, 58], hepatocellular carcinoma [59–66], neuroblastoma as well as breast, pancreatic, and prostate carcinoma [67–69]. 2DE separates proteins according to their isoelectric point and in the second separation according to their molecular weight. PROTEOMEX was developed to identify proteins that raise antibody responses *in vivo* in cancer patients. When applying PROTEOMEX, three to five 2DE gels are run simultaneously under defined conditions with equal amounts of proteins to be separated. Two to three gels are blotted onto nitrocellulose or PVDF membranes and probed with serum from a cancer patient or a healthy individual, respectively (Figure 4.1). The remaining gel (i.e. the preparative

PROTEOMEX

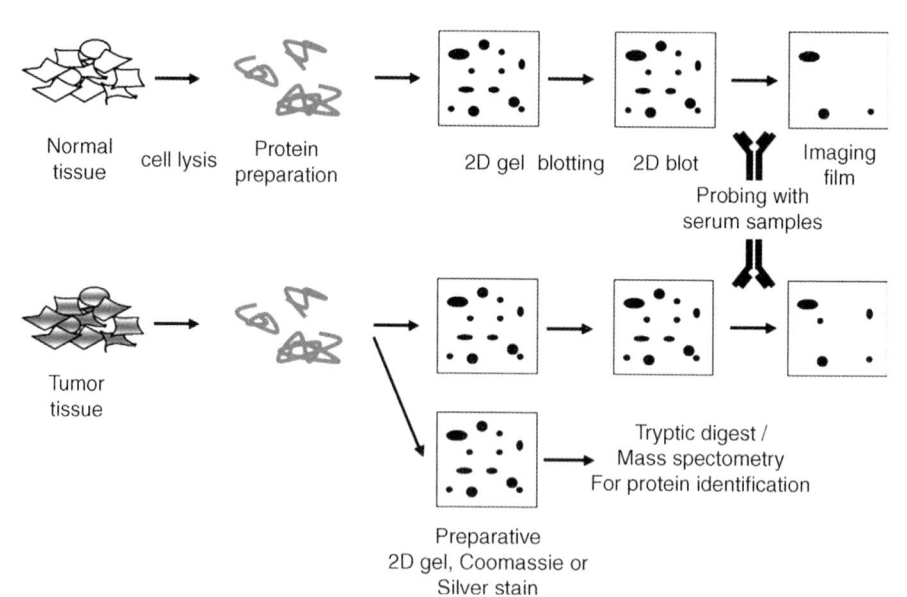

Figure 4.1 Schematic view of PROTEOMEX. Adapted from Rauch and Gires, Proteomics Clinical Application (in press).

gel) is stained with Coomassie blue. Immunoreactive spots from immunoblots of cancer patients are subsequently compared to those of controls. Differentially expressed spots are excised from the preparative gel, heated with trypsin within gel slices, and peptides thereoff subjected to mass spectrometry.

4.2.1.1 PROTEOMEX Technology and its 'Pros' and 'Cons'

When compared to SEREX, PROTEOMEX experiments require less time for completion, are simpler to carry out, and it is more straightforward to implement specificity controls, in contrast to SEREX [12, 56]. In addition, PROTEOMEX allows the identification of post-translational modifications in TAAs, thereby providing a larger number of antigenic determinants for serological testing. Another advantage is the global antibody/target interaction, which is detected by the use of 2DE immunoblots in PROTEOMEX. However, this technology also exhibits some disadvantages, which are inherent to 2DE technology [12, 42]. Generally 2DE is biased towards the identification of relatively abundant proteins despite recent improvements in sensitivity resulting from the use of specific fluorogenic dyes. Another disadvantage is the lack of separation of different proteins that co-migrate on 2DE gels due to their post-translational modifications. In other words, one spot on a given 2DE gel may represent several proteins with similar migration properties, thus complicating the visualization of distinct spots. Despite these limitations PROTEOMEX is a valuable technique for the identification of TAA candidates.

4.2.1.2 Candidate Biomarkers Identified by PROTEOMEX

As a proof of principle, PROTEOMEX was employed for mapping the expression pattern of key components of the MHC class I antigen processing and presentation machinery (APM) in RCC [70]. Results thereof clearly confirmed the identification of APM-related components and allowed their mapping on 2DE with the sole exception of the peptide transporters TAP1 and TAP2, which represent ER-resident membrane proteins and hence can hardly be separated by 2DE.

Based on these results the PROTEOMEX technology has been successfully applied to the identification of autoantibodies to different tumor antigens in cancers of distinct histology [14, 42]. TAAs as identified by PROTEOMEX include differentiation antigens and proteins that are over-expressed in tumors such as annexin I and II, heat shock proteins, various metabolic enzymes, and structural proteins as well as the ubiquitin carboxyl terminal hydrolase (UCHL) 1 [54, 61, 71]. In hepatocellular carcinoma (HCC) autoantibodies to eight proteins were detected in sera from more than 10% of patients, but not in sera from healthy individuals [59]. These include autoantibodies of calreticulin, cytokeratin 8, hsp 60, β-tubulin, nucleoside diphosphate kinase A, and the F_1-ATP synthase β-subunit [59]. In breast cancer the RS/DJ-1 protein was found in the serum of 37% of newly diagnosed mammary carcinoma patients [67]. The RS/DJ-1 gene has been proposed to play an important role in the control of gene expression through a variety of post-transcriptional mechanisms that control mRNA stability, localization, and translation. Furthermore, in human lung squamous cell carcinoma PROTEOMEX was used to identify six out of 14 proteins, which were also up-regulated in lung squamous cell carcinoma using comparative proteomics studies. These include enolase, pre-B cell-enhancing factor precursor, triosephosphate isomerase, phosphoglycerate mutase 1, fructose biphosphate aldolase A, and the guanine nucleotide binding protein β-subunit-like protein [72].

So far, RCC and lung carcinoma have been the major tumor types analyzed using this technology. Various research groups have detected a number of unique, but also overlapping antigens in both tumor types; these include annexins I and IV, thymidine phosphorylase, carbonic anhydrase 1, major vault protein, and superoxide dismutase, but in most cases only small number of tumor samples have been analyzed [54, 61, 71]. In addition, autoantibodies have also been defined in pancreatic carcinoma and neuroblastoma [66, 68]. In pancreatic carcinoma the cognate self-antigens belong to the family of metabolic enzymes and cytoskeletal proteins [68], whereas in neuroblastoma the cytoskeletal component β-tubulin in particular has been identified [66].

4.2.1.3 Implementation of Candidate Biomarkers in the Clinic

Despite the detection of a number of TAAs in a variety of studies it is noteworthy that several proteins were also recognized by serum obtained from healthy individuals. Thus, the value of the markers has still to be validated. However, taking the high UCHL1 levels in different tumor lesions such as lung carcinoma, pancreas carcinoma, and RCC into account, UCHL1 might represent a potential marker for diagnosis and prognosis. A comparable situation occurs for annexin I and II which have been demonstrated to induce autoantibodies in different cancers. With regard to hepatocellular carcinoma (HCC) a distinct repertoire of autoantibodies was detected in tumor patients when compared with samples from chronic hepatitis patients. Based on the published data, it has been suggested that the truncated form of calreticulin may be a suitable marker and should be monitored for the early diagnosis of HCC in high-risk subjects. Thus, the detection of autoantibodies directed against HCC-associated antigens may be of value in HCC screening, diagnosis of follow-up and in monitoring the impact of antigen-based immunotherapy. Notably, RS/DJ-1 appears to be a robust marker since it was detected at high frequency in the sera of breast cancer patients. This circulating protein and/or the cognate autoantibody in the serum might therefore have potential clinical utility. However, the identification of novel markers using PROTEOMEX demonstrated the usefulness of this technology in the analyses of cell type-independent and -dependent gene expression in addition to monitoring large series of cellular proteins, which will provide complementary information to transcription analyses [12, 42].

4.2.2
AMIDA

First described in 2004, AMIDA is the acronym for Autoantibody-Mediated IDentification of Antigens. The first generation of AMIDA technology was carried out in an autologous fashion with serum antibodies and tumor samples from the same donor [13]. The second generation was able to analyze allogeneic samples and incorporated a library of immunoglobulins from healthy volunteers as a control [14].

4.2.2.1 Autologous AMIDA

Using the autologous version of AMIDA, serum antibodies are employed to specifically immunoprecipitate potential TAAs from tumor lysates from the same donor. Antibody–antigen complexes are subsequently separated by high-resolution 2DE according to the isoelectric point and molecular weight (M_r) of proteins. Thereafter, immunoprecipitates from tumor lysates and those from non-transformed autologous cells (e.g. leukocytes) are visualized with Coomassie blue or silver nitrate and staining patterns subjected to computer-assisted comparison. By this means, it is feasible to distinguish potential TAAs from proteins that are similarly immunogenic in normal tissues and most likely reflect the presence of so-called naturally-occurring autoantibodies (NAAs) [52]. Thus, only proteins retrieved from carcinoma cells are eligible for mass spectrometry analysis. Two additional controls can be implemented in the autologous version of AMIDA: (i) Immunoglobulins can be omitted during the process of immunoprecipitation in order to visualize proteins that bind non-specifically to protein A/G sepharose. These protein spots will obviously be excluded from the pool of proteins to be analyzed. (ii) Immunoglobulins from patients' sera can be covalently coupled to sepharose beads pre-coated with anti-human Ig secondary antibodies before immunoprecipitation. Beads coated with secondary antibody will only then serve as a control for non-specific binding of proteins. Protein spots that match these selection criteria are eligible for mass spectrometry technologies of choice, such as Quadrupole, ESI-ToF, LTQ-FT and MALDI-ToF/ToF.

4.2.2.2 Allo-AMIDA

AMIDA was primarily designed as an autologous method with the aim of circumventing problems of false-positive results owing to allogeneic cross-reactivity when using non-autologous samples. However, in particular, primary material as a source of proteins to be immunoprecipitated, unlike serum as a source of immunoglobulins, is difficult to handle in the laboratory and the routine clinical environment. The availability of primary tumor samples and the size of tumor biopsies dictate the number of experiments that can be undertaken. An estimated cell number of $5 \times 10^7 - 10^8$ at least is required to ensure that low abundant proteins are sufficiently represented in cell lysates. For a given 60-kD protein of which there are 10 000 copies/cell, a minimum of 10^7 cells are needed in order to obtain 10 ng of this protein[1]. A plethora of potentially significant proteins such as components of the signal

---

1) $10\,\text{ng} = 167\,\text{fmol} \times \text{Avogadro } 6{,}022 \ast 10^{23} = 10^{11}$ molecules → for a given protein of 60 kD of which there are 10 000 copies per cell, 10^7 cells are needed in order to yield 10 ng.

transduction pathways are present in even lower copy numbers, requiring an even higher number of cells to produce an adequate concentration of the protein. Thus, some difficulties arise when analyzing autologous samples owing to the small number of tumor cells available from primary biopsies and tumor heterogeneity. With these restrictions in mind, Rauch et al. developed an allogeneic version of AMIDA, known as allo-AMIDA [16]. In this approach tumor cell lines served as a source of proteins to be incubated with allogeneic, purified serum antibodies. Cell lines represent a fairly steady and unlimited supply of antigens, thus facilitating experimental replication, and the use of cell lines does not comprise proteins from undefined and varying numbers of infiltrating cells. Since immunoglobulins and lysates are of different origin, a pool of immunoglobulins from healthy donors ($n = 100$) can be used for a control immunoprecipitation in parallel to antibodies from the respective cancer patient. Additionally, purification of tumor-derived antibodies allows for a stringent quantification of the immunoglobulins used in every experiment. When performing allo-AMIDA, the comparison of immunoprecipitates between healthy and tumor immunoglobulins pinpoints potential TAAs. By using control immunoglobulins, proteins immunoprecipitated with naturally-occurring autoantibodies from healthy individuals and with allogeneic-reactive antibodies, can be excluded from the analysis. Following incubation of cell lysates and antibodies, immunoprecipitates are separated by 2DE, proteins are stained with Coomassie blue or silver nitrate and differential proteins excised in order to be processed for mass spectrometry analysis. A schematic representation of AMIDA is given in Figure 4.2. It can be estimated that potential TAAs represented <5% of all protein spots on 2DE gels from individual patient samples when applying this optimized version of allogeneic AMIDA [14].

AMIDA

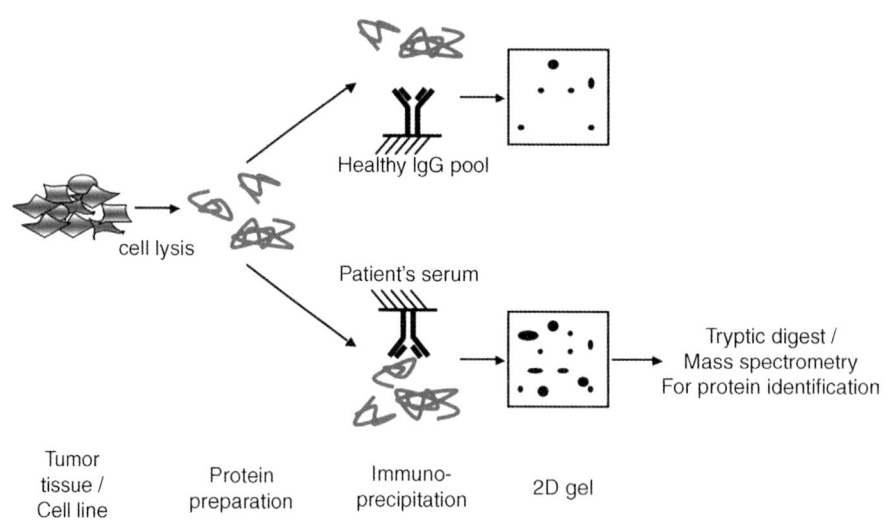

Figure 4.2 Schematic view of AMIDA. Adapted from Rauch and Gires, Proteomics Clinical Application (in press).

4.2.3
Advantages and Disadvantages of AMIDA

AMIDA, which is chronologically the most recently developed strategy, was derived from experience gained with SEREX and PROTEOMEX, and was developed to circumvent their limitations to some degree. In the present section the advantages and disadvantages of AMIDA will be discussed in a comparative manner.

AMIDA is a screening technology that can be used for proteins in their native state, which is an advantage over PROTEOMEX where antigens first need to be separated by 2DE and immobilized on membranes before detection, and are hence denatured. For this reason AMIDA can be used to isolate antigens with antibodies that recognize structural epitopes. However, due to the use of silver nitrate staining instead of immunoblotting, AMIDA technology is associated with decreased sensitivity and efficiency of protein identification by mass spectrometry. The use of serum antibodies for immunoprecipitation of potential TAAs before their separation can be viewed as a biological filter, which greatly decreases the complexity of the proteins to be separated and analyzed. In light of the separation capacity of standard two-dimensional gels, which at best cover 10% of the proteome, such a selection seems desirable.

4.3
AMIDA Antigens and Clinical Application

4.3.1
Diagnostic TAAs Detected with AMIDA Screening

TAAs may have several functions with respect to their clinical use. Prostate-specific antigen (PSA) and the receptor-tyrosine kinase HER2/neu are excellent examples of TAAs which are of great importance in clinical routine [73–75]. PSA is certainly the most popular marker for the diagnosis of prostate carcinoma, while overexpression of HER2/neu is not only useful in diagnostics, but is also a therapeutic target for monoclonal antibodies such as Herceptin® (Trastuzumab) and selective small drugs, such as tyrosine kinase inhibitors.

An increase in specific serum antibodies to the AMIDA antigen cytokeratin 8 (CK8) is a very reliable indicator of head and neck malignancies [13]. Unfortunately, determination of the antibody titre failed to show any differences between benign and malignant diseases [76]. Nevertheless, CK8 itself is a valuable diagnostic marker whose expression characterizes pre-malignant and dysplastic, but not hyperplastic specimens [77]. Other AMIDA antigens that are substantially over-expressed in malignant cells are hnRNP H, eFABP, TOB3, and eIF3i [14, 77–79]. Molecular analysis revealed a function for TOB3 in cytokinesis. Accordingly, siRNA-mediated knock-down of TOB3 resulted in large, multinuclear cells [78]. EIF3i was involved in mTOR-mediated growth and proliferation [79]. Hence, AMIDA antigens not only provide new candidate markers for diagnostics, but in addition are able to provide information regarding the molecular biology of carcinomas.

4.3.2
Therapeutic Markers

A number of different immunotherapeutic strategies are available including stimulation of $CD8^+$ cytotoxic T Lymphoytes (CTL), $CD4^+$ T-helper cells and dendritic cells (DC) by various cytokine and antibody-based therapies. Consequently, two general scenarios become obvious when defining therapeutic antigens: (i) TAAs are targets for monoclonal, bi-specific, or derivative antibody constructs, and/or (ii) targets are processed, presented at the cell surface in the context of MHC molecules, and recognized by $CD8^+$ CTL and $CD4^+$ T-cells.

With regard to AMIDA antigens, it is solely CK8 that emerges as a potential target for monoclonal antibodies. CK8 has the remarkable feature of associating with plasma membranes in tumor cells [80–82]. Interestingly, CK8 serves as a cell surface receptor for plasminogen and tissue-type plasminogen activator in hepatocytes, HepG2 cells, and breast cancer cells [81–83]. It is therefore anticipated that CK8 generates a plasma membrane-associated complex that increases plasmin formation, consequently enhancing the invasion capacity of tumor cells. In support of this notion, Chu and colleagues demonstrated that *de novo* expression of CK8/18 in mouse fibroblasts augmented cell migration and invasion [84]. However, the potential of antibodies directed against CK8 thereby blocking $CK8^+$ tumors has yet to be experimentally determined.

4.4
Conclusions

In summary the field of cancer immunomics with the identification of novel biomarkers is intimately tied to advances in proteomics technologies [85, 86]. SEREX, SELDI, protein microarray, PROTEOMEX, and AMIDA technologies represent the most powerful techniques in the discovery phase. However, the discovery thus far has not provided a solution to the problem of early tumor detection. PROTEOMEX and/or the AMIDA technology incorporates different advantages such as the screening of native proteins, the use of serum antibodies as a biological filter for tumor proteins that elicit an immune response *in vivo*, but also has disadvantages with respect to the sensitivity of TAA recognition.

As for other target identification technologies, PROTEOMEX and AMIDA led to the identification of a considerable number of TAAs, although not as many as those achieved with SEREX, which are awaiting further validation in a pre-clinical setting. To this end, the frequency and selectivity of the TAA will be required to attain significance in large cohorts of patients in order to define its diagnostic and prognostic value. A good biomarker would be a protein that is detectable in serum, blood and/or urine with demonstrable robustness coupled with high specificity and sensitivity. However, it would be expect that a series of TAAs would display increased sensitivity and specificity for a particular tumor type. A combination of extensive screening, pre-clinical validation including animal models, and thorough functional analysis of the TAAs of choice is therefore highly desirable. Lastly, TAAs need further validation as potential biomarkers in the context of early detection, prognosis and response to therapy.

Acknowledgments

B.S. is supported by grants from the BMBF, 'Deutsche Krebshilfe', and Else-Kröner-Fresenius Stiftung.

O.G. is supported by grants from 'Deutsche Forschungsgemeinschaft' (GI-540/1-1), the GSF-National Center Health and Environment, and 'Deutsche Krebshilfe'.

References

1 Von Hansemann, D. (1890) Ueber asymmetrische Zellteilung in Epithelkrebsen und deren biologische Bedeutung. *Virchows Archiv A: Pathological Anatomy*, **199**, 299–326.

2 Bignold, L.P., Coghlan, B.L. and Jersmann, H.P. (2006) Hansemann, Boveri, chromosomes and the gametogenesis-related theories of tumours. *Cell Biology International*, **30**, 640–644.

3 Koretzky, G.A. (2007) The legacy of the Philadelphia chromosome. *The Journal of Clinical Investigation*, **117**, 2030–2032.

4 Nowell, P.C. (2007) Discovery of the Philadelphia chromosome: a personal perspective. *The Journal of Clinical Investigation*, **117**, 2033–2035.

5 Sherbenou, D.W. and Druker, B.J. (2007) Applying the discovery of the Philadelphia chromosome. *The Journal of Clinical Investigation*, **117**, 2067–2074.

6 Schiffer, C.A. (2007) BCR-ABL tyrosine kinase inhibitors for chronic myelogenous leukemia. *The New England Journal of Medicine*, **357**, 258–265.

7 Pfreundschuh, M. (2000) Exploitation of the B cell repertoire for the identification of human tumour antigens. *Cancer Chemotherapy and Pharmacology*, **46**, S3–S7.

8 Sahin, U., Tureci, O. and Pfreundschuh, M. (1997) Serological identification of human tumour antigens. *Current Opinion in Immunology*, **9**, 709–716.

9 Sahin, U., Tureci, O., Schmitt, H., Cochlovius, B., Johannes, T., Schmits, R., Stenner, F., Luo, G., Schobert, I. and Pfreundschuh, M. (1995) Human neoplasms elicit multiple specific immune responses in the autologous host. *Proceedings of the National Academy of Sciences of the United States of America*, **92**, 11810–11813.

10 Tureci, O., Chen, Y.T., Sahin, U., Gure, A.O., Zwick, C., Villena, C., Tsang, S., Seitz, G., Old, L.J. and Pfreundschuh, M. (1998) Expression of SSX genes in human tumours. *International Journal of Cancer*, **77**, 19–23.

11 Kellner, R., Lichtenfels, R., Atkins, D., Bukur, J., Ackermann, A., Beck, J., Brenner, W., Melchior, S. and Seliger, B. (2002) Targeting of tumour associated antigens in renal cell carcinoma using proteome-based analysis and their

clinical significance. *Proteomics*, **2**, 1743–1751.

12 Seliger, B. and Kellner, R. (2002) Design of proteome-based studies in combination with serology for the identification of biomarkers and novel targets. *Proteomics*, **2**, 1641–1651.

13 Lichtenfels, R., Kellner, R., Atkins, D., Bukur, J., Ackermann, A., Beck, J., Brenner, W., Melchior, S. and Seliger, B. (2003) Identification of metabolic enzymes in renal cell carcinoma utilizing PROTEOMEX analyses. *Biochimica et Biophysica Acta*, **1646** (1–2), 21–31.

14 Klade, C.S., Voss, T., Krystek, E., Ahorn, H., Zatloukal, K., Pummer, K. and Adolf, G.R. (2001) Identification of tumour antigens in renal cell carcinoma by serological proteome analysis. *Proteomics*, **1** (7), 890–898.

15 Gires, O., Munz, M., Schaffrik, M., Kieu, C., Rauch, J., Ahlemann, M., Eberle, D., Mack, B., Wollenberg, B., Lang, S., Hofmann, T., Hammerschmidt, W. and Zeidler, R. (2004) Profile identification of disease-associated humoral antigens using AMIDA, a novel proteomics-based technology. *Cellular and Molecular Life Sciences*, **61**, 1198–1207.

16 Rauch, J., Ahlemann, M., Schaffrik, M., Mack, B., Ertongur, S., Andratschke, M., Zeidler, R., Lang, S. and Gires, O. (2004) Allogenic antibody-mediated identification of head and neck cancer antigens. *Biochemical and Biophysical Research Communications*, **323**, 156–162.

17 von Boehmer, H., Aifantis, I., Gounari, F., Azogui, O., Haughn, L., Apostolou, I., Jaeckel, E., Grassi, F. and Klein, L. (2003) Thymic selection revisited: how essential is it? *Immunological Reviews*, **191**, 62–78.

18 von Boehmer, H. and Kisielow, P. (2006) Negative selection of the T-cell repertoire: where and when does it occur? *Immunological Reviews*, **209**, 284–289.

19 von Boehmer, H., Teh, H.S. and Kisielow, P. (1989) The thymus selects the useful, neglects the useless and destroys the harmful. *Immunology Today*, **10**, 57–61.

20 Bannish, G., Fuentes-Panana, E.M., Cambier, J.C., Pear, W.S. and Monroe, J.G. (2001) Ligand-independent signaling functions for the B lymphocyte antigen receptor and their role in positive selection during B lymphopoiesis. *The Journal of Experimental Medicine*, **194**, 1583–1596.

21 Liu, H., Schmidt-Supprian, M., Shi, Y., Hobeika, E., Barteneva, N., Jumaa, H., Pelanda, R., Reth, M., Skok, J., Rajewsky, K. and Shi, Y. (2007) Yin Yang 1 is a critical regulator of B-cell development. *Genes and Development*, **21**, 1179–1189.

22 Sasaki, Y., Schmidt-Supprian, M., Derudder, E. and Rajewsky, K. (2007) Role of NFκB signaling in normal and malignant B cell development. *Advances in Experimental Medicine and Biology*, **596**, 149–154.

23 Lindsley, R.C., Thomas, M., Srivastava, B. and Allman, D. (2007) Generation of peripheral B cells occurs via two spatially and temporally distinct pathways. *Blood*, **109**, 2521–2528.

24 Thomas, M.D., Srivastava, B. and Allman, D. (2006) Regulation of peripheral B cell maturation. *Cellular Immunology*, **239**, 92–102.

25 LeBien, T.W. (1990) Developmental regulation of normal and leukemic human B cell precursors. *Progress in Clinical and Biological Research*, **352**, 417–422.

26 LeBien, T.W. (1998) B-cell lymphopoiesis in mouse and man. *Current Opinion in Immunology*, **10**, 188–195.

27 LeBien, T.W. (2000) Fates of human B-cell precursors. *Blood*, **96**, 9–23.

28 LeBien, T.W. and Villablanca, J.G. (1990) Ontogeny of normal human B-cell and T-cell precursors and its relation to leukemogenesis. *Hematology/Oncology Clinics of North America*, **4**, 835–847.

29 Schmidt, D., Amrani, A., Verdaguer, J., Bou, S. and Santamaria, P. (1999) Autoantigen-independent deletion of diabetogenic CD4+ thymocytes by

protective MHC class II molecules. *Journal of Immunology (Baltimore, Md: 1950)*, **162**, 4627–4636.
30 Stinissen, P., Medaer, R. and Raus, J. (1998) Myelin reactive T cells in the autoimmune pathogenesis of multiple sclerosis. *Multiple Sclerosis (Houndmills, Basingstoke, England)*, **4**, 203–211.
31 Root-Bernstein, R. (2007) Antigenic complementarity in the induction of autoimmunity: a general theory and review. *Autoimmun Rev*, **6**, 272–277.
32 Bettelli, E., Oukka, M. and Kuchroo, V.K. (2007) T(H)-17 cells in the circle of immunity and autoimmunity. *Nature Immunology*, **8**, 345–350.
33 Foster, M.H. (2007) T cells and B cells in lupus nephritis. *Seminars in Nephrology*, **27**, 47–58.
34 Thedrez, A., Sabourin, C., Gertner, J., Devilder, M.C., Allain-Maillet, S., Fournie, J.J., Scotet, E. and Bonneville, M. (2007) Self/non-self discrimination by human gammadelta T cells: simple solutions for a complex issue? *Immunological Reviews*, **215**, 123–135.
35 Zanetti, M. (1988) Self-immunity and the autoimmune network: a molecular perspective to ontogeny and regulation of the immune system. *Annales de l'Institut Pasteur. Immunologie*, **139**, 619–631.
36 Zinkernagel, R.M. and Hengartner, H. (2001) Regulation of the immune response by antigen. *Science*, **293**, 251–253.
37 Spiotto, M.T., Fu, Y.X. and Schreiber, H. (2003) Tumour immunity meets autoimmunity: antigen levels and dendritic cell maturation. *Current Opinion in Immunology*, **15**, 725–730.
38 Spiotto, M.T., Yu, P., Rowley, D.A., Nishimura, M.I., Meredith, S.C., Gajewski, T.F., Fu, Y.X. and Schreiber, H. (2002) Increasing tumour antigen expression overcomes 'ignorance' to solid tumours via crosspresentation by bone marrow-derived stromal cells. *Immunity*, **17**, 737–747.
39 Preuss, K.D., Zwick, C., Bormann, C., Neumann, F. and Pfreundschuh, M. (2002) Analysis of the B-cell repertoire against antigens expressed by human neoplasms. *Immunological Reviews*, **188**, 43–50.
40 Sahin, U., Tureci, O., Schmitt, H., Cochlovius, B., Johannes, T., Schmits, R., Stenner, F., Luo, G., Schobert, I. and Pfreundschuh, M. (1995) Human neoplasms elicit multiple specific immune responses in the autologous host. *Proceedings of the National Academy of Sciences of the United States of America*, **92**, 11810–11813.
41 Nzula, S., Going, J.J. and Stott, D.I. (2003) Antigen-driven clonal proliferation, somatic hypermutation, and selection of B lymphocytes infiltrating human ductal breast carcinomas. *Cancer Research*, **63**, 3275–3280.
42 Caron, M., Choquet-Kastylevsky, G. and Joubert-Caron, R. (2007) Cancer immunomics using autoantibody signatures for biomarker discovery. *Molecular & Cellular Proteomics*, **6**, 1115–1122.
43 Cho-Chung, Y.S. (2006) Autoantibody biomarkers in the detection of cancer. *Biochimica et Biophysica Acta*, **1762**, 587–591.
44 Finn, O.J. (2005) Immune response as a biomarker for cancer detection and a lot more. *The New England Journal of Medicine*, **353**, 1288–1290.
45 Kohler, G. and Milstein, C. (1975) Continuous cultures of fused cells secreting antibody of predefined specificity. *Nature*, **256**, 495–497.
46 Shiku, H., Takahashi, T., Resnick, L.A., Oettgen, H.F. and Old, L.J. (1977) Cell surface antigens of human malignant melanoma. III. Recognition of autoantibodies with unusual characteristics. *The Journal of Experimental Medicine*, **145**, 784–789.
47 Old, L.J. and Chen, Y.T. (1998) New paths in human cancer serology. *The Journal of Experimental Medicine*, **187**, 1163–1167.

48 Old, L.J. (1992) Tumour immunology: the first century. *Current Opinion in Immunology*, **4**, 603–607.

49 Herlyn, M., Steplewski, Z., Herlyn, D. and Koprowski, H. (1979) Colorectal carcinoma-specific antigen: detection by means of monoclonal antibodies. *Proceedings of the National Academy of Sciences of the United States of America*, **76**, 1438–1442.

50 DeLeo, A.B., Jay, G., Appella, E., Dubois, G.C., Law, L.W. and Old, L.J. (1979) Detection of a transformation-related antigen in chemically induced sarcomas and other transformed cells of the mouse. *Proceedings of the National Academy of Sciences of the United States of America*, **76**, 2420–2424.

51 Tedder, T.F. and Engel, P. (1994) CD20: a regulator of cell-cycle progression of B lymphocytes. *Immunology Today*, **15**, 450–454.

52 Baeuerle, P.A. and Gires, O. (2007) EpCAM (CD326) finding its role in cancer. *British Journal of Cancer*, **96**, 417–423.

53 Armstrong, A. and Eck, S.L. (2003) EpCAM: A new therapeutic target for an old cancer antigen. *Cancer Biology & Therapy*, **2**, 320–326.

54 Unwin, R.D., Harnden, P., Pappin, D., Rahman, D., Whelan, P., Craven, R.A., Selby, P.J. and Banks, R.E. (2003) Serological and proteomic evaluation of antibody responses in the identification of tumour antigens in renal cell carcinoma. *Proteomics*, **3** (1), 45–55.

55 Le Naour, F. (2001) Contribution of proteomics to tumour immunology. *Proteomics*, **1** (10), 1295–1302.

56 Gunawardana, C.G. and Diamandis, E.P. (2007) High throughput proteomic strategies for identifying tumour-associated antigens. *Cancer Letters*, **249** (1), 110–119.

57 Klade, C.S., Dohnal, A., Furst, W., Sommergruber, W., Heider, K.H., Gharwan, H., Ratschek, M. and Adolf, G.R. (2002) Identification and characterization of 9D7, a novel human protein overexpressed in renal cell carcinoma. *International Journal of Cancer*, **97** (2), 217–224.

58 Seliger, B., Menig, M., Lichtenfels, R., Atkins, D., Bukur, J., Halder, T.M., Kersten, M., Harder, A., Ackermann, A., Beck, J., Muehlenweg, B., Brenner, W., Melchior, S., Kellner, R. and Lottspeich, F. (2003) Identification of markers for the selection of patients undergoing renal cell carcinoma-specific immunotherapy, *Proteomics*, **3**, 979–990.

59 Le Naour, F., Brichory, F., Misek, D.E., Brechot, C., Hanash, S.M. and Beretta, L. (2002) A distinct repertoire of autoantibodies in hepatocellular carcinoma identified by proteomic analysis. *Molecular & Cellular Proteomics*, **1**, 197–203.

60 Ward, D.G., Cheng, Y., N'Kontchou, G., Thar, T.T., Barget, N., Wei, W., Billingham, L.J., Martin, A., Beaugrand, M. and Johnson, P.J. (2006) Changes in the serum proteome associated with the development of hepatocellular carcinoma in hepatitis C-related cirrhosis. *British Journal of Cancer*, **94** (2), 287–292.

61 Brichory, F.M., Misek, D.E., Yim, A.M., Krause, M.C., Giordano, T.J., Beer, D.G. and Hanash, S.M. (2001) An immune response manifested by the common occurrence of annexins I and II autoantibodies and high circulating levels of IL-6 in lung cancer. *Proceedings of the National Academy of Sciences of the United States of America*, **98**, 9824–9829.

62 Brichory, F., Beer, D., Le Naour, F., Giordano, T. and Hanash, S. (2001) Proteomics-based identification of protein gene product 9.5 as a tumour antigen that induces a humoral immune response in lung cancer. *Cancer Research*, **61**, 7908–7912.

63 Yang, F., Xiao, Z.Q., Zhang, X.Z., Li, C., Zhang, P.F., Li, M.Y., Chen, Y., Zhu, G.Q., Sun, Y., Liu, Y.F. and Chen, Z.C. (2007) Identification of tumour antigens in human lung squamous carcinoma by

64 Hanash, S., Brichory, F. and Beer, D. (2001) A proteomic approach to the identification of lung cancer markers. *Disease Markers*, **17** (4), 295–300.

65 Huang, L.J., Chen, S.X., Huang, Y., Luo, W.J., Jiang, H.H., Hu, Q.H., Zhang, P.F. and Yi, H. (2006) Proteomics-based identification of secreted protein dihydrodiol dehydrogenase as a novel serum markers of non-small cell lung cancer. *Lung Cancer (Amsterdam, Netherlands)*, **54** (1), 87–94.

66 Prasannan, L., Misek, D.E., Hinderer, R., Michon, J., Geiger, J.D. and Hanash, S.M. (2000) Identification of beta-tubulin isoforms as tumour antigens in neuroblastoma. *Clinical Cancer Research*, **6**, 3949–3956.

67 Le Naour, F., Misek, D.E., Krause, M.C., Deneux, L., Giordano, T.J., Scholl, S. and Hanash, S.M. (2001) Proteomics-based identification of RS/DJ-1 as a novel circulating tumour antigen in breast cancer. *Clinical Cancer Research*, **7**, 3328–3335.

68 Tomaino, B., Cappello, P., Capello, M., Fredolini, C., Ponzetto, A., Novarino, A., Ciuffreda, L., Bertetto, O., De Angelis, C., Gaia, E., Salacone, P., Milella, M., Nistico, P., Alessio, M., Chiarle, R., Giuffrida, M.G., Giovarelli, M. and Novelli, F. (2007) Autoantibody signature in human ductal pancreatic adenocarcinoma. *Journal of Proteome Research*, **6**, 4025–4031.

69 Mobley, J.A., Lam, Y.W., Lau, K.M., Pais, V.M., L'Esperance, J.O., Steadman, B., Fuster, L.M., Blute, R.D., Taplin, M.E. and Ho, S.M. (2004) Monitoring the serological proteome: the latest modality in prostate cancer detection. *The Journal of Urology*, **172** (1), 331–337.

70 Lichtenfels, R., Ackermann, A., Kellner, R. and Seliger, B. (2001) Mapping and expression pattern analysis of key components of the major histocompatibility complex class I antigen processing and presentation pathway in a representative human renal cell carcinoma cell line. *Electrophoresis*, **22**, 1801–1809.

71 Seliger, B., Fedorushchenko, A., Brenner, W., Ackermann, A., Atkins, D., Hanash, S. and Lichtenfels, R. (2007) Ubiquitin COOH-terminal hydrolase 1: a biomarker of renal cell carcinoma associated with enhanced tumour cell proliferation and migration. *Clinical Cancer Research*, **13**, 27–37.

72 Li, C., Xiao, Z., Chen, Z., Zhang, X., Li, J., Wu, X., Li, X., Yi, H., Li, M., Zhu, G. and Liang, S. (2006) Proteome analysis of human lung squamous carcinoma. *Proteomics*, **6**, 547–558.

73 Stephan, C., Cammann, H., Meyer, H.A., Lein, M. and Jung, K. (2007) PSA and new biomarkers within multivariate models to improve early detection of prostate cancer. *Cancer Letters*, **249**, 18–29.

74 Cooke, T., Reeves, J., Lanigan, A. and Stanton, P. (2001) HER2 as a prognostic and predictive marker for breast cancer. *Annals of Oncology*, **12**, S23–S28.

75 Menard, S., Pupa, S.M., Campiglio, M. and Tagliabue, E. (2003) Biologic and therapeutic role of HER2 in cancer. *Oncogene*, **22**, 6570–6578. the selection of patients undergoing renal cell carcinoma-specific immunotherapy, *Proteomics*, **3**, 979–90.

76 Ahlemann, M., Schmitt, B., Stieber, P., Gires, O., Lang, S. and Zeidler, R. (2006) Evaluation of CK8-specific autoantibodies in carcinomas of distinct localisations. *Anticancer Research*, **26**, 783–789.

77 Gires, O., Mack, B., Rauch, J. and Matthias, C. (2006) CK8 correlates with malignancy in leukoplakia and carcinomas of the head and neck. *Biochemical and Biophysical Research Communications*, **343**, 252–259.

78 Schaffrik, M., Mack, B., Matthias, C., Rauch, J. and Gires, O. (2006) Molecular characterization of the tumour-associated

antigen AAA-TOB3. *Cellular and Molecular Life Sciences*, **63**, 2162–2174.
79 Ahlemann, M., Zeidler, R., Lang, S., Mack, B., Munz, M. and Gires, O. (2006) Carcinoma-associated eIF3i overexpression facilitates mTOR-dependent growth transformation. *Molecular Carcinogenesis*, **45**, 957–967.
80 Gires, O., Andratschke, M., Schmitt, B., Mack, B. and Schaffrik, M. (2005) Cytokeratin 8 associates with the external leaflet of plasma membranes in tumour cells. *Biochemical and Biophysical Research Communications*, **328**, 1154–1162.
81 Hembrough, T.A., Li, L. and Gonias, S.L. (1996) Cell-surface cytokeratin 8 is the major plasminogen receptor on breast cancer cells and is required for the accelerated activation of cell-associated plasminogen by tissue-type plasminogen activator. *The Journal of Biological Chemistry*, **271**, 25684–25691.
82 Hembrough, T.A., Vasudevan, J., Allietta, M.M., Glass, W.F. and Gonias, S.L. (1995) A cytokeratin 8-like protein with plasminogen-binding activity is present on the external surfaces of hepatocytes, HepG2 cells and breast carcinoma cell lines. *Journal of Cell Science*, **108**, 1071–1082.
83 Hembrough, T.A., Kralovich, K.R., Li, L. and Gonias, S.L. (1996) Cytokeratin 8 released by breast carcinoma cells *in vitro* binds plasminogen and tissue-type plasminogen activator and promotes plasminogen activation. *The Biochemical Journal*, **317**, 763–769.
84 Chu, Y.W., Runyan, R.B., Oshima, R.G. and Hendrix, M.J. (1993) Expression of complete keratin filaments in mouse L cells augments cell migration and invasion. *Proceedings of the National Academy of Sciences of the United States of America*, **90**, 4261–4265.
85 Hardouin, J., Lasserre, J.P., Sylvius, L., Joubert-Caron, R. and Caron, M. (2007) Cancer immunomics: from serological proteome analysis to multiple affinity protein profiling. *Annals of the New York Academy of Sciences*, **1107**, 223–230.
86 Poon, T.C. and Johnson, P.J. (2001) Proteome analysis and its impact on the discovery of serological tumour markers. *Clinica Chimica Acta; International Journal of Clinical Chemistry*, **313** (1–2), 231–239.

5
cDNA and Microarray-based Technologies

Ena Wang, Ping Jin, Hui Lui Liu, David F. Stroncek, and Francesco M. Marincola

5.1
Introduction

The molecular characterization of tumor-associated antigens (TAA) recognized by T cells [1] revolutionized the field of tumor immune biology providing conclusive evidence that CD8+ cytotoxic T cells (CTLs) specifically recognize and kill autologous cancer through recognition of molecularly-defined cancer-specific elements. Since then a myriad of TAA have been identified that has triggered their utilization as anti-cancer vaccines [2–8]. These, in turn, have provided a powerful tool to analyze the dynamics of developing immune responses in cancer-bearing patients [9]. Assessment of the clinical results clearly demonstrates that TAA-based active immunization can consistently achieve its biological end point [10] but this is not sufficient to induce tumor rejection. It remains to be clarified whether this failure is due to a poor selection of TAA that may not represent, for various reasons, ideal targets for CTLs or whether additional biological interactions between the host's CTLs and cancer cells are necessary within the tumor micro-environment to induce full activation of the CTL effector mechanism independent of their TAA target [11]. Among the various reasons which may explain why the current TAA may not constitute ideal targets are their shared expression by normal cells that may have induced negative selection of CTLs with high affinity for them during embryogenesis. Furthermore, as several TAA are expressed by cancer cells as remnants of their derivation from specialized tissues, such as for example melanoma differentiation antigens which are part of the functional repertoire of normal melanocyte differentiation, their expression bears no relevance to cancer cell survival and, therefore, they can be easily dispensed with during cancer progression allowing cancer cells to escape immune recognition [12]. Comparison of the transcriptional profile of renal cell cancer (RCC) and paired normal tissues, demonstrated in fact, that the large majority of genes expressed by early cancers are related to their differentiation from kidney tissues suggesting that most genes expressed by cancer are not related to the oncogenic process and their expression can become progressively lost during cancer progression and by selection pressure applied by the immune system [13].

Those who believe instead that the identity and level of expression of the currently used TAA is not the primary reason for the lack of effectiveness of vaccine-activated CTLs, argue that tumor rejection can occur when the same TAA are used as vaccines. Most pertinently, activation of TAA-specific CTL *ex vivo* and their adoptive transfer to the cancer-bearing host can induce long-lasting complete cancer regressions [14–16]. In addition, the combined administration of immune-modulators such as interleukin (IL)-2 can enhance the clinical effectiveness of TAA-specific CTLs suggesting that other factors are required *in vivo* for their full activation independent of their antigen specificity [17].

Whatever individual investigators' beliefs may be, the actual truth remains unknown. For this reason, we will summarize in this chapter, high throughput strategies that could be applied to the real-time study of clinical samples with the dual purpose of identifying novel TAA that may constitute better targets for active and specific immunization, and simultaneously understand basic concepts of human tumor immune biology. In particular, we will emphasize the description of tools that we or others have developed for the real-time analysis of tumor/host interactions *ex vivo*, using genome-wide high throughput technology [18].

5.2
Technical Aspects

Let us begin with a simplified description of array technology and how it can be applied to the study of human samples. Gene profiling arrays consist of gene-specific cDNA strands (600–2000 bp) or synthetic oligonucleotide probes (50–70 nucleotides) separately spotted on a solid surface. For the dual color system, total messenger RNA or antisense RNA (aRNA) from test and reference samples can be converted into more stable cDNA for direct or indirect differential labeling or direct chemical labeling with fluorescent dyes emitting colors of easily discernible wavelength such as Cy5 (red) and Cy3 (green) fluorochromes. After labeling, the test and reference sample cDNA or aRNA are co-hybridized to the same array slide and the slides are read by a high-resolution scanner. Individual gene expression is portrayed as Cy5/Cy3 fluorescence intensity ratios. Since the ratio is a continuous parameter, array data provide a semi-quantitative estimate of the actual differences in gene expression. Alternatively, single channel array platforms compare test samples from different sources by individual hybridization of each sample to a single array and subsequent normalization of the data gathered in different experiments using bioinformatic algorithms. Most frequently, normalizations are based on entire signal intensity comparisons between the expression of test versus reference sample or on 'house keeping genes' presumed to be similarly expressed in any tissue or other cellular material studied independent of various physiological, pathological or therapeutic conditions.

Although each strategy has its own pros and cons, we prefer the dual color system [19]; indeed, gene profiling using a consistent internal reference offers an important advantage over other gene expression measurement tools including quantitative PCR which estimates gene abundance relative to that of a reference (housekeeping) gene which is assumed to be constantly expressed. Gene profiling compares the expression of each gene between test and reference sample. By maintaining the reference constant it is possible to indirectly but accurately estimate the relative expression of each gene among test samples. In addition, the inter-experimental stability of gene expression in reference samples (reference concordance) can be accurately measured by periodically repeating the same test/reference sample combination [19]. This is critical because the stability of expression of housekeeping genes is questionable [19] and questions as to which housekeeping gene is to be used may arise depending on the origin of the tissues to be analyzed. Pooled healthy donor PBMCs represent a good reference since their individual fluctuation in gene expression is minimal [20].

Quantification of gene expression is a powerful tool for the global understanding of the biology underlying complex patho-physiological conditions. Advances in gene profiling analysis using cDNA or oligo-based microarray systems uncovered genes critically important in disease development, progression, and response to treatment [13, 21–32].

5.2.1
Handling of Samples and the Need for Consistent Messenger RNA Amplification

A major problem, however, in the study of human samples is the limited amount of material that can be obtained, particularly when time-dependent mechanistic studies are needed and multiple biopsies of the same lesions are required [18]. We therefore devoted extensive efforts to developing and validating technologies that allow amplification of messenger RNA (mRNA) from small samples into quantities that could be used for high-fidelity gene profiling studies. The following sections are extracted from a previous manuscript summarizing our experience in the preparation and analysis of human samples, in particular, the consistent adoption of mRNA amplification strategies which provide better quality material for transcriptional analysis [33, 34]. In fact, we consistently recommend mRNA amplification even in samples in which adequate starting material is available to maintain comparability of results and in general to obtain better quality hybridization results. For more details about mRNA amplification the reader is referred to the original article published in the *Journal of Translational Medicine* [34].

While the expression of a limited number of genes can be readily estimated using minimum amounts of mRNA from experimental or clinical samples, gene profiling requires larger amounts, which can only be generated by global RNA amplification. At total of least 10–50 μg RNA (T-RNA) are generally necessary for microarray studies. These amounts of RNA are frequently not obtainable from clinical specimens and are only relevant for experimental endeavors where cultured cell lines or tissues from large excisional biopsies are available [21]. However, most biological specimens directly obtained *ex vivo* for diagnostic or prognostic purposes or for clinical monitoring of treatment are too scarce to yield sufficient RNA for high throughput gene expression analysis. Needle or punch biopsies provide the opportunity to serially sample lesions during treatment in order to study the mechanism of action of that treatment. Such biopsies also provide the opportunity to sample lesions before treatment, so that potential biomarkers that may predict treatment outcome can be identified by directly correlating the biological properties evaluated in the biopsies with the fate of the rest of the sampled lesion that remained in the patient and received a given treatment. In addition, the simplicity of the storage procedure associated with the collection of small samples which can be carried out at the bedside provides superior quality of RNA with minimum degradation [18]. Finally, hypoxia, which follows ligation of tumor-feeding vessels before excision, is avoided with these minimally invasive methods, therefore obtaining a true snapshot of the *in vivo* transcriptional program. However, these minimally invasive biopsies may yield only a few micrograms of total RNA and most frequently even less [18, 35]. Similarly, breast and nasal lavages and cervical brush biopsies, routinely used for pathological diagnosis, generate insufficient material which is far below the detection limits of most current assays. Acquisition of cell subsets by fluorescent or magnetic sorting or laser capture micro-dissection (LCM) for a more accurate portrayal of individual cell interactions in a pathological process, generate even less material, in most cases, nanograms of total RNA [36–39].

Efforts have been made to broaden the utilization of cDNA microarrays using two main strategies: intensifying fluorescence signals [40–43] or amplifying RNA. Signal intensification approaches have reduced the RNA requirement by several fold but cannot extend the utilization of microarray to sub-microgram levels. RNA amplification in turn has gained extreme popularity based on amplification efficiency, linearity, and reproducibility thus lowering the amount of total RNA needed for microarray analysis to nanograms without introducing significant biases. Methods aimed at the amplification of poly(A)-RNA [44] via *in vitro* transcription (IVT) [45] or cDNA amplification via polymerase chain reaction (PCR) [46], have reduced the material needed for cDNA microarray application and extended the spectrum of clinic samples that can be studied. Nanograms of total RNA have been successfully amplified into micrograms of pure mRNA for the screening of the entire transcriptome without losing the proportionality of gene expression displayed by the source material. Curiously, the most important advances were made by Eberwine whose initial goal was not to use clinical material for high-throughput studies but rather to amplify enough material from single cells for individual or multiple gene analysis [47, 48]. Modifications, optimizations, and validations of RNA amplification technology based on Eberwine's pioneering work are still being actively explored.

In this chapter, we will summarize the attempts to optimize mRNA preparation and amplification and describe in detail current amplification procedures that have been validated and applied to cDNA microarray analysis.

5.2.1.1 Collection of Source Material and RNA Isolation

In principle, samples collected from the clinics should keep the fingerprint of the *in vivo* transcriptional activity. Because of the fragility of mRNA, prevention of RNA degradation and RNAase contamination are key to maintaining these fingerprints. For needle biopsies, material should be collected at the patient's bedside in 5 ml of ice-cold RPMI solution without additional supplement material to cool down the cellular metabolism rapidly and minimize RNA degradation. ACK lysing buffer should be added with an equal volume of 1 × PBS (2.5 ml) and incubated for 5 min on ice to lyse red blood cells in the case of excessive contamination. Ideally, excisional biopsies should be handled within 20 min and snap frozen at −80 °C for long-term storage. The use of an RNA protection reagent, such as RNAlater™ (Ambion, Austin, TX) for solid tissue samples and RNAprotect® Cell Reagent (Qiagen, Rockville, MD, USA) for cell suspensions, makes sample collection and long-term RNA preservation easier and safer compared with storage in cell lysis buffer, for example RLT buffer with fresh addition of 2-mercapital ethanol (Qiagen) or Trizol reagent (Invitrogen, Carlsbad, CA, USA) according to our unpublished data.

The method of RNA isolation strongly affects the quality and quantity of RNA that can be isolated using commercially available isolation kits. The T-RNA content per mammalian cell ranges between 20 and 40 pg of which only 0.5–1.0 pg are mRNA [49, 50]. Sample condition, viability, functional status, and phenotype of the cells are the major reasons for variable RNA yield. Sample handling with precautions for RNase contamination always improves the quality and quantity of the RNA obtained. Good quality T-RNA is indicated by an $OD_{260/280}$ ratio above 1.8. It is preferable to estimate RNA quality using Agilent Bioanalyzer or RNA gels. Clear 28S and 18S ribosomal RNA bands with a 28S/18S rRNA ratio equal or close to 2 suggest good RNA quality. Since degradation of 28S rRNA occurs earlier than that of 18S and mRNA degradation in most cases correlates with 28S ribosomal RNA, the ratio of 28S versus 18S rRNA is a good indicator of mRNA quality [51].

5.2.1.2 Single Strand cDNA Synthesis

A critical step in RNA or cDNA amplification is the generation of double stranded cDNA (ds-cDNA) templates. First strand cDNAs are reverse-transcribed from mRNA using oligo dT or random primers. In order to generate full length first strand cDNA, oligo dT(15–24nt) attached to a bacterial phage T7 promoter sequence is commonly used to initiate the cDNA synthesis [34, 44, 48, 52–55]. In the case of degraded RNA [56], random primers attached to a T3 RNA polymerase promoter (T3N9) have been used for first and second strand cDNA synthesis [57]. To prevent RNA degradation during denaturation and the reverse transcription (RT) reaction, it is useful to denature the RNA in the presence of RNasin® Plus RNase Inhibitor, as it forms a stable complex with RNases and inactivates RNase at temperatures up to 70 °C.

To enhance the efficiency of the RT reaction and reduce incorporation errors, the temperature of the RT reaction should be maintained at 50 °C [58, 59] instead of 42 °C to avoid the formation of secondary mRNA structures. This can be done by using thermo-stable reverse transcriptase or regular RTase [60] in the presence of disaccharide trehalose [61–63]. Disaccharide trehalose enhances the thermo-stability of RTase and possesses thermo-activation properties. This modification greatly enhances the accuracy and efficiency of RT with minimum impact on the DNA polymerase activity [58]. The utilization of DNA binding protein T4gp32 in RT reactions also improves cDNA synthesis [59, 60, 64, 65]. T4gp32 protein may essentially contribute to the qualitative and quantitative efficiency of the RT reaction by reducing higher order structures of RNA molecules and hence reducing the pause sites during cDNA synthesis.

In Van Gelder and Eberwine's T7-based RNA amplification [47], the amount of oligo dT-T7 primer used in the first strand cDNA synthesis can affect the amplified RNA in both quantity and quality. Excessive oligo dT-T7 in the RT reaction could lead to template-independent amplification [66]. This phenomenon is not observed when the template switch approach is combined with *in vitro* transcription.

5.2.1.3 Double-stranded cDNA (ds-cDNA) Synthesis

RNA amplification methods differ according to the strategies used for the generation of ds-cDNA as templates for *in vitro* transcription or PCR amplification. There are two basic strategies that have been extensively validated and applied to high throughput transcriptional analysis. The first is based on Gubler–Hoffman's [67] ds-cDNA synthesis subsequently optimized by Van Gelder and Eberwine [47, 48]. This technology utilizes RNase H digestion to create short fragments of RNA as primers to initiate the second-strand cDNA elongation using DNA polymerase I. Fragments of second-strand cDNA are then sequentially ligated to each other using *E. Coli* DNA ligase followed by fine-tuning using T4 DNA polymerase to eliminate loops and form blunt ends. Amplifications based on this methods have been widely used in samples obtained in physiological or pathological conditions and extensively validated for their fidelity, reproducibility, and linearity compared to non-amplified RNA from the same source materials [48, 53, 66–71].

The alternative ds-cDNA synthesis approach utilizes retroviral RNA recombination as a mechanism for template switch to generate full-length ds-cDNA. The method was initially developed for full-length cDNA cloning and, therefore, the main targets of this method are undegraded transcripts. Gubler–Hoffman's ds-cDNA synthesis has the potential of introducing amplification biases because of a possible 5′ under-representation. In addition, the low stringency of the temperature at which ds-cDNA synthesis occurs may introduce additional biases [53]. Although 5′ under-representation could, in theory, be overcome by hairpin loop second-strand synthesis [72], the multiple enzymes used in the reaction could also in turn cause errors. To ensure generation of full-length ds-cDNA, [73] synthesis is carried out by taking advantage of the intrinsic terminal transferase activity and template switch ability of Moloney Murine Leukemia Virus RTase [74]. This enzyme adds non-template nucleotides at the 3′ end of the first strand cDNA, preferentially dCTP oligo nucleotides. A template-switch oligonucleotide (TS primer) containing a short string of dG residues at the 3′ end is added to the reaction to anneal to the dC string of the newly synthesized cDNA. This produces an overhang that allows the RTase to switch template and extend the cDNA beyond the dC to create a short segment of ds-cDNA duplex. After treatment with RNase H to remove the original mRNA, the TS primer initiates the second stranded cDNA synthesis by PCR. Since the terminal transferase activity of the RTase is triggered only when the cDNA synthesis is complete, only full-length single-stranded cDNA will be tailed with the TS primer and converted into ds-cDNA. Using the TS primer, second strand cDNA synthesis is carried at 75 °C after a 95 °C denaturing and a 65°C annealing step in the presence of single DNA polymerase [34, 54]. This technique, in theory, overcomes the bias generated by amplification methods which depend only on 3' nucleotide synthesis and hence it is, in theory, superior to the Gubler–Hoffman ds-cDNA synthesis. However, no significant differences in correlation coefficients of amplified versus non-amplified RNA were observed when the Gubler–Hoffman ds-cDNA method was compared with the TS ds-cDNA amplification using high throughput analysis [59, 75]. The fidelity of template switch-based amplification methods has been assessed by numerous gene profiling analyses on different types of microarray platform, and by real-time PCR and sophisticated statistical analyses and its reliability has been proven for high throughput transcriptome studies [33, 34, 55].

5.2.2
RNA Amplifications

5.2.2.1 Linear Amplification

Linear (non-logarithmic as per PCR) amplification methods have been developed which in theory should maintain the proportionality of each RNA species present in the original sample. IVT using ds-cDNA equipped with a bacteriophage T7 promoter [48] provides an efficient way to amplify mRNA sequences and thereby generate templates for the synthesis of fluorescently-labeled single-stranded cDNA [44, 45, 47, 48, 53, 72]. Depending on the T7 or other (T3 or SP6) promoter sequence position on the ds-cDNA, amplified RNA can be either in sense or antisense orientation. Oligo-dT attachments to the promoter sequence, for example oligo-dT-T7, prime first-strand cDNA positioned at the promoter at the 3' end of genes (5' end of cDNA) and, therefore, lead to the amplification of antisense RNA (aRNA) or complement RNA (cRNA). Promoters positioned at the 5' end of genes by random [76] or TS primers generate sense RNA (sRNA). Amplified sRNA can also be produced by tailing oligo-dT to the 3' of the cDNA followed by oligo-dA-T7 priming for double-stranded T7 promoter generation at the 5' end of genes [77]. The uniqueness of this approach resides in the utilization of a DNA polymerase blocker at the 3' end of the oligo-dA-T7 primer, which prevents the elongation of second-strand cDNA synthesis while priming for the elongation of the double-stranded promoter. In this way, only sense amplification can be achieved by the presence of the 5' ds-T7 promoter followed by single-strand cDNA templates.

IVT using DNA-dependent RNA polymerase is an isothermal reaction with linear kinetics. The input ds-cDNA templates are the only source of template for the complete amplification and, therefore, any errors created on the newly synthesized RNA will not be carried or amplified in the following reactions. Overall, RNA polymerase makes an error at a frequency of about one in every 10 000 nucleotides corresponding to about once per RNA strand created. This contrasts with DNA-dependent DNA polymerases, which incorporate an error once in every 400 nucleotides. Most importantly, these errors are exponentially amplified in the following reaction since the amplicons serve as templates. Thus, RNA polymerase catalyzes transcription robotically and efficiently without sequence-dependent bias. Recombinant RNA polymerases have been engineered to enhance the stability of the enzyme interacting with templates and reduce the abortive tendency [78] of the wild-type RNA polymerase which in turn improved the elongation phase resulting in complete mRNA transcripts. The length of amplified RNA ranges from 200 to 6000 nucleotides for the first round of amplification and 100 to 3000 nucleotides for the second round when random primers are used [33, 55]. The amplification efficiency is greater than 2000-fold in the first round and 100 000-fold in the second round [33].

Two rounds of IVT are commonly required when sub-microgram levels of input total RNA are used. It has been estimated that after two rounds of amplification the frequency of only 10% of the genes in a specimen is reduced [79] and more than two rounds of amplification may still retain at least in part the proportionality of gene expression among different RNA populations [54]. However, we do not generally recommend going over two rounds of amplification unless necessary for extremely scant specimens such as when processing specimens containing one or only a few cells, to avoid unnecessary biases related to amplification. The fidelity of IVT has been extensively assessed by gene profiling analysis, quantitative real-time PCR, and statistical testing by comparing estimates of gene expression in amplified versus non-amplified RNA [19, 33, 34, 54].

The fidelity of the first round amplification decreases when the input starting material is less than 100 ng because of the intrinsic low abundance of transcripts (particularly those under-represented in the biological specimen). This can be rescued by two rounds of IVT if sufficient RNA species are present in the input material [54]. In addition, two rounds of amplification tend to introduce a 3′ bias due to the use of random primers in the cDNA synthesis for ds-cDNA template creation. This should not affect the usefulness of the technique for high throughput gene profiling analysis since cloned cDNA arrays are 3′ biased and even oligonucleotide arrays are designed to target the 3′ end of each gene. Sequence-specific biases introduced during amplification are generally reproducible and, although negligible, could mislead data interpretation only when amplified RNA is directly compared with non-amplified RNA on the same array platform. This type of error can be easily circumvented by using samples processed using identical conditions. Degradation of amplified RNA during prolonged IVT may result in a lower average size of the aRNA and decreased yields [56]. This results from residual RNase in the enzyme mixture used for the IVT reaction and can be prevented by the addition of RNase inhibitor to the reaction if a prolonged amplification is needed.

5.2.2.2 PCR-based Exponential Amplification

IVT is burdensome, time consuming and may, theoretically, produce a 3′ bias especially when two rounds of amplification are undertaken. Exponential amplification (PCR-based) may avoid these drawback and it has shown promise since, contrary to the IVT, it is simple and efficient. However, PCR-based amplification has its own drawbacks.

The limitations of PCR-based amplification stem from the characteristics of the DNA-dependent DNA polymerase enzymatic function. The function of this enzyme is biased towards a lower efficiency in the amplification of GC-rich sequences compared with AT-rich sequences. In addition, as previously discussed, not only does it create errors more frequently than RNA polymerase but it also amplifies these errors because the reaction utilizes the amplicons as templates for subsequent amplification [80]. In addition, due to the exponential amplification, the reaction could reach saturation under conditions in which excess quantities of input template are used or the substrate becomes exhausted. This would favor the amplification of high-abundance transcripts, which would compete more efficiently for substrate in the earlier cycles of the amplification process resulting in the loss of proportionality of the amplification process. Optimization of the number of PCR cycles to avoid reaching the saturation cycle and adjustments in the amount of template input may solve the problems [81]. The utilization of DNA polymerase with a proofreading function may eradicate errors created in the cDNA amplification [82]. This approach preserves the relative abundance of the transcript [83] and it may outperform IVT when less than 50 ng of input RNA are available as starting material [84, 85].

PCR-based cDNA amplification can be categorized as template switching (TS)-PCR [68, 86, 87], random PCR [88], and 3' tailing with 5' adaptor ligation PCR [89] based on the generation of a 5' anchor sequence which provides a platform for 5' primer annealing. TS-PCR employs the same template switch mechanism in the generation of ds-cDNA and the amplification of ds-cDNA using 5' TS primer II (truncated TS primer) and 3' oligo dT or dT-T7 primers (depending upon the primer used in the first-strand cDNA synthesis). Random PCR utilizes modified oligo dT primers (dT-T7 or dT-TAS (Target Amplification Sequence)) or random primers with an adaptor sequence for the first-strand cDNA initiation and random primers attached to the same adaptor, for example dN10-TAS [88], for second-strand cDNA synthesis. The attached sequence, such as TAS, generates a 5' anchor on the cDNA for subsequent PCR amplification with a single TAS-PCR primer. This approach is more suitable for partly degraded RNA which may be under-represented at the 5' end. The third exponential amplification utilizes the terminal deoxynucleotidyl transferase function to add a polymonomer tail, for example poly dA, to the 5' end of the gene. The tailed poly dA provides an annealing position for the oligo dT primer which leads the second-strand cDNA synthesis. Ds-cDNA can then be amplified under one oligo dT primer or dT-adaptor primer if an adaptor sequence is attached [84]. Direct adaptor ligation is another alternative for generating ds-cDNA with a known anchor sequence at the 5' end [89]. In this way, single-strand cDNA is generated using oligo dT primers immobilized onto magnetic beads and second-strand cDNA is completed using Van Gelder and Eberwine's ds-cDNA generation method. A ds-T7 promoter-linker is then unidirectionally ligated to the blunted ds-cDNA at the 5' end. PCR amplification can then be performed using the 5' promoter primer and the 3' oligo dT or dT-adaptor primer if an adaptor is attached. PCR amplified ds-cDNA is suitable for either sense or antisense probe arrays.

The combination of PCR amplification to generate sufficient ds-cDNA template followed by IVT [88, 89] is an attractive strategy to amplify minimal starting material since it takes advantage of the efficiency of the PCR reaction and the linear kinetics of IVT while minimizing the disadvantages discussed above. There have been fewer validations for PCR-based RNA amplification methods than for IVT but have so far been persuasive in spite of the prevalent expectations. Skepticism concerning the reproducibility and linearity are still one of the key factors preventing the extensive application of this approach.

5.2.2.3 Target Labeling for cDNA Microarray using Amplified RNA

The generation of high quality cDNA microarray data depends not only on a sufficient quantity of a highly representative amplified target, but also on the efficacy and reproducibility of target labeling. Steps involved in target preparation such as RNA amplification, target labeling, pre-hybridization, hybridization, and slide washing are imperative to enhance the foreground signal to background noise ratios. A linear spectrum of signal intensity that correlates with gene copy numbers without compromising the detection sensitivity is one of the key factors in high quality cDNA analysis. Therefore, target labeling is a critical step in achieving consistently high signal images.

Fluorescence-labeled cDNA is generated by incorporation of conjugated nucleotide analogs during the reverse transcription process. Depending upon the detection system, labeled markers can be either radioactive, color matrix or fluorescent. Fluorescence labeling outperforms the other labeling methods because of the versatile excitation and emission wavelengths. In addition, it has the advantage of not being hazardous. Among the fluorochromes, Cy3 and Cy5 are most commonly used in cDNA microarray applications due to their distinctive emissions (510 and 664 nm, respectively). Cy5-labeled dUTP and dCTP are less efficiently incorporated during the labeling reaction compared to Cy3-labeled dUTP or dCTP and they are more sensitive to photobleaching because of their chemical structure. Therefore, the labeling bias needs to be accurately analyzed and the results should be normalized according to standard normalization procedures.

Target labeling can be divided into two major categories: direct and indirect fluorescence incorporation. The first category utilizes fluorescence-labeled dUTP or dCTP to partially substitute unlabeled dTTP or dCTP in the RT reaction to generate Cy dye-labeled cDNA. This label incorporation method is suitable for cDNA clone microarray using amplified aRNA as the template or oligo array using amplified sRNA as the template. One limitation of direct labeling is that fluorescent nucleotides are not the normal substrates for polymerases and some may be particularly sensitive to the structural diversity of these artificial oligonucleotides. The fluorescent moieties associated with these nucleotides are often quite bulky and, therefore, the efficiency of incorporation of such nucleotides by polymerases tends to be much lower than that of natural substrates. An alternative is to incorporate, either by synthesis or by enzymatic activity, a nucleotide analog similar to the natural nucleotide in structure featuring a chemically reactive group, such as 5-(3-aminoallyl)-2′-deoxyuridine 5′-triphosphate (aa-dUTP), to which a fluorescent dye such as a Cy dye, may then be attached [90]. The reactive amine of the aa-dUTP can be incorporated by a variety of RNA-dependent and DNA-dependent DNA polymerases. After removing free nucleotides, the aminoallyl-labeled samples can be coupled to the dye, purified again, and then applied to a microarray [91]. The optimized ratio of aa-dUTP versus dTTP in the labeling reaction should be 2 to 3, respectively.

In theory, indirect outperforms direct labeling by reducing the cost and maximizing signal intensity through increases in incorporation of fluorochrome or through signal amplification using fluorescence-labeled antibody or biotin–streptavidin complexes. However, more steps are involved in the purification of the indirectly-labeled target prior to hybridization, thus these strategies are used less frequently.

5.2.2.4 Bioinformatics Tools

Gene expression data are subjected to preliminary normalization to avoid any bias due to the fluorescent dye bias and to filtration in order to exclude non-specific spots, eliminate background noise and spots with extremely low signal. Normalization strategies are based on global or intensity parameters. Normalization is more effectively achieved by adjusting channel intensity to a median CY5/Cy3 ratio of 1 with the assumption that the majority of genes are similarly expressed between two samples. Locally weighted linear regression (*lowess*) fit can be applied to adjust intensity-dependent biases [92].

Hierarchical clustering, K-mean clustering, and self-organizing maps are most commonly used in class discovery. Unsupervised clustering analysis calculates the similarity among individual data series comparing gene-to-gene expression patterns and/or similarities in gene expression among various experiments. As a result, coordinate patterns of gene expression (signatures) or molecularly defined disease subclasses can be identified. Class comparison uses pre-existing knowledge about the experimental settings to cross-examine genes which are differentially expressed between experimental groups, estimating significance with parametric or non-parametric tools. The chance that multiple comparisons of large numbers of genes may result by chance in *p*-values below the set threshold of significance can be estimated by a step-wise multivariant permutation test. Class prediction consists of a multivariate analysis based on genes identified by class comparison. The accuracy of class prediction can be estimated by (1) splitting the samples into a training and test set, (2) using the training set to build the predictor, and (3) predicting the phenotype in the test set. A good example of this strategy was used to identify and validate common cancer biomarkers [93].

Several bioinformatics systems have been developed to manage high throughput biological data [94]. Significance Analysis of Microarrays (SAM) is a type of supervised learning software for mining genomic expression data (http://www-stat.stanford.edu/~tibs/SAM/). BioConductor array analysis includes tools for statistical normalization, differential expression testing and genomic visualization. BRB ArrayTools is an integrated package for the visualization and statistical analysis of DNA microarray gene expression data which utilizes the Excel program (http://linus.nci.nih.gov/BRB-ArrayTools.html). Most often, the function of a given gene and/or its participation in various biological processes is poorly understood. GoMiner (http://discover.nci.nih.gov/gominer) is biological interpretation software that is applicable to genomic and proteomic data [95] to fulfil the intellectual gap between statistical identification of significant data and their biological interpretation. This package includes a freely available computer resource that incorporates the hierarchical structure defined by the Gene Ontology (GO) Consortium to automate the functional categorization of the gene lists generated by individual studies. This software links the following sites: LocusLink, PubMed, MedMiner, GeneCards, the NCBI's Structure Database, BioCarta and KEGG pathway maps offered by the NCI Cancer Genome Anatomy Project.

5.2.2.5 Limitations of Transcriptional Profiling

Gene expression profiling addresses mRNA expression and it is not meant to provide information about the expression of the corresponding protein. Thus, we can equate transcriptional profiling to a test that identifies the reaction of biological systems under particular environmental conditions rather than providing information about their actual functional status; in other words transcriptional profiling provides information about the potential of a cell rather than the actual function of that cell. For instance, detection of IFN-γ mRNA expression in T cells is an accurate marker of antigen-specific stimulation independent of the actual expression of the protein. Although it is not clear to what degree mRNA expression correlates with the expression of the respective protein, it can, however, be assumed that it may at least be indicative. In any case, validation is required and we emphasize that the power of high throughput technologies rests more on the efficiency with which human samples can be tested in a hypothesis-generating frame rather than as an hypothesis testing tool [96].

For validation, we suggest that quantitative real-time PCR, although often used [97], is not particularly useful since at best it provides the same information as the array methods. If feasible, we prefer to validate array data by examining the level of the corresponding protein or by undertaking functional assays that support, refine or disprove the hypothesis generated by the array data [98–101]. Therefore, a well-designed clinical study should prospectively set aside frozen material for cell-specific validation by molecular or immunological methods. *In situ* hybridization can be applied to frozen sections to identify the cell type responsible for the expression of a given gene. In addition, it is possible to amplify material from selected cell types by magnetic cell sorting [99] or laser-guided micro-dissection [102]. Finally, it is possible to confirm protein expression through tissue arrays consisting of matrixes of relevant histopathological material for immunohistochemical or other imaging techniques [103].

5.2.2.6 Usefulness of Transcriptional Profiling for Antigen Discovery and the Understanding of Tumor–Host Interactions

As previously stated transcriptional profiling can assist tumor immunologists in various ways; while on one hand it can help identify novel targets for immunotherapy, on the other it allows the simultaneous analysis of parameters related to immune responsiveness and their relationships in the tumor microenvironment at the genome-wide level.

Identification of Novel Antigens Better Suited as Targets of Immunotherapy: Need for Novel TAA Most TAA that have been extensively used in clinic are tumor differentiation antigens or cancer testis antigens. The former represent remnants of the tumor ontogeny [104] and are very rapidly lost since they are not necessary for tumor cell survival [12, 105, 106]. The latter are expressed in common by some cancers and germinal cells [107] and although their expression is related to the oncogenic process, it remains unclear whether their expression remains stable throughout the disease process. There has been recent interest in the identification of novel TAA which has been initiated by two specific requirements. First, it is important to identify molecules that are necessary for the survival of cancer cells, the expression of which is therefore less likely to be lost during the progression of disease and, second, it is important to identify TAA that are expressed by tumors other than the classically studied melanoma.

For instance, β-catenin is mutated in a subset of melanomas and possibly other cancers [108] and has been shown to contain T cell epitopes that are associated with tumor regression [109]. Because the expression of the mutated protein is strongly related to the oncogenic potential of the cancer cells [108], it is likely that their expression persists indefinitely. Indeed, a recent analysis of a long-term patient who experienced five recurrences of metastatic melanoma over a period spanning 12 years, demonstrated that the original mutation and the expression of the protein persisted throughout the patient's life-time [110, 111]. This example, demonstrates a TAA that may represent a long-term target of immunotherapy. Unfortunately, however, these TAA and their mutations are patient-specific and, therefore, they may not be ideal candidates for immunotherapy strategies applicable to a broad patient population.

Proteins associated with the oncogenic process but non-mutated would be ideal. Among them are those that are over-expressed in cancer compared with normal cells. Survivin represents a good example [112–115] and its expression is associated with a poor prognosis [116]. For this reason, some researchers have started a systematic search for potential tumor-associated proteins that may also serve as TAA using high throughput technologies [117]. This strategy makes good sense since it is clear that tumors rapidly evolve from a phase in which they resemble their normal progenitor cells such as normal epithelial melanocytes [29, 118] or normal kidney tissues [13], to an anaplastic phase where they converge into a neoplastic phenotype that is shared by cancers of different etiology [13]. In particular, by comparing the transcriptional patterns of a large number of cancers of different etiology with that of a broad selection of normal tissues, we identified several genes that were selectively and consistently expressed by all cancers [93]. These genes could potentially constitute novel targets for immunotherapy. In addition, comparing the transcriptional profile of melanoma and renal cell cancer with that of other cancers less susceptible to immunotherapy, we identified transcriptional signatures that are unique to these two cancers and may be relevant to their immune responsiveness [118]. Among them, several genes were identified such as the microphtalmia transcription factor which has several regulatory activities in the biology of melanoma and enolase-2. Finally, macrophage migration inhibiting factor was found to be commonly expressed and this may be a putative modulator of these two cancers' immune microenvironments. It is possible, though unlikely, that in part this could be due to the specific antigenic potential of proteins whose expression is shared by these two cancers compared to less immunogenic tumors.

Study of Immunological Signatures within the Tumor Microenvironment We have repeatedly emphasized the need to simultaneously study the expression of TAA and the function of the immune response in the tumor microenvironment [18, 32, 119–121]. High throughput technology, can definitely improve the interpretation of monogenic studies. For instance, it has been proposed that the over-expression of carbonic anhydrase IX by RCC is an accurate biomarker of its immune responsiveness to interleukin-2 therapy [122]. Transcriptional profiling analyzing the co-ordinate expression of this gene in melanoma and RCC demonstrated that carbonic anhydrase IX is not expressed in melanoma (another cancer that is highly immuno-responsive to interleukin-2 therapy) suggesting that it may rather be a biomarker of sensitivity to immunotherapy rather that representing the causative agent of responsiveness. In fact, even in RCC we observed that carbonic anhydrase expression is associated with an increased expression of genes associated with immune activation [123] suggesting that a broader panel of genes is related to this biological phenomenon. We also identified immunological signatures restricted to melanoma and/or RCC that might explain their propensity to respond to immunotherapy. The large majority of melanoma-restricted genes associated with immune function are tightly clustered together [118] and they are most likely co-regulated resulting in the co-ordinated expression of growth factors, chemokines, cytokines, and interferon stimulated genes (ISGs) [106]. Most of the immunologically related genes that are typically expressed by melanomas are not only the TAA related to their melanocytic lineage but several are markers of immune activation and regulate effector natural killer and T cell function. In particular, NK4/IL32 is associated with *in vitro* activation of immunization-induced T cells [99], regression of a melanoma during interleukin-2 therapy [25], basal cell cancer in response to treatment with Toll-like receptor agonists [100] and the rejection of allografts [124]. The constitutive expression of CX3CR1 by natural killer cells makes them sensitive to chemo attraction by CXCL12 and CXC3L1 [125] and this may explain the preferential localization of these effector cells in melanoma lesions. In addition, the presence of these and other natural killer cell-related genes suggested potent chemo-attraction toward natural killer cells by the tumor microenvironment of subcutaneous and cutaneous melanoma metastases. The constitutive expression of interferon regulatory factor-7 implicated in the amplification of the innate immune response [126] through interactions with the NF-*k*B pathway [127], may lead to the activation of various forms of type I interferons [128].

Thus, immune reactivity is not only a function of TAA expression which is a questionable predictor of immune responsiveness [129, 130], but it may be related to the status of immune activation of individual cancer microenvironments. This may not only apply to metastatic melanoma [29], but also to ovarian [131] and colon [132] carcinoma. We propose that future studies should continue to investigate the relationship between cancer cells and the host cells in real-time taking into account the complexity of the biology underlying the phenomenon of immunologically-mediated tumor rejection.

5.3
Summary

In summary, the present chapter provides an overview of array technology applied to the study of tumor immunology emphasizing the need to obtain material of good quality sampled at appropriate and clinically-relevant time points. Several suggestions have been made regarding the technical issues that must be addressed in order to obtain high-quality results. For more details, we refer the reader to a summary recently published elsewhere [34].

References

1 van der Bruggen, P., Traversari, C., Chomez, P., Lurquin, C., De Plaen, E., Van den Eynde, B., Knuth, A. and Boon, T. (1991) A gene encoding an antigen recognized by cytolytic T lymphocytes on a human melanoma. *Science*, **254**, 1643.

2 Boon, T., Coulie, P.G. and Van den Eynde, B. (1997) Tumor antigens recognized by T cells. *Immunology Today*, **18**, 267.

3 Kawakami, Y. and Rosenberg, S.A. (1996) T-cell recognition of self peptides as tumor rejection antigens. *Immunologic Research*, **15**, 179.

4 Parmiani, G., Anichini, A. and Castelli, C. (1997) New tumour-restricted melanoma antigens as defined by cytotoxic T-cell responses. *Melanoma Research*, **7** (Suppl 2), S95–S98.

5 Parmiani, G., Castelli, C., Dalerba, P., Mortarini, R., Rivoltini, L., Marincola, F.M. and Anichini, A. (2002) Cancer immunotherapy with peptide-based vaccines: what have we achieved? Where are we going? *Journal of the National Cancer Institute*, **94**, 805.

6 Parmiani, G., De, F.A., Novellino, L. and Castelli, C. (2007) Unique human tumor antigens: immunobiology and use in clinical trials. *Journal of Immunology (Baltimore, Md: 1950)*, **178**, 1975.

7 Dalgleish, A. and Pandha, H. (2007) Tumor antigens as surrogate markers and targets for therapy and vaccines. *Advances in Cancer Research*, **96**, 175.

8 Jager, D. (2007) Potential target antigens for immunotherapy identified by serological expression cloning (SEREX). *Methods in Molecular Biology*, **360**, 319.

9 Marincola, F.M. and Ferrone, S. (2003) Immunotherapy of melanoma: the good news, the bad news and what to do next. *Seminars in Cancer Biology*, **13**, 387.

10 Slingluff, C.L., Jr. and Speiser, D.E. (2005) Progress and controversies in developing cancer vaccines. *Journal of Translational Medicine*, **3**, 18.

11 Monsurro', V., Wang, E., Panelli, M.C., Nagorsen, D., Jin, P., Smith, K., Ngalame, Y., Even, J. and Marincola, F.M. (2003) Active-specific immunization against melanoma: is the problem at the receiving end? *Seminars in Cancer Biology*, **13**, 473.

12 Marincola, F.M., Hijazi, Y.M., Fetsch, P., Salgaller, M.L., Rivoltini, L., Cormier, J.N., Simonis, T.B., Duray, P.H., Herlyn, M., Kawakami, Y. and Rosenberg, S.A. (1996) Analysis of expression of the melanoma associated antigens MART-1 and gp100 in metastatic melanoma cell lines and in *in situ* lesions. *Journal of Immunotherapy*, **19**, 192.

13 Wang, E., Lichtenfels, R., Bukur, J., Ngalame, Y., Panelli, M.C., Seliger, B. and Marincola, F.M. (2004) Ontogeny and oncogenesis balance the transcriptional profile of renal cell cancer. *Cancer Research*, **64**, 7279.

14 Rosenberg, S.A., Yang, J.C. and Restifo, N.P. (2004) Cancer immunotherapy: moving beyond current vaccines. *Nature Medicine*, **10**, 909.

15 Mocellin, S., Mandruzzato, S., Bronte, V. and Marincola, F.M. (2004) Correspondence 1: Cancer vaccines: pessimism in check. *Nature Medicine*, **10**, 1278.

16 Wang, E., Selleri, S., Sabatino, M., Monaco, A., Pos, Z., Stroncek, D.F. and Marincola, F.M., (2008) Spontaneous and tumor-induced cancer rejection in humans. *Expert Opinion on Biological Therapy*, **8**, 337.

17 Rosenberg, S.A., Yang, J.C., Schwartzentruber, D., Hwu, P., Marincola, F.M., Topalian, S.L., Restifo, N.P., Dufour, E., Schwartzberg, L., Spiess, P., Wunderlich, J., Parkhurst, M.R., Kawakami, Y., Seipp, C., Einhorn, J.H. and White, D. (1998) Immunologic and therapeutic evaluation of a synthetic tumor associated peptide vaccine for the treatment of patients with metastatic melanoma. *Nature Medicine*, **4**, 321.

18 Wang, E. and Marincola, F.M. (2000) A natural history of melanoma: serial gene expression analysis. *Immunology Today*, **21**, 619.

19 Jin, P., Zhao, Y., Ngalame, Y., Panelli, M.C., Nagorsen, D., Monsurro', V., Smith, K., Hu, N., Su, H., Taylor, P.R., Marincola, F.M. and Wang, E. (2004) Selection and validation of endogenous reference genes using a high throughput approach. *BMC Genomics*, **5**, 55.

20 Whitney, A.R., Diehn, M., Popper, S.J., Alizadeh, A.A., Boldrick, J.C., Relman, D.A. and Brown, P.O. (2003) Individuality and variation in gene expression patterns in human blood. *Proceedings of the National Academy of Sciences of the United States of America*, **100**, 1896.

21 DeRisi, J., Penland, L., Brown, P.O., Bittner, M., Meltzer, P.S., Ray, M., Chen, Y., Su, Y.A. and Trent, J.M. (1996) Use of cDNA microarray to analyse gene expression patterns in human cancer. *Nature Genetics*, **14**, 457.

22 Bittner, M., Meltzer, P., Chen, Y., Jiang, E., Seftor, E., Hendrix, M., Radmacher, M., Simon, R., Yakhini, Z., Ben-Dor, A., Dougherty, E., Wang, E., Marincola, F.M., Gooden, C., Lueders, J., Glatfelter, A., Pollock, P., Gillanders, E., Dietrich, K., Alberts, D., Sondak, V.K., Hayward, N. and Trent, J.M. (2000) Molecular classification of cutaneous malignant melanoma by gene expression: shifting from a continuous spectrum to distinct biologic entities. *Nature*, **406**, 536.

23 Islam, T.C., Lindvall, J., Wennborg, A., Branden, L.J., Rabbani, H. and Smith, C.I. (2002) Expression profiling in transformed human B cells: influence of Btk mutations and comparison to B cell lymphomas using filter and oligonucleotide arrays. *European Journal of Immunology*, **32**, 982.

24 Mellick, A.S., Day, C.J., Weinstein, S.R., Griffiths, L.R. and Morrison, N.A. (2002) Differential gene expression in breast cancer cell lines and stroma-tumor differences in microdissected breast cancer biopsies by display array analysis. *International Journal of Cancer*, **100**, 172.

25 Panelli, M.C., Wang, E., Phan, G., Puhlman, M., Miller, L., Ohnmacht, G.A., Klein, H. and Marincola, F.M. (2002) Genetic profiling of peripheral mononuclear cells and melanoma metastases in response to systemic interleukin-2 administration. *Genome Biology*, **3**, RESEARCH0035.

26 Sasaki, H., Ide, N., Fukai, I., Kiriyama, M., Yamakawa, Y. and Fujii, Y. (2002) Gene expression analysis of human thymoma correlates with tumor stage. *International Journal of Cancer*, **101**, 342.

27 Singh, D., Febbo, P.G., Ross, K., Jackson, D.G., Manola, J., Ladd, C., Tamayo, P., Renshaw, A.A., D'Amico, A.V., Richie, J.P., Lander, E.S., Loda, M., Kantoff, P.W., Golub, T.R. and Sellers, W.R. (2002) Gene expression correlates of clinical prostate cancer behavior. *Cancer Cell*, **1**, 203.

28 Skotheim, R.I., Monni, O., Mousses, S., Fossa, S.D., Kallioniemi, O.P., Lothe, R.A. and Kallioniemi, A. (2002) New insights into testicular germ cell tumorigenesis from gene expression profiling. *Cancer Research*, **62**, 2359.

29 Wang, E., Miller, L.D., Ohnmacht, G.A., Mocellin, S., Petersen, D., Zhao, Y., Simon, R., Powell, J.I., Asaki, E., Alexander, H.R., Duray, P.H., Herlyn, M., Restifo, N.P., Liu, E.T., Rosenberg, S.A. and Marincola, F.M. (2002) Prospective molecular profiling of subcutaneous melanoma metastases suggests classifiers of immune responsiveness. *Cancer Research*, **62**, 3581.

30 Yeoh, E.-J., Ross, M.E., Shurtleff, S.A., Williams, W.K., Patel, D., Mahfouz, R., Behm, F.G., Raimondi, S.C., Relling, M.V., Patel, A., Cheng, C., Campana, D., Wilkins, D., Zhou, X., Li, J., Liu, H., Pui, C.-H., Evans, W.E., Naeve, C., Wong, L. and Downing, J.R. (2002) Classification, subtype discovery and prediction of outcome in pediatric acute lymphoblastic leukemia by gene expression profiling. *Cancer Cell*, **1**, 133.

31 Wang, E., Panelli, M.C. and Marincola, F.M. (2003) Genomic analysis of cancer. *Principles and Practice of Oncology Updates*, **17**, 1.

32 Wang, E., Ngalame, Y., Panelli, M.C., Deavers, M., Mueller, P., Ju, W., Savary, C., Nguyen-Jackson, H., Freedman, R.S. and Marincola, F.M. (2005) Peritoneal and sub-peritoneal stroma may facilitate regional spread of ovarian cancer. *Clinical Cancer Research*, **11**, 113.

33 Feldman, A.L., Costouros, N.G., Wang, E., Qian, M., Marincola, F.M., Alexander, H.R. and Libutti, S.K. (2002) Advantages of mRNA amplification for microarray analysis. *Biotechniques*, **33**, 906.

34 Wang, E. (2005) RNA amplification for successful gene profiling analysis. *Journal of Translational Medicine*, **3**, 28.

35 Wang, E. and Marincola, F.M. (2001) cDNA microarrays and the enigma of melanoma immune responsiveness. *Cancer Journal from Scientific American*, **7**, 16.

36 Bonner, R.F., Emmert-Buck, M., Cole, K., Pohida, T., Chuaqui, R., Goldstein, S. and Liotta, L.A. (1997) Laser capture microdissection: molecular analysis of tissue. *Science*, **278**, 1481–1483.

37 Pappalardo, P.A., Bonner, R., Krizman, D.B., Emmert-Buck, M.R. and Liotta, L.A. (1998) Microdissection, microchip arrays, and molecular analysis of tumor cells (primary and metastases). *Semin Radiat Oncol*, **8**, 217.

38 St Croix, B., Rago, C., Velculescu, V., Traverso, G., Romans, K.E., Montgomery, E., Lal, A., Riggins, G.J., Lengauer, C., Vogelstein, B. and Kinzler, K.W. (2000) Genes expressed in human tumor endothelium. *Science*, **289**, 1121.

39 Tsuda, H., Birrer, M.J., Ito, Y.M., Ohashi, Y., Lin, M., Lee, C., Wong, W.H., Rao, P.H., Lau, C.C., Berkowitz, R.S., Wong, K.K. and Mok, S.C. (2004) Identification of DNA copy number changes in microdissected serous ovarian cancer tissue using a cDNA microarray platform. *Cancer Genetics and Cytogenetics*, **155**, 97.

40 Chen, J.J., Wu, R., Yang, P.C., Huang, J.Y., Sher, Y.P., Han, M.H., Kao, W.C., Lee, P.J., Chiu, T.F., Chang, F., Chu, Y.W., Wu, C.W. and Peck, K. (1998) Profiling expression patterns and isolating differentially expressed genes by cDNA microarray system with colorimetry detection. *Genomics*, **51**, 313.

41 Zejie, Y., Xiaoyi, W., Yu, T. and Huan, H. (1998) The method of microdisplacement measurement to improve the space resolution of array detector. *Medical Engineering & Physics*, **20**, 149.

42 Rajeevan, M.S., Dimulescu, I.M., Unger, E.R. and Vernon, S.D. (1999) Chemiluminescent analysis of gene expression on high-density filter arrays. *The Journal of Histochemistry and Cytochemistry*, **47**, 337.

43 Yu, J., Othman, M.I., Farjo, R., Zareparsi, S., MacNee, S.P., Yoshida, S.

and Swaroop, A. (2002) Evaluation and optimization of procedures for target labeling and hybridization of cDNA microarrays. *Molecular Vision*, **8**, 130.

44 Lockhart, D.J., Dong, H., Byrne, M.C., Folliette, M.T., Gallo, M.V., Chee, M.S., Mittmann, M., Wang, C., Kabayashi, M., Horton, H. and Brown, E.L. (1996) Expression monitoring of hybridization to high-density oligonucleotide arrays. *Nature Biotechnol*, **14**, 1675.

45 Luo, L., Salunga, R.C., Guo, H., Bittner, A., Joy, K.C., Galindo, J.E., Xiao, H., Rogers, K.E., Wan, J.S., Jackson, M.R. and Erlander, M.G. (1999) Gene expression profiles of laser-captured adjacent neuronal subtypes. *Nature Medicine*, **5**, 117.

46 Trenkle, T., Welsh, J., Jung, B., Mathieu-Daude, F. and McClelland, M. (1998) Non-stoichiometric reduced complexity probes for cDNA arrays. *Nucleic Acids Research*, **26**, 3883.

47 Van Gelder, R.N., von Zastrow, M.E., Yool, A., Dement, W.C., Barchas, J.D. and Eberwine, J.H. (1990) Amplified RNA synthesized from limited quantities of heterogeneous cDNA. *Proceedings of the National Academy of Sciences of the United States of America*, **87**, 1663.

48 Eberwine, J.H., Yeh, H., Miyashiro, K., Cao, Y., Nair, S., Finnell, R., Zettel, M. and Coleman, P. (1992) Analysis of gene expression in single live neurons. *Proceedings of the National Academy of Sciences of the United States of America*, **89**, 3010.

49 Roozemond, R.C. (1976) Ultramicrochemical determination of nucleic acids in individual cells using the Zeiss UMSP-I microspectrophotometer. Application to isolated rat hepatocytes of different ploidy classes. *The Histochemical Journal*, **8**, 625.

50 Uemura, E. (1980) Age-related changes in neuronal RNA content in rhesus monkeys (*Macaca mulatta*). *Brain Research Bulletin*, **5**, 117.

51 Skrypina, N.A., Timofeeva, A.V., Khaspekov, G.L., Savochkina, L.P. and Bcabealashvilli, R.S. (2003) Total RNA suitable for molecular biology analysis. *Journal of Biotechnology*, **105**, 1.

52 Eberwine, J.H. (1996) Amplification of mRNA populations using aRNA generated from immobilized oligo(dT)-T7 primed cDNA. *Biotechniques*, **20**, 584.

53 Phillips, J. and Eberwine, J.H. (1996) Antisense RNA amplification: a linear amplification method for analyzing the mRNA population from single living cells. *Methods*, **10**, 283.

54 Wang, E., Miller, L., Ohnmacht, G.A., Liu, E. and Marincola, F.M. (2000) High fidelity mRNA amplification for gene profiling using cDNA microarrays. *Nature Biotech*, **17**, 457.

55 Wang, E. and Marincola, F.M. (2002) in Amplification of small quantities of mRNA for transcript analysis, (eds Bowtell, D. and Sambrook, J.), in *DNA Arrays – A Molecular Cloning Manual*, Cold Spring Harbor Laboratory Press, Cold Springs Harbor, NY. p. 204.

56 Spiess, A.N., Mueller, N. and Ivell, R. (2003) Amplified RNA degradation in T7-amplification methods results in biased microarray hybridizations. *BMC Genomics*, **4**, 44.

57 Xiang, C.C., Chen, M., Ma, L., Phan, Q.N., Inman, J.M., Kozhich, O.A. and Brownstein, M.J. (2003) A new strategy to amplify degraded RNA from small tissue samples for microarray studies. *Nucleic Acids Research*, **31**, e53.

58 Malboeuf, C.M., Isaacs, S.J., Tran, N.H. and Kim, B. (2001) Thermal effects on reverse transcription: improvement of accuracy and processivity in cDNA synthesis. *Biotechniques*, **30**, 1074, 1080, 1082, passim.

59 Kenzelmann, M., Klaren, R., Hergenhahn, M., Bonrouhi, M., Grone, H.J., Schmid, W. and Schutz, G. (2004) High-accuracy amplification of nanogram total RNA amounts for gene profiling. *Genomics*, **83**, 550.

60 Schlingemann, J., Thuerigen, O., Ittrich, C., Toedt, G., Kramer, H., Hahn, M. and Lichter, P. (2005) Effective transcriptome amplification for expression profiling on sense-oriented oligonucleotide microarrays. *Nucleic Acids Research*, **33**, e29.

61 Carninci, P., Nishiyama, Y., Westover, A., Itoh, M., Nagaoka, S., Sasaki, N., Okazaki, Y., Muramatsu, M. and Hayashizaki, Y. (1998) Thermostabilization and thermoactivation of thermolabile enzymes by trehalose and its application for the synthesis of full length cDNA. *Proceedings of the National Academy of Sciences of the United States of America*, **95**, 520.

62 Mizuno, Y., Carninci, P., Okazaki, Y., Tateno, M., Kawai, J., Amanuma, H., Muramatsu, M. and Hayashizaki, Y. (1999) Increased specificity of reverse transcription priming by trehalose and oligo-blockers allows high-efficiency window separation of mRNA display. *Nucleic Acids Research*, **27**, 1345.

63 Spiess, A.N., Mueller, N. and Ivell, R. (2004) Trehalose is a potent PCR enhancer: lowering of DNA melting temperature and thermal stabilization of taq polymerase by the disaccharide trehalose. *Clinical Chemistry*, **50**, 1256.

64 Rapley, R. (1994) Enhancing PCR amplification and sequencing using DNA-binding proteins. *Molecular Biotechnology*, **2**, 295.

65 Villalva, C., Touriol, C., Seurat, P., Trempat, P., Delsol, G. and Brousset, P. (2001) Increased yield of PCR products by addition of T4 gene 32 protein to the SMART PCR cDNA synthesis system. *Biotechniques*, **31**, 81–86.

66 Baugh, L.R., Hill, A.A., Brown, E.L. and Hunter, C.P. (2000) Quantitative analysis of mRNA amplification by *in vitro* transcription. *Nucleic Acids Research*, **29**, E29–38.

67 Gubler, U. and Hoffman, B.J. (1983) A simple and very efficient method for generating cDNA libraries. *Gene*, **25**, 263.

68 Li, Y., Ali, S., Philip, P.A. and Sarkar, F.H. (2003) Direct comparison of microarray gene expression profiles between non-amplification and a modified cDNA amplification procedure applicable for needle biopsy tissues. *Cancer Detection and Prevention*, **27**, 405.

69 Li, Y., Li, T., Liu, S., Qiu, M., Han, Z., Jiang, Z., Li, R., Ying, K., Xie, Y. and Mao, Y. (2004) Systematic comparison of the fidelity of aRNA, mRNA and T-RNA on gene expression profiling using cDNA microarray. *Journal of Biotechnology*, **107**, 19.

70 Park, J.Y., Kim, S.Y., Lee, J.H., Song, J., Noh, J.H., Lee, S.H., Park, W.S., Yoo, N.J., Lee, J.Y. and Nam, S.W. (2004) Application of amplified RNA and evaluation of cRNA targets for spotted-oligonucleotide microarray. *Biochemical and Biophysical Research Communications*, **325**, 1346.

71 Rudnicki, M., Eder, S., Schratzberger, G., Mayer, B., Meyer, T.W., Tonko, M. and Mayer, G. (2004) Reliability of t7-based mRNA linear amplification validated by gene expression analysis of human kidney cells using cDNA microarrays. *Nephron. Experimental Nephrology*, **97**, e86–95.

72 Kacharmina, J.E., Crino, P.B. and Eberwine, J.H. (1999) Preparation of cDNA from single cells and subcellular regions. *Methods in Enzymology*, **303**, 3.

73 Matz, M., Shagin, D., Bogdanova, E., Britanova, O., Lukyanov, S., Diatchenko, L. and Chenchik, A. (1999) Amplification of cDNA ends based on template-switching effect and step-out PCR. *Nucleic Acids Research*, **27**, 1558.

74 Chenchik, A., Zhu, Y.Y., Diatchenko, L., Li, R., Hill, J. and Siebert, P.D. (1998) Generation and use of high-quality cDNA from small amounts of total RNA by SMART^{TM} PCR, in *Gene Cloning and Analysis by RT-PCR*, (eds Siebert, P. and Larrick, J.), Biotechniques Books, Natick, MA. p. 305.

75 Zhao, H., Hastie, T., Whitfield, M.L., Borresen-Dale, A.L. and Jeffrey, S.S.

(2002) Optimization and evaluation of T7 based RNA linear amplification protocols for cDNA microarray analysis. *BMC Genomics*, **3**, 31.

76 Marko, N.F., Frank, B., Quackenbush, J. and Lee, N.H. (2005) A robust method for the amplification of RNA in the sense orientation. *BMC Genomics*, **6**, 27.

77 Goff, L.A., Bowers, J., Schwalm, J., Howerton, K., Getts, R.C. and Hart, R.P. (2004) Evaluation of sense-strand mRNA amplification by comparative quantitative PCR. *BMC Genomics*, **5**, 76.

78 Cheetham, G.M., Jeruzalmi, D. and Steitz, T.A. (1998) Transcription regulation, initiation, and 'DNA scrunching' by T7 RNA polymerase. *Cold Spring Harbor Symposia on Quantitative Biology*, **63**, 263.

79 Gold, D., Coombes, K., Medhane, D., Ramaswamy, A., Ju, Z., Strong, L., Koo, J.S. and Kapoor, M. (2004) A comparative analysis of data generated using two different target preparation methods for hybridization to high-density oligonucleotide microarrays. *BMC Genomics*, **5**, 2.

80 Polz, M.F. and Cavanaugh, C.M. (1998) Bias in template-to-product ratios in multitemplate PCR. *Applied and Environmental Microbiology*, **64**, 3724.

81 Seth, D., Gorrell, M.D., McGuinness, P.H., Leo, M.A., Lieber, C.S., McCaughan, G.W. and Haber, P.S. (2003) SMART amplification maintains representation of relative gene expression: quantitative validation by real time PCR and application to studies of alcoholic liver disease in primates. *Journal of Biochemical and Biophysical Methods*, **55**, 53.

82 Smith, L., Underhill, P., Pritchard, C., Tymowska-Lalanne, Z., Abdul-Hussein, S., Hilton, H., Winchester, L., Williams, D., Freeman, T., Webb, S. and Greenfield, A. (2003) Single primer amplification (SPA) of cDNA for microarray expression analysis. *Nucleic Acids Research*, **31**, e9.

83 Bettinotti, M.P., Panelli, M.C., Ruppe, E., Mocellin, S., Phan, G.Q., White, D.E. and Marincola, F.M. (2003) Clinical and immunological evaluation of patients with metastatic melanoma undergoing immunization with the HLA-C2*0702 associated epitope MAGE-A12:170–178. *International Journal of Cancer*, **105**, 210.

84 Iscove, N.N., Barbara, M., Gu, M., Gibson, M., Modi, C. and Winegarden, N. (2002) Representation is faithfully preserved in global cDNA amplified exponentially from sub-picogram quantities of mRNA. *Nature Biotech*, **20**, 940.

85 Stirewalt, D.L., Pogosova-Agadjanyan, E.L., Khalid, N., Hare, D.R., Ladne, P.A., Sala-Torra, O., Zhao, L.P. and Radich, J.P. (2004) Single-stranded linear amplification protocol results in reproducible and reliable microarray data from nanogram amounts of starting RNA. *Genomics*, **83**, 321.

86 Petalidis, L., Bhattacharyya, S., Morris, G.A., Collins, V.P., Freeman, T.C. and Lyons, P.A. (2003) Global amplification of mRNA by template-switching PCR: linearity and application to microarray analysis. *Nucleic Acids Research*, **31**, e142.

87 Rox, J.M., Bugert, P., Muller, J., Schorr, A., Hanfland, P., Madlener, K., Kluter, H. and Potzsch, B. (2004) Gene expression analysis in platelets from a single donor: evaluation of a PCR-based amplification technique. *Clinical Chemistry*, **50**, 2271.

88 Klur, S., Toy, K., Williams, M.P. and Certa, U. (2004) Evaluation of procedures for amplification of small-size samples for hybridization on microarrays. *Genomics*, **83**, 508.

89 Ohtsuka, S., Iwase, K., Kato, M., Seki, N., Shimizu-Yabe, A., Miyauchi, O., Sakao, E., Kanazawa, M., Yamamoto, S., Kohno, Y. and Takiguchi, M. (2004) An mRNA amplification procedure with directional cDNA cloning and strand-specific cRNA synthesis for comprehensive gene expression analysis. *Genomics*, **84**, 715.

90 Randolph, J.B. and Waggoner, A.S. (1997) Stability, specificity and fluorescence

brightness of multiply-labeled fluorescent DNA probes. *Nucleic Acids Research*, **25**, 2923.
91 DeRisi, J. (2002) Stability, specificity and fluorescence brightness of mulitply-labeled fluorescent DNA probes, in *DNA Arrays – A Molecular Cloning Manual*, (eds Bowtell, D., and Sambrook, J.), Cold Spring Harbor Laboratory Press, Cold Spring Harbor, NY. p. 204.
92 Quackenbush, J. (2002) Microarray data normalization and transformation. *Nature Genetics*, **32s**, 496.
93 Basil, C.F., Zhao, Y., Zavaglia, K., Jin, P., Panelli, M.C., Voiculescu, S., Mandruzzato, S., Lee, H.M., Seliger, B., Freedman, R.S., Taylor, P.R., Hu, N., Zanovello, P., Marincola, F.M. and Wang, E. (2006) Common cancer biomarkers. *Cancer Research*, **66**, 2953.
94 Broberg, P. (2003) Statistical methods for ranking differentially expressed genes. *Genome Biology*, **4**, R41.
95 Zeeberg, B.R., Feng, W., Wang, G., Wang, M.D., Fojo, A.T., Sunshine, M., Narasimhan, S., Kane, D.W., Reinhold, W., Lababidi, S., Bussey, K.J., Riss, J., Barrett, J.C. and Weinstein, J.N. (2003) GoMiner: a resource for biological interpretation of genomic and proteomic data. *Genome Biology*, **4**, R28.
96 Marincola, F.M. (2007) In support of descriptive studies: relevance to translational research. *Journal of Translational Medicine*, **5**, 21.
97 Rajeevan, M.S., Ranamukhaarachchi, D.G., Vernon, S.D. and Unger, E.R. (2001) Use of real-time quantitative PCR to validate the results of cDNA array and differential display PCR technologies. *Methods*, **25**, 443.
98 Panelli, M.C., Martin, B., Nagorsen, D., Wang, E., Smith, K., Monsurro', V. and Marincola, F.M. (2003) A genomic and proteomic-based hypothesis on the eclectic effects of systemic interleukin-2 administration in the context of melanoma-specific immunization. *Cells, Tissues, Organs*, **177**, 124.
99 Monsurro', V., Wang, E., Yamano, Y., Migueles, S.A., Panelli, M.C., Smith, K., Nagorsen, D., Connors, M., Jacobson, S. and Marincola, F.M. (2004) Quiescent phenotype of tumor-specific CD8 + T cells following immunization. *Blood*, **104**, 1970.
100 Panelli, M.C., Stashower, M., Slade, H.B., Smith, K., Norwood, C., Abati, A., Fetsch, P.A., Filie, A., Walters, S.A., Astry, C., Arico, E., Zhao, Y., Selleri, S., Wang, E. and Marincola, F.M. (2006) Sequential gene profiling of basal cell carcinomas treated with Imiquimod in a placebo-controlled study defines the requirements for tissue rejection. *Genome Biology*, **8**, R8.
101 Deola, S., Panelli, M.C., Maric, D., Selleri, S., Dmitrieva, N.I., Voss, C.Y., Klein, H.G., Stroncek, D.F., Wang, E. and Marincola, F.M. (2008) 'Helper' B cells promote cytotoxic T cell survival and proliferation independently of antigen presentation through CD27-CD70 interactions. *Journal of Immunology*, **180**, 1362.
102 Crnogorac-Jurcevic, T., Efthimiou, E., Nielsen, T., Loader, J., Terris, B., Stamp, G., Baron, A., Scarpa, A. and Lemoine, N. (2002) Expression profiling of microdissected pancreatic carcinomas. *Oncogene*, **21**, 4587.
103 Ahram, M., Best, C.J., Flaig, M.J., Gillespie, J.W., Leiva, I.M., Chuaqui, R.F., Zhou, G., Shu, H., Duray, P.H., Linehan, W.M., Raffeld, M., Ornstein, D.K., Zhao, Y., Petricoin, E.F. and Emmert-Buck, M.R. (2002) Proteomic analysis of human prostate cancer. *Molecular Carcinogenesis*, **33**, 9.
104 Kawakami, Y., Robbins, P., Wang, R.F., Parkhurst, M.R., Kang, X. and Rosenberg, S.A. (1998) Tumor antigens recognized by T cells. The use of melanosomal proteins in the immunotherapy of melanoma. *Journal of Immunotherapy*, **21**, 237.
105 Marincola, F.M., Jaffe, E.M., Hicklin, D.J. and Ferrone, S. (2000) Escape of human solid tumors from T cell recognition:

molecular mechanisms and functional significance. *Advances in Immunology*, **74**, 181.

106 Marincola, F.M., Wang, E., Herlyn, M., Seliger, B. and Ferrone, S. (2003) Tumors as elusive targets of T cell-based active immunotherapy. *Trends in Immunology*, **24**, 335.

107 Suri, A. (2006) Cancer testis antigens-- their importance in immunotherapy and in the early detection of cancer. *Expert Opinion on Biological Therapy*, **6**, 379.

108 Rubinfeld, B., Robbins, P., el Gamil, M., Albert, I., Porfiri, E. and Polakis, P. (1997) Stabilization of beta-catenin by genetic defects in melanoma cell lines. *Science*, **275**, 1790.

109 Robbins, P.F., el-Gamil, M., Li, Y.F., Kawakami, Y., Loftus, D., Appella, E. and Rosenberg, S.A. (1996) A mutated beta-catenin gene encodes a melanoma-specific antigen recognized by tumor infiltrating lymphocytes. *J. Exp. Med*, **183**, 1185.

110 Wang, E., Voiculescu, S., Le Poole, I.C., el Gamil, M., Li, X., Sabatino, M., Robbins, P.F., Nickoloff, B.J. and Marincola, F.M. (2006) Clonal persistence and evolution during a decade of recurrent melanoma. *The Journal of Investigative Dermatology*, **126**, 1372.

111 Sabatino, M., Zhao, Y., Voiculescu, S., Monaco, A., Robbins, P.F., Nickoloff, B.J., Karai, L., Selleri, S., Maio, M., Selleri, S., Marincola, F.M. and Wang, E. (2008) Conservation of a core of genetic alterations over a decade of recurrent melanoma supports the melanoma stem cell hypothesis. *Cancer Research*, **68**, 222.

112 Andersen, M.H., Pedersen, L.O., Capeller, B., Brocker, E.B., Becker, J.C. and Thor Straten, P., (2001) Spontaneous cytotoxic T-cell responses against survivin-derived MHC class I-restricted T-cell epitopes in situ as well as *ex vivo* in cancer patients. *Cancer Research*, **61**, 5964.

113 Andersen, M.H. and Thor Straten, P. (2002) Survivin – A universal tumor antigen. *Histology and Histopathology*, **17**, 669.

114 Tsuruma, T., Hata, F., Torigoe, T., Furuhata, T., Idenoue, S., Kurotaki, T., Yamamoto, M., Yagihashi, A., Ohmura, T., Yamaguchi, K., Katsuramaki, T., Yasoshima, T., Sasaki, K., Mizushima, Y., Minamida, H., Kimura, H., Akiyama, M., Hirohashi, Y., Asanuma, H., Tamura, Y., Shimozawa, K., Sato, N. and Hirata, K. (2004) Phase I clinical study of anti-apoptosis protein, survivin-derived peptide vaccine therapy for patients with advanced or recurrent colorectal cancer. *Journal of Translational Medicine*, **2** 19.

115 Andersen, M.H., Soerensen, R.B., Becker, J.C. and Thor Straten, P., (2006) HLA-A24 and survivin: possibilities in therapeutic vaccination against cancer. *Journal of Translational Medicine*, **4**, 38.

116 Li, Y.H., Hu, C.F., Shao, Q., Huang, M.Y., Hou, J.H., Xie, D., Zeng, Y.X. and Shao, J.Y. (2008) Elevated expressions of survivin and VEGF protein are strong independent predictors of survival in advanced nasopharyngeal carcinoma. *Journal of Translational Medicine*, **6**, 1.

117 Matsuzaki, Y., Hashimoto, S., Fujita, T., Suzuki, T., Sakurai, T., Matsushima, K. and Kawakami, Y., (2005) Systematic identification of human melanoma antigens using serial analysis of gene expression (SAGE). *Journal of Immunother*, **28**, 10.

118 Wang, E., Panelli, M.C., Zavaglia, K., Mandruzzato, S., Hu, N., Taylor, P.R., Seliger, B., Zanovello, P., Freedman, R.S. and Marincola, F.M. (2004) Melanoma-restricted genes. *Journal of Translational Medicine*, **2**, 34.

119 Wang, E., Panelli, M.C., Smith, K. and Marincola, F.M. (2002) Applications of functional genomics in oncology. *Principles and Practice of Oncology Updates*, **16**, 1.

120 Wang, E., Panelli, M.C., Monsurro', V. and Marincola, F.M. (2004) Gene expression profiling of anti-cancer

immune responses. *Current Opinion in Molecular Therapeutics*, **6**, 288.
121 Wang, E., Panelli, M. and Marincola, F.M. (2006) Autologous tumor rejection in humans: trimming the myths. *Immunological Investigations*, **35**, 437.
122 Atkins, M.B., Regan, M., McDermott, D., Mier, J., Stanbridge, E., Youmans, A., Febbo, P., Upton, M., Lechpammer, M. and Signoretti, S. (2005) Carbonic anhydrase IX expression predicts outcome in interleukin-2 therapy of renal cancer. *Clinical Cancer Research*, **11**, 3714.
123 Panelli, M.C., Wang, E. and Marincola, F.M. (2005) The pathway to biomarker discovery: carbonic anhydrase IX and the prediction of immune responsiveness. *Clinical Cancer Research*, **11**, 3601.
124 Sarwal, M., Chua, M.S., Kambham, N., Hsieh, S.C., Satterwhite, T., Masek, M. and Salvatierra, Jr. O. (2003) Molecular heterogeneity in acute renal allograft rejection identified by DNA microarray profiling. *The New England Journal of Medicine*, **349**, 125.
125 Robertson, M.J. (2002) Role of chemokines in the biology of natural killer cells. *Journal of Leukocyte Biology*, **71**, 173.
126 Zhang, L. and Pagano, J.S. (2002) Structure and function of IRF-7. *Journal of Interferon and Cytokine Research*, **22**, 95.
127 Hiscott, J., Grandvaux, N., Sharma, S., Tenoever, B.R., Servant, M.J. and Lin, R. (2003) Convergence of the NF-kappaB and interferon signaling pathways in the regulation of antiviral defense and apoptosis. *Annals of the New York Academy of Sciences*, 1010: 237.
128 Levy, D.E., Marie, I., Smith, E. and Prakash, A. (2002) Enhancement and diversification of IFN induction by IRF-7-mediated positive feedback. *Journal of Interferon and Cytokine Research*, **22**, 87.
129 Kammula, U.S., Lee, K.-H., Riker, A., Wang, E., Ohnmacht, G.A., Rosenberg, S.A. and Marincola, F.M. (1999) Functional analysis of antigen-specific T lymphocytes by serial measurement of gene expression in peripheral blood mononuclear cells and tumor specimens. *Journal of Immunology (Baltimore, Md: 1950)*, **163**, 6867.
130 Ohnmacht, G.A., Wang, E., Mocellin, S., Abati, A., Filie, A., Fetsch, P.A., Riker, A., Kammula, U.S., Rosenberg, S.A. and Marincola, F.M. (2001) Short term kinetics of tumor antigen expression in response to vaccination. *Journal of Immunology (Baltimore, Md: 1950)*, 167: 1809.
131 Zhang, L., Conejo-Garcia, J.R., Katsaros, D., Gimotty, P.A., Massobrio, M., Regnani, G., Makrigiannakis, A., Gray, H., Schlienger, K., Liebman, M.N., Rubin, S.C. and Coukos, G. (2003) Intratumoral T cells, recurrence, and survival in epithelial ovarian cancer. *New England Journal of Medicine*, **348**, 203.
132 Galon, J., Costes, A., Sanchez-Cabo, F., Kirilovsky, A., Mlecnik, B., Lagorce-Pages, C., Tosolini, M., Camus, M., Berger, A., Wind, P., Zinzindohoue, F., Bruneval, P., Cugnenc, P.H., Trajanoski, Z., Fridman, W.H. and Pages, F. (2006) Type, density, and location of immune cells within human colorectal tumors predict clinical outcome. *Science*, **313**, 1960.

6
Detection and Identification of TAA by SELDI-TOF

Ferdinand von Eggeling and Christian Melle

6.1
Introduction

In recent years, the search for new cancer biomarkers, especially tumor-associated antigens (TAA), has been advanced by genomic and proteomic high-throughput techniques. Biomarkers or biomarker patterns should enable scientists or medical staff to make a more reliable early diagnosis of malignant tumors, and to predict their progression, which would facilitate the design of more differentiated individually-orientated tumor therapies. However, up until now relevant markers have been established for only a few tumor diseases. Prostate-specific antigen (PSA), prostatic acid phosphatase (PAP), CA 125, carcinoembryonic antigen (CEA), alpha-fetoprotein (AFP), human chorionic gonadotropin (HCG), CA 19–9, CA 15-3, CA 27–29, lactate dehydrogenase (LDH), and neuron-specific enolase (NSE) are known examples. Most of these were first described a long time ago (e.g. PSA in 1971, CEA in 1955) [1]. These markers are mainly used for prognostic purposes, and their application in tumor diagnosis is very limited, especially for early-stage cancer [2]. Furthermore, the survival rates of patients diagnosed late in disease progression and suffering from regionally or distantly spread tumors have shown little improvement over the last 30 years.

In the 1990s genomic methods in particular were preferred for biomarker research, but these methods provided no information as to whether or not the gene products, i.e. proteins, had really been expressed [3]. The gap in information regarding changes in protein expression levels, post-translational modifications or degradations prevents a real understanding of the pathway networks responsible for the regulation of physiological and pathological processes.

Some of the proteomic techniques available now are newly developed, although most of them are based on already existing procedures developed more than 30 years ago but ignored by the majority of the scientific community. The work-horse of the proteomic techniques is two-dimensional polyacrylamide gel electrophoresis, the 2-D PAGE [4]. This technique has been developed to a level that allows high-resolution

Tumor-Associated Antigens. Edited by Olivier Gires and Barbara Seliger
Copyright © 2009 WILEY-VCH Verlag GmbH & Co. KGaA, Weinheim
ISBN: 978-3-527-32084-4

separation of proteins, good reproducibility, and adequate sensitivity, but it is neither a large-scale nor a high-throughput technology. Nevertheless, in combination with mass spectrometry (MS), 2-D PAGE can be a powerful tool for the separation and subsequent identification of proteins of interest. In particular, matrix assisted laser desorption and ionization time-of-flight mass spectrometry (MALDI-TOF) [5] is commonly used for the accurate measurement of the molecular masses of peptides derived from in-gel digested proteins and subsequent identification by peptide mass fingerprinting (PMF) [6, 7]. In contrast, the direct analysis of non pre-fractionated biological samples is beyond the scope of traditional MALDI-TOF, as it is limited by the complexity of the analyte, the presence of buffer components such as salts or detergents, and the presence of non-protein components such as lipids and carbohydrates.

Nevertheless, today most proteomic techniques are based on or linked to MS-approaches [8]. As reviewed by Srinivas [9], this is true for the 'Isotope Affinity Tags', where different isotopes are used to label two different cell states [10], MALDI-derived techniques allowing the ionization of larger proteins [11], 'reverse liquid chromatography tandem mass spectrometry' (LC-MS/MS) for the separation and sequencing of peptides in the low femto-molar range [12], and 'Imaging MS', by which proteins on the surface of tissue sections or individual cells can be mapped directly [13, 14]. But it has to be kept in mind that proteome research is significantly more complex and requires a greater amount of work than the deciphering of genomic information. Among the 1 billion possible protein molecules, about 20 000 different entities are present in one cell. For the yeast proteins for example it is known that 84% of the proteins are present in multi-protein complexes [15]. Thus, the detection of proteomic biomarkers – particularly TAA – the identification and elucidation of the biological role of those molecules will be one of the most exciting challenges for the future.

6.2
SELDI (ProteinChip) Technology

6.2.1
The Procedural Method

The SELDI (ProteinChip) technology was developed in the early 1990s and first described by Hutchens and Yip [16]. It utilizes chips with affinity surface coatings to specifically retain proteins based on their physico-chemical characteristics prior to time-of-flight (TOF) MS analysis (Figure 6.1). Crude protein lysates or catapulted specimens can be applied directly onto the chip surface. The desired proteins are retained at the chromatographic surface and the contaminants such as buffer salts or detergents are washed off. Therefore, this technique is quantitative and displays greater sensitivity than comparable MALDI systems [17, 18], however this is at the expense of the resolution which is lower. Because this instrument is not primarily used for the exact measurement of peptide masses but rather for the comparative

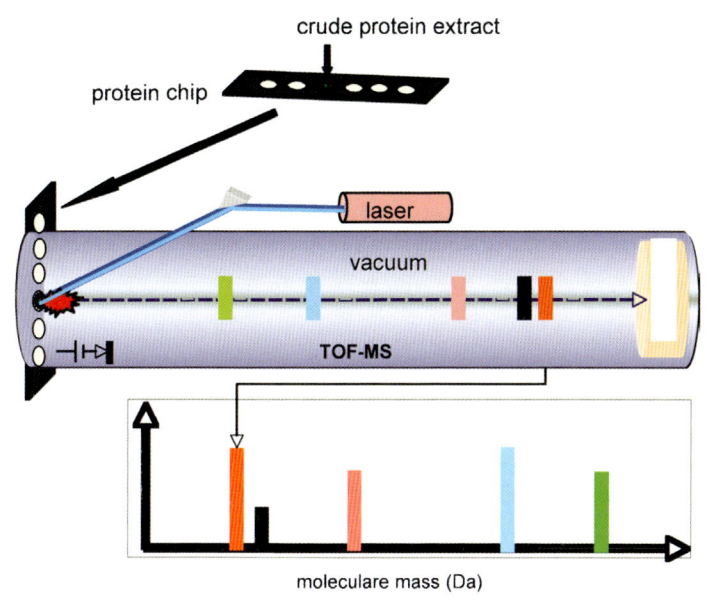

Figure 6.1 Schematic representation of the SELDI-MS technology. Analysis of samples using SELDI-TOF: after placing the ProteinChip Array to which specific proteins are bound into the reader, a laser beam is directed onto the sample on the spot. In this way protons are transferred onto the peptides and proteins that are subsequently accelerated by electromagnetic fields through a flight tunnel. The time-of-flight measured at the detector corresponds inversely to the molecular mass.

analysis of two different protein lysates (e.g. derived from normal vs. tumor tissue) this is not a real disadvantage. SELDI-MS can be used as a complementary method to 2-DE for biomarker research, because small proteins, and also proteins with extreme pI are detectable. For a current review see Engwegen et al. [19]. As only a small number of cells is required for SELDI-MS analysis, this technology is ideal for small biopsies or microdissected tissue samples. Consequently, from its inception SELDI-MS was combined with laser capture microdissection assessment of tumor specimens [20–22]. Moreover, the use of microdissected samples enables the selection of tumor cells which is an advantage over the use of whole biopsies which incorporate varying amounts of non-tumor cells such as infiltrating immune cells.

Lysates of the microdissected specimen can be directly applied onto the chip surface. The desired proteins are retained at the chromatographic surface and the contaminants such as buffer salts or detergents are simply washed off. This easy purification step eliminates the need for pre-separation techniques and allows quantitative evaluations. Its compatibility with laser microdissection and catapulting procedures has been established in various studies [23–28] and was referred to in a review by von Eggeling et al. as 'Microdissecting the proteome' [29].

6.2.2
SELDI-MS in TAA Identification

6.2.2.1 Tumor Samples

Our group published the first extended study on head and neck squamous cell carcinomas (HNSCC) [30]. We used ProteinChip technology as a sensitive tool for the comparison of normal and tumor tissue, 2-DE for the separation of isolated biomarkers, and Tandem MS and immunodepletion for their identification. Further characterization of the protein found was achieved with immunhistochemistry while annexin A5-positive tissue areas were microdissected once more and re-analyzed on ProteinChip arrays to confirm the identity of this protein. This strategy was refined in an additional study on HNSCC and referred to as a 'technical triade' for proteomic identification and characterization of cancer biomarkers [31]. Using this refinement S100A8 (calgranulin A) and S100A9 (calgranulin B) were identified as potential markers.

In further studies, we analysed hepatocellular carcinoma [32], gastric cancer [33], and colorectal carcinoma [34, 35] for the presence of selective TAAs. In the latter study it was shown that tumor-specific markers (in this case α-Defensines, also known as HNP 1–3) detected in tissue could be also traced in serum. Further on we were able to prove that these markers (HNP1–3) would not have been found in serum without the information gained through the analysis of microdissected tissue. In another study we were able to show for the first time at the proteomic level that changes in the histological features of tumors as compared to the tissues from which they arise, are reflected in the convergence of proteomic patterns during the development of cancer [28].

In summary, the results from these studies corroborated that the combination of laser assisted microdissection with SELDI-ProteinChip technology and immunohistochemistry potentiates the identification and characterization of new specific biomarkers. This 'Technical Triade' provides further opportunities to identify tumor-specific proteins, which will significantly improve the detection and treatment of cancer [31, 36].

6.2.2.2 Body Fluids

Until now, biomarker discovery with the ProteinChip System was mostly achieved by analyzing body fluids such as serum or urine [37–41]. In contrast to reviewed work with microdissected tissues, body fluid analyses are rapid and easy to carry out by direct application on the arrays. Nevertheless, it is known that there is a wide range of intra-individual differences in serum. For instance sex hormone levels, nutritional status, or inflammation can markedly change the protein profile. Hence, biomarkers or TAAs responsible for the genesis and progression of cancer must be present at a high level to be observed above normal changes. Nevertheless, this research on body fluids is fascinating because it offers the potential to detect biomarkers in material obtained by non- or minimally invasive procedures such as venipuncture. This is especially true for body fluids like nipple aspirate fluid (NAF) or urine, which are in more direct contact with the tumor and therefore contain a higher concentration of potential biomarkers than serum. Such markers in near-to-tumor body fluids which can be obtained by non-invasive screening would be ideal for screening high-risk individuals or even individuals without an elevated risk, provided that the costs are affordable. However, this goal can only be attained for a few tumor entities (e.g. transitional cell carcinoma; TCC), whereas in many other cases, detectable markers will be present only at an advanced tumor stage, and hence too late for an early and preventive diagnostic assay.

6.2.2.3 SELDI-MS TAAs and Clinical Potential

SELDI was especially designed to identify TAAs for the diagnosis of different cancer entities. Cavallo *et al.* [42] used an integrated approach of immunogenomics and bioinformatics to identify new TAAs for mammary cancer vaccination. In a meta-analysis they attempted to identify a set of new TAA targets, which could be used instead of or in conjunction with Her2. Five TAAs were identified (Tes, Rcn2, Rnf4, Cradd, Galnt3) and found to be expressed linearly in relation to the tumor mass thus making them useful in the design of preclinical immunopreventive vaccines. Spruessel *et al.* [43] showed that pre-analytical factors, such as the period for which the tissue was ischemic, could dramatically influence the concentration of TAAs (i.e. CEA). Wright and co-worker [44] reviewed TAAs and other serum markers for hepatocellular cancer. They reported that several studies had documented the presence of tumor-derived autoantibodies in the sera of HCC patients. These may be early (anti-p53) and newly discovered ones (anti-telomerase immunoglobulins). But most studies that employed SELDI-MS used carcinoembryonic antigen CEA as the 'gold standard' with which to compare their findings.

Zheng *et al.* [45] endeavored to develop a proteomic pattern for the distinction of individuals with colorectal cancer from healthy controls and for monitoring micrometastases using SELDI-TOF-MS. As a result they found a four-peak model, i.e. a signature of four distinct peaks that discriminated cancer from non-cancer samples with higher sensitivity and specificity than the combined use of CEA, CA199 and CA242 in the early detection of colorectal cancer. Su and co-workers [46] analyzed serum samples from individuals with gastric cancer, healthy individuals, and nine patients with benign gastric lesions using SELDI-MS. The sensitivity and specificity of the resulting fingerprints were all higher than those achieved in an analysis conducted in parallel by measuring serum carcinoembryonic antigen (CEA) and carbohydrate antigen (CA)19-9 together.

Scarlett *et al.* [47] studied cholangiocarcinoma (CC) in both tissue and serum to find new biomarkers using SELDI-MS. Univariate analysis revealed 14 differentially expressed individual peaks. One peak found in serum had better discriminatory ability than carbohydrate antigen 19.9 (CA19.9) and carcinoembryonic antigen (CEA). Nevertheless, the results from serum were further improved with the inclusion of CA19.9 and CEA. Qian and co-workers [48] studied SELDI-MS serum protein patterns in relation to the detection of gastric cancer. Sixteen mass peaks were found to be potential biomarkers. Among them, nine mass peaks showed increased expression in patients with gastric cancer. Thus, the sensitivity, specificity, and accuracy were higher than those achieved using the clinical serum biomarkers CEA, CA19-9, and CA72-4.

6.3
Conclusions and Future Perspectives

In conclusion, the SELDI-based ProteinChip technology completes the panoply of powerful tools dedicated to the isolation and identification of TAA and other biomarkers. The aim for the next decade will be further research into these markers so that a more differentiated individual tumor diagnosis and therapy become a possibility. The improved estimation of the biological importance of biomarkers with regard to the progression from pre-neoplastic alterations in the tissue to malignant tumors and the prediction of the potential metastases formation, will be necessary prerequisites for providing a more detailed insight and understanding of tumor progression and adequate therapeutic strategies.

References

1 Bidart, J.M., Thuillier, F., Augereau, C., Chalas, J. et al. (1999) Kinetics of serum tumor marker concentrations and usefulness in clinical monitoring. *Clinical Chemistry*, **45**, 1695–1707.

2 Etzioni, R., Urban, N., Ramsey, S., McIntosh, M. et al. (2003) The case for early detection. *Nature Reviews Cancer*, **3**, 243–252.

3 Gygi, S.P., Rochon, Y., Franza, B.R. and Aebersold, R. (1999) Correlation between protein and mRNA abundance in yeast. *Molecular and Cellular Biology*, **19**, 1720–1730.

4 Klose, J. (1975) Protein mapping by combined isoelectric focusing and electrophoresis of mouse tissues. A novel approach to testing for induced point mutations in mammals. *Humangenetik*, **26**, 231–243.

5 Siuzdak, G. (1996) *Mass Spectrometry for Biotechnology*, Academic Press.

6 Loo, J.A., Brown, J., Critchley, G., Mitchell, C. et al. (1999) High sensitivity mass spectrometric methods for obtaining intact molecular weights from gel-separated proteins. *Electrophoresis*, **20**, 743–748.

7 Lahm, H.W. and Langen, H. (2000) Mass spectrometry: A tool for the identification of proteins separated by gels. *Electrophoresis*, **21**, 2105–2114.

8 Pandey, A. and Mann, M. (2000) Proteomics to study genes and genomes. *Nature*, **405**, 837–846.

9 Srinivas, P.R., Verma, M., Zhao, Y.M. and Srivastava, S. (2002) Proteomics for cancer biomarker discovery. *Clinical Chemistry*, **48**, 1160–1169.

10 Gygi, S.P., Rist, B., Gerber, S.A., Turecek, F. et al. (1999) Quantitative analysis of complex protein mixtures using isotope-coded affinity tags. *Nature Biotechnology*, **17**, 994–999.

11 Karas, M. and Hillenkamp, F. (1988) Laser desorption ionization of proteins with molecular masses exceeding 10000 Daltons. *Analytical Chemistry*, **60**, 2299–2301.

12 Gygi, S.P., Han, D.K.M., Gingras, A.C., Sonenberg, N. and Aebersold, R. (1999) Protein analysis by mass spectrometry and sequence database searching: Tools for cancer research in the post-genomic era. *Electrophoresis*, **20**, 310–319.

13 Stoeckli, M., Farmer, T.B. and Caprioli, R.M. (1999) Automated mass spectrometry imaging with a matrix-assisted laser desorption ionization time-of-flight instrument. *Journal of the American Society for Mass Spectrometry*, **10**, 67–71.

14 Ernst, G., Melle, C., Schimmel, B., Bleul, A. and von Eggeling, F. (2006)

Proteohistography – direct analysis of tissue with high sensitivity and high spatial resolution using proteinchip technology. *The Journal of Histochemistry and Cytochemistry*, **54**, 13–17.

15 Gavin, A.C., Bosche, M., Krause, R., Grandi, P. *et al.* (2002) Functional organization of the yeast proteome by systematic analysis of protein complexes. *Nature*, **415**, 141–147.

16 Hutchens, T.W. and Yip, T.T. (1993) New desorption strategies for the mass spectrometric analysis of macromolecules. *Rapid Communications in Mass Spectrometry*, **7**, 576–580.

17 Semmes, O.J., Feng, Z., Adam, B.L., Banez, L.L. *et al.* (2005) Evaluation of serum protein profiling by surface-enhanced laser desorption/ionization time-of-flight mass spectrometry for the detection of prostate cancer: I. Assessment of platform reproducibility. *Clinical Chemistry*, **51**, 102–112.

18 Vorderwulbecke, S., Cleverly, S., Weinberger, S.R. and Wiesner, A. (2005) Protein quantification by SELDI-TOF-MS-based ProteinChip System. *Nature Methods*, **2**, 393–395.

19 Engwegen, J.Y., Gast, M.C., Schellens, J.H. and Beijnen, J.H. (2006) Clinical proteomics: searching for better tumour markers with SELDI-TOF mass spectrometry. *Trends in Pharmacological Sciences*, **27**, 251–259.

20 Wright, G.L., Cazares, L.H., Leung, S.M., Nasim, S. *et al.* (1999) Proteinchip(R) surface enhanced laser desorption/ionization (SELDI) mass spectrometry: a novel protein biochip technology for detection of prostate cancer biomarkers in complex protein mixtures. *Prostate Cancer and Prostatic Diseases*, **2**, 264–276.

21 von Eggeling, F., Davies, H., Lomas, L., Fiedler, W. *et al.* (2000) Tissue-specific microdissection coupled with proteinchip array technologies: applications in cancer research. *Biotechniques*, **29**, 1066–1070.

22 von Eggeling, F., Junker, K., Fiedler, W., Wollscheid, V. *et al.* (2001) Mass spectrometry meets chip technology: A new proteomic tool in cancer research? *Electrophoresis*, **22**, 2898–2902.

23 Krieg, R.C., Fogt, F., Braunschweig, T., Herrmann, P.C. *et al.* (2004) ProteinChip Array analysis of microdissected colorectal carcinoma and associated tumor stroma shows specific protein bands in the 3.4 to 3.6 kDa range. *Anticancer Research*, **24**, 1791–1796.

24 Krieg, R.C., Gaisa, N.T., Paweletz, C.P. and Knuechel, R. (2005) Proteomic analysis of human bladder tissue using SELDI approach following microdissection techniques. *Methods in Molecular Biology (Clifton, NJ)*, **293**, 255–267.

25 Nakagawa, T., Huang, S.K., Martinez, S.R., Tran, A.N. *et al.* (2006) Proteomic profiling of primary breast cancer predicts axillary lymph node metastasis. *Cancer Research*, **66**, 11825–11830.

26 Guedj, N., Dargere, D., Degos, F., Janneau, J.L. *et al.* (2006) Global proteomic analysis of microdissected cirrhotic nodules reveals significant biomarkers associated with clonal expansion. *Laboratory Investigation; A Journal of Technical Methods and Pathology*, **86**, 951–958.

27 Melle, C., Ernst, G., Scheibner, O., Kaufmann, R. *et al.* (2007) Identification of specific protein markers in microdissected hepatocellular carcinoma. *Journal of Proteome Research*, **6**, 306–315.

28 Muller, U., Ernst, G., Melle, C., Guthke, R. and von Eggeling, F. (2006) Convergence of the proteomic pattern in cancer. *Bioinformatics (Oxford, England)*, **22**, 1293–1296.

29 von Eggeling, F., Melle, C. and Ernst, G. (2007) Microdissecting the proteome. *Proteomics*, **7**, 2729–2737.

30 Melle, C., Ernst, G., Schimmel, B. Bleul, A. *et al.* (2003) Biomarker discovery and identification in laser microdissected head and neck squamous cell carcinoma with ProteinChip(R) technology, two-dimensional gel electrophoresis,

tandem mass spectrometry, and immunohistochemistry. *Molecular & Cellular Proteomics*, **2**, 443–452.

31 Melle, C., Ernst, G., Schimmel, B., Bleul, A. et al. (2004) A Technical Triade for proteomic identification and characterization of cancer biomarkers. *Cancer Research*, **64**, 4099–4104.

32 Melle, C., Kaufmann, R., Hommann, M., Bleul, A. et al. (2004) Proteomic profiling in microdissected hepatocellular carcinoma tissue using ProteinChip technology. *International Journal of Oncology*, **24**, 885–891.

33 Melle, C., Ernst, G., Schimmel, B., Bleul, A. et al. (2005) Characterization of pepsinogen C as a potential biomarker for gastric cancer using a histo-proteomic approach. *Journal of Proteome Research*, **4**, 1799–1804.

34 Melle, C., Ernst, G., Schimmel, B., Bleul, A. et al. (2006) Different expression of calgizzarin (S100A11) in normal colonic epithelium, adenoma and colorectal carcinoma. *International Journal of Oncology*, **28**, 195–200.

35 Melle, C., Ernst, G., Schimmel, B., Bleul, A. et al. (2005) Discovery and identification of alpha-defensins as low abundant, tumor-derived serum markers in colorectal cancer. *Gastroenterology*, **129**, 66–73.

36 von Eggeling, F. and Ernst, G. (2007) Microdissected tissue: an underestimated source for biomarker discovery. *Biomarkers in Medicine*, **1**, 217–219.

37 Petricoin, E.F., Ardekani, A.M., Hitt, B.A., Levine, P.J. et al. (2002) Use of proteomic patterns in serum to identify ovarian cancer. *Lancet*, **359**, 572–577.

38 Petricoin, E.F., Mills, G.B., Kohn, E.C. and Liotta, L.A. (2002) Proteomic patterns in serum and identification of ovarian cancer – Reply. *Lancet*, **360**, 170–171.

39 Vlahou, A., Schellhammer, P.F., Mendrinos, S., Patel, K. et al. (2001) Development of a novel proteomic approach for the detection of transitional cell carcinoma of the bladder in urine. *The American Journal of Pathology*, **158**, 1491–1502.

40 Paweletz, C.P., Trock, B., Pennanen, M., Tsangaris, T. et al. (2001) Proteomic patterns of nipple aspirate fluids obtained by SELDI-TOF: potential for new biomarkers to aid in the diagnosis of breast cancer. *Disease Markers*, **17**, 301–307.

41 Rosty, C., Christa, L., Kuzdzal, S., Baldwin, W.M. et al. (2002) Identification of hepatocarcinoma-intestine-pancreas/pancreatitis-associated protein I as a biomarker for pancreatic ductal adenocarcinoma by protein biochip technology. *Cancer Research*, **62**, 1868–1875.

42 Cavallo, F., Astolfi, A., Iezzi, M., Cordero, F. et al. (2005) An integrated approach of immunogenomics and bioinformatics to identify new Tumor Associated Antigens (TAA) for mammary cancer immunological prevention *BMC. Bioinformatics (Oxford, England)*, **6** (Suppl 4), S7.

43 Spruessel, A., Steimann, G., Jung, M., Lee, S.A. et al. (2004) Tissue ischemia time affects gene and protein expression patterns within minutes following surgical tumor excision. *Biotechniques*, **36**, 1030–1037.

44 Wright, L.M., Kreikemeier, J.T. and Fimmel, C.J. (2007) A concise review of serum markers for hepatocellular cancer. *Cancer Detection and Prevention*, **31**, 35–44.

45 Zheng, G.X., Wang, C.X., Qu, X., Deng, X.M. et al. (2006) Establishment of serum protein pattern for screening colorectal cancer using SELDI-TOF-MS. *Experimental Oncology*, **28**, 282–287.

46 Su, Y., Shen, J., Qian, H., Ma, H. et al. (2007) Diagnosis of gastric cancer using decision tree classification of mass spectral data. *Cancer Science*, **98**, 37–43.

47 Scarlett, C.J., Saxby, A.J., Nielsen, A., Bell, C. *et al.* (2006) Proteomic profiling of cholangiocarcinoma: diagnostic potential of SELDI-TOF MS in malignant bile duct stricture. *Hepatology (Baltimore, Md)*, **44**, 658–666.

48 Qian, H.G., Shen, J., Ma, H., Ma, H.C. *et al.* (2005) Preliminary study on proteomics of gastric carcinoma and its clinical significance. *World Journal of Gastroenterology*, **11**, 6249–6253.

Part Three
TAAs and Their Usefulness

7
Tumor-associated Antigens in Childhood Cancer

Uta Behrends and Josef Mautner

7.1
Introduction to Childhood Cancer

7.1.1
Incidence, Etiology and Types of Cancer in Children

Childhood cancer accounts for only about 1% of all malignancies [1], but represents the leading cause of disease-related death in childhood. In the US and Germany, nearly 10 400 and 1800 children (<15 years) are newly diagnosed with cancer each year, respectively [2, 3]. For largely unknown reasons, the incidence of childhood cancer has been increasing slightly, but not significantly, over the last several decades [2, 4]. Confirmed risk factors for childhood malignancy are few, but a growing body of evidence suggests that genetic and environmental factors play a role [1, 2], with ionizing radiation being recognized as a significant risk factor for pediatric thyroid cancer [5] and leukemia [6].

The types of cancer commonly found in children are rarely the same as those reported in adult cancer patients. Acute leukemias account for about one-third (27–34%) of cancer diagnoses, followed by intracranial and intraspinal tumors (18–22%), lymphomas (12–13%) and various extracranial malignancies, including tumors of the peripheral nervous system (8–10%), soft tissue and other extraosseous sarcomas (6–8%), renal tumors (4–6%), bone malignancies (4–5%), germ cell tumors (3–7%), retinoblastoma (2%), hepatic tumors (1%), and other rare, malignant neoplasms (<1%) such as thyroid cancer, carcinoma and melanoma [2, 3].

7.1.2
Cure Rates, Treatment Failure and Toxicity

Similar to the treatment of cancer in adults, various disease-specific multimodal treatment protocols have been developed for affected children. Depending on the type of cancer, the therapeutic regimens include surgery, chemotherapy,

Tumor-Associated Antigens. Edited by Olivier Gires and Barbara Seliger
Copyright © 2009 WILEY-VCH Verlag GmbH & Co. KGaA, Weinheim
ISBN: 978-3-527-32084-4

radiotherapy, hematopoietic stem cell transplantation (HSCT), cytokines and/or targeted therapy with antibodies or small molecules in some cases. The 5-year survival rate among children with cancer increased from less than 25% in the 1960s to about 60% in the late 1970s, and the current 5-year survival probability is approximately 80% [3, 7].

Successful treatment of pediatric cancer requires maintaining a fine balance between the dual problems of treatment failure and toxicity. On the one hand, a failure to eliminate minimal residual disease (MRD) leads to cancer relapse and disease-related death in a significant number of cases. On the other hand, cure of cancer in the developing organism of a child is paid for dearly. The treatment's toxicity not only manifests in many acute side effects, but also results in various late complications that significantly compromise the health and quality of life of many childhood cancer survivors. Depending on the treatment protocol, late effects include damage to the developing brain, bones and endocrine organs as well as progressive diseases of the heart and kidney, adding to a significantly increased risk for secondary cancer [8]. Pediatric patients that underwent allogeneic HSCT are at a specifically high risk for severe late effects of cancer treatment [9]. Reducing the incidence of therapy-related complications and treatment failure is one of pediatric oncology's major challenges. Novel individualized treatment protocols that allow for a more detailed risk-stratification and for additional, well-tolerated treatment options are needed.

7.2
TAA for Pediatric Cancer Therapy

7.2.1
Potential Clinical Impact of Childhood TAA and Strategies for their Identification

Since tumor-associated antigens (TAA) can provide novel diagnostic markers as well as therapeutic targets, they may contribute to both better risk-stratification and more effective treatment of cancer in children. Targeted immunotherapy seems particularly promising in childhood cancers, since very low levels of MRD can be achieved in the majority of affected children, providing the most favorable conditions for an immune attack. Moreover, since targeted tumor immunotherapy is considered to be largely non-toxic, it might allow for intensified treatment of children with high-risk cancer as well as for the reduction of treatment toxicity in those with low- or intermediate-risk disease.

Due to a different spectrum of tumor entities in childhood and adult life, targeted immunotherapies developed for adult patients might not be appropriate in treating aggressive cancer in children, and diagnostic markers might not significantly overlap. The repertoire of TAA expressed in pediatric cancer as well as childhood immune responses against TAA need to be carefully assessed before reaching any conclusions regarding the possible clinical meanings of TAA in the pediatric setting.

In the late 1990s evidence accumulated that TAA originally identified in carcinoma tissues of adult patients also occur in common childhood malignancies [10]. Since then, a number of publications have documented the expression of cancer-testis (CT) antigens in childhood cancer types [11, 12] but only a few groups have searched for novel antigens that are uniquely expressed in childhood cancer [13–16]. A large part of the reason for this may be that the majority of childhood cancer types do not grow well *in vitro*, and moreover, the number of T cells that can be collected from a young child is limited. Both issues significantly hamper TAA-specific T-cell cloning.

The SEREX method first described by Sahin and colleagues [17], facilitates the systematic screening for antigens in childhood cancer, as recently demonstrated by our group for the two most common embryonal childhood malignancies [13, 14]. This method has two main advantages. First, the amount of tumor tissue and serum needed for the screening is small enough to allow analyses of cancer in infants and very young children. Second, since fresh frozen tumor biopsies are sufficient, the SEREX approach can be applied to tumor entities that fail to grow *in vitro*.

Because chromosomal translocations are common in pediatric tumors, and peptides spanning the break-point region might represent novel T cell epitopes, genetic approaches to the identification of novel diagnostic markers and immunotherapeutic targets are particularly promising [18, 19].

In this chapter we will discuss current knowledge relating to the expression and possible clinical meanings of TAA in the most common types of aggressive pediatric cancer.

7.2.2
TAA in Childhood Leukemia

Leukemia is the most common form of pediatric cancer and accounts for almost one-third of neoplastic diseases in children [2, 3]. Approximately 80% of the affected children are diagnosed with acute lymphoblastic leukemia (ALL), 15% with acute myelogenous leukemia (AML), and 5% with myelodysplastic/myeloproliferative diseases (MDS/MPS), chronic myelogenous leukemia (CML), or other types of leukemia [2, 3]. The 10-year survival probabilities for pediatric patients with leukemia, ALL or AML has increased to about 80, 85 and 60%, respectively [2, 3]. A major element in the treatment of childhood leukemia is polychemotherapy, complemented by craniospinal radiation, allogeneic HSCT and/or targeted therapy with antibodies or small molecules in selected cases [20, 21]. Optimal treatment of leukemia requires exact classification of the subtype at diagnosis and is guided by precise quantification of MRD during complete hematological remission [22]. Subtype-specific chromosomal translocations such as t(4;11), t(9;22), t(8;21) or t(15;17) play a central role in both leukemia subtyping and MRD diagnosis [21, 22]. Individual fusion genes generated by clone-specific B- and T-cell receptor rearrangements serve as genetic markers in successive quantification of MRD in ALL patients [23].

The graft-versus-leukemia (GVL) effect after allogeneic HSCT suggested that T cells play a major role in the rejection of leukemic cells [24]. This was further supported by the observations that donor lymphocyte infusions efficiently prevented promising candidate for immunotherapy of DLBCL [91], and SSX antigens might serve as targets in HL [92].

Viral antigens are attractive immunotherapeutic targets in EBV-associated malignancies, including nasopharynx carcinoma as well as EBV-positive BL, EBV-positive HL and EBV-associated post-transplantation lymphoproliferative disease (PTLD) [93]. These tumors consistently express at least one of nine latency proteins of the virus, and spontaneous virus replication, which leads to the expression of up to 80 lytic genes of EBV, occurs in a low percentage of tumor cells [93]. Latent as well as lytic cycle gene products are recognized by CD8+ and CD4+ T cells [94], and EBV-associated tumors in HSCT recipients have been successfully treated with donor-derived EBV-specific T-cell preparations [95]. Treatment of patients with HL or PTLD after solid organ transplantation using autologous EBV-specific T-cell lines, however, still remains challenging [95]. With the definition of the immunodominant CD4+ and CD8+ T-cell epitopes, T-cell lines enriched in these specificities may show improved clinical efficacy [94, 96].

7.2.5
TAA in Pediatric Neuroblastoma

Neuroblastoma (NB) is an embryonal tumor of the primordial neural crest and the most common extracranial solid tumor of childhood, comprising 8–10% of all childhood cancers. It shows diverse biological characteristics and a broad spectrum of clinical behavior. Some tumors may regress spontaneously, possibly reflecting immune recognition, induction of apoptosis and/or restored differentiation. The majority of tumors, however, grows very aggressively and requires intensive treatment [21]. The age of the patient at diagnosis, the stage of the disease and distinct molecular makers such as the amplification of the *MYCN* oncogene and a loss of heterozygocity on chromosome 1p are the most important risk factors currently used to stratify children with NB for the most appropriate treatment [97]. About 40% of the patients are stratified as high-risk and undergo an intensive multimodal treatment including high-dose chemotherapy and maintenance therapy with either antibodies to eradicate MRD [98, 99] or retinoic acid to induce differentiation [100]. Although the 10-year survival probability of children with NB has increased to approximately 75%, only about 40% of the high-risk patients older than 1 year survive longer than 5 years [3, 21, 101]. To improve clinical outcome a number of novel diagnostic and therapeutic strategies are currently under investigation [102].

Many aspects of the tumor's biological behavior, including lymphocytic infiltration and spontaneous regression in children less than 1 year of age, have led to the concept that harnessing immune recognition of NB cells may eradicate MRD in high-risk patients and prevent tumor relapse. Despite multiple defects in components of the antigen-processing machinery [103], NB cells were shown to be susceptible to cytotoxic effector mechanisms both *in vitro* and *in vivo* [104, 105]. Owing to the paucity of known NB-associated antigens, several cell-based vaccines have been evaluated. Although they proved to be safe in phase I clinical studies, they showed little clinical benefit in patients with active advanced or recurrent NB [106]. The induction of a graft-versus-tumor effect by allogeneic HSCT is a more recent approach to NB immunotherapy [107].

The identification of NB-associated antigens has paved the way towards targeted therapies which are expected to be more efficacious and to bear less risk of autoimmune responses against normal tissues. The disialoganglioside (GD2), a surface antigen of normal brain and peripheral neurons, is highly expressed on NB cells. Anti-GD2 monoclonal antibodies currently form the mainstay of NB immunotherapy in patients with primary refractory disease in the US [99], while analyses of other patient cohorts did not show any clear advantage of anti-GD2 antibody-based consolidation therapy [108]. Besides conferring direct antibody-mediated cytotoxicity to tumor cells, anti-GD2 antibodies have recently been engineered as vehicles to deliver anticancer agents, or to direct T-cell responses to the site of the tumor [99, 109, 110]. The correlation of *MYCN* oncogene amplification with rapid tumor progression and the requirement of high-level MYCN expression for the maintenance of the malignant phenotype, moreover, have suggested that MYCN might be an ideal target for vaccine therapy. Experiments demonstrating that MYCN-specific cytotoxic T cells are able to specifically lyse HLA-matched, *MYCN*-amplified tumor cells implied that MYCN-derived peptides can indeed serve as tumor-specific antigens [111]. Furthermore, the neural cell adhesion molecules NCAM and L1 (CD171) are highly expressed on NB cells and are currently being evaluated as targets for antibody-based and/or T cell-mediated immunotherapy [112, 113]. Last but not least, several CT-antigens have been shown to be expressed in NB [12, 114–116], which could serve as potential diagnostic markers [117, 118] and/or as targets for immunotherapeutic approaches [119]. However, humoral responses against the CT-antigen NY-ESO-1 were detected with only low frequency in patients with NB, which might indicate that this group of antigens is poorly immunogenic in NB patients [119].

By serological screening of a tumor proteome library novel isoforms of β-tubulin were identified and proposed as diagnostic or therapeutic targets [16]. In our own screening of NB cDNA libraries with autologous serum, several antigens were identified that might contribute to the detection of MRD (HuD antigens), to serological disease monitoring (Hu antigens and 018INX), to a better understanding of NB-associated paraneoplastic neuropathy (Hu antigens), and/or to NB immunotherapy (018INX) [13]. By allogeneic SEREX screening of a NB cell line library with serum from a NB patient that had received an autologous tumor vaccine, Lloyd Old's group identified 11 genes. Most of the serum-reactive clones coded for topoisomerase II alpha (TOP2A), suggesting that this gene is highly expressed in this tumor. The abundance of TOP2A transcripts in NB as well as the clinical effectiveness of TOP2A inhibitors on NB point to the possibility that TOP2A may be a biologically crucial gene in NB and a potential immune target [16]. More recently, serum from mice immunized with a NB cell-based vaccine after regulatory T cell depletion was used to screen a cDNA expression library constructed from the parental NB tumor cell line. Interestingly, a number of potentially oncogenic transcripts were identified including the *DEK* oncogene. The authors suggested that identification of protective immunogenic tumor antigens may require the use of serum from post-treatment or vaccinated subjects rather than serum collected at the time of diagnosis [120].

7.2.6
TAA in Rhabdomyosarcoma and other Soft Tissue Sarcomas

The heterogeneous group of soft tissue sarcomas accounts for about 6–8% of all childhood malignancies. Initial risk stratification is carried out by taking into account the histological subtype, site and size of the tumor as well as the age of the patient and the stage of the disease. Without any specific serological markers available, disease monitoring is largely limited to diagnostic imaging. Improved multimodal treatment results in an actual 10-year survival probability of about 70%, with remarkably high cure rates in patients with completely resectable low-risk tumors (> 90%) and dismal outcomes in the case of disseminated disease (<30% survival) [3, 21, 121, 122].

Approximately half of the children with soft tissue sarcoma are diagnosed with rhabdomyosarcoma (RMS), which is dominated by the embryonal type (80%). HLA class I and in some cases class II expression has been mainly observed in this less aggressive subtype [123] but *in vitro* treatment with interferon-γ was reported to restore HLA class I expression in HLA-negative RMS cell lines [124], and TAA were shown to be naturally processed and presented in RMS cells [125]. The use of cell-based RMS vaccines and/or RMS-specific adoptive T-cell transfer has therefore been proposed [126–128]. Allogeneic HSCT is currently undertaken in select cases and aims at a possible graft-versus-tumor effect [107, 129]. Since RMS have been shown to express members of the CT-antigen families MAGE, BAGE, GAGE, XAGE, SAGE, HAGE and/or NY-ESO-1/LAGE [12, 115, 125, 130, 131], and to overexpress growth factor receptors such as EGF-R, HER2 and HER3 [132], some patients might benefit from TAA-based immunotherapeutic strategies. *In vitro* studies indicated that antibody-mediated targeting of EGF-R can inhibit growth and induce apoptosis of RMS cells [132]. Alveolar RMS, the more aggressive subtype, is characterized by *PAX3/FKHR* or *PAX7/FKHR* fusion genes derived from t(2;13) and (t1;13) chromosomal translocations, respectively, and products of the *PAX3/FKHR* chromosomal break point region were shown to contain potential MHC class I and class II binding motifs [133]. Some of the T cells that recognize PAX/FKHR-derived peptides were demonstrated to lyse antigen-positive tumor cells *in vitro* [134], and T cells specific for PAX/FKHR fusion products can recognize and eradicate tumor cells bearing the PAX3/FKHR fusion protein in a preclinical model [19]. However, only a minority of the peptides corresponding to the predicted PAX3/FKHR-derived epitopes were recognized by T cells [133], and four patients with alveolar RMS, vaccinated in a pilot clinical trial with tumor-specific PAX3/FKHR peptides, did not show any significant immunological or clinical responses [135]. In an alternative T-cell immunotherapeutic approach, genetically modified T cells were recently generated that express a chimeric receptor composed of the antigen-binding domain of an antibody against the fetal acetylcholine receptor (fAChR) and the signaling domain of the human T-cell receptor ζ-chain. Expression of fAChR was found to be increased by chemotherapy *in vivo*, and fAChR-binding T cells were shown to lyse fAChR-positive tumor cells [136]. By serological screening of an embryonal RMS library, we recently identified a novel *PMS1* gene product that was probably generated by *trans*-splicing and that was found to be expressed in various embryonal childhood cancers but not in normal tissues. Therefore, this novel antigen might represent a potential diagnostic marker and/or therapeutic target for the treatment of embryonal childhood malignancies (unpublished data).

Several other sarcomas occurring in adults and children carry characteristic chromosomal translocations which give rise to tumor-specific fusion genes spanning the chromosomal break point regions [21]. As an example *SYT* is fused with a member of the *SSX* CT-gene family in synovial sarcoma [t(X;18)] [137], and epitopes derived from the fusion gene product might provide a basis for future targeted therapy [138]. In addition, many of the soft tissue sarcomas express several CT-antigens including PRAME, NY-ESO-1/LAGE-1, MAGE and GAGE antigens as well as the novel antigen DLG7 and products of the *SSX* genes, independently of any chromosomal translocation [139–141]. Although this group of antigens is highly expressed in sarcoma, they apparently fail to elicit detectable humoral immune responses in a significant number of patients [142]. Whether CT-antigens constitute relevant targets for adjuvant therapy or contribute to improved risk-adapted management of sarcoma patients remains to be seen [143].

7.2.7
TAA in Osteosarcoma

Osteosarcoma (OS) is the most frequent malignant bone tumor in children and adolescents and accounts for about 2–3% of all childhood malignancies [2, 3]. As in other types of sarcoma, diagnosis is largely based on diagnostic imaging and histological examination. Although the 10-year survival probability has increased to almost 70% in recent years by combining polychemotherapy with radical surgery [3, 21], the prognosis for patients with advanced disease has remained grim. Since no molecular prognostic markers have yet been clinically established, risk-stratification is largely based on the initial stage of the disease and on the response to chemotherapy [21]. Recently, an association of CXCR4 expression with initial metastases, P-glycoprotein expression with an increased risk for chemotherapy resistance, and survivin overexpression with poor clinical outcome has been suggested [144, 145].

OS are often infiltrated by T cells and can express HLA class I and class II antigens [146]. Remarkably, patients with HLA class I-positive tumors were reported to show better survival rates than those with HLA class I-negative tumors [147], implying that T cell-mediated immune responses may play an important role in the control of tumor growth. Accordingly, clinical trials are currently underway that use cytokines for immune stimulation and/or tumor cell-pulsed dendritic cell vaccines [148, 149].

Although data concerning the repertoire and function of TAA in OS is still scarce, increasing evidence suggests that these tumors might express several diagnostic and/or therapeutic targets. OS cells were shown to express several known TAA including members of the MAGE family [150], SART1 [151] and SART3 [152]. SART3-derived peptides have been demonstrated to activate HLA-A24-restricted T cells of patients with OS, and SART3-specific T cells were shown to recognize SART3-positive tumors. Moreover, the melanoma antigen MUC18 is expressed in OS [153], and anti-MUC18 antibodies were reported to inhibit the development of OS metastases in a preclinical model [153]. More recently, the heparin-binding growth factor midkine known to be involved in the pathogenesis of many tumor types was found to be overexpressed in OS, and anti-midkine antibodies suppressed the growth of OS cell lines *in vitro* [154].

Several groups have attempted to identify novel targets for immunotherapy by assessing humoral and cellular immune responses against OS [155–157]. Among a number of serologically-defined antigens, OSAA-3 and OSAA-5 were recognized exclusively by serum from OS patients but not by serum from healthy volunteers, suggesting that the immune responses to these two antigens were tumor-associated [157]. Serological screening of an OS cell line led to the identification of the *clusterin-associated protein 1* (CLUAP1) gene which was shown to be overexpressed in OS and other malignancies [156]. An OS-derived cDNA library was screened with autologous tumor-reactive CD8+ T cells and this led to the identification of papillomavirus binding factor (PBF), which is overexpressed in most OS and expressed at much lower levels in some normal tissues including ovary, pancreas, liver and spleen. PBF represents a shared tumor-associated antigen which may contribute to peptide-based vaccination and/or adoptive antigen-specific T-cell therapy of patients with OS and other bone and soft tissue tumors [158]. Whether these patients also harbor robust cellular immune responses to survivin as demonstrated in patients with NB [159] and carcinomas [160], and whether targeting this antigen by immunotherapy will improve clinical outcome, remains to be determined.

7.2.8
TAA in Tumors of the Ewing Family

The Ewing family tumors (EFT) represent the second most common bone cancer in children and adolescents, accounting for approximately 1–2% of all childhood malignancies [2, 3]. EFTs comprise a group of morphologically heterogeneous tumors that are characterized by non-random chromosomal translocations involving the *EWS* gene and one of several members of the ETS family of transcription factors. t(11;22)(q24;q12) is the most common translocation and leads to the formation of the EWS-FLI1 fusion protein, which contributes to EFT pathogenesis by modulating the expression of target genes [161, 162].

Despite combining high-dose chemotherapy for systemic control of disease with advanced surgical and/or radiotherapeutic approaches for local treatment, the actual 10-year survival probability is only about 65% with cure rates of less than 30% for metastatic disease [3, 21, 163]. Most interestingly, interferon-γ sensitizes resistant EFT cells to tumor necrosis factor apoptosis-inducing ligand (TRAIL)-induced apoptosis by up-regulation of caspase-8 without altering chemosensitivity, providing a rationale for the inclusion of interferon-γ in upcoming clinical trials with TRAIL [164]. To achieve graft-versus-tumor effects, allogeneic HSCT is being evaluated in very high-risk EFT, accepting the potentially life-threatening toxicity [107, 129, 165]. Early vaccination with tumor lysate-pulsed dendritic cells demonstrated that this approach is feasible and warrants further evaluation [128, 149]. However, targeted therapeutic strategies might be more effective and less toxic in intensively pretreated EFT patients [10, 165, 166].

There is compelling evidence that insulin-like growth factor (IGF) and its receptors (IGF-R) play a major role in growth and transformation of human neoplasia, including EFT. Therefore, interfering with the tyrosine kinase activity of IGF-R1 using anti-IGF-R-antibodies [167] or small molecules [168] is an attractive strategy for the treatment of EFT [166]. Apart from this, CD99 has been proposed as a potential therapeutic target. Ewing tumor cells consistently express CD99 and engagement of CD99 by agonistic antibodies induces massive apoptosis through caspase-independent mechanisms. However, expression of CD99 is also found on normal cells including T and B cells [166], and anti-CD99 monoclonal antibodies are able to induce apoptosis of T cells [169]. Thus, it is currently unknown whether anti-CD99 antibody treatment will show clinical benefit in patients with EFT. In contrast, the EWS/FLI1 fusion protein generated by chromosomal translocation provides a tumor-specific target for immunotherapeutic approaches and is associated with the oncogenesis of EFT. The unique peptides spanning the translocation breakpoint region contain potential MHC class I and class II binding motifs suggesting that they may serve as novel T-cell antigens [19, 170]. Although EWS/FLI1-specific immune responses were documented in 1/12 patients with EFT enrolled in an early clinical peptide vaccination trial, clinical responses were not significant [135]. Antisense oligonucleotides and siRNAs are thought to inhibit the generation of oncogenic chimeric transcription factors derived from *EWS* fusion genes and both are being evaluated in preclinical studies [171, 172]. Recently, the PBF identified in OS was shown to be overexpressed in EFT and therefore represents a potential therapeutic target. Moreover, PBF overexpression was found to be associated with a poor prognosis [173]. The majority of EFT shows mRNA expression of at least one CT-antigen, including members of the MAGE and GAGE family, but not of LAGE/NY-ESO-1 [12, 174].

7.2.9
TAA in Wilms' Tumor

Nephroblastoma (Wilms' tumor, WT) arises from pluripotent embryonic renal precursors and makes up more than 90% of all renal tumors in children, which account for 4–6% of all childhood cancers [3, 21]. With approximately 90% actual 10-year survival probability, the outcome of patients with WT is generally good [3], but the combined-modality treatment in high-risk patients is associated with severe early and late complications. Improved risk-stratification and reduction of the treatment's toxicity in children with low-risk tumors are major goals of current clinical investigations [21]. According to the histological subtype and stage of the disease, approximately 10% of the patients with WT are stratified as high-risk and currently demonstrate 5-year survival rates of less than 50%, despite highly toxic treatment. For this group of patients novel strategies are needed that aim at amplifying the therapeutic effect with minimal toxicity [175, 176]. Due to the lack of known TAA, clinical trials using tumor lysate-pulsed dendritic cells were undertaken in patients with WT [149].

Little information on the immunogenicity of aggressive WT and its TAA repertoire is currently available. The epidermal growth factor receptor 2 (Her2/neu, erbB2) was found to be expressed in the majority of WT and to contribute to tumor growth by promoting angiogenesis. In an animal model, systemic administration of anti-erbB2 monoclonal antibodies inhibited growth of WT xenografts *in vivo*, suggesting that erbB2 may serve as a therapeutic target [177]. CT-antigens of the MAGE family were shown to be expressed in WT, but nothing is known about their immunogenicity in affected children. In a recent SEREX screening of WT, we identified the nuclear protein NARG2 (unpublished data), which was formerly shown to be expressed in various fetal tissues, adult testis and, at very low levels, in adult lung tissue [178] and therefore may provide a therapeutic target in WT. Whether mutations in the tumor suppressor gene *WT1* will result in neoepitopes that are recognized by the immune system, is largely unknown.

7.3
Conclusion

At present, it is not known which of the many potential targets and immunotherapeutic approaches will prove most effective in the treatment of pediatric malignancies. The limited number of patients available for study will make it difficult to directly compare clinical efficacy of the various strategies. Although some clues may be provided by studies in larger adult patient cohorts, it may be problematic to extrapolate results from adults to children.

While many cancers in adults evolve over years or even decades, tumors in children usually develop over a much shorter period of time. Whether this rapid kinetics of tumor growth impinges on quality and magnitude of the antitumoral immune response especially in a still developing immune system is currently not known. Although CT-antigens are expressed in a variety of pediatric cancers, immune responses against these antigens have rarely been detected in children. Whether this is simply a reflection of the low number of cases analyzed, or the consequence of immunological tolerance against antigens that were expressed rather recently during embryonal development, remains to be elucidated.

With increasing clinical evidence that the immune system truly can eradicate tumor cells, coupled with the identification of TAA and a progressively improving understanding of the molecular and cellular basis of the immune system, immune-mediated targeted therapy will sooner rather than later become a standard tool for the pediatric oncologist.

References

1 Buka, I., Koranteng, S. and Osornio Vargas, A.R. (2007) Trends in childhood cancer incidence: review of environmental linkages. *Pediatric Clinics of North America*, **54**, 177–203.
2 Linabery, A.M. and Ross, J.A. (2008) Trends in childhood cancer incidence in the U.S. (1992–2004). *Cancer*, **112**, 416–432.
3 GCCR , (2006) German Childhood Cancer Registry, http://www.kinderkrebsregister.de/english/.
4 Steliarova-Foucher, E., Stiller, C., Lacour, B. and Kaatsch, P. (2005) International Classification of Childhood Cancer. *Cancer*, **103**, 1457–1467.
5 Vasko, V., Bauer, A.J., Tuttle, R.M. and Francis, G.L. (2007) Papillary and follicular thyroid cancers in children. *Endocrine Development*, **10**, 140–172.
6 Belson, M., Kingsley, B. and Holmes, A. (2007) Risk factors for acute leukemia in children: a review. *Environmental Health Perspectives*, **115**, 138–145.
7 McGregor, L.M., Metzger, M.L., Sanders, R. and Santana, V.M. (2007) Pediatric cancers in the new millennium: dramatic progress, new challenges. *Oncology (Williston Park)*, **21**, 809–820. discussion 820, 823–824.
8 Bhatia, S. (2005) Cancer survivorship – pediatric issues. *Hematology American Society of Hematology Education Program*, 507–515.
9 Leung, W., Ahn, H., Rose, S.R., Phipps, S., Smith, T., Gan, K., O'Connor, M., Hale, G.A., Kasow, K.A., Barfield, R.C., Madden, R.M. and Pui, C.H. (2007) A prospective cohort study of late sequelae of pediatric allogeneic, hematopoietic stem cell transplantation. *Medicine (Baltimore)*, **86**, 215–224.
10 Mackall, C.L. and Helman, L.J. (2000) Targeting pediatric malignancies for T cell-mediated immune responses. *Current Oncology Reports*, **2**, 539–546.
11 Rousseau, R.F. and Brenner, M.K. (2005) Vaccine therapies for pediatric malignancies. *Cancer Journal (Sudbury, Mass)*, **11**, 331–339.
12 Jacobs, J.F., Brasseur, F., Hulsbergen-van de Kaa, C.A., van de Rakt, M.W., Figdor, C.G., Adema, G.J., Hoogerbrugge, P.M., Coulie, P.G. and de Vries, I.J. (2007) Cancer-germline gene expression in pediatric solid tumors using quantitative real-time PCR. *International Journal of Cancer*, **120**, 67–74.
13 Behrends, U., Jandl, T., Golbeck, A., Lechner, B., Muller-Weihrich, S., Schmid, I., Till, H., Berthold, F., Voltz, R. and Mautner, J.M. (2002) Novel products of the HUD, HUC, NNP-1 and alpha-internexin genes identified by autologous antibody screening of a pediatric neuroblastoma library. *International Journal of Cancer*, **100**, 669–677.
14 Behrends, U., Schneider, I., Rossler, S., Frauenknecht, H., Golbeck, A., Lechner, B., Eigenstetter, G., Zobywalski, C., Muller-Weihrich, S., Graubner, U., Schmid, I., Sackerer, D., Spath, M., Goetz, C., Prantl, F., Asmuss, H.P., Bise, K. and Mautner, J. (2003) Novel tumor antigens identified by autologous antibody screening of childhood medulloblastoma cDNA libraries. *International Journal of Cancer*, **106**, 244–251.
15 Prasannan, L., Misek, D.E., Hinderer, R., Michon, J., Geiger, J.D. and Hanash, S.M. (2000) Identification of beta-tubulin isoforms as tumor antigens in neuroblastoma. *Clinical Cancer Research*, **6**, 3949–3956.
16 Chen, Y.T. (2004) Identification of human tumour antigens by serological expression, cloning: an online review on SEREX http://www.cancerimmunity.org/SEREX/.
17 Sahin, U., Tureci, O., Schmitt, H., Cochlovius, B., Johannes, T., Schmits, R., Stenner, F., Luo, G., Schobert, I. and Pfreundschuh, M. (1995) Human

neoplasms elicit multiple specific immune responses in the autologous host. *Proceedings of the National Academy of Sciences of the United States of America*, **92**, 11810–11813.

18 Kawakami, Y., Fujita, T., Matsuzaki, Y., Sakurai, T., Tsukamoto, M., Toda, M. and Sumimoto, H. (2004) Identification of human tumor antigens and its implications for diagnosis and treatment of cancer. *Cancer Science*, **95**, 784–791.

19 Mackall, C., Berzofsky, J. and Helman, L.J. (2000) Targeting tumor specific translocations in sarcomas in pediatric patients for immunotherapy. *Clinical Orthopaedics and Related Research*, 25–31.

20 Kuriakose, P. (2005) Targeted therapy for hematologic malignancies. *Cancer Control*, **12**, 82–90.

21 Pizzo, P.A. and Poplack, D.G. (2005) *Principles and Practice of Pediatric Oncology*, 5th edn, Lippincott Williams and Wilkins, Philadelphia.

22 Haferlach, C., Rieder, H., Lillington, D.M., Dastugue, N., Hagemeijer, A., Harbott, J., Stilgenbauer, S., Knuutila, S., Johansson, B. and Fonatsch, C. (2007) Proposals for standardized protocols for cytogenetic analyses of acute leukemias, chronic lymphocytic leukemia, chronic myeloid leukemia, chronic myeloproliferative disorders, and myelodysplastic syndromes. *Genes, Chromosomes & Cancer*, **46**, 494–499.

23 Stanulla, M., Cario, G., Meissner, B., Schrauder, A., Moricke, A., Riehm, H. and Schrappe, M. (2007) Integrating molecular information into treatment of childhood acute lymphoblastic leukemia – a perspective from the BFM Study Group. *Blood Cells, Molecules & Diseases*, **39**, 160–163.

24 Horowitz, M.M., Gale, R.P., Sondel, P.M., Goldman, J.M., Kersey, J., Kolb, H.J., Rimm, A.A., Ringden, O., Rozman, C., Speck, B. *et al.* (1990) Graft-versus-leukemia reactions after bone marrow transplantation. *Blood*, **75**, 555–562.

25 Kolb, H.J., Schmid, C., Barrett, A.J. and Schendel, D.J. (2004) Graft-versus-leukemia reactions in allogeneic chimeras. *Blood*, **103**, 767–776.

26 Slavin, S., Morecki, S., Weiss, L., Shapira, M.Y., Resnick, I. and Or, R. (2004) Nonmyeloablative stem cell transplantation: reduced-intensity conditioning for cancer immunotherapy–from bench to patient bedside. *Seminars in Oncology*, **31**, 4–21.

27 Blair, A., Goulden, N.J., Libri, N.A., Oakhill, A. and Pamphilon, D.H. (2005) Immunotherapeutic strategies in acute lymphoblastic leukaemia relapsing after stem cell transplantation. *Blood Reviews*, **19**, 289–300.

28 Goulmy, E. (2006) Minor histocompatibility antigens: from transplantation problems to therapy of cancer. *Human Immunology*, **67**, 433–438.

29 Bleakley, M. and Riddell, S.R. (2004) Molecules and mechanisms of the graft-versus-leukaemia effect. *Nature Reviews. Cancer*, **4**, 371–380.

30 Barrett, A.J. and Rezvani, K. (2007) Translational mini-review series on vaccines: Peptide vaccines for myeloid leukaemias. *Clinical and Experimental Immunology*, **148**, 189–198.

31 Van Driessche, A., Gao, L., Stauss, H.J., Ponsaerts, P., Van Bockstaele, D.R., Berneman, Z.N. and Van Tendeloo, V.F. (2005) Antigen-specific cellular immunotherapy of leukemia. *Leukemia*, **19**, 1863–1871.

32 Molldrem, J., Dermime, S., Parker, K., Jiang, Y.Z., Mavroudis, D., Hensel, N., Fukushima, P. and Barrett, A.J. (1996) Targeted T-cell therapy for human leukemia: cytotoxic T lymphocytes specific for a peptide derived from proteinase 3 preferentially lyse human myeloid leukemia cells. *Blood*, **88**, 2450–2457.

33 Molldrem, J.J., Clave, E., Jiang, Y.Z., Mavroudis, D., Raptis, A., Hensel, N., Agarwala, V. and Barrett, A.J. (1997) Cytotoxic T lymphocytes specific for a

nonpolymorphic proteinase 3 peptide preferentially inhibit chronic myeloid leukemia colony-forming units. *Blood*, **90**, 2529–2534.

34 Rezvani, K., Price, D.A., Brenchley, J.M., Kilical, Y., Gostick, E., Sconocchia, G., Hansmann, K., Kurlander, R., Douek, D.C. and Barrett, A.J. (2007) Transfer of PR1-specific T-cell clones from donor to recipient by stem cell transplantation and association with GvL activity. *Cytotherapy*, **9**, 245–251.

35 Hutchings, Y., Osada, T., Woo, C.Y., Clay, T.M., Lyerly, H.K. and Morse, M.A. (2007) Immunotherapeutic targeting of Wilms' tumor protein. *Current Opinion in Molecular Therapeutics*, **9**, 62–69.

36 Ariyaratana, S. and Loeb, D.M. (2007) The role of the Wilms tumour gene (WT1) in normal and malignant haematopoiesis. *Expert Review of Molecular Medical*, **9**, 1–17.

37 Oka, Y., Tsuboi, A., Taguchi, T., Osaki, T., Kyo, T., Nakajima, H., Elisseeva, O.A., Oji, Y., Kawakami, M., Ikegame, K., Hosen, N., Yoshihara, S., Wu, F., Fujiki, F., Murakami, M., Masuda, T., Nishida, S., Shirakata, T., Nakatsuka, S., Sasaki, A., Udaka, K., Dohy, H., Aozasa, K., Noguchi, S., Kawase, I. and Sugiyama, H. (2004) Induction of WT1 (Wilms' tumor gene)-specific cytotoxic T lymphocytes by WT1 peptide vaccine and the resultant cancer regression. *Proceedings of the National Academy of Sciences of the United States of America*, **101**, 13885–13890.

38 Rezvani, K., Yong, A.S., Mielke, S., Savani, B.N., Musse, L., Superata, J., Jafarpour, B., Boss, C. and Barrett, A.J. (2008) Leukemia-associated antigen-specific T-cell responses following combined PR1 and WT1 peptide vaccination in patients with myeloid malignancies. *Blood*, **111**, 236–242.

39 Greiner, J., Ringhoffer, M., Taniguchi, M., Hauser, T., Schmitt, A., Dohner, H. and Schmitt, M. (2003) Characterization of several leukemia-associated antigens inducing humoral immune responses in acute and chronic myeloid leukemia. *International Journal of Cancer*, **106**, 224–231.

40 Schmitt, M., Schmitt, A., Rojewski, M.T., Chen, J., Giannopoulos, K., Fei, F., Yu, Y., Gotz, M., Heyduk, M., Ritter, G., Speiser, D.E., Gnjatic, S., Guillaume, P., Ringhoffer, M., Schlenk, R.F., Liebisch, P., Bunjes, D., Shiku, H., Dohner, H. and Greiner, J. (2008) RHAMM-R3 peptide vaccination in patients with acute myeloid leukemia, myelodysplastic syndrome, and multiple myeloma elicits immunologic and clinical responses. *Blood*, **111**, 1357–1365.

41 Guinn, B.A., Collin, J.F., Li, G., Rees, R.C. and Mufti, G.J. (2002) Optimised SEREX technique for the identification of leukaemia-associated antigens. *Journal of Immunological Methods*, **264**, 207–214.

42 Guinn, B.A., Bland, E.A., Lodi, U., Liggins, A.P., Tobal, K., Petters, S., Wells, J.W., Banham, A.H. and Mufti, G.J. (2005) Humoral detection of leukaemia-associated antigens in presentation acute myeloid leukaemia. *Biochemical and Biophysical Research Communications*, **335**, 1293–1304.

43 Dohnal, A.M., Inthal, A., Felzmann, T., Glatt, S., Sommergruber, W., Mann, G., Gadner, H. and Panzer-Grumayer, E.R. (2006) Leukemia-associated antigenic isoforms induce a specific immune response in children with T-ALL. *International Journal of Cancer*, **119**, 2870–2877.

44 Takahashi, H., Furukawa, T., Yano, T., Sato, N., Takizawa, J., Kurasaki, T., Abe, T., Narita, M., Masuko, M., Koyama, S., Toba, K., Takahashi, M. and Aizawa, Y. (2007) Identification of an overexpressed gene, HSPA4L, the product of which can provoke prevalent humoral immune responses in leukemia patients. *Experimental Hematology*, **35**, 1091–1099.

45 Niemeyer, P., Tureci, O., Eberle, T., Graf, N., Pfreundschuh, M. and Sahin, U.

(2003) Expression of serologically identified tumor antigens in acute leukemias. *Leukemia Research*, **27**, 655–660.

46 Pagano, L., Fianchi, L., Caira, M., Rutella, S. and Leone, G. (2007) The role of Gemtuzumab Ozogamicin in the treatment of acute myeloid leukemia patients. *Oncogene*, **26**, 3679–3690.

47 Bocchia, M., Korontsvit, T., Xu, Q., Mackinnon, S., Yang, S.Y., Sette, A. and Scheinberg, D.A. (1996) Specific human cellular immunity to bcr-abl oncogene-derived peptides. *Blood*, **87**, 3587–3592.

48 Yotnda, P., Firat, H., Garcia-Pons, F., Garcia, Z., Gourru, G., Vernant, J.P., Lemonnier, F.A., Leblond, V. and Langlade-Demoyen, P. (1998) Cytotoxic T cell response against the chimeric p210 BCR-ABL protein in patients with chronic myelogenous leukemia. *The Journal of Clinical Investigation*, **101**, 2290–2296.

49 Bocchia, M., Gentili, S., Abruzzese, E., Fanelli, A., Iuliano, F., Tabilio, A., Amabile, M., Forconi, F., Gozzetti, A., Raspadori, D., Amadori, S. and Lauria, F. (2005) Effect of a p210 multipeptide vaccine associated with imatinib or interferon in patients with chronic myeloid leukaemia and persistent residual disease: a multicentre observational trial. *Lancet*, **365**, 657–662.

50 Campana, D. (2007) Monitoring minimal residual disease in pediatric hematologic malignancies. *Clinical Advances in Hematology & Oncology*, **5**, 876–877. 915.

51 Steinbach, D., Schramm, A., Eggert, A., Onda, M., Dawczynski, K., Rump, A., Pastan, I., Wittig, S., Pfaffendorf, N., Voigt, A., Zintl, F. and Gruhn, B. (2006) Identification of a set of seven genes for the monitoring of minimal residual disease in pediatric acute myeloid leukemia. *Clinical Cancer Research*, **12**, 2434–2441.

52 Greiner, J., Ringhoffer, M., Simikopinko, O., Szmaragowska, A., Huebsch, S., Maurer, U., Bergmann, L. and Schmitt, M. (2000) Simultaneous expression of different immunogenic antigens in acute myeloid leukemia. *Experimental Hematology*, **28**, 1413–1422.

53 Hargrave, D.R. and Zacharoulis, S. (2007) Pediatric CNS tumors: current treatment and future directions. *Expert Reviews of Neuroethics*, **7**, 1029–1042.

54 Walker, P.R., Calzascia, T. and Dietrich, P.Y. (2002) All in the head: obstacles for immune rejection of brain tumours. *Immunology*, **107**, 28–38.

55 Khan-Farooqi, H.R., Prins, R.M. and Liau, L.M. (2005) Tumor immunology, immunomics and targeted immunotherapy for central nervous system malignancies. *Neurological Research*, **27**, 692–702.

56 Yu, J.S., Liu, G., Ying, H., Yong, W.H., Black, K.L. and Wheeler, C.J. (2004) Vaccination with tumor lysate-pulsed dendritic cells elicits antigen-specific, cytotoxic T-cells in patients with malignant glioma. *Cancer Research*, **64**, 4973–4979.

57 Rutkowski, S., De Vleeschouwer, S., Kaempgen, E., Wolff, J.E., Kuhl, J., Demaerel, P., Warmuth-Metz, M., Flamen, P., Van Calenbergh, F., Plets, C., Sorensen, N., Opitz, A. and Van Gool, S.W. (2004) Surgery and adjuvant dendritic cell-based tumour vaccination for patients with relapsed malignant glioma, a feasibility study. *British Journal of Cancer*, **91**, 1656–1662.

58 Fenstermaker, R.A. and Ciesielski, M.J. (2004) Immunotherapeutic strategies for malignant glioma. *Cancer Control*, **11**, 181–191.

59 Herold-Mende, C., Mueller, M.M., Bonsanto, M.M., Schmitt, H.P., Kunze, S. and Steiner, H.H. (2002) Clinical impact and functional aspects of tenascin-C expression during glioma progression. *International Journal of Cancer*, **98**, 362–369.

60 Reardon, D.A., Zalutsky, M.R. and Bigner, D.D. (2007) Antitenascin-C monoclonal antibody radioimmunotherapy for

malignant glioma patients. *Expert Review of Anticancer Therapy*, **7**, 675–687.

61 Martens, T., Schmidt, N.O., Eckerich, C., Fillbrandt, R., Merchant, M., Schwall, R., Westphal, M. and Lamszus, K. (2006) A novel one-armed anti-c-Met antibody inhibits glioblastoma growth *in vivo*. *Clinical Cancer Research*, **12**, 6144–6152.

62 Voelzke, W.R., Petty, W.J. and Lesser, G.J. (2008) Targeting the Epidermal Growth Factor Receptor in High-Grade Astrocytomas. *Current Treatment Options in Oncology*, **9**, 23–31.

63 Wu, A.H., Xiao, J., Anker, L., Hall, W.A., Gregerson, D.S., Cavenee, W.K., Chen, W. and Low, W.C. (2006) Identification of EGFRvIII-derived CTL epitopes restricted by HLA A0201 for dendritic cell based immunotherapy of gliomas. *Journal of Neuro-Oncology*, **76**, 23–30.

64 Yajima, N., Yamanaka, R., Mine, T., Tsuchiya, N., Homma, J., Sano, M., Kuramoto, T., Obata, Y., Komatsu, N., Arima, Y., Yamada, A., Shigemori, M., Itoh, K. and Tanaka, R. (2005) Immunologic evaluation of personalized peptide vaccination for patients with advanced malignant glioma. *Clinical Cancer Research*, **11**, 5900–5911.

65 Sasaki, M., Nakahira, K., Kawano, Y., Katakura, H., Yoshimine, T., Shimizu, K., Kim, S.U. and Ikenaka, K. (2001) MAGE-E1, a new member of the melanoma-associated antigen gene family and its expression in human glioma. *Cancer Research*, **61**, 4809–4814.

66 Sahin, U., Koslowski, M., Tureci, O., Eberle, T., Zwick, C., Romeike, B., Moringlane, J.R., Schwechheimer, K., Feiden, W. and Pfreundschuh, M. (2000) Expression of cancer testis genes in human brain tumors. *Clinical Cancer Research*, **6**, 3916–3922.

67 Scarcella, D.L., Chow, C.W., Gonzales, M.F., Economou, C., Brasseur, F. and Ashley, D.M. (1999) Expression of MAGE and GAGE in high-grade brain tumors: a potential target for specific immunotherapy and diagnostic markers. *Clinical Cancer Research*, **5**, 335–341.

68 Zhang, J.G., Eguchi, J., Kruse, C.A., Gomez, G.G., Fakhrai, H., Schroter, S., Ma, W., Hoa, N., Minev, B., Delgado, C., Wepsic, H.T., Okada, H. and Jadus, M.R. (2007) Antigenic profiling of glioma cells to generate allogeneic vaccines or dendritic cell-based therapeutics. *Clinical Cancer Research*, **13**, 566–575.

69 Schmits, R., Cochlovius, B., Treitz, G., Regitz, E., Ketter, R., Preuss, K.D., Romeike, B.F. and Pfreundschuh, M. (2002) Analysis of the antibody repertoire of astrocytoma patients against antigens expressed by gliomas. *International Journal of Cancer*, **98**, 73–77.

70 Struss, A.K., Romeike, B.F., Munnia, A., Nastainczyk, W., Steudel, W.I., Konig, J., Ohgaki, H., Feiden, W., Fischer, U. and Meese, E. (2001) PHF3-specific antibody responses in over 60% of patients with glioblastoma multiforme. *Oncogene*, **20**, 4107–4114.

71 Fischer, U., Struss, A.K., Hemmer, D., Pallasch, C.P., Steudel, W.I. and Meese, E. (2001) Glioma-expressed antigen 2 (GLEA2): a novel protein that can elicit immune responses in glioblastoma patients and some control. *Clinical and Experimental Immunology*, **126**, 206–213.

72 Crawford, J.R., MacDonald, T.J. and Packer, R.J. (2007) Medulloblastoma in childhood: new biological advances. *Lancet Neurology*, **6**, 1073–1085.

73 Raffaghello, L., Nozza, P., Morandi, F., Camoriano, M., Wang, X., Garre, M.L., Cama, A., Basso, G., Ferrone, S., Gambini, C. and Pistoia, V. (2007) Expression and functional analysis of human leukocyte antigen class I antigen-processing machinery in medulloblastoma. *Cancer Research*, **67**, 5471–5478.

74 Ahmed, N., Ratnayake, M., Savoldo, B., Perlaky, L., Dotti, G., Wels, W.S., Bhattacharjee, M.B., Gilbertson, R.J., Shine, H.D., Weiss, H.L., Rooney, C.M., Heslop, H.E. and Gottschalk, S. (2007)

Regression of experimental medulloblastoma following transfer of HER2-specific Tcells. *Cancer Research*, **67**, 5957–5964.

75 Bodey, B., Bodey, V., Siegel, S.E. and Kaiser, H.E. (2004) Survivin expression in childhood medulloblastomas: a possible diagnostic and prognostic marker. *In Vivo*, **18**, 713–718.

76 Li, X.N., Shu, Q., Su, J.M., Adesina, A.M., Wong, K.K., Perlaky, L., Antalffy, B.A., Blaney, S.M. and Lau, C.C. (2007) Differential expression of survivin splice isoforms in medulloblastomas. *Neuropathology and Applied Neurobiology*, **33**, 67–76.

77 Haberler, C., Slavc, I., Czech, T., Gelpi, E., Heinzl, H., Budka, H., Urban, C., Scarpatetti, M., Ebetsberger-Dachs, G., Schindler, C., Jones, N., Klein-Franke, A., Maier, H., Jauk, B., Kiefer, A. and Hainfellner, J.A. (2006) Histopathological prognostic factors in medulloblastoma: high expression of survivin is related to unfavourable outcome. *European Journal of Cancer (Oxford, England: 1990)*, **42**, 2996–3003.

78 Bradley, M.B. and Cairo, M.S. (2008) Stem cell transplantation for pediatric lymphoma: past, present and future. *Bone Marrow Transplantation*, **41**, 149–158.

79 Miles, R.R., Cairo, M.S., Satwani, P., Zwick, D.L., Lones, M.A., Sposto, R., Abromovitch, M., Tripp, S., Angiolillo, A.L., Roman, E., Davenport, V. and Perkins, S.L. (2007) Immunophenotypic identification of possible therapeutic targets in paediatric non-Hodgkin lymphomas: a children's oncology group report. *British Journal of Haematology*, **138**, 506–512.

80 Fanale, M.A. and Younes, A. (2007) Monoclonal antibodies in the treatment of non-Hodgkin's lymphoma. *Drugs*, **67**, 333–350.

81 Cashen, A.F. and Bartlett, N.L. (2007) Therapy of relapsed Hodgkin lymphoma. *Blood Reviews*, **21**, 233–243.

82 Cooney-Qualter, E., Krailo, M., Angiolillo, A., Fawwaz, R.A., Wiseman, G., Harrison, L., Kohl, V., Adamson, P.C., Ayello, J., vande Ven, C., Perkins, S.L. and Cairo, M.S. (2007) A phase I study of 90yttrium-ibritumomab-tiuxetan in children and adolescents with relapsed/refractory CD20-positive non-Hodgkin's lymphoma: a Children's Oncology Group study. *Clinical Cancer Research*, **13**, 5652s–5660s.

83 Neelapu, S.S. and Kwak, L.W. (2007) Vaccine therapy for B-cell lymphomas: next-generation strategies. *Hematology American Society of Hematology Education Program*, **2007**, 243–249.

84 Veelken, H. (2003) Active immunotherapy in follicular lymphoma. *Seminars in Cancer Biology*, **13**, 241–247.

85 Kanter, G., Yang, J., Voloshin, A., Levy, S., Swartz, J.R. and Levy, R. (2007) Cell-free production of scFv fusion proteins: an efficient approach for personalized lymphoma vaccines. *Blood*, **109**, 3393–3399.

86 Hishizawa, M., Imada, K., Sakai, T., Nishikori, M., Arima, N., Tsudo, M., Ishikawa, T. and Uchiyama, T. (2006) Antibody responses associated with the graft-versus-leukemia effect in adult T-cell leukemia. *International Journal of Hematology*, **83**, 351–355.

87 Liggins, A.P., Guinn, B.A. and Banham, A.H. (2005) Identification of lymphoma-associated antigens using SEREX. *Methods in Molecular Medicine*, **115**, 109–128.

88 Kersten, C., Delabie, J., Gaudernack, G., Smeland, E.B. and Fossa, A. (2004) Analysis of the autoantibody repertoire in Burkitt's lymphoma patients: frequent response against the transcription factor ATF-2. *Cancer Immunology, Immunotherapy*, **53**, 1119–1126.

89 Huang, S., Preuss, K.D., Xie, X., Regitz, E. and Pfreundschuh, M. (2002) Analysis of the antibody repertoire of lymphoma patients. *Cancer Immunology, Immunotherapy*, **51**, 655–662.

90 Eichmuller, S., Usener, D., Dummer, R., Stein, A., Thiel, D. and Schadendorf, D. (2001) Serological detection of cutaneous T-cell lymphoma-associated antigens. *Proceedings of the National Academy of Sciences of the United States of America*, **98**, 629–634.

91 Cooper, C.D., Liggins, A.P., Ait-Tahar, K., Roncador, G., Banham, A.H. and Pulford, K. (2006) PASD1, a DLBCL-associated cancer testis antigen and candidate for lymphoma immunotherapy. *Leukemia*, **20**, 2172–2174.

92 Colleoni, G.W., Capodieci, P., Tickoo, S., Cossman, J., Filippa, D.A. and Ladanyi, M. (2002) Expression of SSX genes in the neoplastic cells of Hodgkin's lymphoma. *Human Pathology*, **33**, 496–502.

93 Rickinson, A.B. and Kieff, E. (2006) Epstein-Barr virus, in *Field's Virology*, 5th edn, Lippincott-Raven, Philadelphia, PA. pp. 2655–2700.

94 Hislop, A.D., Taylor, G.S., Sauce, D. and Rickinson, A.B. (2007) Cellular responses to viral infection in humans: lessons from Epstein-Barr virus. *Annual Review of Immunology*, **25**, 587–617.

95 Gottschalk, S., Heslop, H.E. and Rooney, C.M. (2005) Adoptive immunotherapy for EBV-associated malignancies. *Leukemia & Lymphoma*, **46**, 1–10.

96 Adhikary, D., Behrends, U., Boerschmann, H., Pfunder, A., Burdach, S., Moosmann, A., Witter, K., Bornkamm, G.W. and Mautner, J. (2007) Immunodominance of lytic cycle antigens in Epstein-Barr virus,-specific CD4 + T cell preparations for therapy *PLoS ONE*, **2**, e583.

97 Schwab, M., Westermann, F., Hero, B. and Berthold, F. (2003) Neuroblastoma: biology and molecular and chromosomal pathology. *The Lancet Oncology*, **4**, 472–480.

98 Kushner, B.H., Kramer, K. and Cheung, N.K. (2001) Phase II trial of the anti-G(D)2 monoclonal antibody 3F8 and granulocyte-macrophage colony-stimulating factor for neuroblastoma. *Journal of Clinical Oncology*, **19**, 4189–4194.

99 Modak, S. and Cheung, N.K. (2007) Disialoganglioside directed immunotherapy of neuroblastoma. *Cancer Investigation*, **25**, 67–77.

100 Matthay, K.K., Villablanca, J.G., Seeger, R.C., Stram, D.O., Harris, R.E., Ramsay, N.K., Swift, P., Shimada, H., Black, C.T., Brodeur, G.M., Gerbing, R.B. and Reynolds, C.P. (1999) Treatment of high-risk neuroblastoma with intensive chemotherapy, radiotherapy, autologous bone marrow transplantation, and 13-cis-retinoic acid. Children's Cancer Group *The New England Journal of Medicine*, **341**, 1165–1173.

101 Berthold, F., Hero, B., Kremens, B., Handgretinger, R., Henze, G., Schilling, F.H., Schrappe, M., Simon, T. and Spix, C. (2003) Long-term results and risk profiles of patients in five consecutive trials (1979–1997) with stage 4 neuroblastoma over 1 year of age. *Cancer Letters*, **197**, 11–17.

102 Henry, M.C., Tashjian, D.B. and Breuer, C.K. (2005) Neuroblastoma update. *Current Opinion in Oncology*, **17**, 19–23.

103 Raffaghello, L., Prigione, I., Airoldi, I., Camoriano, M., Morandi, F., Bocca, P., Gambini, C., Ferrone, S. and Pistoia, V. (2005) Mechanisms of immune evasion of human neuroblastoma. *Cancer Letters*, **228**, 155–161.

104 Prigione, I., Corrias, M.V., Airoldi, I., Raffaghello, L., Morandi, F., Bocca, P., Cocco, C., Ferrone, S. and Pistoia, V. (2004) Immunogenicity of human neuroblastoma. *Annals of the New York Academy of Sciences*, **1028**, 69–80.

105 Wolfl, M., Jungbluth, A.A., Garrido, F., Cabrera, T., Meyen-Southard, S., Spitz, R., Ernestus, K. and Berthold, F. (2005) Expression of MHC class I, MHC class II, and cancer germline antigens in neuroblastoma. *Cancer Immunology, Immunotherapy*, **54**, 400–406.

106 Verneris, M.R. and Wagner, J.E. (2007) Recent developments in cell-based immune therapy for neuroblastoma. *J Neuroimmune Pharmacol*, **2**, 134–139.

107 Lang, P., Pfeiffer, M., Muller, I., Schumm, M., Ebinger, M., Koscielniak, E., Feuchtinger, T., Foll, J., Martin, D. and Handgretinger, R. (2006) Haploidentical stem cell transplantation in patients with pediatric solid tumors: preliminary results of a pilot study and analysis of graft versus tumor effects. *Klinische Pädiatrie*, **218**, 321–326.

108 Simon, T., Hero, B., Faldum, A., Handgretinger, R., Schrappe, M., Niethammer, D. and Berthold, F. (2005) Infants with stage 4 neuroblastoma: the impact of the chimeric anti-GD2-antibody ch14.18 consolidation therapy. *Klinische Pädiatrie*, **217**, 147–152.

109 Johnson, E., Dean, S.M. and Sondel, P.M. (2007) Antibody-based immunotherapy in high-risk neuroblastoma. *Expert Reviews in Molecular Medicine*, **9**, 1–21.

110 Manzke, O., Russello, O., Leenen, C., Diehl, V., Bohlen, H. and Berthold, F. (2001) Immunotherapeutic strategies in neuroblastoma: antitumoral activity of deglycosylated Ricin A conjugated anti-GD2 antibodies and anti-CD3xanti-GD2 bispecific antibodies. *Medical and Pediatric Oncology*, **36**, 185–189.

111 Sarkar, A.K. and Nuchtern, J.G. (2000) Lysis of MYCN-amplified neuroblastoma cells by MYCN peptide-specific cytotoxic T lymphocytes. *Cancer Research*, **60**, 1908–1913.

112 Jensen, M. and Berthold, F. (2007) Targeting the neural cell adhesion molecule in cancer. *Cancer Letters*, **258**, 9–21.

113 Park, J.R., Digiusto, D.L., Slovak, M., Wright, C., Naranjo, A., Wagner, J., Meechoovet, H.B., Bautista, C., Chang, W.C., Ostberg, J.R. and Jensen, M.C. (2007) Adoptive transfer of chimeric antigen receptor re-directed cytolytic T lymphocyte clones in patients with neuroblastoma. *Molecular Therapy: The Journal of the American Society of Gene Therapy*, **15**, 825–833.

114 Soling, A., Schurr, P. and Berthold, F. (1999) Expression and clinical relevance of NY-ESO-1, MAGE-1 and MAGE-3 in neuroblastoma. *Anticancer Research*, **19**, 2205–2209.

115 Ishida, H., Matsumura, T., Salgaller, M.L., Ohmizono, Y., Kadono, Y. and Sawada, T. (1996) MAGE-1 and MAGE-3 or -6 expression in neuroblastoma-related pediatric solid tumors. *International Journal of Cancer*, **69**, 375–380.

116 Corrias, M.V., Scaruffi, P., Occhino, M., De Bernardi, B., Tonini, G.P. and Pistoia, V. (1996) Expression of MAGE-1, MAGE-3 and MART-1 genes in neuroblastoma. *International Journal of Cancer*, **69**, 403–407.

117 Oberthuer, A., Hero, B., Spitz, R., Berthold, F. and Fischer, M. (2004) The tumor-associated antigen PRAME is universally expressed in high-stage neuroblastoma and associated with poor outcome. *Clinical Cancer Research*, **10**, 4307–4313.

118 Cheung, I.Y., Barber, D. and Cheung, N.K. (1998) Detection of microscopic neuroblastoma in marrow by histology, immunocytology, and reverse transcription-PCR of multiple molecular markers. *Clinical Cancer Research*, **4**, 2801–2805.

119 Rodolfo, M., Luksch, R., Stockert, E., Chen, Y.T., Collini, P., Ranzani, T., Lombardo, C., Dalerba, P., Rivoltini, L., Arienti, F., Fossati-Bellani, F., Old, L.J., Parmiani, G. and Castelli, C. (2003) Antigen-specific immunity in neuroblastoma patients: antibody and T-cell recognition of NY-ESO-1 tumor antigen. *Cancer Research*, **63**, 6948–6955.

120 Zheng, J., Kohler, M.E., Chen, Q., Weber, J., Khan, J., Johnson, B.D. and Orentas, R.J. (2007) Serum from mice immunized in the context of Treg inhibition identifies DEK as a neuroblastoma tumor antigen. *BMC Immunology*, **8**, 4.

121 Klingebiel, T., Boos, J., Beske, F., Hallmen, E., Int-Veen, C., Dantonello, T., Treuner, J., Gadner, H., Marky, I., Kazanowska, B. and Koscielniak, E. (2008) Treatment of children with metastatic soft tissue sarcoma with oral maintenance compared to high dose chemotherapy: report of the HD CWS-96 trial. *Pediatr Blood Cancer*, **50**, 739–745.

122 Dantonello, T.M., Int-Veen, C., Winkler, P., Leuschner, I., Schuck, A., Schmidt, B.F., Lochbuehler, H., Kirsch, S., Hallmen, E., Veit-Friedrich, I., Bielack, S.S., Niggli, F., Kazanowska, B., Ladenstein, R., Wiebe, T., Klingebiel, T., Treuner, J. and Koscielniak, E. (2008) Initial patient characteristics can predict pattern and risk of relapse in localized rhabdomyosarcoma. *Journal of Clinical Oncology*, **26**, 406–413.

123 Fernandez, J.E., Concha, A., Aranega, A., Ruiz-Cabello, F., Cabrera, T. and Garrido, F. (1991) HLA class I and II expression in rhabdomyosarcomas. *Immunobiology*, **182**, 440–448.

124 De Giovanni, C., Nanni, P., Nicoletti, G., Ceccarelli, C., Scotlandi, K., Landuzzi, L. and Lollini, P.L. (1989) Metastatic ability and differentiative properties of a new cell line of human embryonal rhabdomyosarcoma (CCA). *Anticancer Research*, **9**, 1943–1949.

125 Tanzarella, S., Lionello, I., Valentinis, B., Russo, V., Lollini, P.L. and Traversari, C. (2004) Rhabdomyosarcomas are potential target of MAGE-specific immunotherapies. *Cancer Immunology, Immunotherapy*, **53**, 519–524.

126 Rodeberg, D.A., Erskine, C. and Celis, E. (2007) In vitro induction of immune responses to shared tumor-associated antigens in rhabdomyosarcoma. *Journal of Pediatric Surgery*, **42**, 1396–1402.

127 Belova, O.B., Vinnichuk, U.D., Shlakhovenko, V.A. and Berezhnaya, N.M. (2007) Efficacy of different immunotherapy approaches toward treatment of doxorubicin-resistant and doxorubicin-sensitive transplantable rhabdomyosarcoma. *Experimental Oncology*, **29**, 272–276.

128 Ackermann, B., Troger, A., Glouchkova, L., Korholz, D., Gobel, U. and Dilloo, D. (2004) Characterization of CD34 + progenitor-derived dendritic cells pulsed with tumor cell lysate for a vaccination strategy in children with malignant solid tumors and a poor prognosis. *Klinische Pädiatrie*, **216**, 176–182.

129 Koscielniak, E., Gross-Wieltsch, U., Treuner, J., Winkler, P., Klingebiel, T., Lang, P., Bader, P., Niethammer, D. and Handgretinger, R. (2005) Graft-versus-Ewing sarcoma effect and long-term remission induced by haploidentical stem-cell transplantation in a patient with relapse of metastatic disease. *Journal of Clinical Oncology*, **23**, 242–244.

130 Dalerba, P., Frascella, E., Macino, B., Mandruzzato, S., Zambon, A., Rosolen, A., Carli, M., Ninfo, V. and Zanovello, P. (2001) MAGE, BAGE and GAGE gene expression in human rhabdomyosarcomas. *International Journal of Cancer*, **93**, 85–90.

131 Brinkmann, U., Vasmatzis, G., Lee, B. and Pastan, I. (1999) Novel genes in the PAGE and GAGE family of tumor antigens found by homology walking in the dbEST database. *Cancer Research*, **59**, 1445–1448.

132 Ricci, C., Polito, L., Nanni, P., Landuzzi, L., Astolfi, A., Nicoletti, G., Rossi, I., De Giovanni, C., Bolognesi, A. and Lollini, P.L. (2002) HER/erbB receptors as therapeutic targets of immunotoxins in human rhabdomyosarcoma cells. *Journal of Immunotherapy*, **25** (1997), 314–323.

133 Rodeberg, D.A., Nuss, R.A., Heppelmann, C.J. and Celis, E. (2005) Lack of effective T-lymphocyte response to the PAX3/FKHR translocation area in alveolar rhabdomyosarcoma. *Cancer Immunology, Immunotherapy*, **54**, 526–534.

134 van den Broeke, L.T., Pendleton, C.D., Mackall, C., Helman, L.J. and Berzofsky,

J.A. (2006) Identification and epitope enhancement of a PAX-FKHR fusion protein breakpoint epitope in alveolar rhabdomyosarcoma cells created by a tumorigenic chromosomal translocation inducing CTL capable of lysing human tumors. *Cancer Research*, **66**, 1818–1823.

135 Dagher, R., Long, L.M., Read, E.J., Leitman, S.F., Carter, C.S., Tsokos, M., Goletz, T.J., Avila, N., Berzofsky, J.A., Helman, L.J. and Mackall, C.L. (2002) Pilot trial of tumor-specific peptide vaccination and continuous infusion interleukin-2 in patients with recurrent Ewing sarcoma and alveolar rhabdomyosarcoma: an inter-institute NIH study. *Medical and Pediatric Oncology*, **38**, 158–164.

136 Gattenlohner, S., Marx, A., Markfort, B., Pscherer, S., Landmeier, S., Juergens, H., Muller-Hermelink, H.K., Matthews, I., Beeson, D., Vincent, A. and Rossig, C. (2006) Rhabdomyosarcoma lysis by T cells expressing a human autoantibody-based chimeric receptor targeting the fetal acetylcholine receptor. *Cancer Research*, **66**, 24–28.

137 de Bruijn, D.R., Nap, J.P. and van Kessel, A.G. (2007) The (epi)genetics of human synovial sarcoma. *Genes, Chromosomes & Cancer*, **46**, 107–117.

138 Uren, A. and Toretsky, J.A. (2005) Pediatric malignancies provide unique cancer therapy targets. *Current Opinion in Pediatrics*, **17**, 14–19.

139 Segal, N.H., Blachere, N.E., Shiu, H.Y., Leejee, S., Antonescu, C.R., Lewis, J.J., Wolchok, J.D. and Houghton, A.N. (2005) Antigens recognized by autologous antibodies of patients with soft tissue sarcoma. *Cancer Immunity*, **5**, 4.

140 Ayyoub, M., Taub, R.N., Keohan, M.L., Hesdorffer, M., Metthez, G., Memeo, L., Mansukhani, M., Hibshoosh, H., Hesdorffer, C.S. and Valmori, D. (2004) The frequent expression of cancer/testis antigens provides opportunities for immunotherapeutic targeting of sarcoma. *Cancer Immunity*, **4**, 7.

141 Tureci, O., Chen, Y.T., Sahin, U., Gure, A.O., Zwick, C., Villena, C., Tsang, S., Seitz, G., Old, L.J. and Pfreundschuh, M. (1998) Expression of SSX genes in human tumors. *International Journal of Cancer*, **77**, 19–23.

142 Lee, S.Y., Obata, Y., Yoshida, M., Stockert, E., Williamson, B., Jungbluth, A.A., Chen, Y.T., Old, L.J. and Scanlan, M.J. (2003) Immunomic analysis of human sarcoma. *Proceedings of the National Academy of Sciences of the United States of America*, **100**, 2651–2656.

143 Gallego, S., Llort, A., Roma, J., Sabado, C., Gros, L. and de Toledo, J.S. (2006) Detection of bone marrow micrometastasis and microcirculating disease in rhabdomyosarcoma by a real-time RT-PCR assay. *Journal of Cancer Research and Clinical Oncology*, **132**, 356–362.

144 Clark, J.C., Dass, C.R. and Choong, P.F. (2008) A review of clinical and molecular prognostic factors in osteosarcoma. *Journal of Cancer Research and Clinical Oncology*, **134**, 281–297.

145 Osaka, E., Suzuki, T., Osaka, S., Yoshida, Y., Sugita, H., Asami, S., Tabata, K., Sugitani, M., Nemoto, N. and Ryu, J. (2007) Survivin expression levels as independent predictors of survival for osteosarcoma patients. *Journal of Orthopaedic Research*, **25**, 116–121.

146 Trieb, K., Lechleitner, T., Lang, S., Windhager, R., Kotz, R. and Dirnhofer, S. (1998) Evaluation of HLA-DR expression and T-lymphocyte infiltration in osteosarcoma. *Pathology, Research and Practice*, **194**, 679–684.

147 Tsukahara, T., Kawaguchi, S., Torigoe, T., Asanuma, H., Nakazawa, E., Shimozawa, K., Nabeta, Y., Kimura, S., Kaya, M., Nagoya, S., Wada, T., Yamashita, T. and Sato, N. (2006) Prognostic significance of HLA class I expression in osteosarcoma defined by anti-pan HLA class I monoclonal

antibody, EMR8-5. *Cancer Science*, **97**, 1374–1380.
148 Nagarajan, R., Clohisy, D. and Weigel, B. (2005) New paradigms for therapy for osteosarcoma. *Current Oncology Reports*, **7**, 410–414.
149 Geiger, J.D., Hutchinson, R.J., Hohenkirk, L.F., McKenna, E.A., Yanik, G.A., Levine, J.E., Chang, A.E., Braun, T.M. and Mule, J.J. (2001) Vaccination of pediatric solid tumor patients with tumor lysate-pulsed dendritic cells can expand specific T cells and mediate tumor regression. *Cancer Research*, **61**, 8513–8519.
150 Sudo, T., Kuramoto, T., Komiya, S., Inoue, A. and Itoh, K. (1997) Expression of MAGE genes in osteosarcoma. *Journal of Orthopaedic Research*, **15**, 128–132.
151 Ishida, H., Komiya, S., Inoue, Y., Yutani, S., Inoue, A. and Itoh, K. (2000) Expression of the SART1 tumor-rejection antigen in human osteosarcomas. *International Journal of Oncology*, **17**, 29–32.
152 Tsuda, N., Murayama, K., Ishida, H., Matsunaga, K., Komiya, S., Itoh, K. and Yamada, A. (2001) Expression of a newly defined tumor-rejection antigen SART3 in musculoskeletal tumors and induction of HLA class I-restricted cytotoxic T lymphocytes by SART3-derived peptides. *Journal of Orthopaedic Research*, **19**, 346–351.
153 McGary, E.C., Heimberger, A., Mills, L., Weber, K., Thomas, G.W., Shtivelband, M., Lev, D.C. and Bar-Eli, M. (2003) A fully human antimelanoma cellular adhesion molecule/MUC18 antibody inhibits spontaneous pulmonary metastasis of osteosarcoma cells *in vivo*. *Clinical Cancer Research*, **9**, 6560–6566.
154 Maehara, H., Kaname, T., Yanagi, K., Hanzawa, H., Owan, I., Kinjou, T., Kadomatsu, K., Ikematsu, S., Iwamasa, T., Kanaya, F. and Naritomi, K. (2007) Midkine as a novel target for antibody therapy in osteosarcoma. *Biochemical and Biophysical Research Communications*, **358**, 757–762.
155 Nabeta, Y., Kawaguchi, S., Sahara, H., Ikeda, H., Hirohashi, Y., Goroku, T., Sato, Y., Tsukahara, T., Torigoe, T., Wada, T., Kaya, M., Hiraga, H., Isu, K., Yamawaki, S., Ishii, S., Yamashita, T. and Sato, N. (2003) Recognition by cellular and humoral autologous immunity in a human osteosarcoma cell line. *Journal of Orthopaedic Science*, **8**, 554–559.
156 Ishikura, H., Ikeda, H., Abe, H., Ohkuri, T., Hiraga, H., Isu, K., Tsukahara, T., Sato, N., Kitamura, H., Iwasaki, N., Takeda, N., Minami, A. and Nishimura, T. (2007) Identification of CLUAP1 as a human osteosarcoma tumor-associated antigen recognized by the humoral immune system. *International Journal of Oncology*, **30**, 461–467.
157 Liao, B., Ma, B., Liu, Z., Zhang, H., Long, H., Yang, T. and Fan, Q. (2007) Serological identification and bioinformatics analysis of immunogenic antigens in osteosarcoma. *Cancer Biology & Therapy*, **6**, 1805–1809.
158 Tsukahara, T., Nabeta, Y., Kawaguchi, S., Ikeda, H., Sato, Y., Shimozawa, K., Ida, K., Asanuma, H., Hirohashi, Y., Torigoe, T., Hiraga, H., Nagoya, S., Wada, T., Yamashita, T. and Sato, N. (2004) Identification of human autologous cytotoxic T-lymphocyte-defined osteosarcoma gene that encodes a transcriptional regulator, papillomavirus binding factor. *Cancer Research*, **64**, 5442–5448.
159 Coughlin, C.M., Fleming, M.D., Carroll, R.G., Pawel, B.R., Hogarty, M.D., Shan, X., Vance, B.A., Cohen, J.N., Jairaj, S., Lord, E.M., Wexler, M.H., Danet-Desnoyers, G.A., Pinkus, J.L., Pinkus, G.S., Maris, J.M., Grupp, S.A. and Vonderheide, R.H. (2006) Immunosurveillance and survivin-specific T-cell immunity in children with high-risk neuroblastoma. *Journal of Clinical Oncology*, **24**, 5725–5734.
160 Andersen, M.H., Svane, I.M., Becker, J.C. and a Straten, P.T. (2007) The universal

character of the tumor-associated antigen survivin. *Clinical Cancer Research*, **13**, 5991–5994.
161 Riggi, N. and Stamenkovic, I. (2007) The biology of Ewing sarcoma. *Cancer Letters*, **254**, 1–10.
162 McAllister, N.R. and Lessnick, S.L. (2005) The potential for molecular therapeutic targets in Ewing's sarcoma. *Current Treatment Options in Oncology*, **6**, 461–471.
163 Thacker, M.M., Temple, H.T. and Scully, S.P. (2005) Current treatment for Ewing's sarcoma. *Expert Review of Anticancer Therapy*, **5**, 319–331.
164 Lissat, A., Vraetz, T., Tsokos, M., Klein, R., Braun, M., Koutelia, N., Fisch, P., Romero, M.E., Long, L., Noellke, P., Mackall, C.L., Niemeyer, C.M. and Kontny, U. (2007) Interferon-gamma sensitizes resistant Ewing's sarcoma cells to tumor necrosis factor apoptosis-inducing ligand-induced apoptosis by up-regulation of caspase-8 without altering chemosensitivity. *The American Journal of Pathology*, **170**, 1917–1930.
165 Burdach, S. (2004) Treatment of advanced Ewing tumors by combined radiochemotherapy and engineered cellular transplants. *Pediatric Transplantation*, **8** (Suppl. 5), 67–82.
166 Scotlandi, K. (2006) Targeted therapies in Ewing's sarcoma. *Advances in Experimental Medicine and Biology*, **587**, 13–22.
167 Kolb, E.A., Gorlick, R., Houghton, P.J., Morton, C.L., Lock, R., Carol, H., Reynolds, C.P., Maris, J.M., Keir, S.T., Billups, C.A. and Smith, M.A. (2008) Initial testing (stage 1) of a monoclonal antibody (SCH 717454) against the IGF-1 receptor by the pediatric preclinical testing program. *Pediatr Blood Cancer*, **50**, 1190–1197.
168 Martins, A.S., Mackintosh, C., Martin, D.H., Campos, M., Hernandez, T., Ordonez, J.L. and de Alava, E. (2006) Insulin-like growth factor I receptor pathway inhibition by ADW742, alone or in combination with imatinib, doxorubicin, or vincristine, is a novel therapeutic approach in Ewing tumor. *Clinical Cancer Research*, **12**, 3532–3540.
169 Bernard, G., Breittmayer, J.P., de Matteis, M., Trampont, P., Hofman, P., Senik, A. and Bernard, A. (1997) Apoptosis of immature thymocytes mediated by E2/CD99. *Journal of Immunology (Baltimore, Md: 1950)*, **158**, 2543–2550.
170 Meyer-Wentrup, F., Richter, G. and Burdach, S. (2005) Identification of an immunogenic EWS-FLI1-derived HLA-DR-restricted T helper cell epitope. *Pediatric Hematology and Oncology*, **22**, 297–308.
171 Toub, N., Bertrand, J.R., Tamaddon, A., Elhamess, H., Hillaireau, H., Maksimenko, A., Maccario, J., Malvy, C., Fattal, E. and Couvreur, P. (2006) Efficacy of siRNA nanocapsules targeted against the EWS-Fli1 oncogene in Ewing sarcoma. *Pharmaceutical Research*, **23**, 892–900.
172 Maksimenko, A., Polard, V., Villemeur, M., Elhamess, H., Couvreur, P., Bertrand, J.R., Aboubakar, M., Gottikh, M. and Malvy, C. (2005) *In vivo* potentialities of EWS-Fli-1 targeted antisense oligonucleotides-nanospheres complexes. *Annals of the New York Academy of Sciences*, **1058**, 52–61.
173 Yabe, H., Tsukahara, T., Kawaguchi, S., Wada, T., Sato, N., Morioka, H. and Yabe, H. (2008) Overexpression of papillomavirus binding factor in Ewing's sarcoma family of tumors conferring poor prognosis. *Oncology Reports*, **19**, 129–134.
174 Zendman, A.J., Van Kraats, A.A., Weidle, U.H., Ruiter, D.J. and Van Muijen, G.N. (2002) The XAGE family of cancer/testis-associated genes: alignment and expression profile in normal tissues, melanoma lesions and Ewing's sarcoma. *International Journal of Cancer*, **99**, 361–369.
175 Dome, J.S., Cotton, C.A., Perlman, E.J., Breslow, N.E., Kalapurakal, J.A., Ritchey,

M.L., Grundy, P.E., Malogolowkin, M., Beckwith, J.B., Shamberger, R.C., Haase, G.M., Coppes, M.J., Coccia, P., Kletzel, M., Weetman, R.M., Donaldson, M., Macklis, R.M. and Green, D.M. (2006) Treatment of anaplastic histology Wilms' tumor: results from the fifth National Wilms' Tumor Study. *Journal of Clinical Oncology*, **24**, 2352–2358.

176 Gommersall, L.M., Arya, M., Mushtaq, I. and Duffy, P. (2005) Current challenges in Wilms' tumor management. *Nature Clinical Practice Oncology*, **2**, 298–304 quiz 291 pp. following p. 324.

177 Pinthus, J.H., Fridman, E., Dekel, B., Goldberg, I., Kaufman-Francis, K., Eshhar, Z., Harmelin, A., Rechavi, G., Mor, O., Ramon, J. and Mor, Y. (2004) ErbB2 is a tumor associated antigen and a suitable therapeutic target in Wilms tumor. *The Journal of Urology*, **172**, 1644–1648.

178 Sugiura, N., Dadashev, V. and Corriveau, R.A. (2004) NARG2 encodes a novel nuclear protein with (S/T)PXX motifs that is expressed during development. *European Journal of Biochemistry*, **271**, 4629–4637.

8
Epigenetically-regulated Therapeutic Tumor-associated Antigens

Hugues J. M. Nicolay, Luca Sigalotti, Sandra Coral, Elisabetta Fratta, Alessia Covre, Ester Fonsatti, and Michele Maio

8.1
Introduction

The phenotypic characterization of neoplastic cells and the definition of their functional interactions with the host's immune system have provided a strong rationale for designing and applying new strategies for cancer immunotherapy in the clinic. In particular, the identification of tumor-associated antigens (TAA) which are recognized by antibodies and cytotoxic T lymphocytes (CTL) [1–4] together with new data regarding the molecular mechanisms responsible for the regulation of tumor–host interactions [5], have prompted the design of various pre-clinical and clinical studies of specific passive and active immunotherapy.

The aim of passive immunotherapeutic strategies is to provide an immune response directly by the transfer of TAA-specific antibodies or immune cells into the cancer-bearing host [6, 7]. With this approach, significant clinical results in cancer treatment were obtained with humanized monoclonal antibodies (mAb) directed against CD20 on malignant B cells [8] or HER2 in breast cancer [9]. Additionally, therapeutic transfusion in melanoma patients of *ex vivo* generated T cells which recognize selected TAA (i.e. melanocyte differentiation antigens, Cancer Testis Antigens (CTA) or unique antigens) has recently been reported with promising clinical results [10].

In active specific immunotherapeutic approaches, also defined as therapeutic vaccination, TAA are administered, alone or in combination with various immunological adjuvants in different formulations, with the aim of eliciting an anti-TAA specific immune response in cancer patients. Among others, the cancer vaccines mostly utilized in the clinical setting are based on whole cancer cells, anti-idiotypic antibodies, DNA, recombinant proteins, peptides, autologous heat shock proteins and peptide-/tumor-pulsed dendritic cells [3, 11, 12].

TAA-based immunotherapy is emerging as a promising and additional therapeutic option that is rapidly becoming integrated into the overall treatment of cancer patients. However, the cellular and humoral responses to TAA triggered by immunotherapy do not seem to be sufficiently efficacious to eradicate the tumor mass [11]. Based on this evidence, cancer researchers have mainly focused on understanding the mechanisms that may impair the immunologic efficiency of these therapeutic approaches. Among these, loss and/or down-regulated expression of HLA class I antigens has frequently been detected in cancer specimens [13–15], with or without concurrent abnormalities in the expression and/or function of different components of the antigen processing machinery (i.e. TAP1, TAP2, LMP-2, LMP-7, tapasin) [16–18]. Furthermore, absent, down-regulated, and/or heterogeneous expression of different TAA [19] as well as reduced levels of co-stimulatory/accessory molecules (e.g. CD40, CD54, CD58) have been demonstrated in neoplastic lesions [20, 21]. Occurrence of these events either separately or concomitantly may promote disease progression by enabling the neoplastic cells to evade detection by the host's constitutive or immunotherapy-triggered immune surveillance.

Forthcoming evidence will demonstrate that epigenetic events play a crucial role in determining a poorly immunogenic phenotype in cancer cells leading to the loss or down-regulated expression of various immune molecules such as HLA antigens, accessory/co-stimulatory molecules and TAA [22–30]. In contrast to genetic alterations, epigenetic modifications are readily reversible with the use of epigenetic drugs, which are already available for clinical use. Along these lines, the evidence that the DNA hypomethylating drug 5-aza-2'-deoxycytidine (5-AZA-CdR) persistently up-regulates/induces the concomitant expression of several TAA as well as of selected components of the tumor recognition complex (e.g. HLA class I antigens, CD54) [31–33] provides a rationale for the potential use of epigenetic modifiers in the clinical setting to improve the efficacy of TAA-based vaccination.

8.2
Epigenetics

Epigenetics is the study of heritable changes in gene expression that do not derive from alterations of the nucleotide sequence of DNA [25, 34–36]. In this respect, the most fully characterized epigenetic regulators of gene expression are methylation of genomic DNA in CpG dinucleotides, and post-translational modification of core histone proteins involved in the packing of DNA into chromatin.

8.2.1
DNA Methylation

DNA methylation involves the addition of a methyl group to position C5 of the cytidine ring in a CpG dinucleotide. The reaction is mediated by various DNA methyltransferase (DNMT) enzymes that are characterized by specific substrate affinities. Among these, DNMT1 is considered to be responsible for maintaining DNA methylation patterns and has up to 30-fold more affinity for hemimethylated sites than for unmethylated sites. In contrast, DNMT3a and DNMT3b have a similar affinity for hemimethylated and unmethylated DNA and are likely involved in the generation of new methylation patterns [36–39].

CpG methylation in promoter regions leads to transcriptional repression either by directly blocking the binding of transcriptional activators, or by binding methyl-CpG-binding proteins (MBP) which silence gene expression by recruiting co-repressor complexes [40]. The effect of DNA methylation on gene expression is eventually sustained by the generation of a transcriptionally repressive chromatin structure, driven by the ability of MBP and DNMT to bind histone deacetylase (HDAC) and histone methyltransferase (HMT) complexes [40].

Transcriptional silencing by DNA methylation plays a crucial role in the physiological regulation of cellular processes, and is involved in establishing developmental stage- and tissue-specific patterns of gene expression, in determining chromosome and chromatin structure stability, in silencing chromosome X in females and also in the inactivation of viral sequences and transposable elements integrated into the genome. In cancer, methylation patterns are severely disrupted by aberrant hypo- and hyper-methylation events leading respectively to genomic instability and down-regulation of several tumor suppressor genes, likely having a crucial role in tumor development and progression [37, 41].

8.2.2
Histone Post-translational Modifications

Key transcriptional regulatory mechanisms derive from the packing of DNA into higher order chromatin structures. The building block of the chromatin is the nucleosome which is consists of 146 bp DNA wrapped around a histone octamer core comprising two copies each of histones H2A, H2B, H3 and H4. The flexible histone N-terminal tails protrude from the nucleosome and are targets for extensive post-translational modifications. These modifications constitute the 'histone code' which is translated by multiprotein complexes into chromatin compaction status, ultimately leading to the regulation of gene transcription. Amongst all the possible histone post-translational modifications, acetylation and methylation of histone tails have been most widely investigated and characterized for their role in transcriptional regulation. The acetylation status of histones is controlled by the balanced action of histone acetyltransferases (HAT) which adds acetyl groups to the N-terminal lysine (K) residues, and HDAC which has the opposite role. Transcriptional regulation by acetylation is quite straightforward: acetylation of histones results in chromatin decompaction and gene transcription, while deacetylated chromatin associates with repressed genes. The functional importance of HAT and HDAC is highlighted by the fundamental regulatory roles that they play in developmental processes, and by the evidence that both HAT and HDAC misregulation is invariably associated with cancer. In fact, genes encoding HAT and HDAC enzymes have been found translocated, amplified, over-expressed and/or mutated in different types of cancers [42]. On the other hand, histone methylation is carried out by different HMT and is associated with specific effects depending on the target residue. In fact, while methylation of H3-K4 and H3-K79 is associated with transcriptionally active chromatin, methylation of H4-K20, H3-K9, H3-K27 and H3-K36 marks transcriptionally inactive chromatin [43–45]. HMT are required for correct embryonic development and cellular behavior, and most of them have been closely linked to the development of cancer [46].

8.3
TAA

8.3.1
Classification of TAA

According to their pattern of expression in neoplastic and normal tissues, TAA thus far identified can be classified in nine groups including: (a) *CTA* expressed in tumors of various histological origin but not in normal tissues other than testis and placenta; (b) *Differentiation antigens* expressed in normal and neoplastic cells of the same lineage (e.g. Melan-A/MART-1); (c) *Over-expressed antigens* expressed at higher levels in malignant compared to benign tissues (e.g. HER2/neu); (d) *Tumor-specific unique antigens* arising from point mutations of normal genes such as β-catenin and CDK-4 whose molecular changes often accompany neoplastic transformation or progression; (e) *Fusion proteins* deriving from translocation of chromosomes resulting in the fusion of distant genes which characterize each type of disease (e.g. bcr-abl and pml-RARα in chronic myelogenous leukemia and acute promyelocytic leukemia, respectively); (f) *Oncofetal antigens* expressed in fetal, neoplastic and specific benign adult tissues in quantitatively varying levels, often used as diagnostic or prognostic markers for malignancies (e.g. α-1-fetoprotein and carcinoembryonic antigen) [47, 48] (g) *Glycolipids and glycoproteins* (i.e. high molecular weight melanoma-associated antigen (HMW-MAA), gangliosides and mucins) expressed at high levels and/or with an aberrant glycosylation status in malignancies of different histotype; (h) *Splice variants of known genes*, identified by SEREX and found to be immunogenic in cancer patients (e.g. NY-CO-37/38); (i) *Cancer-related autoantigens* also identified by SEREX, PROTEOMEX and AMIDA are expressed ubiquitously and at a similar level in healthy and malignant tissues (e.g. CEBPγ). The encoding genes are not altered in tumor samples; however, they elicit antibody responses in cancer patients but not in healthy individuals [49, 50].

8.3.2
Epigenetically-regulated TAA

8.3.2.1 HMW-MAA
The HMW-MAA is a cell surface chondroitin sulfate proteoglycan that belongs to a family of adhesion receptors. It is expressed in the majority of melanoma lesions and shows a restricted distribution in normal tissues. An array of evidence indicates that the HMW-MAA plays a role in the intracellular signal cascade involved in cellular

adhesion, spreading and invasion [51]. Anti-idiotypic mAb that mimic the HMW-MAA have been utilized for vaccination of melanoma patients in association with different immune adjuvants [52]. The results of these studies demonstrated that treatment elicits a humoral response associated with an improvement in the clinical course of the disease [52].

In the rare cases of HMW-MAA-negative melanoma lesions, the HMW-MAA protein is not expressed due to its silencing by epigenetic events. In fact, the HMW-MAA promoter was found to be heavily methylated in the HMW-MAA-negative M14#5, 1520 and SK-MEL-5 melanoma cell lines. The involvement of DNA hypermethylation in this silencing was further supported by the observation that treatment with the DNA hypomethylating agent (DHA) 5-AZA-CdR led to a hypomethylation of the promoter that resulted in the re-expression of the HMW-MAA mRNA and protein. Despite these data, DNA methylation analysis of a HMW-MAA-negative acral lentiginous melanoma lesion revealed an unmethylated pattern of the promoter region of the HMW-MAA suggesting that DNA hypomethylation may not be sufficient for HMW-MAA gene expression [53].

8.3.2.2 Mucins

Mucins are high molecular weight glycoproteins secreted by specialized epithelial cells (i.e. normal goblet or columnar cells) which play an important role as a physiological barrier against various attacks on the underlying epithelial tissues. So far, 20 mucin genes (MUC) have been identified and designated as MUC1–2, MUC3A, MUC3B, MUC4, MUC5AC, MUC5B, MUC6–13, MUC15–17, and MUC19–20. The biochemical and immunological properties of the MUC1 mucin which is over-expressed in 90% of all adenocarcinomas (breast, lung, pancreas, prostate, stomach, colon and ovary) make this TAA an attractive target for immunotherapeutic approaches [54, 55].

Mucins, when expressed in normal tissues, are regulated in a DNA methylation-dependent manner. In fact, several studies have shown that the strong expression of MUC2 in normal human goblet cells is associated with the average methylation of about 50% of the CpG sites in its promoter. In contrast, MUC2 promoter in the non-expressing normal columnar cells is methylated to nearly 100% [56, 57].

In cancer, MUC2 and MUC5B present a frequently methylated GC-rich structure in their promoters which is associated with gene silencing. Along these lines, COLO 205 colon carcinoma cells in which methylation of the MUC2 promoter is associated with a non-expressing pattern, were treated with 5-AZA-CdR resulting in a *de novo* expression of the MUC2 gene [57].

The effect of trichostatin A (TSA), a HDAC inhibitor (HDACi), and 5-AZA-CdR was also investigated on KATO-III gastric epithelial cancer cells and OE33 esophageal epithelial cancer cells for mucin expression. The results of these studies demonstrated that the MUC2 gene which is not expressed in KATO-III cells, was strongly induced after 5-AZA-CdR and TSA treatment. Following on from this, treatment of pre- and post-confluent OE33 cells showed that the MUC5B gene which is not expressed in OE33 cells, was strongly induced after 5-AZA-CdR treatment in proliferating cells, while TSA treatment induced MUC5B expression in proliferating and confluent cells. This evidence highlights that MUC2 and MUC5B are regulated by DNA methylation and histone deacetylation [56, 58].

8.3.2.3 CTA

Among known TAA, CTA have been well characterized as suitable targets for cancer vaccination because of their restricted presentation by the HLA machinery on tumor cell surface [59–61], their shared expression by neoplasia of different histotype, and their ability to generate a spontaneous humoral and cellular immune response [1, 62–64].

So far, at least 44 families of CTA have been identified [60, 65]. Among these, recombinant proteins or HLA class I and HLA class II-restricted antigenic peptides of MAGE, NY-ESO and SSX gene families and GAGE/PAGE/XAGE super-families have been utilized for cancer vaccination in clinical trials [60, 65]. The results that emerged from the initial clinical trials indicated that CTA-based vaccination is a safe therapeutic procedure and that complete responses as well as prolonged disease-free survival were observed in various solid tumors [1, 3, 49, 66, 67].

Even if clinical results are encouraging, they suggest that several issues still need to be optimized to improve the efficacy of CTA-based immunotherapy. The demonstration that CTA expression is associated with the methylation of their gene promoter has suggested immunotherapeutic implications for DHA such as 5-AZA-CdR.

CTA Expression in Cancer In normal tissues CTA expression is restricted to germ cells of the testis, ovary and placenta [68] while they are widely and variably distributed among tumors of different histotype. The available data are mainly based on the analysis of their transcripts and demonstrate that members of different families and super-families of CTA are largely expressed in melanoma, bladder and non-small cell lung cancer, moderately expressed in breast and prostate cancer, poorly expressed in renal and colon cancer [60, 65] and in hematologic malignancies [69, 70]. Depending on the specific CTA being investigated and on the tumor histotype being analyzed, the percentages of CTA-positive lesions range from 0 to 66% [71]. In addition, little is known about the incidence of CTA expression in tumor subtypes. In this respect, a study was conducted to assess MAGE-A1 expression in a series of lung neoplasms (i.e. 10 large cell undifferentiated carcinomas, 37 adenocarcinomas, 12 squamous cell carcinomas) both by immunohistochemistry (IHC) and RT-PCR analysis [72]. The results point to the highly heterogeneous expression of this CTA among lung tumors of the same or of different subtypes. Interestingly, both RT-PCR and immunohistochemical staining identified 32% of the investigated tumors as MAGE-A1-positive, suggesting that protein expression of MAGE-A1 correlates with its mRNA expression [72]. Similar results were also obtained in cutaneous melanoma. A study carried out on unrelated metastatic lesions from a large number of melanoma patients demonstrated the highly heterogeneous constitutive expression of MAGE-A2, -A3 and -A4, GAGE1-6, BAGE, NY-ESO-1 and PRAME, ranging from lesions expressing all the CTA analyzed to lesions that were completely CTA-negative [59]. In contrast, the pattern of CTA expression was homogeneously maintained among synchronous and metachronous metastases from each patient [59]. In addition to inter-individual heterogeneity, IHC analyses also revealed that the intratumoral expression of CTA is characteristically heterogeneous: usually less than 50% of cells were stained by anti- MAGE-A1, -NY-ESO-1 or -SSX mAb in the majority of CTA-positive lesions from tumors of different histotype [19, 71–73]. In light of the constraints posed on immunotherapy of cancer patients by the overall heterogeneous expression of CTA, various research groups have investigated their regulation in neoplastic tissues. Initial studies have suggested that MAGE-A1 expression in neoplastic cells of different histotype resulted from genome-wide demethylation events that also affected MAGE-A1 promoter methylation status [74]. Accordingly, the expression of different MAGE-A and NY-ESO members was invariably associated with the hypomethylation status of the respective CTA promoters in the tumors and cell lines investigated [59, 74–76]. Finally, transfections with reporter genes driven by CTA promoters, methylated or not *in vitro*, demonstrated that promoter methylation was the only factor limiting CTA promoter activity in cancer cells [59, 77]. This correlation between promoter methylation and CTA expression has recently been confirmed at the single cell level which enabled researchers to demonstrate that the intratumoral heterogeneity of MAGE-A3 expression is also associated with the differential methylation pattern of specific CpG dinucleotides within the MAGE-A3 promoter [78].

Due to their *in vivo* immunogenicity and their shared expression among tumors, CTA have been established as very attractive targets for immunotherapeutic approaches. This is particularly relevant in view of the foreseeable targeting of cancer stem cells for which a consistent expression of therapeutic CTA has been most recently shown both at molecular and protein levels, in enriched populations of melanoma stem cells [79]. In fact, cancer stem cells are considered to be the most important neoplastic cell population for initiating and sustaining tumor growth and metastasis, concomitantly being the most resistant to conventional therapeutic approaches, thus representing the ultimate target for therapeutic intervention in cancer patients. Despite the above reported advantages of CTA as therapeutic targets, their intra- and inter-tumoral heterogeneous expression might limit the eligibility of cancer patients for treatment and lead to the immune-selection of CTA-negative neoplastic clones in the course of CTA-based vaccination. In addition, low levels of expression of the targeted CTA on neoplastic cells [72, 73] may limit the efficiency of CTA-based vaccination impairing the immune recognition of neoplastic cells.

Modulated Expression of CTA by Epigenetic Drugs The demonstration that the methylation status of CTA promoters determined their inter- and intra-tumoral heterogeneous expression prompted studies on the use of epigenetic drugs, such as 5-AZA-CdR, to widen the eligibility of cancer patients for CTA-based vaccination and to improve the efficacy of treatment by the modulation of CTA expression [78, 80].

Similarly, it has been well established that treatment with 5-AZA-CdR is able to induce and/or to up-regulate the concomitant expression of multiple members of different CTA families (e.g. MAGE-A, NY-ESO, GAGE) in cultured neoplastic cells of hematologic and solid malignancies, including primary effusion lymphoma, acute myeloid leukemia (AML), myelodysplastic syndrome (MDS), chronic myeloid leukemia (CML), lung, breast, colon and renal cell carcinoma, cutaneous melanoma, sarcoma, and mesothelioma [78, 80–84]. It is noteworthy that the observed modifications of the CTA phenotype induced by pharmacologic inhibition of DNMT were persistent, still being detectable both at the mRNA and protein at least 2 months after drug removal [31, 69, 80]. Furthermore, by using an *in vitro* model, 5-AZA-CdR was recently suggested to represent a potent agent able to reverse the intratumoral heterogeneity of CTA expression. In fact, 5-AZA-CdR was able to homogenize the MAGE-A3 expression levels in different single-cell clones that were generated from a metastatic lesion of a melanoma patient and that constitutively expressed different levels of the antigen [78]. Furthermore, the effect of 5-AZA-CdR may be significantly potentiated by the concomitant/sequential treatment with HDACi. Indeed, the combined treatment of cancer cells of different histotype with HDACi TSA or depsipeptide and 5-AZA-CdR resulted in 1.5- to 7-fold higher expression of MAGE-A1, -A2, -A3, -A12, and NY-ESO-1 as compared to 5-AZA-CdR alone [85–88]. Altogether, the above-reported modulations of the CTA phenotype triggered by 5-AZA-CdR alone or combined with HDACi, finally led to an efficient recognition of previously refractory or poorly recognized neoplastic cells by CTA-specific CTL and enabled their homogeneous intratumoral targeting by effector T cells [20, 36, 78, 80].

The immunomodulatory activities of epigenetic drugs are not limited to CTA. In fact, recent evidence has identified epigenetic drugs as more general immunomodulatory agents, being able to revert the down-regulated expression of HLA molecules and accessory/co-stimulatory molecules (i.e. CD80, CD86, CD40, CD54) [89–91] which frequently occur in cancer cells. In particular, the *de novo* expression of HLA class I antigens induced by 5-AZA-CdR was demonstrated to enable the recognition of melanoma cells by MAGE-A-specific CTL [92]. This finding, although relevant, is likely to be restricted to only a minor fraction of HLA class I abnormalities [32]. In this context, a much wider impact can be expected from the observed ability of 5-AZA-CdR to up-regulate the levels of HLA class I antigens that are frequently low in cutaneous melanoma. In fact, by exploiting a target TAA, gp100, that was not modulated by the hypomethylating treatment, it has been recently shown that the up-regulation of HLA class I antigens and the accessory molecule CD54 by 5-AZA-CdR was sufficient to improve the recognition of melanoma cells by gp100-specific CTL [33]. A review of the effects of epigenetic drugs on TAA and the presenting molecules complex expression can be found in Table 8.1.

The demonstrated *in vitro* immunomodulatory activities of 5-AZA-CdR were finally evaluated in preclinical animal models to assess their potential clinical transferability. Along these lines, recent data have reported that systemic administration of 5-AZA-CdR to mice grafted with murine cancer cells of different histotype induced a *de novo* expression of the epigenetically-regulated murine CTA, P1A, in tumors. In addition, the obtained reversion of the P1A-phenotype of tumors resulted in a significant reduction of lung metastases following adoptive transfer of P1A-specific CTL [93]. Furthermore, it has been shown that the systemic administration of 5-AZA-CdR to BALB/c *nu/nu* mice grafted with primary cultures of human melanoma cells

Table 8.1 Use of epigenetic drugs to reverse the mechanisms which enable tumor cells to evade immune recognition.

Mechanisms	Modulation by DHA and/or HDACi	References
Loss, down-regulation or heterogeneity of TAA expression	MAGE-A1, -A2, -A3, -A4, -A6, -A10	[22, 75, 78, 80, 83, 102]
	MAGE-A12	[88]
	BAGE	[109]
	GAGE1-6	[20, 78, 83]
	SSX1-5	[78, 102]
	NY-ESO-1	[78, 83, 84]
	HAGE	[81]
	PRAME	[78, 83]
	RAGE-1	[80]
	XAGE-1	[22]
	MUC2	[58]
	MUC5B	[58]
	HMW-MAA	[53]
Loss or down-regulation of HLA class I expression	HLA class I	[31–33, 91]
	HLA class II	[90, 91]
	CIITA	[90]
Down-regulation of accessory/ co-stimulatory molecule expression	CD54	[30, 31, 33]
	CD58	[31]
	CD80	[110]
	CD86	[110]
	CD40	[91]

induced and/or up-regulated the expression of MAGE-A1, -A2, -A3, -A4, -A10, GAGE 1-6, and NY-ESO-1 and concomitantly up-regulated the levels of expression of HLA class I antigens. In this setting, the phenotypic modifications thus achieved were shown to be long-lasting *in vivo* as they were still detectable in excised xenografts 30 days after the end of treatment, in particular NY-ESO-1 and HLA class I expression [83, 84]. Finally, immunization of BALB/c mice with 5-AZA-CdR-treated melanoma cells generated a high titer of circulating anti-NY-ESO-1 antibodies, demonstrating the immunogenicity of the *de novo* expressed CTA [83].

Altogether, this evidence established the *in vivo* immunomodulatory activity of 5-AZA-CdR, providing the rationale for the clinical use of epigenetic drugs in combined chemo-immunotherapeutic regimens for the treatment of cancer patients.

Clinical Evidence of CTA Modulation by Epigenetic Drugs The prompt pharmacological reversal of epigenetic modifications involved in the mechanism(s) of immune escape of neoplastic cells provided new possibilities for the clinical use of epigenetic drugs in combined therapies with CTA-directed vaccines or with other immunotherapeutic approaches. This possibility is strengthened by the clinical availability of various epigenetic drugs. Among these, DHA 5-azacytidine (Vidaza®) and

5-AZA-CdR (Decitabine, Dacogen™) were approved by the Food and Drug Administration (FDA) in 2004 and 2006 respectively, for clinical use in hematologic malignancies of all MDS [94–97]. The synthetic compound suberoylanilide hydroxilamic acid and vorinostat (Zolinza™), a HDACi, were approved by the FDA in October 2006 for treatment of refractory cutaneous T-cell lymphoma [42].

5-AZA-CdR was first introduced into clinical development because of its cytotoxic effects on neoplastic cells [98–100]. Only recently have various clinical trials been undertaken to identify the potential association between the DNA hypomethylating activity of 5-AZA-CdR and its clinical response. In this respect, Phase I and II studies have shown that low-dose schedules of 5-AZA-CdR were able to induce DNA hypomethylation in peripheral blood mononuclear cells (PBMC) of patients with hemopoietic malignancies [95, 101], and that the percentage of demethylation of Alu repetitive elements correlated with the clinical response, suggesting a direct role of drug-induced demethylation in triggering the clinical response [101]. Furthermore, the observation that a single low-dose course of 5-AZA-CdR was able to induce *de novo* expression of different CTA in AML and MDS patients, suggested the possibility that long-term disease control may be sustained by the activation of a CTA-specific immune response [102]. Following the characterization of its epigenetic effects in hematopoietic malignancies, the clinical application of the DNA hypomethylating activity of 5-AZA-CdR in solid malignancies has begun to be addressed. A phase I trial conducted in patients affected by metastatic solid tumors of different histotype reported that although no objective clinical response was found after administration of 5-AZA-CdR, changes in the methylation patterns of selected genes were observed in tumor biopsies [103]. Along these lines, another phase I trial reported a well-tolerated 5-AZA-CdR administration schedule in patients with recurrent or metastatic cancer (i.e. renal, squamous cell carcinoma head/neck, ovarian and colon) associated with a global inhibition of DNA methylation accompanied by a significant hypomethylation of the MAGE-A1 promoter in patients' blood samples. It also showed that genomic DNA methylation returned to baseline levels by 28 to 35 days after the beginning of therapy, demonstrating the persistent effect of 5-AZA-CdR [104]. Furthermore, a phase I trial involving patients with thoracic malignancies who had been treated with the maximum tolerated dose of 5-AZA-CdR (60 to 75 mg/m^2) demonstrated that although no objective clinical responses were observed with this regimens, tumor biopsies from 36% of patients exhibited induction of NY-ESO-1, MAGE-A3, or p16 expression. Post-therapy antibodies against NY-ESO-1 where detected in three patients exhibiting NY-ESO-1 induction in their tumor tissues [105]. The demonstrated ability of 5-AZA-CdR to modulate the CTA tumor phenotype and to promote a humoral immune response in treated patients, led to the design of a combined immunotherapeutic and epigenetic approach for the treatment of solid cancer. Indeed, a phase I clinical trial, conducted in melanoma and renal carcinoma patients, reported that sequential low-dose 5-AZA-CdR plus high dose intravenous interleukin-2 (IL-2), resulted in global genomic DNA hypomethylation and in changes in the expression of immunomodulatory genes in PBMC, overlapping with the immune activation triggered by high-dose IL-2 [106].

On the other hand, nearly a dozen small-molecule HDACi (i.e. Butyrate, Valproic acid, AN-9, Vorinostat, PXD101, LAQ824, LBH589, Pyroxamide, MS-275, tacedinaline, Depsipeptide, MGCD-0103) are currently in clinical phases I, II and III, all of which have exhibited some degree of antitumor activity. They can be used either as monotherapies or in combination with other agents in refractory cutaneous T-cell lymphoma, leukemias, lung and cervical cancers, melanoma and other refractory solid tumors. As these small molecules are still in the early phases of development, little is known about their clinical activity and benefit [36, 107, 108]. Nevertheless, the notion that *in vitro* treatment of neoplastic cells with HDACi significantly up-regulates the 5-AZA-CdR-induced CTA expression [85–88], represents a convincing rationale for the use of HDACi in combination with 5-AZA-CdR in CTA-based vaccination.

8.4
Perspectives and Conclusion

As a result of our better understanding of the molecular mechanisms regulating tumor–host immune interactions, the clinical potential of immunotherapy for the treatment of human neoplasia is becoming increasingly clear. The identification of novel strategies able to strengthen this interaction will undoubtedly improve our ability to control cancer through immunotherapy. Characterization of the immune properties of aberrant DNA methylation in cancer cells and the ability of DHA to improve the immunogenicity of transformed cells and their functional recognition by the immune system also support the potential of immunotherapy in the clinical setting.

References

1 Gnjatic, S., Nishikawa, H., Jungbluth, A.A., Gure, A.O., Ritter, G., Jager, E. *et al.* (2006) NY-ESO-1: Review of an Immunogenic Tumor Antigen, in *Advances in Cancer Research*, (ed George F.V.), Academic Press, pp. 1–30.

2 Scanlan, M.J., Gout, I., Gordon, C.M., Williamson, B., Stockert, E., Gure, A.O. *et al.* (2001) Humoral immunity to human breast cancer: antigen definition and quantitative analysis of mRNA expression. *Cancer Immunity*, **1**, 4.

3 Boon, T., Coulie, P.G., Van denEynde, B.J.V. and van der Bruggen, P. (2006) Human T cell responses against melanoma. *Annual Review of Immunology*, **24**, 175–208.

4 van der Bruggen, P., Traversari, C., Chomez, P., Lurquin, C., De Plaen, E., Van den Eynde, B. *et al.* (1991) A gene encoding an antigen recognized by cytolytic T lymphocytes on a human melanoma. *Science*, **254**, 1643–1647.

5 Wang, E., Panelli, M.C., Monsurro, V. and Marincola, F.M. (2004) A global approach to tumor immunology. *Cellular & Molecular Immunology*, **1**, 256–265.

6 Harris, M. (2004) Monoclonal antibodies as therapeutic agents for cancer. *The Lancet Oncology*, **5**, 292–302.

7. June, C.H. (2007) Principles of adoptive T cell cancer therapy. *The Journal of Clinical Investigation*, **117**, 1204–1212.
8. Smith, M.R. (2003) Rituximab (monoclonal anti-CD20 antibody): mechanisms of action and resistance. *Oncogene*, **22**, 7359–7368.
9. Menard, S., Casalini, P., Campiglio, M., Pupa, S.M. and Tagliabue, E. (2004) Role of HER2/neu in tumor progression and therapy. *Cellular and Molecular Life Sciences*, **61**, 2965–2978.
10. Yee, C., Thompson, J.A., Byrd, D., Riddell, S.R., Roche, P., Celis, E. et al. (2002) Adoptive T cell therapy using antigen-specific CD8 + T cell clones for the treatment of patients with metastatic melanoma: *in vivo* persistence, migration, and antitumor effect of transferred T cells. *Proceedings of the National Academy of Sciences of the United States of America*, **99**, 16168–16173.
11. Rosenberg, S.A. (2001) Progress in human tumour immunology and immunotherapy. *Nature*, **411**, 380–384.
12. Terando, A.M., Faries, M.B. and Morton, D.L. (2007) Vaccine therapy for melanoma: Current status and future directions. *Vaccine*, **25**, B4–B16.
13. Seliger, B., Cabrera, T., Garrido, F. and Ferrone, S. (2002) HLA class I antigen abnormalities and immune escape by malignant cells. *Seminars in Cancer Biology*, **12**, 3–13.
14. Aptsiauri, N., Cabrera, T., Garcia-Lora, A., Lopez-Nevot, M.A., Ruiz-Cabello, F. and Garrido, F. (2007) MHC Class I Antigens and Immune Surveillance in Transformed Cells, in *International Review of Cytology A Survey of Cell Biology*, (ed Kwang W.J.), Academic Press, pp. 139–189.
15. Algarra, I., Garcia-Lora, A., Cabrera, T., Ruiz-Cabello, F. and Garrido, F. (2004) The selection of tumor variants with altered expression of classical and nonclassical MHC class I molecules: implications for tumor immune escape. *Cancer Immunology, Immunotherapy*, **53**, 904–910.
16. de la Salle, H., Zimmer, J., Fricker, D., Angenieux, C., Cazenave, J.P., Okubo, M. et al. (1999) HLA class I deficiencies due to mutations in subunit 1 of the peptide transporter TAP1. *The Journal of Clinical Investigation*, **103**, R9–R13.
17. Furukawa, H., Murata, S., Yabe, T., Shimbara, N., Keicho, N., Kashiwase, K. et al. (1999) Splice acceptor site mutation of the transporter associated with antigen processing-1 gene in human bare lymphocyte syndrome. *The Journal of Clinical Investigation*, **103**, 755–758.
18. Romero, J.M., Jimenez, P., Cabrera, T., Cozar, J.M., Pedrinaci, S., Tallada, M. et al. (2005) Coordinated downregulation of the antigen presentation machinery and HLA class I/beta2-microglobulin complex is responsible for HLA-ABC loss in bladder cancer. *International Journal of Cancer*, **113**, 605–610.
19. dos Santos, N.R., Torensma, R., de Vries, T.J., Schreurs, M.W.J., de Bruijn, D.R.H., Kater-Baats, E. et al. (2000) Heterogeneous expression of the SSX cancer/testis antigens in human melanoma lesions and cell lines. *Cancer Research*, **60**, 1654–1662.
20. Sigalotti, L., Coral, S., Fratta, E., Lamaj, E., Danielli, R., Di Giacomo, A.M. et al. (2005) Epigenetic modulation of solid tumors as a novel approach for cancer immunotherapy. *Seminars in Oncology*, **32**, 473–478.
21. Arnold, J.M., Cummings, M., Purdie, D. and Chenevix-Trench, G. (2001) Reduced expression of intercellular adhesion molecule-1 in ovarian adenocarcinomas. *British Journal of Cancer*, **85**, 1351–1358.
22. James, S.R., Link, P.A. and Karpf, A.R. (2006) Epigenetic regulation of X-linked cancer/germline antigen genes by DNMT1 and DNMT3b. *Oncogene*, **25**, 6975–6985.

23 Baylin, S.B. (2005) DNA methylation and gene silencing in cancer. *Nature Clinical Practice Oncologyl*, **2** (Suppl. 1), S4–S11.

24 Baylin, S.B. and Ohm, J.E. (2006) Epigenetic gene silencing in cancer—a mechanism for early oncogenic pathway addiction? *Nature Reviews, Cancer*, **6**, 107–116.

25 Gronbaek, K., Hother, C. and Jones, P.A. (2007) Epigenetic changes in cancer. *APMIS: Acta Pathologica, Microbiologica, et Immunologica Scandinavica*, **115**, 1039–1059.

26 Jones, P.A. and Baylin, S.B. (2002) The fundamental role of epigenetic events in cancer. *Nature Reviews Genetics*, **3**, 415–428.

27 Jones, P.A. and Baylin, S.B. (2007) The epigenomics of cancer. *Cell*, **128**, 683–692.

28 Smith, L.T., Otterson, G.A. and Plass, C. (2007) Unraveling the epigenetic code of cancer for therapy. *Trends in Genetics*, **23**, 449–456.

29 Ting, A.H., McGarvey, K.M. and Baylin, S.B. (2006) The cancer epigenome – components and functional correlates. *Genes and Development*, **20**, 3215–3231.

30 Hellebrekers, D.M., Castermans, K., Vire, E., Dings, R.P., Hoebers, N.T., Mayo, K.H. *et al.* (2006) Epigenetic regulation of tumor endothelial cell anergy: silencing of intercellular adhesion molecule-1 by histone modifications. *Cancer Research*, **66**, 10770–10777.

31 Coral, S., Sigalotti, L., Gasparollo, A., Cattarossi, I., Visintin, A., Cattelan, A. *et al.* (1999) Prolonged upregulation of the expression of HLA class I antigens and costimulatory molecules on melanoma cells treated with 5-aza-2′-deoxycytidine (5-AZA-CdR). *Journal of Immunotherapy*, **22**, 16–24.

32 Fonsatti, E., Sigalotti, L., Coral, S., Colizzi, F., Altomonte, M. and Maio, M. (2003) Methylation-regulated expression of HLA class I antigens in melanoma. *International Journal of Cancer*, **105**, 430–431.

33 Fonsatti, E., Nicolay, H.J., Sigalotti, L., Calabro, L., Pezzani, L., Colizzi, F. *et al.* (2007) Functional up-regulation of human leukocyte antigen class I antigens expression by 5-aza-2′-deoxycytidine in cutaneous melanoma: immunotherapeutic implications. *Clinical Cancer Research*, **13**, 3333–3338.

34 Antequera, F. and Bird, A. (1999) CpG islands as genomic footprints of promoters that are associated with replication origins. *Current Biology*, **9**, R661–R667.

35 Bird, A. (2007) Perceptions of epigenetics. *Nature*, **447**, 396–398.

36 Sigalotti, L., Fratta, E., Coral, S., Cortini, E., Covre, A., Nicolay, H.J. *et al.* (2007) Epigenetic drugs as pleiotropic agents in cancer treatment: biomolecular aspects and clinical applications. *Journal of Cellular Physiology*, **212**, 330–334.

37 Hellebrekers, D.M., Griffioen, A.W. and van Engeland, M. (2007) Dual targeting of epigenetic therapy in cancer. *Biochimica et Biophysica Acta*, **1775**, 76–91.

38 Robertson, K.D., Uzvolgyi, E., Liang, G., Talmadge, C., Sumegi, J., Gonzales, F.A. *et al.* (1999) The human DNA methyltransferases (DNMTs), 1, 3a and 3b: coordinate mRNA expression in normal tissues and overexpression in tumors. *Nucleic Acids Research*, **27**, 2291–2298.

39 Tomasi, T.B., Magner, W.J. and Khan, A.N. (2006) Epigenetic regulation of immune escape genes in cancer. *Cancer Immunology, Immunotherapy*, **55**, 1159–1184.

40 Klose, R.J. and Bird, A.P. (2006) Genomic DNA methylation: the mark and its mediators. *Trends in Biochemical Sciences*, **31**, 89–97.

41 Nephew, K.P. and Huang, T.H. (2003) Epigenetic gene silencing in cancer initiation and progression. *Cancer Letters*, **190**, 125–133.

42 Marks, P.A. and Breslow, R. (2007) Dimethyl sulfoxide to vorinostat:

development of this histone deacetylase inhibitor as an anticancer drug. *Nature Biotechnology*, **25**, 84–90.

43 Liang, G., Lin, J.C., Wei, V., Yoo, C., Cheng, J.C., Nguyen, C.T. et al. (2004) Distinct localization of histone H3 acetylation and H3-K4 methylation to the transcription start sites in the human genome. *Proceedings of the National Academy of Sciences of the United States of America*, **101**, 7357–7362.

44 Margueron, R., Trojer, P. and Reinberg, D. (2005) The key to development: interpreting the histone code? *Current Opinion in Genetics & Development*, **15**, 163–176.

45 Schubeler, D., MacAlpine, D.M., Scalzo, D., Wirbelauer, C., Kooperberg, C., van Leeuwen, F. et al. (2004) The histone modification pattern of active genes revealed through genome-wide chromatin analysis of a higher eukaryote. *Genes and Development*, **18**, 1263–1271.

46 Lund, A.H. and van Lohuizen, M. (2004) Epigenetics and cancer. *Genes and Development*, **18**, 2315–2335.

47 Feizi, T. (1985) Demonstration by monoclonal antibodies that carbohydrate structures of glycoproteins and glycolipids are onco-developmental antigens. *Nature*, **314**, 53–57.

48 Jotti, G.S. and Bombardieri, E. (1990) Circulating tumor markers in breast cancer (review). *Anticancer Research*, **10**, 253–258.

49 Renkvist, N., Castelli, C., Robbins, P.F. and Parmiani, G. (2001) A listing of human tumor antigens recognized by T cells. *Cancer Immunology, Immunotherapy*, **50**, 3–15.

50 Li, G., Miles, A., Line, A. and Rees, R.C. (2004) Identification of tumour antigens by serological analysis of cDNA expression cloning. *Cancer Immunology, Immunotherapy*, **53**, 139–143.

51 Campoli, M.R., Chang, C.C., Kageshita, T., Wang, X., McCarthy, J.B. and Ferrone, S. (2004) Human high molecular weight-melanoma-associated antigen (HMW-MAA): a melanoma cell surface chondroitin sulfate proteoglycan (MSCP) with biological and clinical significance. *Critical Reviews in Immunology*, **24**, 267–296.

52 Mittelman, A., Chen, G.Z., Wong, G.Y., Liu, C., Hirai, S. and Ferrone, S. (1995) Human high molecular weight-melanoma associated antigen mimicry by mouse anti-idiotypic monoclonal antibody MK2-23: modulation of the immunogenicity in patients with malignant melanoma. *Clinical Cancer Research*, **1**, 705–713.

53 Luo, W., Wang, X., Kageshita, T., Wakasugi, S., Karpf, A.R. and Ferrone, S. (2006) Regulation of high molecular weight-melanoma associated antigen (HMW-MAA) gene expression by promoter DNA methylation in human melanoma cells. *Oncogene*, **25**, 2873–2884.

54 Mukherjee, P., Ginardi, A.R., Madsen, C.S., Sterner, C.J., Adriance, M.C., Tevethia, M.J. et al. (2000) Mice with spontaneous pancreatic cancer naturally develop MUC-1-specific CTLs that eradicate tumors when adoptively transferred. *Journal of Immunology (Baltimore, Md: 1950)*, **165**, 3451–3460.

55 Singh, A.P., Chauhan, S.C., Bafna, S., Johansson, S.L., Smith, L.M., Moniaux, N. et al. (2006) Aberrant expression of transmembrane mucins, MUC1 and MUC4, in human prostate carcinomas. *Prostate*, **66**, 421–429.

56 Mesquita, P., Peixoto, A.J., Seruca, R., Hanski, C., Almeida, R., Silva, F. et al. (2003) Role of site-specific promoter hypomethylation in aberrant MUC2 mucin expression in mucinous gastric carcinomas. *Cancer Letters*, **189**, 129–136.

57 Gratchev, A., Siedow, A., Bumke-Vogt, C., Hummel, M., Foss, H.D., Hanski, M.L. et al. (2001) Regulation of the intestinal mucin MUC2 gene expression *in vivo*: evidence for the role of promoter methylation. *Cancer Letters*, **168**, 71–80.

58 Vincent, A., Perrais, M., Desseyn, J.L., Aubert, J.P., Pigny, P. and Van Seuningen, I. (2007) Epigenetic regulation (DNA methylation, histone modifications) of the 11p15 mucin genes (MUC2, MUC5AC, MUC5B, MUC6) in epithelial cancer cells. *Oncogene*, **26**, 6566–6576.

59 Sigalotti, L., Coral, S., Nardi, G., Spessotto, A., Cortini, E., Cattarossi, I. *et al.* (2002) Promoter methylation controls the expression of MAGE2, 3 and 4 genes in human cutaneous melanoma. *Journal of Immunotherapy (1997)*, **25**, 16–26.

60 Scanlan, M.J., Simpson, A.J. and Old, L.J. (2004) The cancer/testis genes: review, standardization, and commentary. *Cancer Immunity*, **4**, 1.

61 Simpson, A.J., Caballero, O.L., Jungbluth, A., Chen, Y.T. and Old, L.J. (2005) Cancer/testis antigens, gametogenesis and cancer. *Nature Reviews. Cancer*, **5**, 615–625.

62 Jager, E., Stockert, E., Zidianakis, Z., Chen, Y.T., Karbach, J., Jager, D. *et al.* (1999) Humoral immune responses of cancer patients against 'Cancer-Testis' antigen NY-ESO-1: correlation with clinical events. *International Journal of Cancer*, **84**, 506–510.

63 Stockert, E., Jager, E., Chen, Y.T., Scanlan, M.J., Gout, I., Karbach, J. *et al.* (1998) A survey of the humoral immune response of cancer patients to a panel of human tumor antigens. *The Journal of Experimental Medicine*, **187**, 1349–1354.

64 Jager, E., Chen, Y.T., Drijfhout, J.W., Karbach, J., Ringhoffer, M., Jager, D. *et al.* (1998) Simultaneous humoral and cellular immune response against cancer-testis antigen NY-ESO-1: definition of human histocompatibility leukocyte antigen (HLA)-A2-binding peptide epitopes. *The Journal of Experimental Medicine*, **187**, 265–270.

65 Scanlan, M.J., Gure, A.O., Jungbluth, A.A., Old, L.J. and Chen, Y.T. (2002) Cancer/testis antigens: an expanding family of targets for cancer immunotherapy. *Immunological Reviews*, **188**, 22–32.

66 Terando, A.M., Faries, M.B. and Morton, D.L. (2007) Vaccine therapy for melanoma: Current status and future directions. *Vaccine*, **25**, B4–16.

67 van der Bruggen, P. and Van den Eynde, B.J.V. (2006) Processing and presentation of tumor antigens and vaccination strategies. *Current Opinion in Immunology*, **18**, 98–104.

68 Zendman, A.J., Ruiter, D.J. and van Muijen, G.N. (2003) Cancer/testis-associated genes: identification, expression profile, and putative function. *Journal of Cellular Physiology*, **194**, 272–288.

69 Calabro, L., Fonsatti, E., Altomonte, M., Pezzani, L., Colizzi, F., Nanni, P. *et al.* (2005) Methylation-regulated expression of cancer testis antigens in primary effusion lymphoma: immunotherapeutic implications. *Journal of Cellular Physiology*, **202**, 474–477.

70 Gattei, V., Fonsatti, E., Sigalotti, L., Degan, M., Di Giacomo, A.M., Altomonte, M. *et al.* (2005) Epigenetic immunomodulation of hematopoietic malignancies. *Seminars in Oncology*, **32**, 503–510.

71 Maio, M., Coral, S., Fratta, E., Altomonte, M. and Sigalotti, L. (2003) Epigenetic targets for immune intervention in human malignancies. *Oncogene*, **22**, 6484–6488.

72 Jungbluth, A.A., Stockert, E., Chen, Y.T., Kolb, D., Iversen, K., Coplan, K. *et al.* (2000) Monoclonal antibody MA454 reveals a heterogeneous expression pattern of MAGE-1 antigen in formalin-fixed paraffin embedded lung tumours. *British Journal of Cancer*, **83**, 493–497.

73 Jungbluth, A.A., Chen, Y.T., Stockert, E., Busam, K.J., Kolb, D., Iversen, K. *et al.* (2001) Immunohistochemical analysis of NY-ESO-1 antigen expression in normal and malignant human tissues. *International Journal of Cancer*, **92**, 856–860.

74 De Smet, C., De, B.O., Faraoni, I., Lurquin, C., Brasseur, F. and Boon, T. (1996) The activation of human gene MAGE-1 in tumor cells is correlated with genome-wide demethylation. *Proceedings of the National Academy of Sciences of the United States of America*, **93**, 7149–7153.

75 De Smet, C., Loriot, A. and Boon, T. (2004) Promoter-Dependent Mechanism Leading to Selective Hypomethylation within the 5' Region of Gene MAGE-A1 in Tumor Cells. *Molecular and Cellular Biology*, **24**, 4781–4790.

76 Honda, T., Tamura, G., Waki, T., Kawata, S., Terashima, M., Nishizuka, S. et al. (2004) Demethylation of MAGE promoters during gastric cancer progression. *British Journal of Cancer*, **90**, 838–843.

77 De Smet, C., Courtois, S.J., Faraoni, I., Lurquin, C., Szikora, J.P., De Backer, O. et al. (1995) Involvement of two Ets binding sites in the transcriptional activation of the MAGE1 gene. *Immunogenetics*, **42**, 282–290.

78 Sigalotti, L., Fratta, E., Coral, S., Tanzarella, S., Danielli, R., Colizzi, F. et al. (2004) Intratumor heterogeneity of cancer/testis antigens expression in human cutaneous melanoma is methylation-regulated and functionally reverted by 5-aza-2'-deoxycytidine. *Cancer Research*, **64**, 9167–9171.

79 Sigalotti, L., Covre, A., Zabierowski, S., Himes, B., Colizzi, F., Natali, P.G. et al. (2008) Cancer testis antigens in human melanoma stem cells: expression, distribution, and methylation status. *Journal of Cellular Physiology*, **215**, 287–291.

80 Coral, S., Sigalotti, L., Altomonte, M., Engelsberg, A., Colizzi, F., Cattarossi, I. et al. (2002) 5-aza-2'-Deoxycytidine-induced expression of functional cancer testis antigens in human renal cell carcinoma: immunotherapeutic implications. *Clinical Cancer Research*, **8**, 2690–2695.

81 Roman-Gomez, J., Jimenez-Velasco, A., Agirre, X., Castillejo, J.A., Navarro, G., San Jose-Eneriz, E. et al. (2007) Epigenetic regulation of human cancer/testis antigen gene, HAGE, in chronic myeloid leukemia. *Haematologica*, **92**, 153–162.

82 Roman-Gomez, J., Jimenez-Velasco, A., Agirre, X., Castillejo, J.A., Navarro, G., Jose-Eneriz, E.S. et al. (2007) Epigenetic regulation of PRAME gene in chronic myeloid leukemia. *Leukemia Research*, **31**, 1521–1528.

83 Coral, S., Sigalotti, L., Colizzi, F., Spessotto, A., Nardi, G., Cortini, E. et al. (2006) Phenotypic and functional changes of human melanoma xenografts induced by DNA hypomethylation: immunotherapeutic implications. *Journal of Cellular Physiology*, **207**, 58–66.

84 Coral, S., Sigalotti, L., Covre, A., Nicolay, H.J., Natali, P.G. and Maio, M. (2007) 5-AZA-2'-deoxycytidine in cancer immunotherapy: a mouse to man story. *Cancer Research*, **67**, 2900–2901.

85 Schrump, D.S. and Nguyen, D.M. (2005) Targeting the epigenome for the treatment and prevention of lung cancer. *Seminars in Oncology*, **32**, 488–502.

86 Weiser, T.S., Guo, Z.S., Ohnmacht, G.A., Parkhurst, M.L., Tong-On, P., Marincola, F.M. et al. (2001) Sequential 5-Aza-2 deoxycytidine-depsipeptide FR901228 treatment induces apoptosis preferentially in cancer cells and facilitates their recognition by cytolytic T lymphocytes specific for NY-ESO-1. *J Immunother (1997)*, **24**, 151–161.

87 Weiser, T.S., Ohnmacht, G.A., Guo, Z.S., Fischette, M.R., Chen, G.A., Hong, J.A. et al. (2001) Induction of MAGE-3 expression in lung and esophageal cancer cells. *The Annals of Thoracic Surgery*, **71**, 295–301.

88 Wischnewski, F., Pantel, K. and Schwarzenbach, H. (2006) Promoter demethylation and histone acetylation mediate gene expression of MAGE-A1, -A2, -A3, and -A12 in human cancer cells. *Molecular Cancer Research*, **4**, 339–349.

89 Altomonte, M., Fonsatti, E., Visintin, A. and Maio, M. (2003) Targeted therapy of solid malignancies via HLA class II antigens: a new biotherapeutic approach? *Oncogene*, **22**, 6564–6569.

90 Chou, S.D., Khan, A.N., Magner, W.J. and Tomasi, T.B. (2005) Histone acetylation regulates the cell type specific CIITA promoters, MHC class II expression and antigen presentation in tumor cells. *International Immunology*, **17**, 1483–1494.

91 Magner, W.J., Kazim, A.L., Stewart, C., Romano, M.A., Catalano, G., Grande, C. et al. (2000) Activation of MHC class I, II, and CD40 gene expression by histone deacetylase inhibitors. *Journal of Immunology (Baltimore, Md: 1950)*, **165**, 7017–7024.

92 Serrano, A., Tanzarella, S., Lionello, I., Mendez, R., Traversari, C., Ruiz-Cabello, F. et al. (2001) Rexpression of HLA class I antigens and restoration of antigen-specific CTL response in melanoma cells following 5-aza-2′-deoxycytidine treatment. *International Journal of Cancer*, **94**, 243–251.

93 Guo, Z.S., Hong, J.A., Irvine, K.R., Chen, G.A., Spiess, P.J., Liu, Y. et al. (2006) De novo induction of a cancer/testis antigen by 5-aza-2′-deoxycytidine augments adoptive immunotherapy in a murine tumor model. *Cancer Research*, **66**, 1105–1113.

94 Plimack, E.R., Kantarjian, H.M. and Issa, J.P. (2007) Decitabine and its role in the treatment of hematopoietic malignancies. *Leukemia and Lymphoma*, **48**, 1472–1481.

95 Issa, J.P., Garcia-Manero, G., Giles, F.J., Mannari, R., Thomas, D., Faderl, S. et al. (2004) Phase 1 study of low-dose prolonged exposure schedules of the hypomethylating agent 5-aza-2′-deoxycytidine (decitabine) in hematopoietic malignancies. *Blood*, **103**, 1635–1640.

96 Kantarjian, H., Oki, Y., Garcia-Manero, G., Huang, X., O'Brien, S., Cortes, J. et al. (2007) Results of a randomized study of 3 schedules of low-dose decitabine in higher-risk myelodysplastic syndrome and chronic myelomonocytic leukemia. *Blood*, **109**, 52–57.

97 Oki, Y., Jelinek, J., Shen, L., Kantarjian, H.M. and Issa, J.P. (2008) Induction of hypomethylation and molecular response after decitabine therapy in patients with chronic myelomonocytic leukemia. *Blood*, **111**, 2382–2384.

98 Kantarjian, H.M., O'Brien, S., Cortes, J., Giles, F.J., Faderl, S., Issa, J.P. et al. (2003) Results of decitabine (5-aza-2′deoxycytidine) therapy in 130 patients with chronic myelogenous leukemia. *Cancer*, **98**, 522–528.

99 Wijermans, P., Lubbert, M., Verhoef, G., Bosly, A., Ravoet, C., Andre, M. et al. (2000) Low-dose 5-aza-2′-deoxycytidine, a DNA hypomethylating agent, for the treatment of high-risk myelodysplastic syndrome: a multicenter phase II study in elderly patients. *Journal of Clinical Oncology*, **18**, 956–962.

100 Momparler, R.L., Momparler, L.F. and Samson, J. (1982) Combinational chemotherapy of L1210 and L1210/ARA-C leukemia with 5-AZA-2′-deoxycytidine and beta-2′-deoxythioguanosine. *International Journal of Cancer*, **30**, 361–364.

101 Yang, A.S., Doshi, K.D., Choi, S.W., Mason, J.B., Mannari, R.K., Gharybian, V. et al. (2006) DNA methylation changes after 5-aza-2′-deoxycytidine therapy in patients with leukemia. *Cancer Research*, **66**, 5495–5503.

102 Sigalotti, L., Altomonte, M., Colizzi, F., Degan, M., Rupolo, M., Zagonel, V. et al. (2003) 5-Aza-2′-deoxycytidine (decitabine) treatment of hematopoietic malignancies: a multimechanism therapeutic approach? *Blood*, **101**, 4644–4646.

103 Aparicio, A., Eads, C.A., Leong, L.A., Laird, P.W., Newman, E.M., Synold, T.W. et al. (2003) Phase I trial of continuous infusion 5-aza-2′-deoxycytidine. *Cancer Chemotherapy and Pharmacology*, **51**, 231–239.

104 Samlowski, W.E., Leachman, S.A., Wade, M., Cassidy, P., Porter-Gill, P., Busby, L. et al. (2005) Evaluation of a 7-day continuous intravenous infusion of decitabine: inhibition of promoter-specific and global genomic DNA methylation. *Journal of Clinical Oncology*, **23**, 3897–3905.

105 Schrump, D.S., Fischette, M.R., Nguyen, D.M., Zhao, M., Li, X., Kunst, T.F. et al. (2006) Phase I study of decitabine-mediated gene expression in patients with cancers involving the lungs, esophagus, or pleura. *Clinical Cancer Research*, **12**, 5777–5785.

106 Gollob, J.A., Sciambi, C.J., Peterson, B.L., Richmond, T., Thoreson, M., Moran, K. et al. (2006) Phase I trial of sequential low-dose 5-aza-2′-deoxycytidine plus high-dose intravenous bolus interleukin-2 in patients with melanoma or renal cell carcinoma. *Clinical Cancer Research*, **12**, 4619–4627.

107 Minucci, S. and Pelicci, P.G. (2006) Histone deacetylase inhibitors and the promise of epigenetic (and more) treatments for cancer. *Nature Reviews. Cancer*, **6**, 38–51.

108 Bolden, J.E., Peart, M.J. and Johnstone, R.W. (2006) Anticancer activities of histone deacetylase inhibitors. *Nature Reviews. Drug Discovery*, **5**, 769–784.

109 Fradet, Y., Picard, V., Bergeron, A. and Larue, H. (2005) Cancer-testis antigen expression in bladder cancer. *Progrès en Urologie*, **15**, 1303–1313.

110 Maeda, T., Towatari, M., Kosugi, H. and Saito, H. (2000) Up-regulation of costimulatory/adhesion molecules by histone deacetylase inhibitors in acute myeloid leukemia cells. *Blood*, **96**, 3847–3856.

9
Cancer Testis Antigens

Jonathan Cebon, Otavia Caballero, Thomas John, and Oliver Klein

9.1
Introduction

Shared antigens are critical for the development of pharmaceutical-grade T cell vaccines to treat or prevent cancer. As targets that can be recognized by both autologous and allogeneic T lymphocytes, they have the potential for the treatment of patient populations with standardized vaccines. The first of these antigens to be discovered were intracellular molecules detected in melanoma patients and included differentiation antigens and a family of molecules expressed both in germ cells and in cancer. The prototype was named Melanoma Antigen (MAGE)-1 [1] and represents a member of a substantial family of homologous molecules [2]. Subsequently a large number of antigens was discovered, which did not share homology with the MAGE family, but are also predominantly expressed in the germ line and reactivated in different cancer types. These have been broadly classified as 'Cancer/Testis' (CT) antigens [3]. In addition to sharing a characteristic pattern of tissue distribution, they are often immunogenic, which makes them attractive targets for cancer immunotherapy [4]. Immunohistochemistry revealed that a common feature of CT antigens is their heterogeneous expression [5, 6]. In addition an understanding of their function is starting to emerge. While heterogeneity might suggest that these may be suboptimal targets for cancer treatment, there are early indications that CT antigens may provide a survival advantage to the cancer [7]. Taken together these observations have led to speculation that CT antigen expression may characterize a subset of cells within the tumor that recapitulate the function of gametic cells [8] or are critical for tumor survival such as stem-like cells [9, 10]. If so, they may prove to be ideal targets for immunotherapy. Indeed there are early suggestions that this may be the case. Encouraging trends in a randomized trial with a MAGE-A3 vaccine in the adjuvant setting in lung cancer and promising clinical outcomes in an early NY-ESO-1 trial in melanoma patients [10] both suggest that vaccinating against CT antigens can prevent cancer relapse. This has encouraged the pharmaceutical industry to seriously pursue the development of anti-cancer vaccines based on full-length CT protein antigens.

Tumor-Associated Antigens. Edited by Olivier Gires and Barbara Seliger
Copyright © 2009 WILEY-VCH Verlag GmbH & Co. KGaA, Weinheim
ISBN: 978-3-527-32084-4

9.2
Definitions and Classification

CT antigens share many of the following characteristics: (i) they are expressed in a wide range of both solid organ and hematological malignancies, but their expression in normal tissues is restricted to germ cells within the testis, immature germ cells in the fetal ovary, and also in trophoblasts; (ii) their expression programs are probably tightly regulated by epigenetic mechanisms such as DNA methylation; and (iii) they are immunogenic and capable of eliciting cellular and/or humoral immune responses [3, 8]. So far, more than 100 CT genes or gene families have been described in the literature and detailed information about their expression profiles, immunogenicity, function (where known), gene structure, location, and orthologous groups is currently being assembled in a comprehensive database (http://www.cta.lncc.br) [11].

A selected subset of these CT antigens is shown in Table 9.1. These genes can be divided between those encoded on the X chromosome (CT-X), which are highly

Table 9.1 Selected Cancer/Testis antigens.

	Method of identification	Chromosome location	References (PMID)
CT-X			
MAGEA	T cell epitope cloning	Xq28	1840703
MAGEB	Positional cloning	Xp21-p22	7761436
GAGE	T cell epitope cloning	Xp11	7544395
SSX	SEREX	Xp11	9378559
NY-ESO-1	SEREX	Xq28	9050879
MAGEC1	Representational difference analysis	Xq26, Xq27	9485030
MAGEC2	Representational difference analysis	Xq27	10699956
CTp11/SPANX	Differential display	Xq27	10626816
XAGE1	Database mining	Xp11	10197611
Cancer/testis antigen CT45	Database mining	Xq26	15905330
Cancer/testis antigen CT47	Structural analysis of chromosome X	Xq24	16382448
PASD1	Database mining	Xq28	15905330
	SEREX		15162151
Non-X CTs			
BAGE	T cell epitope cloning	21p11.1	7895173
SYCP1	SEREX	1p13-p12	9560255
DDX43 (HAGE)	Representational difference analysis	6q12-q13	10919659
CTAGE1	SEREX	18p11	11149944
TPTE	Database mining	21p11	12865919
HORMAD1	Database mining	1q21	15999985
COX6B2	Database mining	19q13	15905330
SLCO6A1	SEREX	5q21	15546177
TSPY1	cDNA arrays	Yp11	16106251
ARMC3	SEREX	10p12	16397042
CEP290	SEREX	12q21	17192878
CRT2	Representational difference analysis	19p13	17975137

expressed in spermatogonia and belong to a number of gene families and non-X CT genes, which are encoded on other chromosomes. Non-X CT genes tend to be expressed during later stages of spermatogenesis and are less frequently members of multi-gene families. Other differences between these two groups of CT antigens include the less restricted pattern of expression of most of the non-X CTs, while most of the CT-X antigens are not expressed in normal somatic tissues [11]. Interestingly, the vast majority of the non-X CTs are conserved during evolution, while most of the CT-X are genes exclusive to *Homo sapiens* and primates and represent recently expanded families on the X chromosome [12]. Although the identification of the several CT antigens, including members of the MAGE, GAGE, and SSX families, was based on their immunogenic properties [1, 13, 14], more recently, the identification of newer CT antigens has been mostly undertaken by mining gene expression databases for genes with restricted expression in testis and tumors [15–17]. Therefore, for many of these newer antigens, information about their immunogenic properties is limited.

9.3
Tissue Distribution

9.3.1
Normal tissues

CT antigens are characterized by their limited expression in normal non-gametogenic tissues. However, some CT gene transcripts are present in normal tissues. Their expression patterns in these non-gametogenic tissues has recenty been surveyed and described in detail [18]. This has resulted in both the CT-X and non-X CT genes being divided into three categories: (1) those that are restricted to testis such as MAGEA1, 2, MAGEB1, 3, 4, 5, 6, SSX2, 3, NY-ESO-1, MAGEC1 and SPANX; (2) Testis/brain restricted CT genes such as MAGEC2, 3, MAGEA9, GAGE1, 2, 4, 5, 7 & 8 and PAGE3, where the distribution in non-gametogenic tissue is limited to brain; and (3) testis-selective CT genes such as MAGEA3, 5, 6, 8, 11 & 12, XAGE1, 2 & 3, PLAC1 and SSX4. For these, more widespread expression is observed in non-gametogenic tissues [18]. It is notable that expression in non-gametogenic tissues such as pancreas, spleen, and liver occurs at levels that are far below that observed in germ cells [3].

9.3.2
Tumors

A wide range of tumors express CT antigens. Bladder cancer, non-small cell lung cancer, melanoma, ovarian, and hepatocellular carcinomas frequently express distinct CT genes, whereas they are rarely expressed in lymphoma, leukemia, renal, colon, and gastric carcinomas. Antigen expression, as observed by immunohistochemical analysis, rarely demonstrates uniform patterns of staining [6, 19]. Even in tumors with strong expression of multiple CT antigens, subpopulations of positively staining cells are often observed, with other areas of the tumor remaining negative.

This staining pattern, when viewed in the context of normal tissue expression being restricted to gametogenic cells, has resulted in the hypothesis that the CT genes are expressed in a subpopulation of primitive cells via a process known as 'gametic recapitulation' [8, 9]. This term hypothesizes the ability of some tumors to acquire the multipotent attributes of gametogenic cells in order to facilitate tumor growth and self renewal. A more detailed description of such functions, where known, appears in the following section.

These molecules, particularly the CT-X antigens, are often not just co-expressed in the one tumor but are also coordinately expressed [3, 20, 21]. NY-ESO-1 for example is very rarely found in the absence of MAGEA3, and tumors which express any of the other CT-X antigens almost always also express MAGEA3. Co-ordinate expression of the CT-X antigens is not surprising considering that the CT-X genes studied, which lie in close physical proximity to one another, are epigenetically silenced as a result of promoter methylation [22, 23]. Experimental demethylation has been shown to result in expression of CT antigens in cells that do not normally express them [24, 25]. The process of tumorigenesis is known to result in global DNA and gene-specific hypomethylation, which may explain why CT antigens are not normally found in somatic tissues but may be expressed aberrantly through hypomethylating events in cancers.

The appearance of CT antigens in primitive cell populations has also led to the hypothesis that they are likely to mark cancer stem cells [9] and preliminary evidence supports this in melanoma cell lines [10]. This has raised concerns about the potential for these molecules to serve as targets on non-cancerous somatic stem cells. It is therefore reassuring that none of the vaccine trials to date which have targeted CT antigens have resulted in clinical evidence of toxicity to these cell populations.

There is also emerging evidence that some of these molecules may play a role in early embryogenesis and could therefore be considered to be 'Embryo-Cancer antigens' [26]. One such molecule is Embryo-Cancer Sequence-A (ECSA), which has been entered into public databases as Developmental Pluripotency Associated-2 (DPPA-2). This gene localized on chromosome 3 was initially isolated from pluripotent cells within the pre-implantation blastocyst [27] and later found to be expressed in a variety of tumors as well as being immunogenic [26]. In non-small cell lung cancer, ECSA/DPPA2 marked small subpopulations of putative stem cells. Furthermore, examination of more than 100 lung cancers revealed that the expression of ECSA/DPPA2 significantly increased the likelihood of expression of CT antigens, particularly those coded on the X chromosome. This pattern of distribution in cancer, stem cells, and pre-implantation embryos is shared by other molecules such as OCT3/4 [28–30], SOX-2 [31, 32], and NANOG [33] which might therefore also be viewed as Embryo-Cancer Antigens.

9.4
Function

The CT antigens located on chromosome X (CT-X), which represent about one-half of CT genes described to date [3], remain functionally uncharacterized both in the physiological context and in cancer development. In contrast, those that map to other chromosomes are very frequently involved in various aspects of spermatogenesis [16, 34–36].

In several different tumor types, the expression of CT-X genes is associated with advanced disease and poor outcome, and indicates that their expression might contribute to tumorigenesis. In NSCLC, cancer-testis gene expression, either cumulatively or individually, showed significant associations with advanced tumor type, nodal and pathologic stages as well as pleural invasion [20]. Likewise, expression of *MAGEA3* in pancreatic ductal adenocarcinoma was found to be a prognostic factor for poor survival [37]. In colorectal cancers, the expression of *NY-ESO-1* may serve as a marker for local metastasis and advanced disease, while the expression of *MAGEA4* was significantly associated with the vessel emboli [38]. Similarly, expression of MAGEA1, MAGEA3, MAGEA4, MAGEC1, and NY-ESO-1 in malignant gammopathies correlated with stage and risk status of disease [39–41]. In serous carcinomas of the ovary, MAGEA4 and NY-ESO-1 correlated with the degree of malignancy. Moreover, a significant inverse correlation was found between MAGEA4 expression and patient survival [42]. A direct correlation between CT-X antigen expression and tumor stage has been reported in tumors of the bladder [43, 44] and lung [45]. In melanoma, acquisition of MAGE antigen expression was associated with poorer prognosis (Svobodova S. *et al.* unpublished data), and tumor progression [6, 46]. NY-ESO-1 was found to be more frequently expressed in metastatic than in primary melanomas and its expression was associated with thicker primary lesions and a higher frequency of metastatic disease, indicative of a worse prognosis [47].

The biological role of CT-X in both germ line and tumors remains poorly understood and restricted mostly to members of the MAGE family. Recent functional genomic and proteomic studies have provided evidence that the CT-X genes are related to neoplastic cell processes. Using a yeast two-hybrid assay, the transcriptional regulator SKIP was identified as a MAGEA1 binding partner [48]. SKIP connects DNA-binding proteins to other proteins that either activate or repress transcription, and participates in a range of signaling pathways, including those involving vitamin D, retinoic acid, estrogens, glucocorticoids, Notch1, and transforming growth factor-β. In the Notch1 pathway, MAGEA1 was found to disrupt SKIP-mediated Notch1 signal transduction by binding to SKIP and recruiting histone deacetylase, acting therefore as a transcriptional repressor. Yeast two-hybrid studies using other cancer-related genes as bait have twice identified MAGE proteins: MAGEA11 and MAGEA4 [49, 50]. MAGEA11 was found to have a role in the regulation of androgen-receptor function by modulating its internal domain interactions and was found to have a dual amplifying effect on androgen signaling [49]. MAGEA4 was identified in a search for binding partners of the oncoprotein gankyrin [50]. Gankyrin destabilizes the retinoblastoma tumor suppressor, contributing to unscheduled entry into the cell cycle and escape from cell-cycle arrest and/or apoptosis. MAGEA4 suppresses the oncogenic activity of gankyrin through the action of a peptide that is naturally cleaved from the carboxyl terminus of MAGEA4 and which induces p53-dependent and p53-independent apoptosis. Overexpression of MAGEA4 in human embryonic kidney cells (293 cells) was found to increase apoptosis as measured by the apoptotic index and caspase-3 activity, while MAGE-A4 silencing using a small interfering RNA (siRNA) approach resulted in decreased caspase-3 activity in a squamous cell lung cancer and in 293/MAGE-A4 cells [51]. Yeast two-hybrid screening was used to identify putative MAGE homology-domain (MHD)-interacting proteins. The MHD of MAGE-C1 was used as bait to screen a human testis cDNA library, leading to NY-ESO-1 being identified as a MAGEC1 binding partner and for the first time, a direct interaction between two CT antigens being described. This interaction may be functionally relevant due to the frequently coordinated expression of these proteins [9]. MAGEA2 protein was shown to confer wild-type-p53-sensitive resistance to etoposide (ET) by inducing a novel p53 inhibitory loop involving recruitment of histone deacetylase 3 (HDAC3) to MAGEA2–p53 complex, thus strongly down-regulating the transactivation function of p53 [52]. In addition, combined trichostatin A-ET treatment in melanoma cells expressing high MAGE-A levels reestablishes the p53 response and reverses the chemoresistance [52]. Multiple MAGE proteins including human MAGEA3, human MAGEC2, and murine Mage-b1 (mMage-b) proteins form complexes with Kap-1, a known co-repressor of p53 [7]. SiRNA suppression of MAGE-A3, mMage-b, and MAGE-C2 genes induces apoptosis and causes increased p53 expression *in vitro*. Furthermore, it was shown that treatment with MAGE-specific siRNA suppresses melanoma growth *in vivo* and therefore, MAGE gene expression may protect cells from programmed cell death and may actively contribute to the development of malignancies by promoting survival [7]. In summary, although very little is known of the function(s) of these antigens, recent discoveries indicate that the expression of CT-X may contribute to the proliferative potential of neoplasms and their survival. These molecules may therefore also represent targets for other therapeutic modalities in addition to immunotherapy, such as rational drug design or gene targeting.

9.5
Immunology

Immunogenicity is a hallmark of many CT antigens. The first MAGE family members were discovered by characterizing the molecular specificity of $CD8^+$ T lymphocyte targets in a melanoma patient [1]; many other CT antigens were subsequently identified using immunological responses as 'probes' to identify their targets. These approaches are not restricted to T lymphocytes. Antibodies have also proven to be useful in the identification of anti-cancer immune responses that spontaneously arise in cancer patients (see related chapters 3 and 4) even though these molecules are intracellular and therefore only capable of being targeted using the cellular immune system. One application that has revealed numerous CT antigens among the antibody repertoire in cancer-bearing subjects is the SEREX technique [53]. DNA expression libraries (generally constructed from patient tumor samples) are screened using autologous sera to identify those proteins that are recognized by auto-antibodies in that patient. A comprehensive list of antigens identified by SEREX can be found at http://www.cancerimmunity.org/SEREX/index.htm.

The high degree of immunogenicity directed against CT antigens likely reflects their lack of expression by normal somatic tissues. Germ cells reside in immune-privileged sites, so their antigens are not generally exposed to immune recognition. As a result the immune system remains 'ignorant' rather than 'tolerant' of them so that subsequent expression by tumor cells or vaccination can readily result in an immune response. NY-ESO-1 represents a particularly well studied example [4]. First identified by the SEREX method, it has been shown to induce antibody responses in a substantial proportion of patients whose tumors express the antigen [54]. Subsequently the breadth of cellular responses either occurring spontaneously or after vaccination has been mapped and numerous epitopes identified by both $CD4^+$ and $CD8^+$ T lymphocytes have been characterized [4]. For the other CT antigens the extent of immune characterization varies. Information relating to the CT antigens is regularly updated at http://www.cta.lncc.br.

9.6
Clinical Trials

Clinical vaccine trials targeting CT antigens have been designed to elicit cellular immune responses, mainly mediated by $CD8^+$, but also $CD4^+$ T lymphocytes. These include peptide, protein, dendritic cell (DC) studies, and more recently trials with recombinant viruses expressing the tumor antigens. Most of the clinical experience in vaccine trials with CT antigens relates to NY-ESO-1 and MAGE-3; a few trials used other CT antigens such as MAGE-4 or MAGE-10, mostly in peptide or dendritic cell studies. Table 9.2 shows selected trials with CT antigens. The outstanding feature that is common to all of these trials is the lack of autoimmune toxicity. This presumably reflects the lack of expression of CT antigens on normal tissues.

The first vaccination studies with CT antigens were peptide-based trials, which used synthetic peptides that bind HLA class I molecules and are recognized by $CD8^+$ T lymphocytes. A pioneering study was undertaken with a MAGE-A3 nona-peptide binding to the HLA-A1 molecule [55]. As with many of the other early trials using this approach, most of the patients had advanced malignant melanoma. Thirty-nine patients received three subcutaneous and intradermal vaccinations at monthly intervals. No significant side-effects were observed. Among the 25 patients who received all three vaccinations, tumor regression was observed in seven, mostly with cutaneous metastasis. In three patients, complete responses were observed with a sustained remission in two. Perhaps surprisingly, no T-cell immunological response against MAGE could be found in the peripheral blood of these patients.

The first NY-ESO-1 peptide epitopes recognized by $CD8^+$ T lymphocytes [56] were used in a vaccination study in 12 HLA-A2$^+$ patients with progressive NY-ESO-1-expressing metastatic tumors including melanoma, ovarian and breast cancer [57]. The peptides were injected intradermally, concurrently with granulocyte macrophage colony stimulating factor (GM-CSF) once weekly for 4 weeks and further doses were administered to patients whose cancers did not progress. In contrast to the

Table 9.2 Selected clinical trials with CT antigens.

Antigen	Form	Patients	Comments	Reference
NY-ESO-1				
	Peptide			
	p157-167, p157-165, p155-163; GM-CSF as adjuvants	12 patients with metastatic tumors of different types (melanoma, ovarian cancer and breast cancer)	Eight patients showed disease stabilization or regression of individual metastases	[57]
	p157-165V, p161-180	37 patients with metastatic melanoma	One partial remission	[60]
	p157-170; incomplete Freund's Adjuvant	18 patients with epithelial ovarian cancer; most of them stage IIIC	One complete remission	[75]
	p157-165, p 157-167, in combination with FLT3L alone or together with Imiquimod	18 patients with melanoma, Stage II-IV	One partial remission; one patient showed disease stabilization	[58]
	Matured myeloid blood dendritic cells pulsed with p157-165 together with MAGE-A4 (230-239) and MAGE-A10 (254-262)	Six patients with resected melanoma, Stage II-IV		[63]
	Protein			
	NY-ESO-1/ISCOMATRIX	46 patients with full resected melanoma and high risk of recurrence	Patients receiving NY-ESO-1/ISCOMATRIX relapsed less frequently than patients vaccinated with protein only or placebo	[67]
	CHP-NY-ESO-1	Nine patients with esophageal cancer, prostate cancer and melanoma	Mixed tumor responses in three patients and disease stabilization in four patients	[68]

Recombinant virus construct			
Recombinant vaccinia NY-ESO-1, recombinant fowlpox-NY-ESO-1	16 patients with metastatic disease (melanoma, malignant teratoma, head and neck cancer, prostate cancer, endometrial cancer and sarcoma; seven patients with fully resected disease (melanoma, ovarian cancer, sarcoma and breast cancer)	One complete response, one mixed response and six patients with disease stabilization in the group with metastatic disease. Six patients in group for adjuvant treatment are free of disease	[73]
MAGE-A3			
Peptide			
MAGE-3.A1	39 patients with metastatic melanoma	Tumor regression in seven of 25 patients who completed the vaccination treatment	[55]
Monocyte derived Dendritic cells pulsed with MAGE-3.A1	11 patients with metastatic melanoma	Six mixed responses	[61]
Protein			
MAGE-A3 with adjuvant SBAS-2	57 patients with metastatic disease, mostly melanoma	Five patients with tumor regression including three partial remissions	[65]
MAGE-3	32 patients with metastatic melanoma limited to lymph nodes or skin	One partial remission; four mixed responses	[66]
MAGE-A3 with adjuvant AS02B	Randomized trial in 182 patients with resected Stage IB and II Non-small lung cancer	Improvement in disease-free survival in patients receiving the vaccine	[71]
Recombinant Virus construct			
ALVAC canarypox virus containing minigene coding for a MAGE-A3 and MAGE-A1 antigen	40 patients with advanced cancer (melanoma, NSCLC, head and neck cancer, esophageal cancer or bladder cancer)	One partial response; five mixed responses; two patients with stable disease	[74]

previously-mentioned MAGE-A3 peptide vaccination study, spontaneous and induced immune responses could be detected in the peripheral blood of these patients. These responses were assessed according the absence or presence of spontaneous immunity to NY-ESO-1 as evidenced by the detection of NY-ESO-1-specifc antibody. $CD8^+$ T cell responses were found in four of seven antibody-negative patients and were associated with disease stabilization and regression of single tumor lesions. In the group of antibody-positive patients disease stabilization and regression of some metastases were observed. In a parallel study, a vaccine, which combined the DC growth factor Flt3 ligand with a mixture of peptides, including several different NY-ESO-1 peptides, was tested in a series of melanoma patients. The NY-ESO-1 peptides were highly immunogenic, however when used alone Flt3 ligand was a suboptimal cytokine adjuvant. The effectiveness of this approach was improved by pre-conditioning the cutaneous vaccine site with the TLR7 ligand imiquimod [58]. Polyepitope vaccines may enhance the efficacy of a vaccine by increasing the possibility of targeting tumor cells that have lost one or more of the original TAA epitopes [59]. Several polyepitope vaccines, which included NY-ESO-1 peptides, have been reported (Table 9.2).

Strategies to improve the effectiveness of peptide-based vaccines have included the use of peptide analogs that have a higher affinity for the HLA molecule thereby increasing the immunogenicity of the antigen, and the implementation of peptides recognized by $CD4^+$ T lymphocytes as well as the use of a combination of HLA class I and class II epitopes for vaccination. The rationale behind this approach is to provide $CD4^+$ T lymphocyte 'help' for the induction of $CD8^+$ T cell responses. These approaches were tested in a trial of 37 patients with metastatic melanoma who received either an injection of an NY-ESO-1 analog peptide restricted by HLA-A2 alone or in combination with an NY-ESO epitope that binds to the HLA class II molecule DP4 [60]. Most of the vaccinated patients developed an immune response to the vaccinating peptide, but the T cells were only able to recognize tumor cells in a small minority of patients. In this study only one partial clinical response was seen.

Several studies have used DC as 'cellular adjuvants' pulsed with peptides or with full-length CT antigen alone or in combination with other antigens. One of the early DC trials used matured monocyte-derived DC pulsed with the MAGE-3.A1 peptide in 11 patients with advanced melanoma [61]. In six out of 11 patients tumor regression was observed although this did not translate to an objective clinical response. Subsequent studies have used DC generated from $CD34^+$ hematopoietic progenitor cells [62] or myeloid dendritic cells isolated directly from peripheral blood [63]. The small numbers of patients included in these studies and the different patient populations in which these trials were undertaken, for example fully resected versus advanced disease, make it difficult to draw any conclusion about the superiority of DC from one source over another for the induction of immune and clinical responses. Disappointingly, a recent phase III trial using monocyte-derived dendritic cells pulsed with multiple peptides including different MAGE-A3 and MAGE-1 peptides failed to demonstrate superiority over standard chemotherapy in patients with metastatic melanoma [64].

The use of full-length antigens, such as recombinant proteins for vaccination, offers the potential advantage of inducing a broader immune response involving both $CD8^+$ and $CD4^+$ T-cell responses against multiple epitopes. Also, full-length antigens incorporate every potential epitope of a tumor antigen, while synthetic peptides, even for the case of multiple epitopes only represent a selection. The availability of more epitopes also reduces eligibility constraints imposed by the HLA requirements of individual peptides. Full-length proteins may therefore be better vaccine candidates than shorter peptides. In one of the first protein vaccination trials, the MAGE-A3 protein in combination with the saponin-containing SBAS2 and the bacterial cell wall component MPL as adjuvants, were used to immunize patients with advanced cancer of different histologies such as patients with malignant melanoma, non-small cell lung cancer, head and neck carcinoma, esophageal and bladder cancer [65]. The vaccine was injected intramuscularly four times at 3 weeks intervals. A total number of 59 patients was enrolled, but only 39 patients received all four vaccinations, the majority of which were patients with malignant melanoma. Signs of tumor regression were observed in four patients, but only two experienced an objective clinical response. The vaccine was generally well tolerated. Interestingly, as with the MAGE-A3 peptide studies, most of the responses were found in patients with soft tissue metastases. A subsequent study with the MAGE-A3 protein alone without adjuvant was carried out in 30 patients with malignant melanoma limited to soft tissue and lymph nodes [66]. Tumor regression was reported in five patients with one experiencing a partial response and four a mixed response.

The first vaccination study with the NY-ESO-1 protein was undertaken in patients with resected NY-ESO-1-expressing cancers (predominantly malignant melanoma) and a high risk of recurrence [67]. The protein was injected intramuscularly alone or in the combination with the saponin-based adjuvant ISCOMATRIX. Three injections were administered at 4-week intervals. The vaccine was safe and well tolerated. Twenty patients received the full dose of recombinant protein with adjuvant and these developed a broad cellular and humoral immune response against multiple epitopes. The clinical impression was that vaccinated patients relapsed less frequently than control patients [10]. This has prompted a larger prospective randomized phase II trial to confirm these results. A subsequent trial with the same vaccine in patients with metastatic melanoma demonstrated no clinical responses (unpublished data). A smaller study in patients with advanced esophageal and prostate cancer used the NY-ESO-1 protein in combination with the adjuvant CHP [68]. Immunological responses against NY-ESO-1 were induced in these patients and tumor responses were seen in four patients. Although the numbers are small, the clinical benefit in terms of durable tumor remissions appears to be limited. There are several explanations for this, limited or heterogeneous antigen expression is the most likely reason, although in the setting of advanced disease, immunoregulation and immunosuppression are also likely to be implicated in the inhibition of effective anti-tumor immune responses [69]. Ongoing and future trials include strategies aimed at disrupting this state of immunosuppression in order to make cancer vaccination more effective. An example of this approach is the depletion of regulatory T lymphocytes [70].

In contrast, 'adjuvant' vaccine trials with CT antigen in the case of fully resected cancers demonstrate promise in preventing disease recurrence in patients who remain at high risk of relapse. These include a recent trial with the MAGE-A3 protein together with the immunological adjuvant AS02B in patients with completely resected AJCC stage IB and II non-small cell lung cancer. These patients usually have a relapse risk of approximately 50%. In total 182 patients with MAGE-A3-positive tumors were included in this trial. Patients who received the vaccine had a reduced risk of relapse with a relative improvement in disease-free survival in 33% of patients. A confirmatory double-blind, randomized, placebo-controlled phase III trial is currently underway to confirm these promising results. In parallel a phase III trial will be undertaken with the same vaccine in patients with resected stage III melanoma [71].

An alternative strategy for delivering full-length CT antigens is the use of viral vectors such as pox viruses, encoding the target antigen. These provide added immune activation by stimulating pattern recognition receptors on antigen-presenting cells thereby presenting the tumor antigen in a more inflammatory environment [72]. A study with NY-ESO-1 used recombinant vaccinia and fowl pox constructs as the antigen delivery system in patients with advanced NY-ESO-1 expressing cancers [73]. Thirty-five patients were included in this trial and 23 of these completed four vaccinations. Two groups of patients received only recombinant vaccina virus or fowl pox virus, another group was treated with a prime-boost strategy with an initial injection of vaccina virus followed by fowl pox virus. The vaccine was well tolerated and in most of the patients an immune response against NY-ESO-1 was induced. From the group of patients who completed all four vaccinations 16 had measurable disease and seven had completely resected disease. All responding patients had malignant melanoma. In patients with advanced disease, there was one complete remission, one minor response and one mixed response. Six patients achieved disease stabilization, one lasting over 31 months. Recombinant canary pox virus, encoding antigenic epitopes from MAGE-A3 and MAGE-A1 presented by the HLA-A1 and -B35, respectively, has also been studied [74]. Forty patients including 37 melanoma patients received four priming vaccinations at 3-week intervals with the virus construct followed by three MAGE-A3 and MAGE-A1 peptide 'boost' injections every 3 weeks. Thirty patients completed the trial and were able to be assessed. Only one patient had a partial response and disease stabilization was described in two others.

9.7
Conclusions

The attractiveness of the CT antigens as vaccine targets results from their expression by a variety of different cancer types and a relatively restricted pattern of expression in normal non-malignant tissues. This has resulted in a safety profile in early clinical trials that does not appear to include significant autoimmunity. Furthermore some CT antigens, such as NY-ESO-1, are particularly immunogenic, and robust immune responses were obtained following vaccination.

A substantial unresolved issue is whether or not heterogeneous expression within an individual cancer is an obstacle to effective anti-cancer immunity or not. Ongoing analysis of the CT antigen function in cancer cells and stem-like cells should help to resolve this issue. If they are found to be markers of critical repopulating cells within cancers, they will assume a central role in cancer vaccine strategies in the future.

References

1. van der Bruggen, P. *et al.* (1991) A gene encoding an antigen recognized by cytolytic T lymphocytes on a human melanoma. *Science*, **254** (5038), 1643–1647.
2. Boon, T. *et al.* (2006) Human T cell responses against melanoma. *Annual Review of Immunology*, **24**, 175–208.
3. Scanlan, M.J., Simpson, A.J. and Old, L.J. (2004) The cancer/testis genes: review, standardization, and commentary. *Cancer Immunology*, **4**, p. 1.
4. Nicholaou, T. *et al.* (2006) Directions in the immune targeting of cancer: lessons learned from the cancer-testis Ag NY-ESO-1. *Immunology and Cell Biology*, **84** (3), 303–317.
5. Vaughan, H.A. *et al.* (2004) Immunohistochemical and molecular analysis of human melanomas for expression of the human cancer-testis antigens NY-ESO-1 and LAGE-1. *Clinical Cancer Research*, **10** (24), 8396–8404.
6. Barrow, C. *et al.* (2006) Tumor antigen expression in melanoma varies according to antigen and stage. *Clinical Cancer Research*, **12** (3 Pt 1), 764–771.
7. Yang, B. *et al.* (2007) MAGE-A, mMage-b, and MAGE-C proteins form complexes with KAP1 and suppress p53-dependent apoptosis in MAGE-positive cell lines. *Cancer Research*, **67** (20), 9954–9962.
8. Simpson, A.J. *et al.* (2005) Cancer/testis antigens, gametogenesis and cancer. *Nature Reviews. Cancer*, **5** (8), 615–625.
9. Costa, F.F., Le Blanc, K. and Brodin, B. (2006) Cancer/Testis antigens, stem cells and cancer. *Stem Cells* **25**, 707–711.
10. Gedye, C. *et al.* (2006) NY-ESO-1 is a critical target for immune destruction of clonogenic melanoma stem-like cells, in *Cancer Immunotherapy 2006*, poster p-30.
11. Almeida, L.G. *et al.* (2009) CT database: a knowledge-base of high-throughput and curated data on cancer-testis antigens, *Nucleic Acids Research* **37**: D816–9, doi:10.1093/nar/gkn.673
12. Ross, M.T. *et al.* (2005) The DNA sequence of the human X chromosome. *Nature*, **434** (7031), 325–337.
13. De Backer, O. *et al.* (1999) Characterization of the GAGE genes that are expressed in various human cancers and in normal testis. *Cancer Research*, **59** (13), 3157–3165.
14. Gure, A.O. *et al.* (2002) The SSX gene family: characterization of 9 complete genes. *International Journal of Cancer*, **101** (5), 448–453.
15. Scanlan, M.J. *et al.* (2002) Identification of cancer/testis genes by database mining and mRNA expression analysis. *International Journal of Cancer*, **98** (4), 485–492.
16. Chen, Y.T. *et al.* (2005) Identification of cancer/testis-antigen genes by massively parallel signature sequencing. *Proceedings of the National Academy of Sciences of the United States of America*, **102** (22), 7940–7945.
17. Chen, Y.T. *et al.* (2005) Identification of CT46/HORMAD1, an immunogenic cancer/testis antigen encoding a putative meiosis-related protein. *Cancer Immunology*, **5**, 9.
18. Hofmann, O. *et al.* (2008) Genome-wide analysis of cancer/testis gene expression. *Proceedings of the National Academy of*

19 Jungbluth, A.A. et al. (2000) Expression of MAGE-antigens in normal tissues and cancer. *International Journal of Cancer*, **85** (4), 460–465.

20 Gure, A.O. et al. (2005) Cancer-testis genes are coordinately expressed and are markers of poor outcome in non-small cell lung cancer. *Clinical Cancer Research*, **11** (22), 8055–8062.

21 Sahin, U. et al. (1998) Expression of multiple cancer/testis (CT) antigens in breast cancer and melanoma: basis for polyvalent CT vaccine strategies. *International Journal of Cancer*, **78** (3), 387–389.

22 De Smet, C. et al. (1999) DNA methylation is the primary silencing mechanism for a set of germ line- and tumor-specific genes with a CpG-rich promoter. *Molecular and Cellular Biology*, **19** (11), 7327–7335.

23 Weber, J. et al. (1994) Expression of the MAGE-1 tumor antigen is up-regulated by the demethylating agent 5-aza-2′-deoxycytidine. *Cancer Research*, **54** (7), 1766–1771.

24 De Smet, C. et al. (1996) The activation of human gene MAGE-1 in tumor cells is correlated with genome-wide demethylation. *Proceedings of the National Academy of Sciences of the United States of America*, **93** (14), 7149–7153.

25 Coral, S. et al. (2002) 5-Aza-2′-deoxycytidine-induced expression of functional cancer testis antigens in human renal cell carcinoma: immunotherapeutic implications. *Clinical Cancer Research*, **8** (8), 2690–2695.

26 John, T. et al. (2008) ECSA/DPPA2 is an embryo-cancer antigen that is coexpressed with cancer-testis antigens in non-small cell lung cancer, *Clinical Cancer Research* **14**, 3291–3298.

27 Monk, M. and Holding, C. (2001) Human embryonic genes re-expressed in cancer cells. *Oncogene*, **20** (56), 8085–8091.

28 Ezeh, U.I. et al. (2005) Human embryonic stem cell genes OCT4, NANOG, STELLAR, and GDF3 are expressed in both seminoma and breast carcinoma. *Cancer*, **104** (10), 2255–2265.

29 Jin, T. et al. (1999) Examination of POU homeobox gene expression in human breast cancer cells. *International Journal of Cancer*, **81** (1), 104–112.

30 Atlasi, Y. et al. (2007) OCT-4, an embryonic stem cell marker, is highly expressed in bladder cancer. *International Journal of Cancer*, **120** (7), 1598–1602.

31 Gure, A.O. et al. (2000) Serological identification of embryonic neural proteins as highly immunogenic tumor antigens in small cell lung cancer. *Proceedings of the National Academy of Sciences of the United States*, **97** (8), 4198–4203.

32 Spisek, R. et al. (2007) Frequent and specific immunity to the embryonal stem cell-associated antigen SOX2 in patients with monoclonal gammopathy. *Journal of Experimental Medicine*, 20062387.

33 Hart, A.H. et al. (2005) The pluripotency homeobox gene NANOG is expressed in human germ cell tumors. *Cancer*, **104** (10), 2092–2098.

34 Alsheimer, M. et al. (2005) The cancer/testis antigen CAGE-1 is a component of the acrosome of spermatids and spermatozoa. *European Journal of Cell Biology*, **84** (2–3), 445–452.

35 de Wit, N.J. et al. (2004) Testis-specific human small heat shock protein HSPB9 is a cancer/testis antigen, and potentially interacts with the dynein subunit TCTEL1. *European Journal of Cell Biology*, **83** (7), 337–345.

36 Tureci, O. et al. (1998) Identification of a meiosis-specific protein as a member of the class of cancer/testis antigens. *Proceedings of the National Academy of Sciences of the United States of America*, **95** (9), 5211–5216.

37 Kim, J. et al. (2006) The clinical significance of MAGEA3 expression in pancreatic cancer. *International Journal of Cancer*, **118** (9), 2269–2275.

38 Li, M. et al. (2005) Expression profile of cancer-testis genes in 121 human

colorectal cancer tissue and adjacent normal tissue, *Clinical Cancer Research* **11**, 1809–1814.
39. Condomines, M. *et al.* (2007) Cancer/testis genes in multiple myeloma: expression patterns and prognosis value determined by microarray analysis. *Journal of Immunology (Baltimore, Md: 1950)*, **178** (5), 3307–3315.
40. Taylor, B.J. *et al.* (2005) SSX cancer testis antigens are expressed in most multiple myeloma patients: co-expression of SSX1, 2, 4, and 5 correlates with adverse prognosis and high frequencies of SSX,-positive PCs. *Journal of Immunotherapy*, **28** (6), 564–575.
41. Jungbluth, A.A. *et al.* (2005) The cancer-testis antigens CT7 (MAGE-C1) and MAGE-A3/6 are commonly expressed in multiple myeloma and correlate with plasma-cell proliferation. *Blood*, **106** (1), 167–174.
42. Yakirevich, E. *et al.* (2003) Expression of the MAGE-A4 and NY-ESO-1 cancer-testis antigens in serous ovarian neoplasms. *Clinical Cancer Research*, **9** (17), 6453–6460.
43. Patard, J.J. *et al.* (1995) Expression of MAGE genes in transitional-cell carcinomas of the urinary bladder. *International Journal of Cancer*, **64** (1), 60–64.
44. Kurashige, T. *et al.* (2001) Ny-ESO-1 expression and immunogenicity associated with transitional cell carcinoma: correlation with tumor grade. *Cancer Research*, **61** (12), 4671–4674.
45. Bolli, M. *et al.* (2002) Tissue microarray evaluation of Melanoma antigen E (MAGE) tumor-associated antigen expression: potential indications for specific immunotherapy and prognostic relevance in squamous cell lung carcinoma. *Annals of Surgery*, **236** (6), 785–793, discussion 793.
46. Gibbs, P. *et al.* (2000) MAGE-12 and MAGE-6 are frequently expressed in malignant melanoma. *Melanoma Research*, **10** (3), 259–264.
47. Velazquez, E.F. *et al.* (2007) Expression of the cancer/testis antigen NY-ESO-1 in primary and metastatic malignant melanoma (MM)—correlation with prognostic factors. *Cancer Immunology*, **7**, 11.
48. Laduron, S. *et al.* (2004) MAGE-A1 interacts with adaptor SKIP and the deacetylase HDAC1 to repress transcription. *Nucleic Acids Research*, **32** (14), 4340–4350.
49. Bai, S., He, B. and Wilson, E.M. (2005) Melanoma antigen gene protein MAGE-11 regulates androgen receptor function by modulating the interdomain interaction. *Molecular and Cellular Biology*, **25** (4), 1238–1257.
50. Nagao, T. *et al.* (2003) MAGE-A4 interacts with the liver oncoprotein gankyrin and suppresses its tumorigenic activity. *The Journal of Biological Chemistry*, **278** (12), 10668–10674.
51. Peikert, T. *et al.* (2006) Melanoma antigen A4 is expressed in non-small cell lung cancers and promotes apoptosis. *Cancer Research*, **66** (9), 4693–4700.
52. Monte, M. *et al.* (2006) MAGE-A tumor antigens target p53 transactivation function through histone deacetylase recruitment and confer resistance to chemotherapeutic agents. *Proceedings of the National Academy of Sciences of the United States of America*, **103** (30), 11160–11165.
53. Sahin, U. *et al.* (1995) Human neoplasms elicit multiple specific immune responses in the autologous host. *Proceedings of the National Academy Of Sciences of the United, States of America*, **92** (25), 11810–11813.
54. Stockert, E. *et al.* (1998) A survey of the humoral immune response of cancer patients to a panel of human tumor antigens. *Journal of Experimental Medicine*, **187** (8), 1349–1354.
55. Marchand, M. *et al.* (1999) Tumor regressions observed in patients with metastatic melanoma treated with an antigenic peptide encoded by gene MAGE-3 and presented by HLA-A1. *International Journal of Cancer*, **80** (2), 219–230.
56. Jager, E. *et al.* (1998) Simultaneous humoral and cellular immune response

against cancer-testis antigen NY-ESO-1: definition of human histocompatibility leukocyte antigen (HLA)-A2-binding peptide epitopes. *The Journal of Experimental Medicine*, **187** (2), 265–270.

57 Jager, E. et al. (2000) Induction of primary NY-ESO-1 immunity: CD8+ T lymphocyte and antibody responses in peptide-vaccinated patients with NY-ESO-1+ cancers. *Proceedings of the National Academy of Sciences of the United States of America*, **97** (22), 12198–12203.

58 Shackleton, M. et al. (2004) The impact of imiquimod, a Toll-like receptor-7 ligand (TLR7L), on the immunogenicity of melanoma peptide vaccination with adjuvant Flt3 ligand. *Cancer Immunology*, **4**, 9.

59 Baumgaertner, P. et al. (2006) Ex vivo detectable human CD8 T-cell responses to cancer-testis antigens. *Cancer Research*, **66** (4), 1912–1916.

60 Khong, H.T. et al. (2004) Immunization of HLA-A*0201 and/or HLA-DPbeta1*04 patients with metastatic melanoma using epitopes from the NY-ESO-1 antigen. *Journal of Immunotherapy (1997)*, **27** (6), 472–477.

61 Thurner, B. et al. (1999) Vaccination with mage-3A1 peptide-pulsed mature, monocyte-derived dendritic cells expands specific cytotoxic T cells and induces regression of some metastases in advanced stage IV melanoma. *The Journal of Experimental Medicine*, **190** (11), 1669–1678.

62 Banchereau, J. et al. (2001) Immune and clinical responses in patients with metastatic melanoma to CD34(+) progenitor-derived dendritic cell vaccine. *Cancer Research*, **61** (17), 6451–6458.

63 Davis, I.D. et al. (2006) Blood dendritic cells generated with Flt3 ligand and CD40 ligand prime CD8+ T cells efficiently in cancer patients. *Journal of Immunotherapy*, **29** (5), 499–511.

64 Schadendorf, D. et al. (2006) Dacarbazine (DTIC) versus vaccination with autologous peptide-pulsed dendritic cells (DC) in first-line treatment of patients with metastatic melanoma: a randomized phase III trial of the DC study group of the DeCOG. *Annals of Oncology*, **17** (4), 563–570.

65 Marchand, M. et al. (2003) Immunisation of metastatic cancer patients with MAGE-3 protein combined with adjuvant SBAS-2: a clinical report. *European Journal of Cancer (Oxford, England: 1990)*, **39** (1), 70–77.

66 Kruit, W.H. et al. (2005) Phase 1/2 study of subcutaneous and intradermal immunization with a recombinant MAGE-3 protein in patients with detectable metastatic melanoma. *International Journal of Cancer*, **117** (4), 596–604.

67 Davis, I.D. et al. (2004) Recombinant NY-ESO-1 protein with ISCOMATRIX adjuvant induces broad integrated antibody and CD4(+) and CD8(+) T cell responses in humans. *Proceedings of the National Academy of Sciences of the United States of America*, **101** (29), 10697–10702.

68 Uenaka, A. et al. (2007) T cell immunomonitoring and tumor responses in patients immunized with a complex of cholesterol-bearing hydrophobized pullulan (CHP) and NY-ESO-1 protein. *Cancer Immunology*, **7**, 9.

69 Rabinovich, G.A., Gabrilovich, D. and Sotomayor, E.M. (2007) Immunosuppressive strategies that are mediated by tumor cells. *Annual Review of Immunology*, **25**, 267–296.

70 Colombo, M.P. and Piconese, S. (2007) Regulatory-T-cell inhibition versus depletion: the right choice in cancer immunotherapy. *Nature Reviews. Cancer*, **7** (11), 880–887.

71 Brichard, V.G. and Lejeune, D. GSK's antigen-specific cancer immunotherpay programme: pilot results leading to Phase III clinical development. *Vaccine*, **25**(Suppl 2), B61–B71.

72 van Duin, D., Medzhitov, R. and Shaw, A.C. (2006) Triggering TLR signaling in vaccination. *Trends in Immunology*, **27** (1), 49–55.

73 Jager, E. et al. (2006) Recombinant vaccinia/fowlpox NY-ESO-1 vaccines

induce both humoral and cellular NY-ESO-1-specific immune responses in cancer patients. *Proceedings of the National Academy of Sciences of the United States of America*, **103** (39), 14453–14458.

74 van Baren, N. *et al.* (2005) Tumoral and immunologic response after vaccination of melanoma patients with an ALVAC virus encoding MAGE antigens recognized by T cells. *Journal of Clinical Oncology*, **23** (35), 9008–9021.

75 Odunsi, K. *et al.* (2007) Vaccination with an NY-ESO-1 peptide of HLA class I/II specificities induces integrated humoral and T cell responses in ovarian cancer. *Proceedings of the National Academy of Sciences of the United States of America*, **104** (31), 12837–12842.

10
Rationale for Treatment of Colorectal Cancer with EpCAM Targeting Therapeutics

Patrick A. Baeuerle and Gert Riethmueller

10.1
Introduction

Tumor-associated antigens have been shown to be useful targets for monoclonal antibody-based therapies [1]. Examples are human epidermal growth factor receptors 1 (EGFR) and 2 (HER-2/neu), which are frequently overexpressed on cancer cells relative to normal tissue. Antibodies cetuximab and trastuzumab have shown clinical efficacy and are presently being used in routine cancer therapy. Likewise, certain differentiation antigens represent useful antibody targets, including CD20 (rituximab, ofatumumab), CD22 (epratuzumab), CD33 (gemtuzumab), and CD52 (alemtuzumab). Antibodies with theses specificities show that ablation of an entire normal cell compartment along with blood-borne tumor cells are tolerated in some cases. Examples of tumor-associated antigens (TAA) used for the development of experimental antibody-based therapies are the carcinoembryonal antigen (CEA), CD40, EpCAM, mucin-1 (MUC-1), Lewis-Y, and certain gangliosides. It is noteworthy that in all cases target antigens are also expressed to a certain extent on normal tissues or blood cells demonstrating that 'tumor-associated' antigens are never strictly tumor-specific.

There are various ways that a therapeutic window can emerge with antigens that are not exclusively expressed on tumor cells: (i) the overexpression of the antigen on the tumor relative to normal tissue; (ii) the sequestration of antigen or selected epitopes on normal tissue while there is good accessibility to these epitopes on tumor cells and (iii) the temporary depletion of a normal cell population expressing the target antigen, which is repopulated after the end of treatment.

Obviously, selection of a target antigen for treatment of a particular cancer indication critically relies on the presence of the target antigen on tumor cells. However, there are additional requirements for an antigen used in a targeted therapy. The following list defines in their sum the criteria for an ideal TAA. However, as it is evident from existing antibody therapies, not all criteria need to be fulfilled.

(i) High frequency and level of target antigen expression on individual cells of a tumor.
(ii) High frequency of antigen expression within the selected patient population.
(iii) No reduction or loss of antigen with tumor progression.
(iv) Specific tumor cell surface binding.
(v) Absence of substantial antigen shedding.
(vi) No substantial internalization of target antigen upon antibody binding.
(vii) Absence, low-level expression or inaccessibility of target antigen on normal tissues and cells.
(viii) No reduction or loss of antigen expression by concomitant conventional therapies.
(ix) A negative correlation of overexpression with survival parameters.
(x) A functional role of the antigen for tumor biology such that antigen loss is of considerable disadvantage for tumor cells.

More recent research suggests that a target antigen should also be expressed on so-called cancer stem cells, which are the seeds for future metastases and the precursor cells for the bulk of tumor cells. In this chapter, it will become evident that EpCAM indeed fulfils most if not all of the requirements desired for an ideal TAA.

Several review articles cover in detail the biology of EpCAM [2–5]. We will therefore only briefly describe the antigen and focus on data supporting the use of EpCAM as the target in the treatment of CRC. EpCAM (CD326) is also referred to in the literature as 17-1A antigen, TROP1, HEA125, EGP-2, Egp34, KSA, ESA, MK-1, GA733-2, KS1/4 and AUA. This relates to the fact that monoclonal antibodies (mAb) to EpCAM were independently generated by many different laboratories each giving the antigen the name of the respective monoclonal antibody.

EpCAM is a 40-kDa glycoprotein with selective expression on most adult human epithelia [6, 7]. Hepatocytes can express EpCAM during chronic inflammatory and regenerating conditions [8, 9]. During embryogenesis, EpCAM is transiently or *de novo* expressed on certain epithelia [2]. The majority of human adeno and squamous cell carcinoma express EpCAM [10, 11]. Based on these data, EpCAM can be considered as one of the most frequently expressed TAA currently known. As determined by RT-PCR in ovarian and breast cancer, EpCAM expression in primary tumors can exceed expression in the respective normal tissues by > 100-fold [12, 13]. Other cancers, including CRC, apparently do not overexpress EpCAM, but retain the high level of expression as found in parental normal epithelia.

EpCAM appears to function as a calcium-independent homotypic cell adhesion molecule [14]. Overexpression of EpCAM interferes with the function of another epithelial-specific cell adhesion molecule, E-cadherin [15]. Impairment of E-cadherin function was shown to be associated with increased proliferation, migration and invasiveness of tumor cells. The short intracellular domain of EpCAM was found to confer a potent proliferation signal when fused to other membrane proteins leading to c-myc and cyclin expression, growth of cells in soft agar and at low serum concentration [16]. The signaling pathway induced by overexpressed EpCAM has just recently been identified [17]. Nuclear signaling by EpCAM uses elements of both the *notch* and *wnt* signaling pathways. The *wnt* pathway is known to be involved in both oncogenic and stem cell signal transduction.

For breast, ovarian, cholangio, ampullary pancreas, and squamous cell carcinoma of the esophagus, EpCAM overexpression was found to be associated with decreased overall survival (reviewed in [3]). For gastric and renal cancer the opposite was observed. Breast cancer cells were directly shown to benefit from EpCAM overexpression. Specific reduction of EpCAM expression by siRNA in breast cancer cell lines led to reduced proliferation, migration and invasiveness of various breast tumor cell lines [13].

EpCAM is not significantly shed. Low ng/ml concentrations of soluble EpCAM were detected in sera of only 10% of colorectal cancer patients, and not in healthy individuals [18].

10.2
EpCAM Expression in Colorectal Cancer

10.2.1
Frequent and High Level Expression of EpCAM in Primary Tumors of Colon Cancer

A key requirement for any TAA is its presence on cancer cells. We will therefore review in detail the literature that describes the frequency and intensity of EpCAM expression on colon cancer tissue, circulating tumor cells, cancer stem cells and cell lines.

Went *et al.* [10] explored EpCAM expression in 3900 patient samples representing the majority of human cancers. EpCAM expression was semi-quantitated by immunohistochemical (IHC) analysis on tissue microarrays (TMA) using the anti-EpCAM antibody ESA. TMA technology is particularly suited to the standardized comparison of staining intensities because all samples are subject to the identical staining procedure. Staining of tumor samples is recorded by relative intensity (scores 0–3) and frequency, i.e. percentage of tumor cells stained per sample (scores 0–4). The highest possible score is 12 points. Scores >4 are considered 'high-level' expression. The clinical relevance of this particular cut-off score is supported by the highly significant correlations seen between EpCAM expression with scores >4 and overall survival as shown, e.g. for breast [19] and ovarian cancer [20].

Of the 3900 human cancer samples analyzed [10], a total of 191 were taken from primary colon cancer: 47 from adenocarcinoma, 50 from adenoma with severe dysplasia, 49 from adenoma with moderate dysplasia, and 45 from adenoma with mild dysplasia. All (99%) but one patient sample with mild dysplastic adenoma expressed EpCAM. The highest level of EpCAM expression was observed for the group of adenocarcinoma in which 81% ($n = 38$) showed intense staining and only 19% ($n = 9$) weak or moderate staining. The percentage of intensely stained samples was 53% for mild dysplatic adenoma ($n = 24$), 37% for moderate dysplastic adenoma ($n = 18$) and 42% for severe dysplastic adenoma ($n = 21$) while the remaining samples stained with moderate or weak intensity.

In a second study using TMA technology, Went et al. [11] analyzed a total of 1186 CRC patient samples of various histology, tumor grading, nodal and tumor stage of their primary colon cancer for EpCAM expression. Anti-EpCAM antibody ESA was used for TMA staining. The vast majority of patient samples originated from adenocarcinoma (92%; $n = 1086$). Mucinous carcinoma represented the second largest group with 88 samples (7.4%). Medullary adenocarcinoma ($n = 5$), signet ring cell carcinoma ($n = 4$) and other types ($n = 3$) were represented by only a small percentage of samples. Four different tumor stages from pT1 ($n = 47$), pT2 ($n = 171$), pT3 ($n = 766$) to pT4 ($n = 187$) were analyzed. The majority of patient samples were of grade 2 (Dukes B) ($n = 1,001$), followed by 153 patient samples of grade 3 (Dukes C) and only 23 of grade 1. Nodal stages were also analyzed; 603 patient samples were of stage N0, 309 samples of N1, and 241 samples of N2.

On average, 97.7% of patient samples ($n = 1159$) showed strong EpCAM expression with a score >4 and only 1.9% ($n = 23$) exhibited weak or moderate expression. Four colon cancer samples (0.4%) lacked EpCAM expression; one was a medullary carcinoma and three were adenocarcinoma. A total of 98% of the adenocarcinomas ($n = 1064$), 80% of medullary carcinomas ($n = 5$), 95.5% of mucinous carcinomas ($n = 84$) and all seven (100%) cases of signet ring cell carcinoma and other carcinoma types, exhibited considerable EpCAM expression. There was no significant difference in high-level EpCAM expression with tumor staging. The percentage of EpCAM overexpression ranged from 96.8% (pT4; $n = 187$) to 100% (pT1, $n = 47$). Likewise, there was no difference in high-level expression between samples from patients with different nodal status. High-level EpCAM overexpression reached 97.1% at nodal stage II ($n = 241$), 97.4% at stage N1 ($n = 3019$) and 98.2% at stage N0.

A statistically significant difference in the percentage of EpCAM overexpression was found for tumor grading ($p = 0.0001$). While grades 1 and 2 showed 100% ($n = 23$) and 98.6% ($n = 1001$) high-level expression, respectively, 'only' 92.1% of grade 3 colon cancer samples showed similar high-level EpCAM expression.

In a univariate survival analyses in the patient population analyzed, the lymph node status (pN0 versus pN+, $p < 0.0001$), vascular invasion ($p < 0.0001$), and post-operative chemotherapy ($p < 0.0001$) were all of high significance regarding tumor-specific survival. Because EpCAM expression and tumor grading showed a significant association, the different grades were analyzed separately. The subgroup of patients with EpCAM-negative, moderately differentiated colon cancers of grade 2 demonstrated a significantly inferior tumor-specific survival (OR 5.421, 95% CI 1.685–17.442, $p = 0.0014$, $n = 284$), whereas in the other subgroups patients with strongly EpCAM-expressing tumors showed no trend towards a better survival.

Colon cancer is the indication with the highest frequency of high-level EpCAM expression on primary tumors among all the major human cancers that have been analyzed to date [10, 11]. The frequency of high-level expression for any subgroup is typically >90%. Therefore EpCAM pre-screening and selection of patients for clinical trials with EpCAM-directed therapies seems to be unnecessary. On the other hand, a retrospective analysis of tumor tissue for intensity and frequency of EpCAM expression in clinical trials with EpCAM-directed therapies may be indicated to retrospectively correlate differences in efficacy with qualitative and quantitative differences in the expression of the target antigen.

A number of other studies using conventional IHC methodology have analyzed the expression of EpCAM on colon cancer lesions. Although only small sample numbers were tested, the use of anti-EpCAM monoclonal antibodies other than ESA supports the notion that EpCAM overexpression is frequent and that the properties of the analytical antibody may not matter much for the analysis of EpCAM expression levels. Moldenhauer et al. [21] reported a high staining intensity (++) with the anti-EpCAM antibody HEA125 in 20/20 colorectal carcinoma samples. No details were given about the percentage of stained tumor cells or the origin of patient samples. Equal staining intensity was reported in this study for normal colon and small intestine epithelial cells. Using the anti-EpCAM mAb KS1/4, Bumol et al. [22] reported similar results. Of the 13 colon adenocarcinoma samples analyzed all exhibited 'very strong positive' EpCAM staining. 'Strongly positive' staining was also reported for 12/12 normal colon epithelium samples. Finally, Zhang and co-authors [23] analyzed a large panel of antibodies directed against TAA for their reactivity with various tumor tissues. Of the eight colon cancer samples tested, only anti-EpCAM antibody GA733 gave a 4+ staining intensity in all samples in >80% of the cells that stained. In contrast, staining of the same lesions with the anti-mucin 2–7, anti-HER-2 and anti-CEA mAbs demonstrated either a lower frequency or total lack of expression of these markers, as shown for anti-HER-2 and PSMA antibodies.

10.2.2
EpCAM Expression on Colon Cancer Metastases

Hosch et al. [24] analyzed cryosections from 71 colorectal liver metastases for the expression of EpCAM ('17-1A') using the mAb 3B10, CD55 and CD59. CD55 and CD59 expression on tumor tissue was analyzed because co-expression of these complement inhibitory factors may predict resistance of tumors to anti-EpCAM antibodies that mediate complement-dependent cytotoxicity (CDC). Non-specific antibody binding to tissues was tested using irrelevant mouse myeloma IgGs of the same isotypes. Twenty patients (28%) had synchronous and 51 (72%) metachronous hepatic metastases. The majority of metastases showed moderate differentiation ($n = 65$; 92%), two were well differentiated (3%), and four poorly differentiated (6%). Forty patients (56%) had solitary metastases, 13 had two metastases (18%), and 18 patients (25%) had three or more isolated tumors in the liver.

All 71 CRC patient samples expressed EpCAM and at least two-thirds of the tumor cells were stained. Sixty-seven samples (94%) showed homogenous EpCAM staining, while only 6% showed heterogeneous expression. Staining occurred predominantly on membranes with additional cytoplasmic staining patterns.

Expression of CD55 and CD59 was consistently heterogeneous and absent in 89 and 94% of colorectal liver metastases, respectively. Fifty-nine metastatic lesions

(83%) were double-negative for CD55 and CD59. This suggests that the majority of metastases will not be effectively protected by CD55 and CD59 from CDC during EpCAM-specific antibody therapies. Although EpCAM expression levels were not further qualified in this study for their staining intensity, the observed frequency of homogenous EpCAM expression on metastases appeared to be similar to the EpCAM expression pattern on primary tumors.

Normal colon and bile duct tissues showed comparable EpCAM expression on epithelial cells and malignant tissues. In contrast to breast or ovarian cancer, EpCAM in colon cancer is therefore not overexpressed relative to normal tissue but rather high expression levels of EpCAM are inherited by tumors from parental tissue.

10.2.3
EpCAM Expression on Circulating Colon Cancer Cells

Sophisticated technologies have been developed to facilitate the detection and analysis of single circulating tumor cells (CTC) in blood samples from patients [25]. The procedure uses magnetic beads coated with the anti-EpCAM antibody VU.D1 for the initial isolation of CTC from 7.5 ml of blood, followed by microscopic phenotyping of cells for other surface antigens, size and DNA content. It is noteworthy that all CTCs analyzed are by definition EpCAM-positive. Thus, all conclusions about the prognostic value of CTC hold true for EpCAM-positive cells. For breast cancer, it has been well documented that the threshold frequency of such CTCs in blood negatively correlates with overall survival of patients. This threshold can be as low as 5 CTC per 7.5 ml of blood. It is possible that the occurrence of CTC in blood is a read-out for the metastatic potential of a tumor because a requirement of metastases formation is their disengagement from the primary tumor and their transition through various compartments of the body to reach distant locations. CTCs appear particularly susceptible to antibody-based therapies because of their encounter with high levels of antibodies and immune cells in the bloodstream. CTC numbers in CRC patients were used to monitor efficacy in clinical trials of the vaccine ING-101 with their numbers being reduced by treatment [26].

10.2.4
EpCAM Expression on Colon Cancer Stem Cells

Three recent studies have isolated and analyzed the phenotype of cancer stem cells derived from colorectal cancer tissues of patients [27–29]. Cancer stem cells (CSC) are defined as a subpopulation of cells within the tumor that are highly tumorigenic and, when injected at very low numbers into NOD/SCID immunodeficient mice, are capable of generating tumors with the same heterogeneity and low frequency of CSC as the original tumor tissue. Two studies have used anti-EpCAM antibodies to initially isolate cells from single cell suspensions of primary CRC tumors and have identified CD133 as a marker for colon cancer-derived CSC. Dalerba *et al.* [29] have shown that only those cells expressing high levels of EpCAM and the cell adhesion molecule CD44 are tumorigenic. CD166 was identified as an additional surface marker.

Together, these studies suggest that EpCAM is not only present on colon CSC, but has a functional relevance for the high tumorigenicity of these cells. This function may relate to the recent finding that EpCAM is a potent oncogenic signal transducer engaging the *wnt* pathway [17], which has been shown to be important for regulating growth and survival of normal and cancer stem cells [30].

10.2.5
EpCAM Expression on Human Colon Cancer Cell Lines

Moldenhauer *et al.* [21] studied the expression of EpCAM on human colon cancer cell lines SW480, HT-29 and SW1116. All three cell lines showed the maximum staining intensity ('+++') with the mAb HEA125. Of all the 39 cell lines and types tested in this study, colon, breast and cervix cancer cell lines demonstrated the highest scores for EpCAM staining. No EpCAM staining was detectable on cell lines derived from melanoma, neuroblastoma, sarcoma, B and T cell lymphoma, myelocytic leukemia, normal fibroblasts or normal blood cells.

Employing two different human EpCAM-specific mAbs, EpCAM surface expression was tested in two human colon cancer cell lines by using saturation binding at 4 °C and a flow cytometry-based bead assay for quantification (unpublished data). With mAb 3B10, 570 000 EpCAM molecules were detected on HT-29 cells, and 215 000 on SW480 cells, whereas with mAb 5.10, 440 000 and 190 000 molecules were detected in these cell lines, respectively. These results demonstrated (i) a high level of EpCAM expression on colon cancer lines comparable to that on the majority of nine breast cancer cell lines tested [31] and (ii) that distinct anti-EpCAM mAbs produced comparable data.

10.3
EpCAM-directed Therapeutic Approaches

10.3.1
Clinical Results with Anti-EpCAM Murine Antibody Edrecolomab in Stage II and III CRC

Edrecolomab (also known as 17-1A, or Panorex®) is a murine IgG2a mAb with relatively low affinity for EpCAM [32–34]. It has a short serum half-life of 21 h, and repeated administration in humans leads to a strong neutralizing antibody response which inactivates repeatedly administered murine antibody. As a murine IgG2a, the interaction of the antibody with human immune effector cells is also reduced thus translating into suboptimal antibody dependent cellular cytotoxicity (ADCC) when compared to anti-EpCAM human IgG1 in cell-based assays [35]. It is therefore surprising that edrecolomab has shown some signs of clinical activity in CRC as described below.

Four clinical phase II/III studies that tested edrocolomab alone or in combination with chemotherapy in resected stage III (Dukes C) and stage II (Dukes B) colon cancer patients are reviewed here. It is noteworthy that EpCAM expression on resected primary tumors was never used as an inclusion criterion. Likewise, no preclinical data were available to support the dose levels of the murine antibody or to provide a rationale for combination therapy in CRC.

The German Cancer Aid 17-1A Study group began a study in 1989 with 189 patients suffering from stage III colorectal adenocarcinoma [36]. Only patients who had undergone curative surgery and were free of manifest residual tumor were admitted and randomized to an observation regimen ($n=90$) or to post-operative treatment with 500 mg edrecolomab followed by 4-monthly i.v. doses of 100 mg antibody ($n=99$).

After a median follow-up of 5 years, a first study [36] reported that antibody treatment reduced the overall death rate by 30% (Cox's proportional hazard, $p=0.03$, log-rank $p=0.05$), and decreased the recurrence rate by 27% ($p=0.03$, $p=0.05$ respectively). The effect of the antibody was most pronounced in patients who had distant metastasis as the first sign of relapse ($p=0.0014$, $p=0.002$), an effect that was not seen for local relapses ($p=0.74$, $p=0.67$). The 7-year follow-up evaluation confirmed the earlier findings [37]. Overall mortality in the antibody arm was reduced by 32% (Cox's proportional hazard $p<0.01$; log-rank $p=0.01$) and the recurrence rate by 23% ($p<0.04$, $p=0.07$ respectively). The intention-to-treat analysis gave a significant effect for overall survival ($p<0.01$, $p=0.02$) and disease-free survival ($p=0.02$, $p=0.11$). While distant metastases were significantly reduced ($p=0.004$, $p=0.004$), local relapses were not ($p=0.65$, $p=0.52$). The data suggested that edrecolomab treatment after surgery of colorectal cancer patients prevented the development of distant metastases in one-third of patients and that the therapeutic effect of the antibody was maintained after a 7-year follow-up period. In this first study, edrecolomab showed a benign safety profile. After a total of 371 infusions, four anaphylactic reactions were seen that did not require hospitalization.

The published data from the 'Riethmuller study', which led to market approval of Panorex® in Germany, prompted larger trials in Europe and the USA. The outcome of the 'European study' with a total of 2761 resected stage III colon cancer patients was published by Punt et al. [38]. There were two major differences between the 'European' and Riethmuller studies: in the former study no observation arm was used but comparisons were made between the newly-established 5-fluorouracil-folinic acid (5-FU) chemotherapy alone ($n=912$) and the combination of antibody and chemotherapy ($n=922$).

Three-year overall survival on combination therapy was not different from that on chemotherapy alone. Disease-free survival was however significantly lower on edrecolomab monotherapy than on chemotherapy (53% versus 65.5%, $p<0.0001$). The addition of the mAb to chemotherapy in the adjuvant treatment of resected stage III colon cancer did therefore not improve overall or disease-free survival. Moreover, the antibody monotherapy was inferior to chemotherapy alone. Based on these data, the commercial provider took Panorex® off the German market. It is noteworthy that the overall survival rate observed in the antibody arm in the negative European trial of 70% closely matched the observed rate in the antibody arm of the Riethmuller trial which was 72%.

Edrecolomab was well tolerated and did not increase the toxicity of chemotherapy in combination treatment. Hypersensitivity reactions occurred in 25% of patients ($n = 452$) and for 71 patients in this population (4%) treatment was therefore discontinued.

A second large study was carried out in the USA in parallel with the European study. To date, the data has only been published as an abstract of the presentation given at the ASCO meeting in 2002 [39]. This study with a total of 1839 resected stage III colon cancer patients had only two arms: one was 5-FU-based chemotherapy (FCT) alone ($n = 924$), and the other a combination with the standard regimen for edreoclomab (500 mg + 4 × 100 mg monthly; $n = 915$). FCT was institution-specific and either used leucovorin ($n = 1421$) or levamisole ($n = 418$) in combination with 5-FU. Median follow-up was at 31 months.

Patients randomized for the chemo/antibody combination benefited from a significant 2.7% relative improvement in overall survival compared to patients receiving only chemotherapy (log-rank $p = 0.023$; hazard ratio $= 0.785$). Overall survival after 3 years was 81.6% in the combination arm and 78.9% in the arm on chemotherapy alone. There was no statistically significant effect of the combination therapy on disease-free survival ($p = 0.198$).

Major grade 3 and 4 side-effects were diarrhea, nausea, abdominal pain and neutropenia. Anaphylaxis occurred in <1% of patients. The antibody could be safely combined with full dose FCT.

The discrepant results from the European and US trial have been extensively discussed in editorials. Differences in standard-of-care, chemotherapy, and surgical quality between the many centers in the European trial were discussed as reasons for the discrepancy in results between the US and European trials. In the trial conducted by Riethmueller and co-authors, patients received the antibody within the first 2 weeks after surgery, whereas in the larger phase III trials antibody treatment was administered at intervals up to 6 weeks following surgery.

In yet another phase III study in Europe which recruited patients with stage II colon cancer, the effect of adjuvant therapy with edrecolomab was compared to observation alone [40]. From January 1997 to July 2000 a total of 377 patients were postoperatively stratified according to tumor stage (T3 versus T4) and center, and randomly allocated to either treatment with edrecolomab (cohort A, $n = 183$) or observation (cohort B, $n = 194$). Patients in cohort A received a total of 900 mg edrecolomab (400 mg plus 4 × 100 mg monthly). The study was terminated prematurely because production of the drug was discontinued in Germany. Three hundred and five patients were eligible for the primary endpoint of overall survival and 282 patients for disease-free survival. After a median follow-up of 42 months, overall survival and disease-free survival were not significantly different. Toxicity was mild. Postoperative adjuvant treatment with edrecolomab in patients with resected stage II colon cancer did not apparently improve overall or disease-free survival. However, since 5-FU also failed to show an increase in overall survival in resected stage II CRC (IMACT study), and edrecolomab is inferior to 5-FU according to Punt et al. [38], edrecolomab could not be expected to show any efficacy in resected stage II disease.

10.3.2
Clinical Results with Edrecolomab as a Vaccine for Induction of Anti-idiotypic Response

Earlier studies had shown that the EpCAM antibody edrecolomab (CO17-1A) administered at doses and intervals as described above for immunocompetent stage III colorectal cancer patients frequently induced anti-idiotypic antibodies [41]. However, no significant association was found between the occurrence of anti-idiotype antibodies and a therapeutic response.

IGN101 is a vaccine designed for therapeutic vaccination against cancers of epithelial origin and consists of 0.5 mg alum-adsorbed murine monoclonal antibody edrecolomab (17-1A, Panorex®) [42]. The vaccine is supposed to trigger an anti-idiotypic response leading to human anti-EpCAM antibody titers. Feasibility has been shown in 60 patients vaccinated on days 1, 15, 29 and 57 with 0.5 g IGN101 [43]. Sera harvested on day 71 contained between 5 and 40 µg/ml anti-EpCAM human IgG. IgG subtypes and anti-tumor activity of anti-EpCAM antibodies in patient sera has not been reported.

The effect of IGN101 vaccination on the number of circulating EpCAM-positive tumor cells (CTC) in blood has been studied using the method of Cristofanilli et al. [25]. In 16 patients, the median number of CTC was found to be reduced from 10.5 at day 1 to 5.0 at day 71 [26]. This change was not statistically significant, nor controlled by an alum placebo arm.

Igeneon conducted a 1:1 prospective randomized double-blind placebo-controlled phase II trial in 240 patients with epithelial cancers stage III or IV, including colorectal cancer (CRC) [44]. The objectives were to assess the effects of IGN101 on survival of the whole study population, but also to analyze in an exploratory manner the survival effects for the different tested indications. The primary endpoint was overall survival (OS), and secondary endpoints included safety, tolerability, and immunogenicity. Vaccinations with IGN101 or placebo (alum) were administered on days 1, 15, 29 and 57 and in weeks 16, 24, 32, 40 and 68. Concomitant standard therapies were permitted.

On intention-to-treat (ITT), 239 patients were included of which 163 had CRC, and 95 patients stage IV CRC. Thirty-two patients had stage III or IV upper GI tract cancer, 38 patients had stage III or IV NSCLC, and six patients had liver or bile duct carcinoma. The groups were well balanced regarding age, sex, Karnofsky index and concomitant therapies. For the whole ITT population, no difference in overall survival was detected. A trend for prolonged survival was demonstrated in stage IV CRC patients (ITT) during the first year (12 months overall survival 48.8% placebo versus 60% IGN101; $p = 0.28$). A statistically significant survival prolongation was observed in the 53 patients with stage IV rectal cancer (ITT). For the IGN101 group, median survival was 415 days compared to 253 days for placebo ($p = 0.037$). One-year survival was doubled from 29.5% for placebo to 62% for IGN101. Use of concomitant chemotherapy (CT) had no effect on survival. Vaccinations with IGN101 were well tolerated. Almost all patients in the IGN101 group produced an antibody response to the vaccine antigen. Whereas the effect observed in the overall colorectal cancer stage

IV group was not statistically significant (potentially due to sample size), vaccination with IGN101 significantly prolonged survival of metastatic rectal cancer patients.

A key difference between colon and rectal cancer is their different route of metastases formation. While in colon cancer metastases primarily arise in the liver, they primarily form in lungs for rectal cancer. It may be possible that established metastases can be more effectively treated with the anti-EpCAM therapy in the lung than in the liver leading to the observed survival difference between colon and rectal cancer patients vaccinated with IGN101.

A major issue with IGN101 may be the variability in anti-EpCAM response in patients. It is not clear whether sufficiently high serum trough levels of cytotoxic antibodies capable of sustaining an ADCC or CDC response against tumor cells are induced in every vaccinated patient. No data are available that further characterize the cytotoxic activity of human anti-EpCAM antibodies induced by IGN101 in patients. Administration of a defined and well-characterized human anti-EpCAM antibody leading to predictable serum trough levels therefore appears superior to IGN101 vaccination. The effect of IGN101 on overall survival in metastasized rectal cancer patients may be meaningful and should be kept in mind when patients are selected for clinical trials or analyzed retrospectively. It is also interesting to note that IGN101 was well tolerated although it induced at high frequency a robust immune response against the EpCAM autoantigen, which is highly expressed on normal epithelial tissues.

10.3.3
EpCAM Protein as a Vaccine in Colorectal Cancer

A remarkable number of clinical trials have been undertaken and are still ongoing in CRC patients that aim at eliciting humoral or T-cell responses against tumor cells overexpressing the autoantigen EpCAM [45]. A study by Mosolits et al. [46] with nine CRC patients showed that it is feasible to elicit an EpCAM-specific $CD8^+$ T-cell response by either vaccination with recombinant EpCAM protein or an anti-idiotypic (Id) antibody mimicking an EpCAM epitope. Eight of nine patients showed ELISPOT reactivity against recombinant EpCAM, and 9/9 patients against EpCAM-derived peptides. An earlier study by the Mellstedt group [47] with 13 grade II–IV CRC patients showed that vaccination with the EpCAM protein but not with the anti-idiotype (in combination with GM-CSF as adjuvant) induced in all patients a humoral response consisting mainly of IgG1 and IgG3 subtypes. Likewise, all but one patient immunized with EpCAM protein or anti-Id showed an EpCAM-specific proliferative Th1-biased T cell response. Neidhart et al. [48] observed similar T-cell responses with 11 CRC patients, as did Herlyn's group in a study of six CRC and six pancreatic cancer patients [49]. Both used baculovirus-produced soluble EpCAM protein (KSA, GA733-2E). EpCAM expressed by a recombinant canary pox virus had a similar effect in patients [50].

While further well-controlled studies are needed to test whether EpCAM-specific cytotoxic T-cell and antibody responses have therapeutic potential in CRC and other cancers, it is remarkable that the tolerance to the auto-antigen EpCAM can be successfully reversed by standard immunization protocols. EpCAM autoantibodies can even be detected in normal and diseased conditions without prior vaccination [12, 51]. It is noteworthy that both vaccine-induced and physiological humoral and T-cell responses to EpCAM appear to be well tolerated. The observation that immune responses to EpCAM are easily achieved and seem not to be deleterious supports the notion that EpCAM expressed on normal tissue within critical organs may be largely inaccessible to antibodies (and T cells) [52].

10.3.4
Adecatumumab, a Novel Fully Human Anti-EpCAM Antibody

Three obvious shortcomings of the murine antibody edrecolomab – namely short serum half-life, high immunogenicity, and suboptimal ADCC with human effector cells – prompted the development of a fully human anti-EpCAM antibody adecatumumab as an IgG1, the human antibody isotype showing the highest level of ADCC and CDC activity and a fairly long half-life of 2 weeks in serum [35, 53, 54]. Another goal was to retain with adecatumumab the favorable safety profile of edrecolomab. Adecatumumab was generated by using a murine antibody with binding properties similar to edrecolomab for phage display guided selection of fully human variable domains. Human immunoglobulin repertoires were employed that harbor very few somatic mutations. In this way, 88 and 98% similarity was achieved between the variable heavy and light chain domains of adecatumumab respectively, and those in the human germline.

Preclinical studies by Naundorf et al. [35] showed that ADCC initiated by murine edrecolomab is inferior to that of human adecatumumab when human peripheral blood mononuclear cells (PBMC) are used as effector cells. On the other hand, the CDC activity of edrecolomab and adecatumumab was very similar. The half-life of edrecolomab in humans is only 1 day while it is 15 days for adecatumumab [54]. Lastly, all patients eventually develop neutralizing antibodies against edrecolomab (human anti-mouse antibodies = HAMAs), while no neutralizing immune response has thus far been observed for adecatumumab in a study of 168 patients [55]. Suboptimal ADCC, short serum half-life, and high immunogenicity are thus major limitations of edrecolomab vis-à-vis other human antibodies such as adecatumumab. The binding properties of edrecolomab and adecatumumab to EpCAM are nevertheless remarkably similar. Both antibodies bind to recombinant EpCAM on Biacore chips with a K_D of approximately 10^{-7} M, while EpCAM on tumor cells is bound with an avidity of approximately 10^{-8} M. It is noteworthy that a study with HER-2-specific single-chain antibodies of various affinities has shown that a target affinity of 10^{-8} M was optimal for tumor penetration [56]. The affinity of adecatumumab for EpCAM may therefore be optimal for achieving a therapeutic window as well as for penetration of a solid tumor.

So far, four human colon cancer cell lines have thus far been tested for ADCC generated by adecatumumab: HT-29, HCT116, SW480 and LS174T. ADCC was observed with all four cell lines with the following values for half-maximal cell lysis (ED50): HT-29: 93 ± 15 nM, HCT116: 37 ± 22 nM, SW480: 329 ± 354 nM and

LS174T: 56 nM (unpublished data). The results were obtained from different experiments using various human PBMC donors as a source of allogenic effector cells. It can be concluded that human colon cancer cell lines are generally susceptible to ADCC initiated by adecatumumab.

Naundorf et al. [35] described in a mouse model, the activity of adecatumumab against xenografts derived from subcutaneous inoculation of the human colon cancer cell line HT-29. In brief, 10^6 cancer cells in 100 µl PBS were injected into the left flank of nude mice on day 0. Antibody was administered by intravenous injection on days 1, 4 and 7. Controls were injected with a vehicle (PBS) and 30 µg of a human IgG1 isotype (rituximab). Early treatment with adecatumumab significantly prevented tumor outgrowth after a total of three doses of 30 µg ($p = 0.05$ by Mann-Whitney test), but not after three doses of 3 µg antibody. Edrecolomab, a murine IgG2a with similar affinity for human EpCAM as adecatumumab showed a trend for high in-vivo efficacy similar to that of adecatumumab. By contrast, C46, a humanized version of edrecolomab, did not achieve significant tumor inhibition. This suggested that the compatibility of the antibody's Fcγ portion with the murine immune system is a key determinant for in-vivo efficacy and presumably ADCC activity in mice. This notion was proven experimentally. Lutterbuese et al. [57] showed that the inhibition of lung cancer in a syngeneic mouse model was greatly improved by administering a version of adecatumumab in which the human IgG1 Fcγ portion was replaced by murine IgG2a sequences compared to treatment with the human version of the antibody. In-vitro ADCC assays showed that human and murinized versions of adecatumumab showed high compatibility with immune effector cells of their respective species, but not across species.

Adecatumumab has thus far been tested in three clinical trials. A phase 1 study in metastatic prostate cancer patients showed a benign safety profile at doses up to 262 mg/m^2 (ca. 6 mg/kg) given every other week [54]. No maximum tolerated dose was reached. Two phase 2 studies were carried out in metastatic breast cancer and early stage prostate cancer to obtain signs of clinical efficacy. While the prostate cancer trial gave inconclusive results, the trial in metastatic breast cancer patients indicated clinical activity in terms of delaying disease progression [58, 59]. In this trial, the antibody was administered biweekly at dose of 2 or 6 mg/kg. Patients were stratified with regard to a high and low score of EpCAM expression on primary tumors (high and low score) and were randomized into the two dose groups such that in total four treatment groups were analyzed and compared with each other. Treatment was continued until the disease progressed. Patients receiving the higher of the two doses had a significantly longer time-to-progression compared to patients receiving the lower dose [58]. While according to RECIST criteria, adecatumumab as monotherapy had no significant impact on existing lesions, the subgroup of patients expressing high EpCAM had a significantly lower occurrence of new lesions compared to the subgroup of patients with low EpCAM [59]. This finding was unexpected because EpCAM overexpression by itself has been shown to be a negative prognostic factor for overall survival in breast cancer [19]. Despite the lack of a placebo control arm, the trial provided evidence for the clinical activity of adecatumumab, which was both dependent on antibody dose and level of target expression.

Adecatumumab was generally well tolerated in three completed trials ([54, 58, 59], unpublished data). As seen with other mAb therapies, acute infusion-related adverse events such as chills and pyrexia which are mainly grade 1 or 2 were observed. Dose-dependent adverse events associated with the gastrointestinal tract such as diarrhea, nausea, and vomiting of mostly grade 1 or 2, were also observed. High affinity anti-EpCAM antibodies were much less well-tolerated and caused acute pancreatitis as a result of dose-limiting toxicity [60–64]. Adecatumumab did not cause acute pancreatitis, but in some patients there was a transient elevation of pancreatic enzymes in the serum, which normalized under treatment.

Adecatumumab has not yet been explored in the therapy of CRC. However, the clinical success of other mAb therapies in CRC, such as cetuximab, panitumumab, and bevacizumab, supports the proposal that this indication is principally amenable to antibody-based therapies.

10.4
Therapeutic Window of EpCAM-directed Therapies

IHC analyses indicate that EpCAM is expressed on certain normal tissues and stains with the same intensity as that on malignant tissues [6, 7, 21]. For instance, high EpCAM levels can be detected by antibody staining on sliced tissue samples from colonic epithelium, pancreas or bile duct. It is therefore not unexpected that two EpCAM-specific antibodies were found to cause acute pancreatitis in clinical trials [59, 64]. Obviously, these two IgG1 antibodies, which both had rather high binding affinity for EpCAM, gained access to the antigen expressed on pancreas tissue and there mediated an acute cytotoxic reaction. In these early trials dose-limiting toxicity was reached at 1 mg/kg and no signs of efficacy were observed, thus indicating that these antibodies had no therapeutic window.

Two other EpCAM-directed immunotherapies are in the advanced stages of development but are not well tolerated when given systemically. The EpCAM/CD3 'tri-specific' mouse/rat IgG antibody catumaxomab, produced a very significant increase in the puncture-free interval of ascites in late-stage ovarian cancer patients after intraperitoneal administration of low (µg) concentrations [65]. Systemic administration of catumaxomab in a lung cancer trial initiated strong cytokine reactions and liver toxicity, and reached a dose limiting toxicity at 5 µg total dose under co-medication with glucocorticoid [66]. Another EpCAM-directed therapy is VB4-845, a single-chain antibody/pseudomonas exotoxin fusion protein, which is injected locally into lesions of head and neck cancer patients, and has been found to control disease progression in 71% of treated patients [67–69]. The same protein can also be instilled into the bladder for treatment of early-stage bladder cancer, and is showing promising signs of activity [70]. It is unclear whether the systemic toxicities of these two proteins are in part due to the targeting of EpCAM on healthy tissues, or are solely the consequence of non-conditional CD3 stimulation in the first case and systemic exposure to a bacterial toxin in the second. Nonetheless, both drug candidates have reached phase III and II studies, respectively, and validate EpCAM as a target for the treatment of various cancer indications. Their therapeutic index may be improved by local administration although it has been shown that using this route also leads to the build-up of active drug levels in the blood.

The murine anti-EpCAM mAb edrecolomab as well as the human mAb adecatumumab, show cytotoxic activity against EpCAM-expressing tumor cells *in vitro* [35], and, as reviewed above, also appear to be active against tumor cells in patients. Nevertheless, both antibodies have benign safety profiles when systemically administered. No cases of acute pancreatitis have been observed to date. On the other hand, both antibodies induce acute diarrhea, which may result from the targeting of EpCAM on colonic epithelium. On the other hand, the diarrhea is resolved by continued antibody treatment suggesting that it may also have been caused by cytokines released during the first dosing.

Another EpCAM-directed antibody in the early stages of development is MT110 [71] which is a single-chain bispecific antibody construct of the BiTE® class that can engage T cells via its anti-CD3 moiety and thereby elicit potent redirected lysis of EpCAM-expressing cancer cells [72]. Eradication of established human tumors and of metastatic tissue from ovarian cancer patients in mouse models has been demonstrated [71, 73]. A surrogate BiTE® antibody specific for murine EpCAM and CD3 was investigated for its therapeutic window in two mouse cancer models [74]. While tumor inhibition could already be observed at doses of 5–12.5 µg/kg, much higher doses were tolerated by mice without irreversible side-effects. These side-effects were cytokine-mediated and there were no instances of damage to normal tissues which express a high level of EpCAM in a similar fashion to human tissues, as evidenced by extensive post mortem IHC analyses.

As described in the Introduction, some vaccination approaches were successful in inducing T-cell responses against peptides of the autoantigen EpCAM. Likewise, titers of anti-EpCAM antibodies can be easily induced by various means, and can occur naturally in certain patients. This indicates that EpCAM is somehow shielded in normal tissue with little initial exposure to the immune system. The basis for this immune privilege of EpCAM is not well understood.

Studies with transgenic mice expressing human EpCAM under the control of a human EpCAM promoter strongly support the differential accessibility of EpCAM to antibodies *in vivo* [52]. Post-mortem analysis of tissues showed that while strong immunohistochemical staining of slices from both tumor and normal epithelial tissues was achieved using a human EpCAM-specific mAb, the antibody only stained tissue when it was systemically administered to mice.

We assume there are three mechanisms contributing to a therapeutic window of EpCAM-specific therapies. These are illustrated in Figure 10.1.

(i) One may relate to the observation that EpCAM is found associated with other proteins in the plasma membrane, including the tetraspanin CD9, the tight junction protein claudin-7 and the cell adhesion molecule CD44 [75–78]. It is conceivable that within the so-called tetraspanin webs only a few epitopes of the

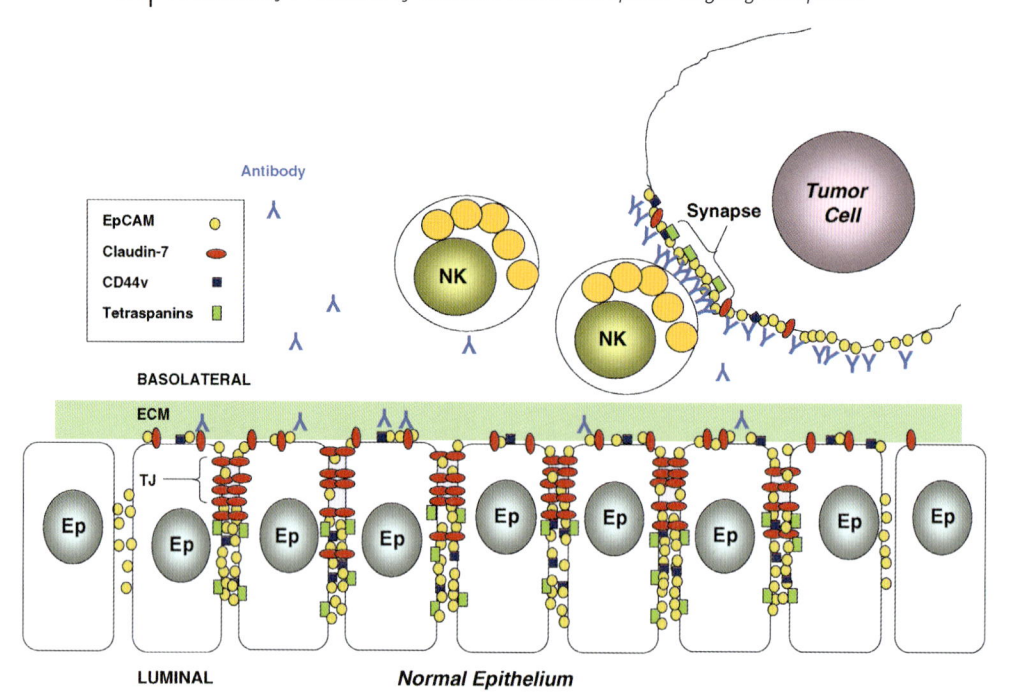

Figure 10.1 Factors that may contribute to a therapeutic window of certain EpCAM-directed therapeutics. As shown, a prerequisite for ADCC is the formation of cytolytic synapses between immune cells and antibody-decorated tumor cells. This requires a certain surface density of antibody coating and direct access of the immune cell to the target cell. In normal epithelia this may be suboptimal for two reasons. One is that only a small percentage of EpCAM target antigen may be accessible to antibody while most is sequestered in intercellular boundaries and by complexation with multiple protein partners. The second is that intact extracellular matrix may reduce access by immune cells. In the proteolyzed environment of tumors, access by both target and immune cells to cancer cells may be highly improved. ECM, extracellular matrix.

complexed EpCAM protein are accessible, while other potential epitopes only become accessible when EpCAM is overexpressed relative to the other partners or when protein partners are lost or disengage from EpCAM during malignant transformation (Figure 10.2). A loss of CD9, the organizer of tetraspanin webs, has indeed been reported during tumor progression [79]. *In-vitro* experiments showed that EpCAM on a normal breast epithelial cell line is much less accessible to certain antibodies than EpCAM on breast cancer cell lines, while other EpCAM antibodies could not differentiate between these cell lines (unpublished observations). This differential binding also translated into a robust difference with respect to redirected lysis between normal and transformed cell lines.

(ii) A second mechanism contributing to a therapeutic window may arise from the sequestration of EpCAM within intact epithelial tissue. IHC studies showed

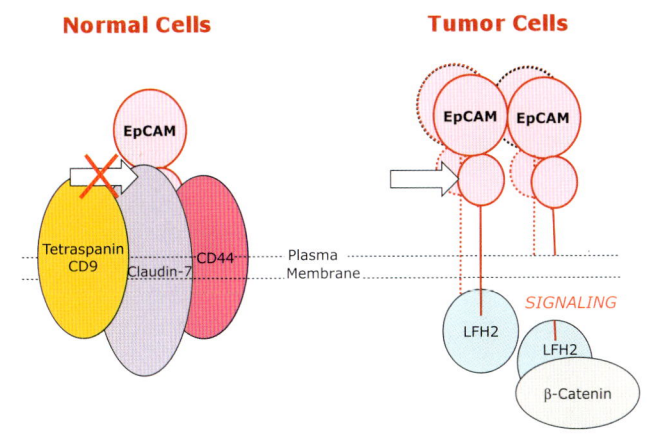

Figure 10.2 The role of accessory proteins in the accessibility of the EpCAM target antigen to antibody-based therapeutics. The left panel shows EpCAM in complex with three of its known protein partners, claudin-7, CD9 and CD44. The arrow indicates the position of an epitope that may not be accessible in such complexes. The right panel shows the presumed situation on tumor cells, where EpCAM overexpression and loss of protein partners or reduced interaction between them provides access to certain epitopes by respective anti-EpCAM antibodies.

that part of the cell surface-associated EpCAM is localized within intercellular boundaries [15] where tight junctions and other dense structures may effectively reduce access by antibodies and the formation of cytolytic synapses with immune cells as is required for ADCC activity. EpCAM exposed on the baso-lateral membrane may still be accessible to certain antibodies provided that particular epitopes protrude from the EpCAM–protein complexes. It is conceivable that antibodies of very high affinity are even capable of stably displacing protein partners from EpCAM while lower affinity antibodies are unable to do so for kinetic reasons, which may explain why the development of pancreatitis was associated with two high affinity anti-EpCAM antibodies while this was not the case for two lower affinity antibodies.

(iii) The third process which contributes to a therapeutic window may relate to the effector mechanism of the targeted therapy. While immunotoxins will kill every cell they can reach and bind to, this may not be the case for therapies that rely for their cytotoxicity on the recruitment of immune effector cells. In the latter case, not only the antibody, but the killer cell requires a close encounter with the target cell. Although this seems feasible in the highly proteolyzed and disintegrated environment of tumors and metastasizing cells, normal epithelial cells are typically embedded in a tight net of intact extracellular matrix and are covered with a basement membrane, the resulting close proximity of the cells enables them to adhere firmly to one another. Access of immune cells to normal tissue may only be possible where malignant (or inflammatory) events have led to a significant degradation of the matrix. We anticipate that all three mechanisms described above ultimately contribute to the therapeutic window of certain but not all EpCAM-directed therapies.

10.5
Conclusions

The following considerations provide in their sum a substantial rationale for treating CRC with EpCAM-directed therapies.

(i) EpCAM is expressed with high frequency on primary and metastatic colon cancer tissues. Of all human tumors analyzed to date, the frequency of high-level EpCAM expression among patients is highest in CRC. EpCAM expression does not appear to decline with tumor progression, and EpCAM expression may not need to be included among the criteria for patient selection for inclusion in clinical trials. Possible issues may arise from two subgroups of CRC patients. One group includes patients with grade 3 tumors showing a statistically significant reduction of ca. 6% in the frequency of high-level EpCAM expression (i.e. 92% compared to >98% in grade 1 and 2 tumors). However, due to the low rate of decrease the number of patients eligible for treatment would not be substantially reduced. Another subgroup to be considered is stage II cancer patients with moderately differentiated primary tumors. Patients in this group whose tissues expressed EpCAM had a statistically significant survival advantage. Stage II patients with moderately differentiated tumors should therefore now be considered for exclusion from treatment with anti-EpCAM therapies until this observation is fully understood.

EpCAM expression levels in normal colon or bile duct epithelium and malignant tissues are not markedly different. In the vast majority of CRC cases, however, EpCAM expression on the tumor is at least as high as that in those subgroups of patients with ovarian or breast cancer showing high-level EpCAM overexpression. The high level of normal EpCAM expression on colon and bile duct epithelium is not a particular issue for CRC as an indication; it would likewise determine the therapeutic window for all other cancer indications.

The very frequent high-level EpCAM expression on CRC at all stages of the disease is probably the strongest argument for selection of colon cancer as an indication for treatment with anti-EpCAM therapies. This expression may reflect a particular need for EpCAM expression in CRC cells.

(ii) EpCAM is suitable to target cancer stem cells derived from CRC. The promise of targeting and eliminating cancer stem cells is cure. This requires that all cancer stem cells carry the target antigen, are all accessible to the targeted therapy and essentially are all eliminated by the treatment. If the target antigen has functional relevance for the tumorigenicity of cancer stem cells then this would be a particular advantage. This seems to be the case for EpCAM in colon cancer. Whether an EpCAM-targeted therapy is capable of reaching and lysing all EpCAM-expressing cancer stem cells remains to be seen. It is noteworthy that antibodies which engage immune effector cells are well suited to a 'seek and destroy' mechanism because this function is built in to most immune effector cells. Moreover, use of a composite of cytotoxic proteins by redirected immune cells for tumor cell lysis appears to be superior to targeted approaches for inhibiting signaling molecules of cancer cells, or delivering just one type of payload.

(iii) Certain EpCAM-specific immunotherapies show signs of clinical activity. Although the anti-EpCAM antibody edrecolomab can be considered as suboptimal for adjuvant treatment of CRC stage III because of its short half-life, high immunogenicity and poor ADCC activity, significant clinical activity of the antibody in CRC has been reported in two trials, but not observed in a third. While these controversial results would disqualify the murine antibody as an efficacious adjuvant treatment for CRC, they may nevertheless support the development of a human anti-EpCAM antibody with much improved pharmacokinetic and -dynamic properties.

The discrepant trial results may be best explained by the borderline and non-robust activity of the murine antibody in colorectal cancer patients, the magnitude of which appears highly susceptible to multiple parameters derived in part from the antibody's short serum half-life and high immunogenicity, but also dependency on the quality and timing of the preceding surgery. Moreover, the murine antibody had very low ADCC activity with human effector cells. Hence, the only anti-tumor activity edrecolomab may still be able to exert during its short exposure time in man is complement-dependent cellular cytotoxicity (CDC). It is not known whether a CDC-based mechanism will profit from combination with 5-FU-based chemotherapies, and the contribution of CDC to the anti-tumor activity of antibodies is also still unknown. The safety profile of edrecolomab in colon cancer patients is generally described as benign and no problems were associated with its combination with a chemotherapeutic regimen. Signs of the clinical activity of edrecolomab in resected stage III colon cancer patients may indicate that a more potent, less immunogenic and more long-lived antibody may produce a more pronounced and robust activity profile in the adjuvant CRC setting post-surgery.

(iv) Some but not all EpCAM-specific therapies have a therapeutic window. There is increasing evidence that EpCAM can be targeted in a tumor-selective fashion despite its expression on normal tissue. This is evident from the benign safety profile of adecatumumab, a therapeutic window seen with EpCAM/CD3-specific BiTE antibodies, the clinical activities of catumaxomab and VB4-845, the ease with which well-tolerated immune responses to the EpCAM autoantigen can be induced through vaccination, and by studies in human EpCAM transgenic mice which indicate little accessibility of antibody to normal EpCAM-expressing tissues. Understanding the basis of this therapeutic window in relation to the binding specificity and affinity of anti-EpCAM antibodies requires further research.

(v) Finally, the anti-EGFR antibodies cetuximab and panitumumab show efficacy as monotherapies in metastatic CRC. This supports the suggestion that this indication is in principle amenable to treatment with monoclonal antibody therapies [80].

References

1 Brekke, O.H. and Sandlie, I. (2003) *Nature Reviews*, **2**, 52–62.
2 Balzar, M., Winter, M.J., de Boer, C.J. and Litvinov, S.V. (1999) *Journal of Molecular Medicine (Berlin, Germany)*, **77**, 699–712.
3 Baeuerle, P.A. and Gires, O. (2007) *British Journal of Cancer*, **96**, 417–423.
4 Trzpis, M., McLaughlin, P.M.J., de Leij, L.M.F.H. and Harmsen, M.C. (2007) *American Journal of Clinical Pathology*, **171**, 1–10.
5 Chaudry, M.A., Sales, K., Ruf, P., Lindhofer, H. and Winslet, M.C. (2007) *British Journal of Cancer*, **10**, 1013–1019.
6 Gottlinger, H.G., Funke, I., Johnson, J.P. et al. (1986) *International Journal of Cancer*, **38**, 47–53.
7 Momburg, F., Moldenhauer, G., Haemmerling, G.J. and Moeller, P. (1987) *Cancer Research*, **47**, 2883–2891.
8 de Boer, C.J., van Krieken, J.H., Janssen van Rhijn, C.M. and Ltvinov, S.V. (1999) *The Journal of Pathology*, **188**, 201–206.
9 Breuhahn, K., Baeuerle, P.A., Peters, M. et al. (2006) *Hepatology Research*, **34**, 50–56.
10 Went, P.T., Lugli, A., Meier, S. et al. (2004) *Human Pathology*, **35**, 122–127.
11 Went, P.T., Kononen, P.J., Simon, R. et al. (2005) *British Journal of Cancer*, **94**, 128–135.
12 Kim, J.H., Herlyn, D., Wong, K.K. et al. (2003) *Clinical Cancer Research*, **9**, 4782–4791.
13 Osta, W.A., Chen, Y., Mikhitarian, K. et al. (2004) *Canc Res*, **64**, 5818–5824.
14 Balzar, M., Briaire-de Bruijn, I.H., Rees-Bakker, H.A. et al. (2001) *Molecular and Cellular Biology*, **21**, 2570–2580.
15 Litvinov, S.V., Balzar, M., Winter, M.J. et al. (1997) *The Journal of Cell Biology*, **139**, 1337–1348.
16 Munz, M., Kieu, C., Mack, B. et al. (2004) *Oncogene*, **23**, 5748–5758.
17 Gires, O., Maetzel, D., Mack, B. et al. (2007) *AACR-NCI-EORTC Meeting*, Abstract No. C175.
18 Abe, H., Kuroki, M. and Imakiire, T. et al. (2002) *Journal of Immunological Methods*, **270**, 227–233.
19 Spizzo, G., Went, P., Dirnhofer, S. et al. (2004) *Breast Cancer Research and Treatment*, **86**, 207–213.
20 Spizzo, G., Went, P., Dirnhofer, S. et al. (2006) *Gynaecol Oncol*, **103**, 483–488.
21 Moldenhauer, G., Momburg, F., Möller, P. et al. (1987) *British Journal of Cancer*, **56**, 714–721.
22 Bumol, T.F., Marder, P., DeHerdt, S.V. et al. (1988) *Hybridoma*, **7**, 407–415.
23 Zhang, S., Zhang, H.S., Cordon-Cardo, C. et al. (1998) *Clinical Cancer Research*, **4**, 2669–2676.
24 Hosch, S.B., Scheunemann, P., Lüth, M. et al. (2001) *Journal of Gastrointestinal Surgery*, **5**, 673–679.
25 Cristofanilli, M., Hayes, D.F., Budd, G.T. et al. (2005) *Journal of Clinical Oncology*, **23**, 1420–1430.
26 Samonigg, H., Himmler, G., Gnant, M. et al. (2005) *ASCO Annual Meeting*, Abstract No. 2557.
27 O'Brien, C.A., Pollett, A., Gallinger, S. and Dick, J.E. (2007) *Nature*, **445**, 106–110.
28 Ricci-Vitiani, L., Lombardi, D.G., Pilozzi, E. et al. (2007) *Nature*, **445**, 111–115.
29 Dalerba, P., Dylla, S.J., Park, I.K. et al. (2007) *Proceedings of the National Academy of Sciences of the United States of America*, **104**, 10158–10163.
30 Dreesen, O. and Brivanlou, A.H. (2007) *Stem Cell Rev*, **3**, 7–17.
31 Prang, N., Preithner, S., Brischwein, K. et al. (2005) *British Journal of Cancer*, **92**, 342–349.
32 Schwartzberg, L.S. (2001) *Critical Reviews in Oncology/Hematology*, **40**, 17–24.
33 Armstrong, A. and Eck, S.L. (2003) *Cancer Biology & Therapy*, **2**, 320–326.
34 Froedin, J.E., Fagerber, J., Hjelm Skog, A.L. et al. (2002) *Hybridoma and Hybridomics*, **21**, 99–101.

35 Naundorf, S., Preithner, S., Mayer, P. et al. (2002) *International Journal of Cancer*, **100**, 101–110.
36 Riethmuller, G., Schneider-Gadicke, E., Schlimok, G. et al. (1994) *Lancet*, **343**, 1177–1183.
37 Riethmuller, G., Holz, E., Schlimok, G. et al. (1998) *Journal of Clinical Oncology*, **16**, 1788–1794.
38 Punt, C.J., Nagy, A., Douillard, J.Y. et al. (2002) *Lancet*, **360**, 671–677.
39 Fields, A.L., Keller, A.M., Schwartzberg, L. et al. (2002) *ASCO Annual Meeting*, Abstract No. 508.
40 Hartung, G., Hofheinz, R.D., Dencausse, Y. et al. (2005) *Onkologie*, **28**, 347–350.
41 Gruber, R., van Haarlem, L.J., Warnaar, S.O., Holz, E. and Riethmueller, G. (2000) *Cancer Research.*, **60**, 1921–1926.
42 Kirman, I. (2006) *Current Opinion in Molecular Therapeutics*, **8**, 358–365.
43 Himmler, G., Loibner, H., Schuster, M. et al. (2003) *ASCO Annual Meeting*, Abstract No. 732.
44 Himmler, G., Settaf, A. Groiss, F. et al. (2005) *ASCO Annual Meeting*, Abstract No. 2555.
45 Mosolits, S., Nilsson, B. and Mellstedt, H. (2005) *Expert Review of Vaccines*, **4**, 329–350.
46 Mosolits, S., Markovic, K., Fagerberg, J. et al. (2005) *Cancer Immunology, Immunotherapy*, **54**, 557–570.
47 Mosolits, S., Markovic, K. and Frodin, J.E. et al. (2004) *Clinical Cancer Research*, **10**, 5391–5402.
48 Neidhart, J., Allen, K.O., Barlow, D.L. et al. (2004) *Vaccine*, **22**, 773–780.
49 Taib, L., Birebent, B., Somasundaram, R. et al. (2001) *International Journal of Cancer*, **92**, 79–87.
50 Ullenhag, G.J., Frodin, J.E. and Mosolits, S. et al. (2003) *Clinical Cancer Research*, **9**, 2447–2456.
51 Furth, E.E., Li, J., Purev, E. et al. (2006) *Cancer Immunology, Immunotherapy*, **55**, 528–537.
52 McLaughlin, P.M., Harmesen, M.C., Dokter, W.H. et al. (2001) *Cancer Research*, **61**, 4105–4111.
53 Kirman, I. and Whelan, R.L. (2007) *Current Opinion in Molecular Therapeutics*, **9**, 190–196.
54 Oberneder, R., Weckermann, D., Ebner, B. et al. (2006) *European Journal of Cancer (Oxford, England: 1990)*, **42**, 2530–2538.
55 Wolf, A., Raum, T. and Rumpler, S. et al. (2007) *AACR-NCI-EORTC Annual Meeting*, Abstract No. B49.
56 Adams, G.P., Schier, R., McCall, A.M. et al. (2001) *Cancer Research*, **61**, 4750–4755.
57 Lutterbuese, P., Brischwein, K., Hofmeister, R. et al. (2007) *Cancer Immunology, Immunotherapy*, **56**, 459–468.
58 Dittrich, C., Schmidt, M. and Awada, A. et al. (2007) *AACR-NCI-EORTC Annual Meeting*, Abstract No. A71.
59 Awada, A., Schmidt, M., Scheulen, M.E. et al. (2006) *ASCO Annual Meeting*, Abstract No. 3588.
60 LoBuglio, A.F., Saleh, M.N., Braddock, J.M. et al. (1997) *ASCO Annual Meeting*, Abstract No. 1562.
61 Saleh, M.N., Posey, J.A. and Khazaeli, L.M. et al. (1997) *ASCO Annual Meeting*, Abstract No. 1680.
62 Khor, S.P., Lampkin, T.A. and Saleh, M.N. et al. (1997) *ASCO Annual Meeting*, Abstract No. 847.
63 de Bono, J.S., Forero, A., Hammond, L.A. et al. (2002) *ASCO Annual Meeting*, Abstract No. 35.
64 de Bono, J.S., Tolcher, A.W. Forero, A. et al. (2004) *Clinical Cancer Research*, **10**, 7555–7565.
65 Parsons, S.L., Kutarska, E. and Koralewski, P. et al. (2007) *ASCO Annual Meeting*, Abstract No. 5520.
66 Sebastian, M., Passlick, B. and Friccius-Quecke, H. et al. (2007) *Cancer Immunology, Immunotherapy*, **56**, 1637–1644.
67 Fitsialos, D., Quenneville, J., Rasomoelisolo, M. et al. (2005) *ASCO Annual Meeting*, Abstract No. 5569.
68 Quenneville, J., Fitsialos, D. and Rasomoelisolo, M. et al. (2005) *ASCO Annual Meeting*, Abstract No. 5539.

69 MacDonald, G.C. and Glover, N. (2005) *Current Opinion in Drug Discovery & Development*, **8**, 177–183.

70 Fitsialos, D., Seitz, S. and Wiecek, E. et al. (2006) *ASCO Annual Meeting*, Abstract No. 4580.

71 Brischwein, K., Schlereth, B., Guller, B. et al. (2006) *Molecular Immunology*, **43**, 1129–1143.

72 Wolf, E., Hofmeister, R., Kufer, P. et al. (2005) *Drug Discovery Today*, **10**, 1237–1244.

73 Schlereth, B., Fichtner, I., Lorenczewski, G. et al. (2005) *Cancer Research*, **56**, 2882–2889.

74 Amann, M., Brischwein, K., Lutterbuese, P. et al. (2008) *Cancer Research*, **68**, 143–151.

75 Schmidt, D.-S., Klingbeil, P., Schnölzer, M. and Zöller, M. (2004) *Experimental Cell Research*, **297**, 329–347.

76 Ladwein, M., Pape, U.-F., Schmidt, D.-S. et al. (2005) *Experimental Cell Research*, **309**, 345–357.

77 Le Naour, F., André M., Greco, C. et al. (2006) *Molecular & Cellular Proteomics*, **5**, 845–857.

78 Kuhn, S., Koch, M. and Nübel, T. et al. (2007) *Molecular Cancer Research*, **5**, 553–567.

79 Miyake, M., Nakano, K., Leki, Y. et al. (1995) *Cancer Research*, **55**, 4127–4131.

80 Chung, K.Y. and Saltz, L.B. (2005) *The Oncologist*, **10**, 701–709.

11
Carcinoembryonic Antigen

Wolfgang Zimmermann and Robert Kammerer

11.1
CEA Biology

11.1.1
CEA Gene Family, Genomic Localization, Protein Structure

CEA is encoded by the *CEA-related cell-cell adhesion molecule 5* (*CEACAM5*) gene, which belongs to the CEA gene family and in humans consists of 22 expressed members and 12 pseudogenes [3, 4]. In the past, the CEA family has been subdivided into the CEACAM and pregnancy-specific glycoprotein (PSG) subgroups. The latter group of genes encodes highly similar, secreted, trophoblast-specific proteins. However, with the recent discovery of more distantly related CEA family members (CEACAM16, CEACAM18-CEACAM20), which are conserved between mammals, this division appears less useful. Evolutionary studies favor a subdivision into an evolutionary divergent, probably CEACAM1 ancestor-derived group and a set of family members with orthologs in other mammals. This subdivision is also supported by the chromosomal arrangement of the gene family on the long arm of chromosome 19q13.2. The divergent family members including the *PSG* are clustered around the *CEACAM1* gene. This gene locus is interrupted by a stretch of non-CEA-related genes. The more distantly related CEA family genes are located 1.5 Mb downstream toward the telomere with the exception of *CEACAM18* which has been translocated from this locus further downstream into the SIGLEC gene locus before the divergence of placental mammals from marsupials.

The CEA family is part of the evolutionary successful immunoglobulin (Ig) superfamily, the members of which are composed of stable and functional versatile often heavily N-glycosylated Ig-like domains. The Ig fold is formed by two β-sheets and comes in two 'flavors': the Ig variable (IgV) and Ig constant (IgC) domains. Almost all human CEA family members are composed of one IgV-like or N domain (CEACAM20 contains a truncated version and CEACAM16 two N domains) and

---

Tumor-Associated Antigens. Edited by Olivier Gires and Barbara Seliger
Copyright © 2009 WILEY-VCH Verlag GmbH & Co. KGaA, Weinheim
ISBN: 978-3-527-32084-4

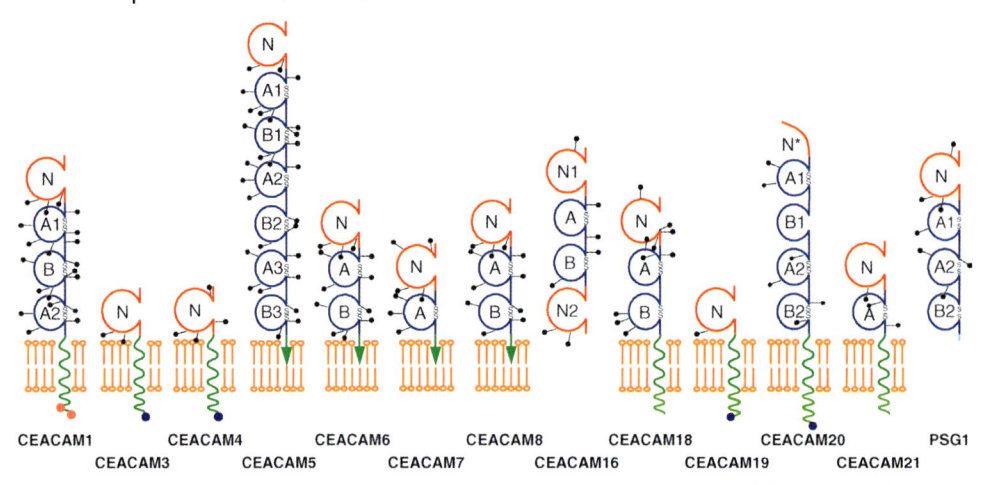

Figure 11.1 Domain organization of the human CEA family. The CEA family contains 12 CEACAM and 11 very closely related PSG members (only one representative is shown). They consist of IgV-like N domains (red) and IgC-like domains of A and B-type (blue) with conserved disulfide bridges (SS). Some of the transmembrane-bound members contain ITAM (blue dots) or ITIM motifs (red dots) in their cytoplasmic domains. Arrows indicate membrane anchorage via a GPI moiety. Potential N-glycosylation sites are depicted as lollipops.

either none or up to six IgC-like domains (A, B domains). The latter composition applies to CEA (Figure 11.1). The N domains have been shown to be indispensable for the interaction with cellular and microbial ligands and receptors. They represent the most similar extracellular domains among the CEACAM1-related members (e.g. 89% amino acid sequence identity between CEA and CEACAM1 or CEACAM6), which results in widespread cross-reactivity of anti-CEACAM antibodies. However, international comparative efforts have allowed identification of a number of molecule-specific monoclonal antibodies [5] some of which are being used for targeting CEA-expressing tumors and metastases in patients (see below). Recent determination of the three-dimensional structure of the N domain of murine and human CEACAM1 by X-ray crystallography facilitates detailed structure–function relationship studies and modeling of antigen–antibody interactions [6, 7]. The human CEA family members are either secreted (CEACAM16, PSG) or membrane bound. Membrane anchorage can be provided by a glycosylphosphatidylinositol (GPI) anchor which tethers CEA, CEACAM6, CEACAM7 and CEACAM8 to the outer half of the cell membrane or by a transmembrane domain as found for CEACAM1, CEACAM3, CEACAM4 and CEACAM18-CEACAM21. GPI-linked members, including CEA, have only been identified in primates. The transmembrane-bound members appear to be involved in opposite types of signaling since they contain (with the exception of CEACAM21) immune receptor tyrosine-based inhibition (ITIM; CEACAM1) or activation consensus motifs (ITAM; CEACAM3, CEACAM4, CEACAM19, CEACAM20) (Figure 11.1).

11.1.2
Evolution

The increasing number of whole genome sequences which have been generated over the past few years is beginning to unveil the divergent evolution of the CEA families in vertebrate species. CEA-related genes have been identified in as distantly related species as teleost fishes (unpublished data), which shared a common ancestor with humans some 450 million years ago. Structural and functional similarities with CEACAM1 (e.g. the presence of putative ITIM elements) imply a CEACAM1-like family ancestor. The group of genes with readily identifiable orthologs within mammalian species can be traced back at least to the egg-laying monotremes (unpublished data). The structure and number of the remaining CEA-related members differ quite significantly even among closely related species [3, 8]. In all species representing different orders which have been analyzed to date, CEA family genes have been identified which are very similar to the species' CEACAM1. They often carry ITAM motifs in their cytoplasmic domains, which in the case of the human granulocyte-specific CEACAM3, is able to convey uptake and destruction of pathogenic Neisseria (see below). On the other hand, allelic CEACAM1 variants exist in various species which have been shown in mice to differ drastically in their capability to serve as (mouse hepatitis) virus receptors [9]. Since CEACAM1 is known or assumed to also be used in other species by viral and bacterial pathogens as a receptor for adherence and cell entry, it has been suggested that the evolution of CEACAM1-related genes (probably with the exception of PSG) has been shaped mostly by host–pathogen interactions [10]. In this context it is interesting to note that soluble or primate-specific GPI-linked CEACAMs which are easily released from the cell surface by phospholipases could serve as decoy receptors to divert pathogens from epithelial layers which often serve as entry points for pathogens [11]. Therefore, pathogen-driven evolution might be responsible for the sole presence of CEA in higher primates.

11.1.3
Expression

Expression of CEA in normal tissues is restricted to apical surfaces of epithelial cells and to squamous epithelial cells. It is found most prominently on mucosal epithelia of the gastrointestinal tract, but also in epithelial cells of the lung, urogenital tract, and in sweat glands. In the colon, CEA has been more specifically localized to vesicular and filamentous structures of the glycocalyx on top of the microvilli of columnar epithelial cells by immune electron microscopy [11]. This localization allows continuous shedding of CEA into the lumen and might explain the lack of targeting of anti-CEA antibodies to normal tissues (see below). CEA is expressed in many tumors of epithelial origin. Most commonly, CEA is found in adenocarcinomas of the colon, rectum, endometrium, and lung, but also in gastric, pancreatic, gall bladder, mucinous ovarian, and small cell lung carcinomas and, less frequently, in breast and serous ovarian carcinomas (reviewed in [11, 12]). Possibly because of its involvement in tumor formation and progression, CEA is also commonly expressed in corresponding metastatic lesions and loss of CEA expression following CEA-targeted therapy has never been reported [13]. An up to 20-fold upregulation of CEA on the cell surface of tumor cells in comparison to normal colonocytes has been demonstrated by flow cytometry analyses of freshly dissociated colorectal tumors [14]. This supports the usefulness of CEA as a target for immune therapies (see below). Since there are widely expressed CEACAMs which are closely related to CEA, cross-reactions might occur during CEA-targeted vaccination and antibody tumor therapies. This might lead to side effects of effective CEA-targeted therapies. The closest relatives of CEA (CEACAM1, CEACAM6) are more widely expressed in healthy adult tissues. Both CEACAM1 and CEACAM6 are expressed on most epithelia and are present on different leukocytes. In addition, CEACAM1 is found in liver, and on endothelial cells [12, 15, 16].

11.1.4
Biological Functions

The members of the CEA family exhibit a wide variety of presumed and proven functions including environmental sensing by homo- and heterotypic cell–cell adhesion (many CEACAMs), modulation of angiogenesis (CEACAM1), tumor suppression (CEACAM1), down-regulation of the epithermal growth factor and insulin receptors as well as T-cell receptor-mediated cytotoxicity (CEACAM1), promotion of metastasis (CEACAM1, CEACAM5), inhibition of differentiation and 'loss of contact'-induced apoptosis (CEACAM5, CEACAM6), regulation of the immune system during pregnancy (PSG), as well as binding and uptake of viral and bacterial pathogens (many CEACAMs). A number of recent review articles cover these functional aspects in depth [12, 17–20].

The primate-specificity of CEA has hampered the elucidation of the *in vivo* function of CEA, because of the lack of appropriate animal models. However, through *in vitro* studies and, more recently, analyses using transgenic mice expressing human CEA, a number of functional roles for CEA have emerged over the years. Intercellular adhesion via homo- and heterophilic interaction with other members of the CEA family based on protein–protein interaction has been shown *in vitro* with CEA-expressing cell lines [21, 22]. Both protein-based as well as carbohydrate–lectin interactions are responsible for the adherence of a number of pathogens to CEA-expressing epithelial cells or intestinal glycocalyx. *Salmonella* and *Escherichia coli* (*E. coli*) strains bind to mannose-rich carbohydrate structures on CEA and related molecules through type I fimbrial lectins [23]. Afa/Dr-I adhesins of diffusely adhering *E. coli* also interact with unidentified target structures on CEACAM1, CEA, and CEACAM6 [24]. Protein–protein interaction is responsible for *in vitro* adhesion of *Neisseria gonorrhoeae* and *Neisseria meningitides* to CEA on epithelial cells allowing adhesion and invasion. Bacterial virulence-associated opacity (Opa) proteins function as adhesins on the pathogen side. The GPI-liked CEA and CEACAM6 as part of the glycocalyx might represent decoy receptors to shield the closely related CEACAM1 from engagement with mucosa-invading pathogens thus protecting the colon and possibly other epithelia from microbial attack [11]. Expression of the highly glycosylated CEA can enhance the metastatic behavior of human colonic tumor cells in immunodeficient mice [25, 26]. The concentration of sialyl-Lewis X (sLe^x) carbohydrate structures correlates with the metastatic potential of tumor cells [27]. Therefore, lectins on endothelial cells recognizing sLe^x carbohydrate moieties, such as E-selectin, are considered to play a role in adhesion and extravasation of tumor cells carrying sLe^x, which is also present on CEA in tumors [28]. Interestingly, soluble CEA might also indirectly enhance metastasis by binding to the CEA receptor, an 80-kDa protein on liver macrophages (Kupffer cells), via a pentapeptide in CEA, which leads to the secretion of proinflammatory cytokines, such as TNF-α [29]. Kupffer cells in turn activate endothelial cells by upregulation of lectins which mediate adhesion of tumor cells as has recently been shown *in vitro* [28, 30]. In addition, it has been suggested that CEA carrying tumor-specific Lewis carbohydrate structures can be taken up by immature dendritic cells (DC) through interaction with the lectin DC-SIGN. Engulfment of glycosylated CEA via the internalizing receptor DC-SIGN might enhance CEA peptide presentation to T cells and thus support CEA-based vaccination strategies [31].

Over-expression of CEA and CEACAM6 suppresses differentiation and thus induces disruption of the tissue architecture and the apoptotic process of anoikis in various *in vitro* and *in vivo* model systems [32–35]. These findings support the notion that CEA and related molecules play an active role in the aberrant behavior of epithelial cells. Suppression of differentiation and anoikis are probably mediated by self-binding of GPI-anchored CEA and CEACAM6 which in turn induces signal transduction by membrane raft clustering. The mechanism of signal transduction by GPI-anchored CEA family members has been elucidated to some degree. CEA has been shown to be capable of initiating signal transduction which leads to the inactivation of components of the intrinsic death pathway such as caspase-8 and caspase-9 [36]. Interestingly, the GPI anchor of CEA appears to contain all the specific information to convey CEA's function once attached to an extracellular adhesive protein domain [37].

In the context of CEA-based immunotherapy it is of special interest that CEA can apparently modulate immune functions. The killing of tumor cells by natural killer (NK) cells is inhibited via heterotypic interaction of CEA on tumor cells with the MHC class I-independent inhibitory NK cell receptor CEACAM1 [38]. A similar inhibitory effect has been reported to result from the transfer of soluble CEA to a still unidentified putative NK cell inhibitory receptor which involves carbohydrate moieties on the CEA molecule [39]. A similar observation has been made previously with lymphokine activated killer (LAK) cells, the antitumor activity of which is negatively correlated with the surface CEA density of tumor cell targets [40]. Transfer of soluble CEA mediated by direct cell–cell contact also appears to be a prerequisite for this inhibition [41]. More recently, induction of mucosal immunity tolerance against luminal antigens by stimulation of CD8$^+$ regulatory T cells through engagement of CEA on antigen-presenting intestinal epithelial cells and CD8 on T cells has been suggested. This interaction involves antigen presentation by the non-polymorphic MHC I molecule CD1d on epithelial cells [42].

Taken together, the functions of CEA seem to help to maintain the dedifferentiated state of cancer cells and allow their escape from the ordered architecture of the tissue thus facilitating invasion and metastasis. This is achieved by supporting survival upon loss of binding to the extracellular matrix. Furthermore, CEA also appears to participate in tumor immune escape processes by negatively modulating tumor killing functions by NK and T cells, and possibly by induction of a state of tolerance as seen in intestinal intraepithelial lymphocytes. These properties make CEA very useful if not indispensable for tumor cells, which in turn might be dependent on continuous expression of this tumor-associated antigen. Thus, the appearance of CEA loss variants under CEA-targeted tumor therapies is not expected to occur.

11.2
Clinical Relevance of CEA

11.2.1
CEA as a Tumor Marker for Prognosis and Post-surgery Follow-up

Apically expressed CEA on normal epithelial cells is shed into the lumen possibly by the action of phospholipases and through the exfoliation of turned-over cells. Thus CEA does not have access the bloodstream. Expression on less differentiated unpolarized tumor cells, however, allows released CEA to enter blood and lymphatic vessels through the intercellular spaces, which can lead to elevated CEA concentrations in the sera of tumor patients. However, as seen for most other tumor markers, CEA measurements are not useful for screening and early detection of cancer due to the lack of sensitivity and specificity [43]. Raised CEA serum levels are also observed in a number of benign conditions, such as inflammation of the lung, intestine and pancreas, which is accompanied by tissue injury and remodeling. Despite these deficiencies, determination of CEA serum levels is useful for pre-operative staging, monitoring treatment responses as well as early detection of disease recurrence in selected groups of patients, especially in those with colorectal and breast carcinomas [44, 45]. It has been agreed by the European Group on Tumor Markers to recommend CEA as a tumor marker in colorectal cancer pre-operatively as a prognostic factor and for early detection of disease in patients with stage II and III tumors who may be candidates for resection of developing liver metastases or systemic treatment. Regular postoperative measurements of CEA serum levels followed by appropriate therapeutic interventions had a significant impact on the survival of patients with colorectal cancers [43]. Elevated serum CEA levels are found in 50–60% of breast cancer patients with metastatic disease. In the case of negativity of other serum tumor markers more commonly elevated in breast tumor patients such as MUC-1 and CA 15-3, measurements of CEA serum concentrations in follow-up patients are recommended [46]. Recent studies demonstrate that peri- or postoperative quantification of CEA mRNA in blood cells, which is believed to be a measurement for circulating epithelial tumor cells, alone or in combination with other marker mRNAs has the potential of early detection of occult recurrences for gastric, colorectal carcinoma, esophageal squamous carcinoma and possibly for non-small cell lung carcinoma patients [47–50].

11.2.2
Targeting CEA for Tumor Localization and Therapy

When targeting CEA for diagnosis and therapy possible cross-reactivity due to the existence of closely related members of the CEA family has to be considered. Antibodies can be designed to be CEA-specific. However, for T cell-mediated therapies it is more difficult to avoid cross-reaction with, for example, CEACAM expressed in leukocytes, when whole CEA protein vaccines are used. On the other hand, the close CEA relative CEACAM6 is highly over-expressed in many adenocarcinomas [51]. Cross-reactivity, therefore, might even enhance the potency of CEA-targeted vaccines or antibodies if potential damage to the hematopoietic system is assumed to be reversible.

11.2.2.1 Animal Models for CEA
Mice are commonly used to analyze the efficacy of tumor therapies and mechanisms of tumor rejection. However, this species cannot be utilized to evaluate CEA-based therapies since no *CEACAM5* ortholog exists in rodents. This problem was circumvented by introducing the human *CEACAM5* gene into the germ line of mice [52–54]. Either cosmid clones containing the *CEACAM5* gene or a bacterial artificial chromosome (BAC) clone, which comprises part of the gene cluster surrounding *CEACAM5*, served as a genetic source. This cluster includes *CEACAM6* and *CEACAM3*, two out of the three genes most similar to *CEACAM5*. In all models, expression patterns of *CEACAM5* and related genes during development and in adult animals mimic those found in humans. These models, therefore, are useful to study approaches aimed at destroying tolerance to the self-antigen CEA and possible side effects caused by autoimmune reactions. Furthermore, the BAC-transgenic mice allow evaluation of possible cross-reactions with related members, which might or might not interfere with CEA-targeted therapies. Indeed, CEA-transgenic mice have been used to demonstrate that no insurmountable central tolerance against CEA exists in these animals by induction of effective antitumor responses using various vaccination strategies [55–64]. So far, antitumor immunity has been induced in models with transplanted CEA-expressing syngeneic tumor cells in both prophylactic and therapeutic settings. Successful treatment of spontaneous autochthonous tumors, expressing transgenic human CEA will be the next challenging step in the development of effective CEA-targeted immune therapies. A number of CEA-positive single and double transgenic murine models predisposed to develop spontaneous CEA-positive breast, stomach, lung, and colonic adenocarcinomas/adenomas have been described [32, 65–68].

11.2.2.2 Tumor Localization and Therapy with Anti-CEA Antibodies
High specificity and affinity of antibodies have been used to concentrate cytotoxic agents at the tumor site thus increasing the efficiency of a given therapy and at the

same time reducing the damage caused to normal tissues. The suitability of CEA as a target antigen for tumor visualization and therapy has been explored for many years. Small tumor lesions expressing CEA can be detected with radioisotope-labeled anti-CEA antibodies. However, other sensitive, more practical technologies (^{18}F-deoxyglucose-based positron emission tomography (PET) and magnetic resonance imaging) are being increasingly used for the imaging of small tumor lesions [69]. Major obstacles like poor tissue penetration in solid tumors and immunogenicity of antibodies and toxins as well as hematologic toxicity of radioisotope-conjugates have been tackled and at least partly resolved, for example by using small recombinant antibody fragments and fully humanized antibodies [70–73]. An unconjugated mouse monoclonal anti-CEA antibody has been tested in patients with advanced colorectal carcinomas in a phase I trial. The treatment was associated with minimal toxicity and unfortunately, little if any clinical effects, as seen in many other trials which have made use of other target antigens [74].

Antibody Delivery of Radionuclides, Drugs and Effector Molecules Delivery of radionuclides to tumors using murine and human anti-CEA antibodies has been studied for many years [70]. The use of low molecular weight single chain antibody fragments and pre-targeting has been found to enhance the sensitivity of tumor visualization as well as increasing the delivered therapeutic dose by separating the antibody targeting to the tumor from the subsequent delivery of the therapeutic radionuclide that binds to the tumor-localized antibody. Initial promising results have been obtained using bispecific anti-CEA/anti-hapten antibodies and hapten–radionuclide conjugates for the radioimmunotherapy of patients with CEA-positive solid tumors, such as medullary thyroid and small-cell lung cancers [70]. Co-delivery of an apoptosis-inducing agent with a bispecific anti-CEA/anti-TNF-α antibody might also enhance radiocurability of CEA-expressing tumors as shown in a CEA-transgenic mouse model [75]. Multistep procedures have been applied to accumulate conjugates consisting of an anti-CEA antibody and a bacterial prodrug-activating enzyme in the tumor followed by clearance of the antibody–enzyme complex from the blood and the systemic application of an inactive prodrug (ADEPT). Enzymatic activation at the cancer sites leads to the formation of highly toxic compounds. Due to their small size the drugs can easily spread locally by diffusion and also inhibit untargeted tumor cells. However, a number of problems like the immunogenicity of prodrug-activating enzymes and the pharmacological complexity of this therapy make it difficult to transfer ADEPT into the clinic [76].

Redirection of Viral and Bacterial Vectors The presence of CEA on tumor cells has been exploited to change the tropism of adenoviral, retroviral, and bacterial vectors as well as of measles viruses to allow a more tumor-selective delivery of suicide and other therapeutic genes and functions. This has been achieved by displaying either anti-CEA single chain fragment variable (scFv) antibody-Moloney murine leukemia virus envelope or measles virus hemagglutinin protein chimera on the virus surface [77, 78]. Engineering an Fc-binding motif from Staphylococcus protein A into the HI loop of the adenovirus fiber protein and addition of an anti-CEA antibody allowed selective targeting of an adenoviral vector with a therapeutic gene to CEA-expressing gastric carcinoma cells which resulted in superior therapeutic efficacy in a nude mouse model [79]. In a different approach, adenovirus retargeting to CEA-positive tumor cells in hepatic tumor grafts was achieved by systemic application of virus together with a bispecific adaptor consisting of the coxsackie/adenovirus receptor ectodomain and an anti-CEA scFv antibody [80]. Furthermore, a *Salmonella typhimurium* strain engineered to display an anti-CEA scFv on its surface showed preferential homing to CEA-expressing tumors in mice which resulted in accumulation of macrophages and $CD3^+$ lymphocytes in the tumor and in prolonged survival [81].

Redirection of Immune Effector Cells Active vaccination approaches against TAA always face the problem of being inadequate owing to the lack of immunogenicity of self-antigens. Only moderate cellular immune responses mostly involving low affinity immune receptors can be induced. Furthermore, downregulation of MHC I molecule expression and lack of consistent expression of NK cell ligands have been commonly observed as a means by which tumor cells can escape from immune surveillance. Therefore, strategies involving redirection of potent effector T cells and NK cells have been employed. This is accomplished by uncoupling target cell recognition from the classical MHC I-peptide/T cell receptor (TCR) and NK cell receptor/ligand interaction thus overriding tolerance. Introduction of chimeric T and NK cell receptors has been used to retarget T and NK cells to tumor cells via CEA [82]. Both anti-CD3/anti-CEA bispecific antibodies or diabodies and anti-CEA antibody/interleukin 2 (IL-2) fusion proteins have been used to target tumor infiltrating lymphocytes and tumor cells via TCR and IL-2 receptors, respectively, resulting in successful treatment of CEA-positive transplanted human and murine tumors [83, 84]. Chimeric TCR have been designed which consist of an anti-CEA scFv extracellular component hooked via a CD28 transmembrane domain to both CD28 and CD3zeta intracellular signaling modules which allows integration of the two signals needed for optimal T-cell stimulation. Similarly, chimeric anti-CEA scFv/Fc-γ receptors have been constructed. In most cases, these constructs were transduced into primary $CD4^+$ and $CD8^+$ T cells or monocytes using lentiviral vectors and were found to reduce tumor growth in *in-vitro* and *in-vivo* murine models [85–87]. Furthermore, Cesson and co-workers have exploited the immune system's propensity to efficiently mount cytotoxic virus-specific T-cell responses against viral infections by redirecting their cytotoxicity to CEA-expressing tumor cells, using a 95-kDa anti-CEA F_{ab} antibody fragment/MHC I- lymphochoriomeningitis virus (LCMV) peptide conjugate. Infection of mice with LCMV abolished growth of grafted MC38-CEA tumor cells [88].

11.2.2.3 CEA-based Vaccines

As stated before, tolerance towards the self-antigen CEA must be overcome in order to render vaccination strategies successful. There is ample evidence that peripheral and not central tolerance exists in humans and in CEA-transgenic mice (see above). Peripheral tolerance can be destroyed more easily. CEA appears to represent a general tumor antigen since colorectal carcinoma patients, who respond clinically to an cytokine staining, and cytokine and chemokine multiplexing [106]. The best *in vitro* assay would quantify the CEA vaccine-induced effector cells and antibodies responsible for clinically measurable anti-tumor responses. A number of MHC I and MHC II epitopes have been identified by applying prediction algorithms or experimentally by peptide elution from purified MHC/peptide complexes and peptide scanning (Table 11.1; [107]). In addition, modification of a CEA-derived T-cell epitope (CAP-1) at a central, non-anchor position which is responsible for T-cell receptor interaction

Table 11.1 CEA T-cell epitopes.

Name	Sequence	Domain (position)	MHC restriction	Reference
Human			**MHC I**	
CEA(9-18)	HRWCIPWQRL	L (9–18)	HLA-B27[1]	[110]
CEA(61)	HLFGYSWYK	N (61–69)	HLA-A3	[111]
CEA268	QYSWFVNGTF	B1 (268–277)	HLA-A24	[112]
CEA(CAP1)	YLSGANLNL	B3 (605–613)	HLA-A2	[113]
CEA(CAP1-6D)[2]	YLSGADLNL			[114]
CEA9(632)	IPQQHTQVL	B3 (632–640)	HLA-B7	[115]
CEA(652)	TYACFVSNL	B3 (652–660)	HLA-A2402	[116]
CEA(691)	IMIGVLVGV	M (691–699)	HLA-A2	[117]
CEA(691-1Y5M)[2]	YMIGMLVGV		HLA-A2[1]	[110]
CEA(691-1Y7L)[2]	YMIGVLLGV		HLA-A2[1]	[110]
CEA(694-702)	GVLVGVALI	M (694–702)	HLA-A2	[118]
			MHC II	
CEA(116)	DTGFYTLHVIKSDLV NEEATGQFRV	N (116–140)	HLA-DR4[1]	[119]
CEA(177–189/ 355–367)	LWWVNNQSLPVSP	A1 (177–189) A2 (355–367)	n.d.	[120]
CEA(625-639)	YSWRINGIPQQHTQV	B3 (625–639)	HLA-DR4[1] HLA-DR53	[121]
CEA(653-667)	YACFVSNLATGRNNS	B3 (653–667)	HLA-DR4 HLA-DR7 HLA-DR9	[122]
Mouse			**MHC I**	
CEA(526-533)	EAQNTTYL	A3(526–533)	H-2D^b	[123]
CEA(572-579)	CGIQNSVSA	A3 (572–579)	H-2D^b	[56]
			MHC II	
1	TYLWWVNNQSLPVS	A1 (175–188)	H-2^b	[124]
	TYLWWVNNQSLPVS	A2 (353–366)	H-2^b	[124]
	TYLWWVNGQSLPVS	A3 (531–544)	H-2^b	[124]
2	SPSYTYYRPGVNLSL	B2 (421–435)	H-2^b	[124]
3	CGIQNSVSANRSDPV	A3 (571-585)	H-2^b	[124]

[1] Identified in HLA-transgenic mice.
[2] 'enhanced' epitope.
L, leader peptide; N, IgV-like domain; A, B, IgC-like domains; M, GPI anchorage signal sequence.

has yielded an agonistic peptide which is able to sensitize CAP-1 T cells 100–1000 times more efficiently than the original peptide CAP-1 [108]. Such enhanced epitope peptides are invaluable for vaccination and monitoring anti-CEA immune responses. Furthermore, CEA CTL epitopes can be used to monitor the induction of antitumor immune responses after vaccination with complex CEA-expressing tumor cell-based vaccines [109].

11.3
Conclusion

CEA is one of the first tumor markers discovered, second to α-fetoprotein. It is still one of the most valuable serum tumor markers to date, especially for the follow-up of patients with various adenocarcinomas, in particular colorectal carcinomas. Indeed, worldwide CEA is the most commonly used tumor marker. More recently, antibody-mediated and cell-based immunotherapies targeting CEA are being evaluated. Due to the high and consistent cell surface expression of CEA in certain tumors it appears to be an ideal target antigen for such immunotherapies. Many preclinical experiments with human CEA-transgenic mice and clinical trials have clearly demonstrated that there is no insurmountable tolerance towards the self-antigen CEA. So far, no autoimmune reactions provoked by CEA-targeted vaccinations have been observed. However, once highly reactive T cells recognizing CEA-derived epitopes are generated, possible cross-reactions with closely related CEA family members have to be considered and circumvented when designing potent vaccines in the future.

Acknowledgments

We would like to thank Petra Stieber and Heike Pohla for their valuable suggestions.

References

1 Gold, P. and Freedman, S.O. (1965) *The Journal of Experimental Medicine*, **122**, 467–481.
2 von Kleist, S. and Burtin, P. (1966) *CR Académie des sciences (Paris)*, **263**, 1543–1546.
3 Zebhauser, R., Kammerer, R., Eisenried, A., McLellan, A., Moore, T. and Zimmermann, W. (2005) *Genomics*, **86**, 566–580.
4 Beauchemin, N., Draber, P., Dveksler, G., Gold, P., Gray-Owen, S., Grunert, F., Hammarstrom, S., Holmes, K.V., Karlsson, A., Kuroki, M., Lin, S.H., Lucka, L., Najjar, S.M., Neumaier, M., Obrink, B., Shively, J.E., Skubitz, K.M., Stanners, C.P., Thomas, P., Thompson, J.A., Virji, M., von Kleist, S., Wagener, C., Watt, S. and Zimmermann, W. (1999) *Experimental Cell Research*, **252**, 243–249.
5 Nap, M., Hammarstrom, M.L., Bormer, O., Hammarstrom, S., Wagener, C., Handt, S., Schreyer, M., Mach, J.P., Buchegger, F. and von Kleistm, S. (1992) *Cancer Research*, **52**, 2329–2339.

6 Tan, K., Zelus, B.D., Meijers, R., Liu, J.H., Bergelson, J.M., Duke, N., Zhang, R., Joachimiak, A., Holmes, K.V. and Wang, J.H. (2002) *The EMBO Journal*, **21**, 2076–2086.

7 Fedarovich, A., Tomberg, J., Nicholas, R.A. and Davies, C. (2006) *Acta Crystallographica. Section D, Biological Crystallography*, **62**, 971–979.

8 Kammerer, R., Popp, T., Hartle, S., Singer, B.B. and Zimmermann, W. (2007) *BMC Evolutionary Biology*, **7**, 196.

9 Dveksler, G.S., Pensiero, M.N., Cardellichio, C.B., Williams, R.K., Jiang, G.S., Holmes, K.V. and Dieffenbach, C.W. (1991) *Journal of Virology*, **65**, 6881–6891.

10 Kammerer, R., Popp, T., Singer, B.B., Schlender, J. and Zimmermann, W. (2004) *Gene*, **339**, 99–109.

11 Hammarstrom, S. (1999) *Seminars in Cancer Biology*, **9**, 67–81.

12 Horst, A. and Wagener, C. (2004) *Handbook of Experimental Pharmacology, Cell Adhesion*, Springer, Berlin, Heidelberg.

13 Gautherot, E., Kraeber-Bodere, F., Daniel, L., Fiche, M., Rouvier, E., Sai-Maurel, C., Thedrez, P., Chatal, J.F. and Barbet, J. (1999) *Clinical Cancer Research*, **5**, 3177s–3182s.

14 Ilantzis, C., Jothy, S., Alpert, L.C., Draber, P. and Stanners, C.P. (1997) *Laboratory Investigation; A Journal of Technical Methods and Pathology*, **76**, 703–716.

15 Prall, F., Nollau, P., Neumaier, M., Haubeck, H.D., Drzeniek, Z., Helmchen, U., Loning, T. and Wagener, C. (1996) *The Journal of Histochemistry and Cytochemistry*, **44**, 35–41.

16 Scholzel, S., Zimmermann, W., Schwarzkopf, G., Grunert, F., Rogaczewski, B. and Thompson, J. (2000) *The American Journal of Pathology*, **156**, 595–605.

17 Gray-Owen, S.D. and Blumberg, R.S. (2006) *Nature Reviews. Immunology*, **6**, 433–446.

18 Kuespert, K., Pils, S. and Hauck, C.R. (2006) *Current Opinion in Cell Biology*, **18**, 565–571.

19 Najjar, S.M. (2002) *Trends in Endocrinology and Metabolism*, **13**, 240–245.

20 Obrink, B. (1997) *Current Opinion in Cell Biology*, **9**, 616–626.

21 Oikawa, S., Inuzuka, C., Kuroki, M., Matsuoka, Y., Kosaki, G. and Nakazato, H. (1989) *Biochemical and Biophysical Research Communications*, **164**, 39–45.

22 Benchimol, S., Fuks, A., Jothy, S., Beauchemin, N., Shirota, K. and Stanners, C.P. (1989) *Cell*, **57**, 327–334.

23 Leusch, H.G., Drzeniek, Z., Markos-Pusztai, Z. and Wagener, C. (1991) *Infection and Immunity*, **59**, 2051–2057.

24 Berger, C.N., Billker, O., Meyer, T.F., Servin, A.L. and Kansau, I. (2004) *Molecular Microbiology*, **52**, 963–983.

25 Hashino, J., Fukuda, Y., Oikawa, S., Nakazato, H. and Nakanishi, T. (1994) *Clinical & Experimental Metastasis*, **12**, 324–328.

26 Thomas, P., Gangopadhyay, A., Steele, G.., Jr. Andrews, C., Nakazato, H., Oikawa, S. and Jessup, J.M. (1995) *Cancer Letters*, **92**, 59–66.

27 Yogeeswaran, G. and Salk, P.L. (1981) *Science*, **212**, 1514–1516.

28 Minami, S., Furui, J. and Kanematsu, T. (2001) *Cancer Research*, **61**, 2732–2735.

29 Bajenova, O.V., Zimmer, R., Stolper, E., Salisbury-Rowswell, J., Nanji, A. and Thomas, P. (2001) *The Journal of Biological Chemistry*, **276**, 31067–31073.

30 Aarons, C.B., Bajenova, O., Andrews, C., Heydrick, S., Bushell, K.N., Reed, K.L., Thomas, P., Becker, J.M. and Stucchi, A.F. (2007) *Clinical & Experimental Metastasis*, **24**, 201–209.

31 van Gisbergen, K.P., Aarnoudse, C.A., Meijer, G.A., Geijtenbeek, T.B. and van Kooyk, Y. (2005) *Cancer Research*, **65**, 5935–5944.

32 Chan, C.H., Cook, D. and Stanners, C.P. (2006) *Carcinogenesis*, **27**, 1909–1916.

33. Ilantzis, C., DeMarte, L., Screaton, R.A. and Stanners, C.P. (2007) *Neoplasia (New York, NY)*, **4**, 151–163.
34. Ordonez, C., Screaton, R.A., Ilantzis, C. and Stanners, C.P. (2000) *Cancer Research*, **60**, 3419–3424.
35. Soeth, E., Wirth, T., List, H.J., Kumbhani, S., Petersen, A., Neumaier, M., Czubayko, F. and Juhl, H. (2001) *Clinical Cancer Research*, **7**, 2022–2030.
36. Camacho-Leal, P. and Stanners, C.P., (2007) *Oncogene*.
37. Screaton, R.A., DeMarte, L., Draber, P. and Stanners, C.P. (2000) *The Journal of Cell Biology*, **150**, 613–626.
38. Stern, N., Markel, G., Arnon, T.I., Gruda, R., Wong, H., Gray-Owen, S.D. and Mandelboim, O. (2005) *Journal of Immunology (Baltimore, Md: 1950)*, **174**, 6692–6701.
39. Stern-Ginossar, N., Nedvetzki, S., Markel, G., Gazit, R., Betser-Cohen, G., Achdout, H., Aker, M., Blumberg, R.S., Davis, D.M., Appelmelk, B. and Mandelboim, O. (2007) *Journal of Immunology (Baltimore, Md: 1950)*, **179**, 4424–4434.
40. Kammerer, R. and von Kleist, S. (1994) *International Journal of Cancer*, **57**, 341–347.
41. Kammerer, R. and von Kleist, S. (1996) *International Journal of Cancer*, **68**, 457–463.
42. Allez, M., Brimnes, J., Shao, L., Dotan, I., Nakazawa, A. and Mayer, L. (2004) *Annals of the New York Academy of Sciences*, **1029**, 22–35.
43. Duffy, M.J., van Dalen, A., Haglund, C., Hansson, L., Holinski-Feder, E., Klapdor, R., Lamerz, R., Peltomaki, P., Sturgeon, C. and Topolcan, O. (2007) *European Journal of Cancer (Oxford, England: 1990)*, **43**, 1348–1360.
44. Goldstein, M.J. and Mitchell, E.P. (2005) *Cancer Investigation*, **23**, 338–351.
45. Molina, R., Barak, V., van Dalen, A., Duffy, M.J., Einarsson, R., Gion, M., Goike, H., Lamerz, R., Nap, M., Soletormos, G. and Stieber, P. (2005) *Tumor Biology*, **26**, 281–293.
46. Harris, L., Fritsche, H., Mennel, R., Norton, L., Ravdin, P., Taube, S., Somerfield, M.R., Hayes, D.F. and Bast, R.C. Jr. (2007) *Journal of Clinical Oncology*, **25**, 5287–5312.
47. Setoyama, T., Natsugoe, S., Okumura, H., Matsumoto, M., Uchikado, Y., Ishigami, S., Owaki, T., Takao, S. and Aikou, T. (2006) *Clinical Cancer Research*, **12**, 5972–5977.
48. Sheu, C.C., Chang, M.Y., Chang, H.C., Tsai, J.R., Lin, S.R., Chang, S.J., Hwang, J.J., Huang, M.S. and Chong, I.W. (2006) *Oncology*, **70**, 203 211.
49. Wang, J.Y., Lin, S.R., Wu, D.C., Lu, C.Y., Yu, F.J., Hsieh, J.S., Cheng, T.L., Koay, L.B. and Uen, Y.H. (2007) *Clinical Cancer Research*, **13**, 2406–2413.
50. Wu, C.H., Lin, S.R., Hsieh, J.S., Chen, F.M., Lu, C.Y., Yu, F.J., Cheng, T.L., Huang, T.J., Huang, S.Y. and Wang, J.Y. (2006) *Disease Markers*, **22**, 103–109.
51. Blumenthal, R.D., Leon, E., Hansen, H.J. and Goldenberg, D.M. (2007) *BMC. Cancer*, **7**, 2.
52. Clarke, P., Mann, J., Simpson, J.F., Rickard-Dickson, K. and Primus, F.J. (1998) *Cancer Research*, **58**, 1469–1477.
53. Eades-Perner, A.M., van Der, P.H., Hirth, A., Thompson, J., Neumaier, M., von Kleist, S. and Zimmermann, W. (1994) *Cancer Research*, **54**, 4169–4176.
54. Chan, C.H. and Stanners, C.P. (2004) *Molecular Therapy: The Journal of the American Society of Gene Therapy*, **9**, 775–785.
55. Huang, Y., Fayad, R., Smock, A., Ullrich, A.M. and Qiao, L. (2005) *Cancer Research*, **65**, 6990–6999.
56. Mennuni, C., Calvaruso, F., Facciabene, A., Aurisicchio, L., Storto, M., Scarselli, E., Ciliberto, G. and La Monica, N. (2005) *International Journal of Cancer*, **117**, 444–455.
57. Niethammer, A.G., Primus, F.J., Xiang, R., Dolman, C.S., Ruehlmann, J.M., Ba, Y., Gillies, S.D. and Reisfeld, R.A. (2001) *Vaccine*, **20**, 421–429.

58 Ojima, T., Iwahashi, M., Nakamura, M., Matsuda, K., Nakamori, M., Ueda, K., Naka, T., Ishida, K., Primus, F.J. and Yamaue, H. (2007) *International Journal of Cancer*, **120**, 585–593.

59 Salucci, V., Mennuni, C., Calvaruso, F., Cerino, R., Neuner, P., Ciliberto, G., La Monica, N. and Scarselli, E. (2006) *Scandinavian Journal of Immunology*, **63**, 35–41.

60 Schwegler, C., Dorn-Beineke, A., Nittka, S., Stocking, C. and Neumaier, M. (2005) *Cancer Research*, **65**, 1925–1933.

61 Xiang, R., Silletti, S., Lode, H.N., Dolman, C.S., Ruehlmann, J.M., Niethammer, A.G., Pertl, U., Gillies, S.D., Primus, F.J. and Reisfeld, R.A. (2001) *Clinical Cancer Research*, **7**, 856s–864s.

62 Zhou, H., Luo, Y., Mizutani, M., Mizutani, N., Becker, J.C., Primus, F.J., Xiang, R. and Reisfeld, R.A. (2004) *The Journal of Clinical Investigation*, **113**, 1792–1798.

63 Saha, A., Chatterjee, S.K., Foon, K.A., Celis, E. and Bhattacharya-Chatterjee, M. (2007) *Cancer Research*, **67**, 2881–2892.

64 Facciabene, A., Aurisicchio, L., Elia, L., Palombo, F., Mennuni, C., Ciliberto, G. and La Monica, N. (2006) *Human Gene Therapy*, **17**, 81–92.

65 Hance, K.W., Zeytin, H.E. and Greiner, J.W. (2005) *Mutation Research*, **576**, 132–154.

66 Horig, H., Wainstein, A., Long, L., Kahn, D., Soni, S., Marcus, A., Edelmann, W., Kucherlapati, R. and Kaufman, H.L. (2001) *Cancer Research*, **61**, 8520–8526.

67 Thompson, J., Epting, T., Schwarzkopf, G., Singhofen, A., Eades-Perner, A.M., van Der, P.H. and Zimmermann, W. (2000) *International Journal of Cancer*, **86**, 863–869.

68 Thompson, J.A., Eades-Perner, A.M., Ditter, M., Muller, W.J. and Zimmermann, W. (1997) *International Journal of Cancer*, **72**, 197–202.

69 Oudoux, A., Salaun, P.Y., Bournaud, C., Campion, L., Ansquer, C., Rousseau, C., Bardet, S., Borson-Chazot, F., Vuillez, J.P., Murat, A., Mirallie, E., Barbet, J., Goldenberg, D.M., Chatal, J.F. and Kraeber-Bodere, F. (2007) *The Journal of Clinical Endocrinology and Metabolism*, **92**, 4590–4597.

70 Goldenberg, D.M., Sharkey, R.M., Paganelli, G., Barbet, J. and Chatal, J.F. (2006) *Journal of Clinical Oncology*, **24**, 823–834.

71 Imakiire, T., Kuroki, M., Shibaguchi, H., Abe, H., Yamauchi, Y., Ueno, A., Hirose, Y., Yamada, H., Yamashita, Y., Shirakusa, T., Ishida, I. and Kuroki, M. (2004) *International Journal of Cancer*, **108**, 564–570.

72 Sharkey, R.M., Cardillo, T.M., Rossi, E.A., Chang, C.H., Karacay, H., McBride, W.J., Hansen, H.J., Horak, I.D. and Goldenberg, D.M. (2005) *Nature Medicine*, **11**, 1250–1255.

73 Pastan, I., Hassan, R., FitzGerald, D.J. and Kreitman, R.J. (2007) *Annual Review of Medicine*, **58**, 221–237.

74 Zbar, A.P., Thomas, H., Wilkinson, R.W., Wadhwa, M., Syrigos, K.N., Ross, E.L., Dilger, P., Allen-Mersh, T.G., Kmiot, W.A., Epenetos, A.A., Snary, D. and Bodmer, W.F. (2005) *International Journal of Colorectal Disease*, **20**, 403–414.

75 Larbouret, C., Robert, B., Linard, C., Teulon, I., Gourgou, S., Bibeau, F., Martineau, P., Santoro, L., Pouget, J.P., Pelegrin, A. and Azria, D. (2007) *International Journal of Radiation Oncology, Biology, Physics*, **69**, 1231–1237.

76 Bagshawe, K.D. (2006) *Expert Review of Anticancer Therapy*, **6**, 1421–1431.

77 Khare, P.D., Shao-Xi, L., Kuroki, M., Hirose, Y., Arakawa, F., Nakamura, K., Tomita, Y. and Kuroki, M. (2001) *Cancer Research*, **61**, 370–375.

78 Hammond, A.L., Plemper, R.K., Zhang, J., Schneider, U., Russell, S.J. and Cattaneo, R. (2001) *Journal of Virology*, **75**, 2087–2096.

79 Tanaka, T., Huang, J., Hirai, S., Kuroki, M., Kuroki, M., Watanabe, N., Tomihara, K., Kato, K. and Hamada, H. (2006) *Clinical Cancer Research*, **12**, 3803–3813.

80 Li, H.J., Everts, M., Pereboeva, L., Komarova, S., Idan, A., Curiel, D.T. and Herschman, H.R. (2007) *Cancer Research*, **67**, 5354–5361.

81 Bereta, M., Hayhurst, A., Gajda, M., Chorobik, P., Targosz, M., Marcinkiewicz, J. and Kaufman, H.L. (2007) *Vaccine*, **25**, 4183–4192.

82 Kuroki, M., Kuroki, M., Shibaguchi, H., Badran, A., Hachimine, K., Zhang, J. and Kinugasa, T. (2004) *Tumour Biology: The Journal of the International Society for Oncodevelopmental Biology and Medicine*, **25**, 208–216.

83 Xu, X., Clarke, P., Szalai, G., Shively, J.E., Williams, L.E., Shyr, Y., Shi, E. and Primus, F.J. (2000) *Cancer Research*, **60**, 4475–4484.

84 Blanco, B., Holliger, P., Vile, R.G. and Alvarez-Vallina, L. (2003) *Journal of Immunology (Baltimore, Md: 1950)*, **171**, 1070–1077.

85 Biglari, A., Southgate, T.D., Fairbairn, L.J. and Gilham, D.E. (2006) *Gene Therapy*, **13**, 602–610.

86 Hombach, A., Kohler, H., Rappl, G. and Abken, H. (2006) *Journal of Immunology (Baltimore, Md: 1950)*, **177**, 5668–5675.

87 Sasaki, T., Ikeda, H., Sato, M., Ohkuri, T., Abe, H., Kuroki, M., Onodera, M., Miyamoto, M., Kondo, S. and Nishimura, T. (2006) *Cancer Science*, **97**, 920–927.

88 Cesson, V., Stirnemann, K., Robert, B., Luescher, I., Filleron, T., Corradin, G., Mach, J.P. and Donda, A. (2006) *Clinical Cancer Research*, **12**, 7422–7430.

89 Correale, P., Cusi, M.G., Tsang, K.Y., Del Vecchio, M.T., Marsili, S., Placa, M.L., Intrivici, C., Aquino, A., Micheli, L., Nencini, C., Ferrari, F., Giorgi, G., Bonmassar, E. and Francini, G. (2005) *Journal of Clinical Oncology*, **23**, 8950–8958.

90 Saha, A., Baral, R.N., Chatterjee, S.K., Mohanty, K., Pal, S., Foon, K.A., Primus, F.J., Krieg, A.M., Weiner, G.J. and Bhattacharya-Chatterjee, M. (2006) *Cancer Immunology, Immunotherapy*, **55**, 515–527.

91 Saha, A., Chatterjee, S.K., Foon, K.A., Primus, F.J., Sreedharan, S., Mohanty, K. and Bhattacharya-Chatterjee, M. (2004) *Cancer Research*, **64**, 4995–5003.

92 Chong, G., Bhatnagar, A., Cunningham, D., Cosgriff, T.M., Harper, P.G., Steward, W., Bridgewater, J., Moore, M., Cassidy, J., Coleman, R., Coxon, F., Redfern, C.H., Jones, J.J., Hawkins, R., Northfelt, D., Sreedharan, S., Valone, F. and Carmichael, J. (2006) *Annals of Oncology*, **17**, 437–442.

93 Bramswig, K.H., Knittelfelder, R., Gruber, S., Untersmayr, E., Riemer, A.B., Szalai, K., Horvat, R., Kammerer, R., Zimmermann, W., Zielinski, C.C., Scheiner, O. and Jensen-Jarolim, E. (2007) *Clinical Cancer Research*, **13**, 6501–6508.

94 Gilboa, E. (2007) *The Journal of Clinical Investigation*, **117**, 1195–1203.

95 Eppler, E., Horig, H., Kaufman, H.L., Groscurth, P. and Filgueira, L. (2002) *European Journal of Cancer (Oxford, England: 1990)*, **38**, 184–193.

96 Morse, M.A., Clay, T.M., Hobeika, A.C., Osada, T., Khan, S., Chui, S., Niedzwiecki, D., Panicali, D., Schlom, J. and Lyerly, H.K. (2005) *Clinical Cancer Research*, **11**, 3017–3024.

97 Babatz, J., Rollig, C., Lobel, B., Folprecht, G., Haack, M., Gunther, H., Kohne, C.H., Ehninger, G., Schmitz, M. and Bornhauser, M. (2006) *Cancer Immunology, Immunotherapy*, **55**, 268–276.

98 Weihrauch, M.R., Ansen, S., Jurkiewicz, E., Geisen, C., Xia, Z., Anderson, K.S., Gracien, E., Schmidt, M., Wittig, B., Diehl, V., Wolf, J., Bohlen, H. and Nadler, L.M. (2005) *Clinical Cancer Research*, **11**, 5993–6001.

99 Liu, K.J., Wang, C.C., Chen, L.T., Cheng, A.L., Lin, D.T., Wu, Y.C., Yu, W.L., Hung, Y.M., Yang, H.Y., Juang, S.H. and Whang-Peng, J. (2004) *Clinical Cancer Research*, **10**, 2645–2651.

100 Correale, P., Cusi, M.G., Del Vecchio, M.T., Aquino, A., Prete, S.P., Tsang, K.Y., Micheli, L., Nencini, C., La Placa, M., Montagnani, F., Terrosi, C., Caraglia, M.,

Formica, V., Giorgi, G., Bonmassar, E. and Francini, G. (2005) *Journal of Immunology (Baltimore, Md: 1950)*, **175**, 820–828.

101 Tacken, P.J., de Vries, I.J., Torensma, R. and Figdor, C.G. (2007) *Nature Reviews. Immunology*, **7**, 790–802.

102 Huang, E.H. and Kaufman, H.L. (2002) *Expert Review of Vaccines*, **1**, 49–63.

103 Arlen, P.M., Kaufman, H.L. and DiPaola, R.S. (2005) *Seminars in Oncology*, **32**, 549–555.

104 Aarts, W.M., Schlom, J. and Hodge, J.W. (2002) *Cancer Research*, **62**, 5770–5777.

105 Marshall, J.L., Gulley, J.L., Arlen, P.M., Beetham, P.K., Tsang, K.Y., Slack, R., Hodge, J.W., Doren, S., Grosenbach, D.W., Hwang, J., Fox, E., Odogwu, L., Park, S., Panicali, D. and Schlom, J. (2005) *Journal of Clinical Oncology*, **23**, 720–731.

106 Keilholz, U., Martus, P. and Scheibenbogen, C. (2006) *Clinical Cancer Research*, **12**, 2346s–2352s.

107 Pickford, W.J., Watson, A.J. and Barker, R.N. (2007) *Clinical Cancer Research*, **13**, 4528–4537.

108 Fong, L., Hou, Y., Rivas, A., Benike, C., Yuen, A., Fisher, G.A., Davis, M.M. and Engleman, E.G. (2001) *Proceedings of the National Academy of Sciences of the United States of America*, **98**, 8809–8814.

109 Koido, S., Hara, E., Homma, S., Torii, A., Toyama, Y., Kawahara, H., Watanabe, M., Yanaga, K., Fujise, K., Tajiri, H., Gong, J. and Toda, G. (2005) *Clinical Cancer Research*, **11**, 7891–7900.

110 Huarte, E., Sarobe, P., Lasarte, J.J., Brem, G., Weiss, E.H., Prieto, J. and Borras-Cuesta, F. (2002) *International Journal of Cancer*, **97**, 58–63.

111 Kawashima, I., Tsai, V., Southwood, S., Takesako, K., Sette, A. and Celis, E. (1999) *Cancer Research*, **59**, 431–435.

112 Nukaya, I., Yasumoto, M., Iwasaki, T., Ideno, M., Sette, A., Celis, E., Takesako, K. and Kato, I. (1999) *International Journal of Cancer*, **80**, 92–97.

113 Tsang, K.Y., Zaremba, S., Nieroda, C.A., Zhu, M.Z., Hamilton, J.M. and Schlom, J. (1995) *Journal of the National Cancer Institute*, **87**, 982–990.

114 Zaremba, S., Barzaga, E., Zhu, M., Soares, N., Tsang, K.Y. and Schlom, J. (1997) *Cancer Research*, **57**, 4570–4577.

115 Lu, J. and Celis, E. (2000) *Cancer Research*, **60**, 5223–5227.

116 Kim, C., Matsumura, M., Saijo, K. and Ohno, T. (1998) *Cancer Immunology, Immunotherapy*, **47**, 90–96.

117 Kawashima, I., Hudson, S.J., Tsai, V., Southwood, S., Takesako, K., Appella, E., Sette, A. and Celis, E. (1998) *Human Immunology*, **59**, 1–14.

118 Schirle, M., Keilholz, W., Weber, B., Gouttefangeas, C., Dumrese, T., Becker, H.D., Stevanovic, S. and Rammensee, H.G. (2000) *European Journal of Immunology*, **30**, 2216–2225.

119 Shen, L., Schroers, R., Hammer, J., Huang, X.F. and Chen, S.Y. (2004) *Cancer Immunology, Immunotherapy*, **53**, 391–403.

120 Campi, G., Crosti, M., Consogno, G., Facchinetti, V., Conti-Fine, B.M., Longhi, R., Casorati, G., Dellabona, P. and Protti, M.P. (2003) *Cancer Research*, **63**, 8481–8486.

121 Ruiz, M., Kobayashi, H., Lasarte, J.J., Prieto, J., Borras-Cuesta, F., Celis, E. and Sarobe, P. (2004) *Clinical Cancer Research*, **10**, 2860–2867.

122 Kobayashi, H., Omiya, R., Ruiz, M., Huarte, E., Sarobe, P., Lasarte, J.J., Herraiz, M., Sangro, B., Prieto, J., Borras-Cuesta, F. and Celis, E. (2002) *Clinical Cancer Research*, **8**, 3219–3225.

123 Schmitz, J., Reali, E., Hodge, J.W., Patel, A., Davis, G., Schlom, J. and Greiner, J.W. (2002) *Cancer Research*, **62**, 5058–5064.

124 Bos, R., van Duikeren, S., van Hall, T., Kaaijk, P., Taubert, R., Kyewski, B., Klein, L., Melief, C.J. and Offringa, R. (2005) *Cancer Research*, **65**, 6443–6449.

12
HER-2 as a Tumor Antigen

Barbara Seliger

12.1
Introduction

The HER2 oncogene, a member of the HER family of receptor tyrosine kinases, represents a potential target for cancer immunotherapies as well as for new biomolecular treatment modalities of solid tumors. HER2 overexpression has been found in many human tumors of epithelial origin and is associated with tumor aggressiveness and reduced patient survival. Furthermore, HER2 appears to be immunogenic since B and T cell-mediated responses directed against this oncoprotein have been detected in tumor patients. Therefore HER2 is a valuable target for both passive and active immunotherapeutic strategies. In addition to a brief description of the expression and function of HER2, which have been reviewed elsewhere in detail, this chapter summarizes HER2 as an immunogenic self-antigen, the existence of MHC class I- and class II-restricted HER2-specific T-cell epitopes, the different immunotherapeutic approaches directed against HER2 and the monitoring of HER2-specific immune responses, but also discusses problems of HER2-directed immunotherapies.

12.2
Biology of HER2

12.2.1
Features of HER2

The human epidermal growth factor receptor HER2 is a 185-kDa protein, which is encoded by the HER2 proto-oncogene mapped to chromosome 17p21. HER2 represents a member of the transmembrane tyrosine kinase receptor family designated as type I growth factor receptors HER1 to HER4. These different receptors confer growth stimulatory activity and are involved both in the normal development and differentiation of mammary epithelial cells. In general HER receptors exist as monomers in the cell membrane, but upon by binding of different ligands, including the epidermal growth factor (EGF), the transforming growth factor-α (TGF-α) and heregulin, they form an active heterodimers. The ligands of HER receptors are classified into three distinct groups based upon their receptor specificity. The epidermal growth factor (EGF), the transforming growth factor α (TGF)-α, amphiregulin and ephigen, exclusively bind to HER1 and belong to the first group. The second group includes β-celluline, the heparin-binding EGF and epiregulin, which have a dual capacity and bind to HER1 as well as HER4, whereas the third group is composed of the neuroregulin family, which bind either to HER3 or to HER4 [1–3]. However, it is noteworthy that the ligand for HER2 has not yet been identified. HER2 is the preferred heterodimerization partner of all HER family members with the strongest affinity to HER1.

In normal cells, activation of HER triggers a network of signaling cascades that control cell growth, cell cycle, differentiation, apoptosis, motility and cell adhesion [4]. In addition, a number of *in vitro* and *in vivo* animal studies also demonstrated the direct role of HER2 in the oncogenic transformation and during tumorigenesis, which is due to increased HER2 expression levels [5].

12.2.2
The Self-antigen HER2 as a Potential Therapeutic Target

As reviewed elsewhere in this book many newly discovered tumor antigens (TA) are proteins that are overexpressed in tumor cells. HER2 represents a member of this family of TA. In a number of different human cancers of epithelial origin including mammary, non-small cell lung (NSCLC), ovarian, uterus, prostate, colon and renal cell carcinoma (RCC, Table 12.1), but also in *in vitro* models of oncogenic transformation, HER2 amplification and/or overexpression is accompanied by an increase in the surface expression of HER2. This overexpression results in the dysregulated homeostasis of the HER signal cascade, which is advantageous to both the growth and survival of the tumor cells. Using a tetracylin-controlled HER2 mouse model the inhibition of HER2 leads to tumor cell apoptosis as well as a tumor size-dependent remission [6]. HER2 expression is

Table 12.1 Human tumors showing HER2 amplification and/or overexpression.

- Breast carcinoma
- Colon carcinoma
- Lung carcinoma
- Mammary carcinoma
- Ovarian carcinoma
- Prostate carcinoma
- Renal cell carcinoma
- Uterus carcinoma

often associated with HER2 gene amplification, transcriptional and/or post-transcriptional dysregulation. The clinical consequences of HER2 overexpression have been mainly studied in lung, ovarian and breast carcinoma. The HER2 receptor is overexpressed in approximately 20 to 40% of mammary and 30% ovarian cancer patients and is predictive of a worse prognosis and poor clinical outcome for these patients. Furthermore, HER2 overexpression is associated with an increased resistance to therapy with hormonal (Tamoxifen) or certain cytotoxic chemotherapeutic agents (doxorubicin). Despite the existence of circulating HER2-specific antibodies in tumor patients the relationship between overexpression of HER2 as a self-protein and its immunogenicity in human cancers has not been well defined. However, self-antigens can be immunogenic in cancer patients when a protein is normally silent or expressed at low levels, but is aberrantly overexpressed or abundant in tumor cells. It has recently been demonstrated that patients with HER2-overexpressing tumors may have pre-existing antibody immunity to HER2 [7]. Thus, HER2 represents an ideal candidate for therapy due to (i) its extensive overexpression mainly occurring in epithelial tumors compared to its low expression in normal tissues, (ii) its amplification as a cause of oncogenic transformation, (iii) its immunogenicity as demonstrated by the presence of antibodies and HER2-specific T cells in HER2-positive tumor patients [8, 9] and (iv) its extracellular domain, which is available for potential antibody binding (Table 12.2). Based on this information HER2 is a valuable target for specific anti-cancer agents such as monoclonal antibodies (mAbs), HER2-specific T cells as well as small targeted molecules like tyrosine kinase inhibitors (TKI). In many studies the growth of tumor cell lines *in vitro* and tumor xenografts *in vivo* were inhibited by TKIs and by various anti-HER2 mAb. In order to overcome the immunogenicity of HER2-specific antibodies they have been humanized and trastuzumab (Herceptin) has been approved by the FDA for clinical use in patients with mammary carcinoma.

However, the efficacy of HER2-specific immune responses is only low or nonexistent in the majority of patients who overexpress this oncoprotein, whilst some patients have high levels of immunity. This suggests that immunotherapies should be developed which will generate and enhance effective HER2-specific B and T cell responses with cytotoxic potential in both animal models and clinical trials [10–12]. In addition, *in vivo* expansion of HER2-specific T cells, which could also be used in adoptive therapy, might be implemented for the treatment of established disease.

Table 12.2 Features of HER2 which make it an ideal candidate for immunotherapies.

- Extensive HER2 overexpression in different tumor types
- Amplification of HER in some tumor entities
- Immunogenic potential
- Membrane bound with an extracellular antibody binding domain

HER2-specific $CD8^+$ and/or $CD4^+$ T lymphocytes of patients, respectively. A number of these known peptides have been demonstrated to be presented by various tumors including breast, ovarian, colorectal, pancreatic, lung, renal, gastric and prostate cancer as well as glioblastoma.

Employing reverse immunology, synthetic peptides derived from the HER2 protein sequence have been selected based on the binding motives of different HLA class I antigens and screened for candidate T-cell epitopes by evaluating the $CD8^+$ T-cell recognition of the respective epitope and antigen-specific T-cell responses generated against the respective HLA-restricted HER2 peptide antigen. With this approach a number of immunodominant epitopes have been identified including the HLA-A2-restricted $HER2_{369-377}$ peptide, which is frequently recognized by T cells infiltrating $HER2^+$ tumors. Other peptides identified are presented in association with HLA-A2, -A3, -A11 and -A24. $CD4^+$ T cell-mediated HER2-specific tumor rejection has been found in the absence of B cells [26]. Furthermore, there is clear evidence that HER2-specific peptides could be also presented by MHC class II antigens. HER2 represents a relevant target for eliciting humoral responses in patients as well as T cell-mediated immunity since both HER2-specific class I- and class II-restricted T-cell epitopes were detected. Indeed several of the HER2 peptides identified are presented by different human tumors including breast, ovarian, colorectal, pancreatic, lung, renal, gastric, prostate cancer and even glioblastoma. Modification of HER2 epitopes enhanced both tumor cell binding as well as antitumor activities [27].

12.4.2
T Cell-mediated Immunity to HER2 in Cancer Patients

In most tumors HER2 represents a non-mutated self protein against which an immunologic tolerance usually is expected. However, there appears to be B and T cell-mediated immunity to the HER2 protein even in the early stages of this disease [28]. HER2-specific T cells have been shown to proliferate *in vitro* and secrete Th1 cytokines in response to HER2 peptides [29, 30]. However, this response was independent of HER2 overexpression suggesting that T cell tolerance can be reversed even in the absence of HER2 overexpression. Recently different mutations have been identified in the HER2 kinase domain. The role of these mutations for T-cell responses has yet to be defined. However, it might be speculated that these mutations may be involved in the resistance of patients to HER2-mediated immune responses.

12.5
Passive Immunotherapy Targeting HER2

Based on the isolation and functional activity of the murine HER2-specific antibody 4D5, this antibody was humanized and still retained its anti-tumor activity. This led to the development of the anti-HER2 antibody trastuzumab (Herceptin), which has been approved by the FDA for the treatment of metastatic

HER2-overexpressing breast cancer. The IgG1 class mAb trastuzumab has significant ADCC-activating functions and a body of evidence implicates an immunological mechanism in the action of trastuzumab. Both trastuzumab alone or in combination with chemotherapy improved the response of patients and reduced the risk of breast cancer recurrence in women with HER2-positive tumors. In addition to trastuzumab, a second generation of agents has been developed [31]. The anti-HER mAb pertuzumab targets the dimerization interface of HER2 and interferes with the heterodimerization of HER2/HER3. Pertuzumab is currently under clinical development and preliminary clinical results look promising [32]. However, in order to increase the efficacy of both antibodies a better knowledge of the molecular mechanisms of action of trastuzumab as well as of pertuzumab is required. So far, several different mechanisms for trastuzumab activity have been demonstrated. In addition to the described antibody-induced HER2 downregulation, a novel mechanism for trastuzumab action has been suggested involving the recycling of a dynamic membrane pool of HER2 receptors without lowering the receptor levels. Furthermore the antibody can downregulate HER signaling activity *in vitro* and *in vivo* in HER2-overexpressing breast cancer patients. Last but not least, trastuzumab binding may mask the protein activity thereby blocking the active domain of the receptor. Blockade of receptors is directly correlated with the response of HER2-positive patients to trastuzumab therapies. Receptor degradation induced by trastuzumab led to enhanced HER2-specific cytotoxic T-cell function [33].

Humanized HER2 mAb vaccines in combination with doxorubicin result in an increased survival of women with HER2$^+$ metastatic breast cancer. Indeed, trastuzumab was active against HER2-overexpressing early stage breast cancer. However, only 20% of patients with HER2-overexpressing tumor lesions exhibit an increased overall survival suggesting that markers predictive of its efficacy have still to be identified. With regard to the action of pertuzumab, the HER2/HER3 heterodimerization signal transduction is blocked which is particularly important since many human tumors express both HER2 and HER3 receptors in their respective lesions. It has therefore been suggested that pertuzumab has a broad spectrum of efficacy in the treatment of tumor patients which is currently being analyzed.

12.6
HER2 Effects on Immunogenicity: Both Sides of the Coin

Although there is a large body of evidence suggesting that HER2 induces immunogenicity, it may have also inhibitory properties such as downregulating the T-cell responses [34]. This is in line with reports that tumor cells can negatively interfere with immune responses. In a number of model systems constitutive or inducible HER2/neu overexpression resulted in a downregulation of MHC class I surface antigens [35]. This is due to a lack or suppression of components in the antigen processing machinery (APM) including the transporters associated with antigen processing (TAP), tapasin and the IFN-γ-inducible proteasome subunits [34]. Deficiencies in APM components are directly associated with a reduced

CTL-mediated recognition of HER2$^+$ tumor cells [36]. Thus, the optimal regimen needs to overcome such immune suppression by the tumor, which might even predict the clinical outcome for patients. Based on these results it might be postulated that HER2-specific immunotherapies need to be revisited, suggesting that the efficacy of responses to HER2-specific vaccination might be increased by the proper selection of MHC class I-positive patients. Furthermore, a loss of the HER2 antigen may be induced by T cell-mediated immune responses in mammary carcinoma, which leads to a relapse in tumor growth. Thus, immune selection may occur resulting in variants which are able to evade the immune system [37].

12.7 Conclusions

Although self-proteins like HER2 have on the one hand been demonstrated to enhance tumor immunogenicity [38–42], a number of open questions still remain. What are the underlying molecular mechanisms by which the abundance of HER2 in tumor cells enhances immunogenicity or immune evasion? Is there an endogenous antibody-immunity to HER2 and can this humoral immunity be used as a diagnostic marker?

Despite reports of a dose-dependent response, what is the relationship between the level of HER2 overexpression and the presence of pre-existing HER2-specific immunity? Does this knowledge lead to the identification of novel immune-relevant targets for the therapeutic development?

Abbreviations

APC	antigen presenting cell
APM	antigen processing machinery
β_2-m	β_2-microglobulin
CISH	chromogenic *in situ* hybridization
CTL	cytotoxic T lymphocytes
DC	dendritic cells
EGF	epidermal growth factor
EGF-R	epidermal growth factor receptor
ER	endoplasmic reticulum
ERAP	ER aminopeptidase associated with antigen processing
FISH	fluorescence *in situ* hybridization
HC	heavy chain
IFN	interferon
IL	interleukin
LMP	low molecular weight proteins
LOH	loss of heterozygosity
mAb	monoclonal antibody

MHC   major histocompatibility complex
NK   natural killer
NSCLC   non-small cell lung carcinoma
PLC   peptide loading complex
RCC   renal cell carcinoma
TA   tumor antigen
TAA   tumor-associated antigen
TAP   transporter associated with antigen processing
TGF   transforming growth factor
TKI   tyrosin kinase inhibitors
Treg   regulatory T cells

References

1 Sergina, N.V. and Moasser, M.M. (2007) The HER family and cancer: emerging molecular mechanisms and therapeutic targets. *Trends in Molecular Medicine*, **13**, 527–534.

2 Kochupurakkal, B.S., Harari, D., Di-Segni, A., Maik-Rachline, G., Lyass, L., Gur, G. et al. (2005) Epigen, the last ligand of ErbB receptors, reveals intricate relationships between affinity and mitogenicity. *The Journal of Biological Chemistry*, **280**, 8503–8512.

3 Stove, C. and Bracke, M. (2004) Roles for neuregulins in human cancer. *Clinical & Experimental Metastasis*, **21**, 665–684.

4 Press, M.F., Cordon-Cardo, C. and Slamon, D.J. (1990) Expression of the HER-2/neu proto-oncogene in normal human adult and fetal tissues. *Oncogene*, **5**, 953–962.

5 Pierce, J.H., Arnstein, P., DiMarco, E., Artrip, J., Kraus, M.H., Lonardo, F. et al. (1991) Oncogenic potential of erbB-2 in human mammary epithelial cells. *Oncogene*, **6**, 1189–1194.

6 Schiffer, I.B., Gebhard, S., Heimerdinger, C.K., Heling, A., Hast, J., Wollscheid, U. et al. (2003) Switching off HER-2/neu in a tetracycline-controlled mouse tumor model leads to apoptosis and tumor-size-dependent remission. *Cancer Research*, **63**, 7221–7231.

7 Goodell, V., Waisman, J., Salazar, L.G., Dela Rosa, C., Link, J., Coveler, A.L., et al. (2008) Level of HER-2/neu protein expression in breast cancer may affect the development of endogenous HER-2/neu-specific immunity. *Molecular Cancer Therapeutics*, **7**, 449–454.

8 Pupa, S.M., Iezzi, M., Di Carlo, E., Invernizzi, A., Cavallo, F., Meazza, R. et al. (2005) Inhibition of mammary carcinoma development in HER-2/neu transgenic mice through induction of autoimmunity by xenogeneic DNA vaccination. *Cancer Research*, **65**, 1071–1078.

9 Disis, M.L., Pupa, S.M., Gralow, J.R., Dittadi, R., Menard, S. and Cheever, M.A. (1997) High-titer HER-2/neu protein-specific antibody can be detected in patients with early-stage breast cancer. *Journal of Clinical Oncology*, **15**, 3363–3367.

10 Finn, O.J. (2003) Cancer vaccines: between the idea and the reality. *Nature Reviews. Immunology*, **3**, 630–641.

11 Mina, L. and Sledge, G.W. (2006) Twenty years of systemic therapy for breast cancer. *Oncology*, **20**, 25–32.

12 Seliger, B. and Kiessling, R. (2004) Take two – get more: optimization strategies for HER-2/neu-based immunotherapies inhibiting carcinogenesis. *Gene Therapy*, **12**, 1549–1550.

13 Di Palma, S., Collins, N., Bilous, M., Sapino, A., Mottolese, M., Kapranos, N., et al. (2008) A quality assurance exercise to evaluate the accuracy and reproducibility of CISH for HER2 analysis in breast cancer. *Journal of Clinical Pathology*, **61**, 757–760.

14 Nielsen, D.L., Andersson, M. and Kamby, C. (2008) HER2-targeted therapy in breast cancer. Monoclonal antibodies and tyrosine kinase inhibitors. *Cancer Treat Rev*.

15 Moasser, M.M. (2007) Targeting the function of the HER2 oncogene in human cancer therapeutics. *Oncogene*, **26**, 6577–6592.

16 Choudhury, A. and Kiessling, R. (2004) Her-2/neu as a paradigm of a tumor-specific target for therapy. *Breast Disease*, **20**, 25–31.

17 Curigliano, G., Spitaleri, G., Dettori, M., Locatelli, M., Scarano, E. and Goldhirsch, A. (2007) Vaccine immunotherapy in breast cancer treatment: promising, but still early. *Expert Review of Anticancer Therapy*, **7**, 1225–1241.

18 Ostrand-Rosenberg, S. (2004) Animal models of tumor immunity, immunotherapy and cancer vaccines. *Current Opinion in Immunology*, **16**, 143–150.

19 Lollini, P.L. and Forni, G. (2002) Antitumor vaccines: is it possible to prevent a tumor? *Cancer Immunology, Immunotherapy*, **51**, 409–416.

20 Baxevanis, C.N., Sotiropoulou, P.A., Sotiriadou, N.N. and Papamichail, M. (2004) Immunobiology of HER-2/neu oncoprotein and its potential application in cancer immunotherapy. *Cancer Immunology, Immunotherapy*, **53**, 166–175.

21 Bernhard, H., Salazar, L., Schiffman, K., Smorlesi, A., Schmidt, B., Knutson, K.L. et al. (2002) Vaccination against the HER-2/neu oncogenic protein. *Endocrine-Related Cancer*, **9**, 33–44.

22 Kiessling, R., Wei, W.Z., Herrmann, F., Lindencrona, J.A., Choudhury, A., Kono, K. et al. (2002) Cellular immunity to the Her-2/neu protooncogene. *Advances in Cancer Research*, **85**, 101–144.

23 Cavallo, F., Offringa, R., van der Burg, S.H., Forni, G. and Melief, C.J. (2006) Vaccination for treatment and prevention of cancer in animal models. *Advances in Immunology*, **90**, 175–213.

24 Calogero, R.A., Musiani, P., Forni, G. and Cavallo, F. (2004) Towards a long-lasting immune prevention of HER2 mammary carcinomas: directions from transgenic mice. *Cell Cycle (Georgetown, Tex)*, **3**, 704–706.

25 Brinkman, J.A., Fausch, S.C., Weber, J.S. and Kast, W.M. (2004) Peptide-based vaccines for cancer immunotherapy. *Expert Opinion on Biological Therapy*, **4**, 181–198.

26 Lindencrona, J.A., Preiss, S., Kammertoens, T., Schuler, T., Piechocki, M., Wei, W.Z. et al. (2004) $CD4^+$ T cell-mediated HER-2/neu-specific tumor rejection in the absence of B cells. *International Journal of Cancer*, **109**, 259–264.

27 Dakappagari, N.K., Lute, K.D., Rawale, S., Steele, J.T., Allen, S.D., Phillips, G. et al. (2005) Conformational HER-2/neu B-cell epitope peptide vaccine designed to incorporate two native disulfide bonds enhances tumor cell binding and antitumor activities. *The Journal of Biological Chemistry*, **280**, 54–63.

28 Disis, M.L., Knutson, K.L., Schiffman, K., Rinn, K. and McNeel, D.G. (2000) Pre-existent immunity to the HER-2/neu oncogenic protein in patients with HER-2/neu-overexpressing breast and ovarian cancer. *Breast Cancer Research and Treatment*, **62**, 245–252.

29 Disis, M.L., Gooley, T.A., Rinn, K., Davis, D., Piepkorn, M., Cheever, M.A. et al. (2002) Generation of T-cell immunity to the HER-2/neu protein after active immunization with HER-2/neu peptide-based vaccines. *Journal of Clinical Oncology*, **20**, 2624–2632.

30 Inokuma, M., dela Rosa, C., Schmitt, C., Haaland, P., Siebert, J., Petry, D. et al.

(2007) Functional T cell responses to tumor antigens in breast cancer patients have a distinct phenotype and cytokine signature. *Journal of Immunology (Baltimore, Md: 1950)*, **179**, 2627–2633.

31 Franklin, M.C., Carey, K.D., Vajdos, F.F., Leahy, D.J., de Vos, A.M. and Sliwkowski, M.X. (2004) Insights into ErbB signaling from the structure of the ErbB2-pertuzumab complex. *Cancer Cell*, **5**, 317–328.

32 Johnson, B.E. and Jänne, P.A. (2006) Rationale for a phase II trial of pertuzumab, a HER-2 dimerization inhibitor, in patients with non-small cell lung cancer. *Clin Cancer Res.*, **12**, 4436s–4440s.

33 zum Buschenfelde, C.M., Hermann, C., Schmidt, B., Peschel, C. and Bernhard, H. (2002) Antihuman epidermal growth factor receptor 2 (HER2) monoclonal antibody trastuzumab enhances cytolytic activity of class I-restricted HER2-specific T lymphocytes against HER2-overexpressing tumor cells. *Cancer Research*, **62**, 2244–2247.

34 Herrmann, F., Lehr, H.A., Drexler, I., Sutter, G., Hengstler, J., Wollscheid, U. et al. (2004) HER-2/neu-mediated regulation of components of the MHC class I antigen-processing pathway. *Cancer Research*, **64**, 215–220.

35 Choudhury, A., Charo, J., Parapuram, S.K., Hunt, R.C., Hunt, D.M., Seliger, B. and Kiessling, R. (2004) Small interfering RNA (siRNA) inhibits the expression of the Her2/neu gene, upregulates HLA class I and induces apoptosis of Her2/neu positive tumor cell lines. *International Journal of Cancer*, **108**, 71–77.

36 Norell, H., Carlsten, M., Ohlum, T., Malmberg, K.J., Masucci, G., Schedvins, K. et al. (2006) Frequent loss of HLA-A2 expression in metastasizing ovarian carcinomas associated with genomic haplotype loss and HLA-A2-restricted HER-2/neu-specific immunity. *Cancer Research*, **66**, 6387–6394.

37 Kmieciak, M., Knutson, K.L., Dumur, C.I. and Manjili, M.H., (2007) HER-2/neu antigen loss and relapse of mammary carcinoma are actively induced by T cell-mediated anti-tumor immune responses. *European Journal of Immunology*, **37**, 675–685.

38 Curcio, C., Di Carlo, E., Clynes, R., Smyth, M.J., Boggio, K., Quaglino, E. et al. (2003) Nonredundant roles of antibody, cytokines, and perforin in the eradication of established Her-2/neu carcinomas. *The Journal of Clinical Investigation*, **111**, 1161–1170.

39 Menard, S., Casalini, P., Campiglio, M., Pupa, S.M. and Tagliabue, E. (2004) Role of HER2/neu in tumor progression and therapy. *Cellular and Molecular Life Sciences*, **61**, 2965–2978.

40 Pupa, S.M., Tagliabue, E., Menard, S. and Anichini, A. (2005) HER-2: a biomarker at the crossroads of breast cancer immunotherapy and molecular medicine. *Journal of Cellular Physiology*, **205**, 10–18.

41 Rescigno, M., Avogadri, F. and Curigliano, G. (2007) Challenges and prospects of immunotherapy as cancer treatment. *Biochimica et Biophysica Acta*, **1776**, 108–123.

42 Revillion, F., Lhotellier, V., Hornez, L., Bonneterre, J. and Peyrat, J.P. (2008) ErbB/HER ligands in human breast cancer, and relationships with their receptors, the bio-pathological features and prognosis. *Annals of Oncology*, **19**, 73–80.

13
Epstein-Barr Virus-associated Antigens
Christoph Mancao and Wolfgang Hammerschmidt

13.1
Introduction

Epstein-Barr virus (EBV) is a ubiquitous virus, which is present in the majority of the adult population for a lifetime but generally without clinical symptoms [1, 2]. This human γ-herpes virus was found in Burkitt's lymphoma cells in the early 1960s [3]; a finding which together with epidemiological and molecular data identified EBV as the first known human tumor virus [4]. Today, EBV is regarded as an important infectious agent, which is a class 1 carcinogen according to the World Health Organization. Given its very high prevalence in the human population EBV-associated cancers are infrequent in the Western world but of importance in other geographic areas [4].

EBV infects most individuals in early childhood, resulting in an asymptomatic infection [5]. Primary infection later in life leads to infectious mononucleosis (IM) in about half of the cases. IM is a self-limiting disease, which nevertheless can cause severe clinical symptoms including fever, sore throat, and lymphadenopathy. The reasons for these different clinical courses are not understood. Permissive epithelial cells of the oropharynx represent the first site of infection, in which the virus can replicate lytically leading to virus progeny [6, 7]. The propagated virus then infects mucosal B-lymphocytes via the viral glycoproteins gp350 and gp42, which bind to the cellular surface receptor CD21 [8] and human leukocyte antigen (HLA) class II [9], respectively. In the newly infected B cells EBV exclusively establishes a latent state of infection, which prevents initial *de novo* synthesis of progeny virus. In IM patients up to 10% of total B cells can be infected with EBV during this phase of disease. The latently infected B cells express viral antigens, which induce an immediate and strong immune response. As a consequence the majority of latently infected B cells will be eliminated by the immune system during the clinical course of IM but a few infected cells down-regulate viral gene expression and therefore escape elimination. This tiny population of EBV-infected cells (on average one in 10^5 to 10^6 peripheral B cells) are predominantly memory B cells

Tumor-Associated Antigens. Edited by Olivier Gires and Barbara Seliger
Copyright © 2009 WILEY-VCH Verlag GmbH & Co. KGaA, Weinheim
ISBN: 978-3-527-32084-4

which are resting and long-lived. From thereon, the virus becomes reactivated from this quiescent state only occasionally by external stimuli such as antigen-mediated B-cell stimulation [10, 11].

The infected host is at risk of developing EBV-associated tumors probably because EBV can induce proliferation in latently infected cells *in vivo* (as well as *in vitro*). Accordingly, there are several malignancies causally linked to EBV infection such as Burkitt's lymphoma (BL), Hodgkin lymphoma (HL), nasopharyngeal carcinoma (NPC), and gastric carcinoma, among others [12]. EBV has also been linked to several diseases such as hairy leukoplakia, immunoblastic lymphomas, and post transplant proliferative disease (PTLD) [13] in immuno-compromised patients, indicating that an intact immune system controls and limits the proliferation of EBV-infected B cells in healthy individuals, which is mandatory for a stable virus–host relationship.

The first part of this chapter will discuss the viral genes which play a potentially causal role in EBV pathogenesis and are critical to the latent state of EBV. The second part focuses on how the immune system reacts to and controls an EBV infection by mounting T-cell responses to EBV-encoded antigenic gene products. The final part covers three EBV-associated malignancies, post-transplant lymphoproliferative disease (PTLD), Burkitt's and Hodgkin lymphoma, and novel therapeutic concepts, which involve the adoptive transfer of T cells directed against EBV encoded epitopes to fight these potentially fatal cancers.

13.2
Functions of EBV Antigens in Latently Infected Cells

EBV-associated tumor cells carry the virus latently similar to tumor-derived cell lines or *in vitro* established lymphoblastoid cell lines (LCLs). A unique characteristic of EBV is the expression pattern of its genes, which can be detected during latency. These patterns differ depending on the type of tumor cell. *In vitro,* EBV-infected cells display a so-called type III latency program (also known as the 'growth program') in which a set of nine viral proteins, two small non-coding viral RNAs (EBERs), and a number of viral micro RNAs are expressed from the viral genome. Six of these latently expressed proteins are known as Epstein-Barr nuclear antigens (EBNAs: EBNA1, -2, -3A, -3B, -3C, LP-); the remaining viral genes encode three latent membrane proteins (LMPs: LMP1, LMP2A, LMP2B). Five of these viral products are thought to be essential for *in vitro* B-cell transformation (EBNA1, -2, -3A, -3C and LMP1) whereas two gene products (EBNA-LP and LMP2A) increase the efficiency of transformation (for a recent review see [1]). EBV-infected B cells from patients suffering from PTLD display such latency, whereas the viral expression pattern observed in BL is limited to EBNA1 only (in addition to the non-coding small viral RNAs), a latency called type I. In latency type II (also called 'default program') the latent membrane proteins LMP1 and LMP2 (A and B) are expressed in addition to EBNA1, as seen in NPC, HL, and certain NK/T-cell lymphomas [14]. In latency type 0, only LMP2A (together with the non-coding small viral RNAs) is expressed as seen in memory B cells in the latently infected host [15].

The following section will provide an overview of the latent EBV proteins with a more detailed look at EBNA1, LMP1 and LMP2A because they are expressed in the majority of EBV-associated lymphoproliferative diseases during latency type II [16]. The epitopes of viral genes expressed during latency and detected by EBV-specific T cells are summarized in a recent review by Hislop *et al.* [11].

13.2.1
EBV Nuclear Antigen 1 (EBNA1)

EBNA1 is essential for extrachromosomal replication of the EBV genome in latently infected cells (for a recent review see [17]). EBNA1 tethers the latent viral origin of replication, termed *oriP*, to cellular chromatin to bring about DNA replication and equal distribution of the viral genome between the daughter cells [18, 19]. A specified sequence within *oriP*, known as the 'family of repeats' owing to its repetitive composition of 20 EBNA1 binding sites, is mandatory for nuclear retention of plasmid DNA. In addition, EBNA1 is directly involved in the extrachromosomal DNA replication of the viral genome by the cellular machinery [20, 21]. For this function, a second independent DNA motif within *oriP* is required, this motif is known as the 'dyad symmetry element' due to its bipartite composition comprising two pairs of two EBNA1 binding sites each located at critical spacing.

EBNA1 can be considered to function as the molecular glue needed to tether the EBV genomic DNA to cellular chromatin and mediate viral DNA replication. However, EBNA1 has additional functions. It also acts as an indispensable transcriptional activator of viral latent genes [22]. Although the subject of controversial discussion [23], expression of EBNA1 under the control of the immunoglobulin heavy chain intron enhancer, induced neoplasia of B cells in transgenic mice [24]. Together these findings suggest that EBNA1 might regulate expression of cellular genes as well, which may contribute to its presumed oncogenic characteristics.

13.2.2
EBV Nuclear Antigen 2 and Leader Protein (EBNA2 and EBNA-LP)

EBNA2 together with EBNA-LP are among the first latency genes of EBV expressed in resting primary B cells after *in vitro* infection. EBNA-LP is found predominantly in nucleoli or associated with promyelocytic leukemia nuclear bodies ([25] and references therein) which indicates a role in cell-cycle regulation [26]. Both EBNA2 and EBNA-LP in concert activate the latent membrane genes of EBV as well as cellular genes at the transcriptional level. EBNA2 indirectly binds to specific DNA motifs by interacting with cellular DNA-binding proteins such as RBP-Jκ (also known as CBF-1) or PU.1 in promoter regions of EBNA2-responsive genes [1]. In the physiological context, RBP-Jκ is the nuclear receptor of the cleaved intracellular domain of NOTCH-1 and is known as NOTCH-ICD. NOTCH-1 in turn is a cellular surface receptor, which upon binding of its ligand is subject to activation via regulated intramembrane proteolysis (RIP) by two transmembrane proteases, TACE and presenilin-1 [27]. EBNA2 and NOTCH-ICD bind to similar regions on RBP-Jκ, which led to the assumption that EBNA2 acts as a constitutively active homolog of NOTCH that hijacks cellular components of NOTCH signaling to foster cell proliferation. EBNA2 or NOTCH-ICD have similar target genes in B lymphocytes, for example the cell entry receptors CD21 and CD23 [28], which corroborate this finding. The functional analogy of EBNA2 and NOTCH-ICD, which share hardly any homology at the primary amino acid sequence level, is a hallmark of EBV because several viral gene products mimic the function of cellular proteins as discussed below.

13.2.3
EBV Nuclear Antigen Family 3 (EBNA3)

This protein family consists of three members namely EBNA3A, 3B and 3C. Several sometimes redundant functions have been attributed to individual members of this gene family. They modulate or interfere with the formation of the EBNA2–RBP–Jκ complex and thereby dampen the activation of EBNA2-responsive promoters. Therefore, EBNA3 proteins are functionally regarded as mediators that counterbalance or fine-tune EBNA2 [29, 30]. EBNA3A and 3C appear to be essential for growth transformation of primary resting B cells, whereas EBNA3B is dispensable [31]. This is of particular interest since it is the EBNA3 family member that is predominantly recognized by cytotoxic CD8+ T lymphocytes (CTLs) suggesting that they govern the latent state of EBV *in vivo*. Because EBNA3C is presumably dispensable for maintaining latency *in vivo*, the dominance of EBNA3C-specific CTLs may select EBNA3C-negative EBV variants, which can be fatal [32]. EBNA3C in particular has also been found to interact with a number of cellular targets, including the oncogene c-Myc and the tumor suppressor protein Rb [33, 34].

13.2.4
Latent Membrane Protein 1 (LMP1)

LMP1 is the only EBV gene, which scores as an oncogene *in vitro* and *in vivo*. It can antagonize senescence and apoptosis, and fosters cell proliferation and survival. LMP1 acts in a similar manner to the CD40 receptor, which induces numerous signaling pathways upon ligand-induced activation. These pathways include the canonical and alternative NF-κB signaling cascades, c-Jun N-terminal kinase (JNK), p38 mitogen-activated protein kinase (MAPK), the JAK/STAT and phosphatidylinositol 3-kinase (PI3K) pathway, the tumor necrosis factor-receptor (TNF-R) and the Toll-like/interleukin-1-receptor (TIR) family pathway [16, 35]. In contrast to the CD40 receptor, which depends on its cognate ligand CD40L for activation, LMP1 acts ligand-independently because its transmembrane domain fosters multimerization and recruitment to lipid rafts. Therefore LMP1 is constitutively active [36–39] (for a recent review see [35]).

Several studies have confirmed LMP1's tumorigenic potential. For instance, in established rodent fibroblast cell lines LMP1 is sufficient to induce a transformed state [40–42]. When it is expressed in transgenic mice LMP1 can partially substitute for CD40 [43] and induces B cell lymphomas [44].

LMP1 consists of a short amino-terminus, six transmembrane domains, and a 200 amino acids (aa) long carboxy-terminus, which serves as the signaling domain for LMP1. It consists of two critical subdomains: transformation effector site 1 (TES1 or C-terminal activation region CTAR1) and TES2 (or CTAR2). The two subdomains provide different outcomes to the infected cell. TES1/CTAR1 is required for initial cellular transformation by binding of TNF-R-associated factor (TRAF) 1, 2, 3, and 5, and thereby induces signaling of the alternative, non-canonical NF-κB pathway. Additionally, this subdomain can also activate the PI3K pathway, which leads to AKT activation and remodeling of the actin cytoskeleton and stress fiber formation. The second domain TES2/CTAR2 is linked to continued maintenance of the transformed status of LCLs. Signaling from this subdomain is mainly accomplished by indirect binding of TRAF6, which leads to activation of the classical NF-κB pathway. Binding of TRAF6 and TRAF3 to TES2/CTAR2 will lead to phosphorylation of either p38 or JNK1 and terminates in the activation of AP1-induced target genes (see [35] and references therein). *In vitro*, withdrawal of LMP1 from lymphoblastoid cell lines results in a substantial loss of proliferation, while its re-expression allows for re-entry into the cell cycle [45]. Hence, LMP1 uses and misuses versatile cellular signaling components and can be regarded as the prototype for a herpesviral oncogene.

13.2.5
Latent Membrane Protein 2 (LMP2)

Similarly to LMP1, LMP2A is expressed in LCLs but its role in *in vitro* in B-cell transformation is controversial [46]. Because LMP2A is the most consistently expressed viral gene *in vivo* [47] it is thought to play a role in maintaining the latent state of EBV and is considered to be an important factor in EBV-associated malignancies. The protein consists of a short carboxy-terminal and a 119 aa long amino-terminal domain separated by 12 membrane-spanning domains. The striking feature of LMP2A is its signaling amino-terminus, which its isogenic twin LMP2B lacks. Several tyrosine residues create a potential binding site for protein tyrosine kinases (PTKs): Y23, Y31, Y60, Y74, Y85, Y101 and Y112. Members of this protein family contain *src*-homology-2-domains (SH2), which promote binding to phosphorylated tyrosine residues. Accordingly, LMP2A binds PI3K and phospholipase-γ-2 (PLC-γ-2), ABL, CRK, and NCK [48]. The residues Y74 and Y85 play a prominent role in the signaling capacity of LMP2A because they exhibit strong homology with the immunoreceptor-tyrosine-based-activation-motif (ITAM). This motif exists in the B cell receptor (BCR) co-receptors CD79a and CD79b (Igα and Igβ) and provides the binding platform for ZAP70/SYK (spleen tyrosine kinase). Similar to LMP1 but in contrast to the BCR, LMP2A presumably clusters autonomously, forms homo-oligomers, and homes to lipid rafts in the absence of a known ligand. These characteristics deliver a constitutively active signal, which is functionally similar to BCR-derived signals. LMP2A induces proliferative signal cascades via calcium release and JNK induction as well as anti-apoptotic pathways via AKT and BAD. LMP2A also has repressive effects on cellular genes and B-cell differentiation [49] but expression of LMP2A in transgenic mice permits BCR-deficient B cells to mature and migrate into the periphery [50]. These findings together with additional observations strengthen the hypothesis that LMP2A can functionally replace the BCR by mimicking a genuine and 'tonic' BCR-derived signal which is essential for B cell survival *in vivo* [51]. *In vitro* pro-apoptotic BCR⁻ cells can be infected with EBV giving rise to growth transformed B-cell lines [52–54] but an LMP2A-deficient mutant EBV did not rescue BCR-deficient cells *in vitro* [46]. The constitutive expression of LMP2A in established and growth-transformed LCLs is absolutely essential for cell survival indicating an indispensable role for LMP2A in maintaining the transformed state [46].

13.3
T-cell Responses to EBV Antigens

13.3.1
CD8$^+$ T-cell Responses

EBV-encoded gene products constitute *de novo* antigens *in vivo*, which become targets for a specific and strong immunological response. Together with certain viral genes, which have been categorized as lytic genes [55–57] EBNA2 and EBNA-LP are among the first EBV genes expressed after infection of primary B cells [1]. Many T-cell epitopes of EBV-encoded proteins are known but the catalog is probably far from complete [11].

Primary infection with EBV is controlled by the induction of EBV-specific cytotoxic T lymphocytes (CTL). These CD8$^+$ T cells exhibit reactivity against a wide range of both lytically and latently infected cells. The hallmark of this immunological response is a marked hierarchy of immunodominance among immediate early (IE), early (E) and late (L) viral proteins [58]. In acute IM up to 40% of the total CD8$^+$ T-cell population is able to display reactivity against IE and E antigens, whereas only about 0.1–5% of T cells detect latent proteins [11]. Only two viral IE proteins are found to be responsible for this initial prominent boost of CD8$^+$ T cells: BZLF1 and BRLF1. These transcription factors directly regulate the expression of other viral genes such as BMLF1, BMRF1, BALF2 and BALF5, which represent the group of detected E antigens. Together with BALF4 and BGLF1 these four viral proteins are the major lytic targets, against which CD8$^+$ T-cell clones can be identified. T cells reactive against other E or L protein-derived epitopes were found only in a small number of patients. It seems that viral proteins, which are expressed late during the lytic phase of the EBV life cycle, are less prominent targets of a specific antiviral T-cell response. EBV-encoded, immune-evasive proteins may be responsible for this phenomenon, which might inhibit proper antigen processing to interfere with T-cell recognition and the killing of virus-producing cells. Two viral candidates, which block or inactivate the HLA loading machinery, are likely candidates for this type of immune evasion: the viral IL-10 homolog BCRF1 was shown to inactivate TAP1 (transporter associated with antigen processing) and LMP2 (low mass protein), as does the cellular IL-10 [57]. A more recent publication identified BNLF2A as an inhibitor of TAP1 and TAP2 [59].

13.3.2
CD4$^+$ T-cell Response

The situation is less clear regarding the response of CD4$^+$ T cells, which do not show the same hierarchic immunodominance as CD8$^+$ T cells [60]. The IE protein BZLF1, certain E proteins (BMLF1, BHRF1 and BNLF2b), L proteins (e.g. BCF1, BDLF1, BALF4 and BLLF1), and a few structural proteins (e.g. gp350) can be detected by CD4$^+$ T cells [61–63]. Stronger responses by CD4$^+$ T cells were detected against epitopes from latent proteins. EBNA1, which induces relatively weak responses by CD8+T cells, is a strong antigen for CD4$^+$ T cells, followed by EBNA2. The family of EBNA3 proteins is immunodominant for CD8$^+$ but not for CD4$^+$ T cells; details can be found in a recent review dedicated to this theme [11].

13.4
EBV-associated Malignancies

13.4.1
Post-Transplant Lymphoproliferative Disease (PTLD)

As indicated above, EBV infection even in immunocompetent and healthy carriers can be closely associated with several malignancies. B lymphocytes are the natural reservoir of latent EBV infections and therefore B-cell lymphomas predominate among EBV-associated malignancies [16]. A striking feature of EBV-infected tumor cells is their heterogeneous viral expression patterns which vary across cancer types.

The situation in post-transplant lymphoproliferative disease (PTLD) is probably the simplest, because the complete set of all viral latency proteins are commonly expressed. PTLD is a life-threatening complication that can arise in immunocompromised patients, namely transplant recipients and HIV carriers. The lesions can range from mild lymphoid hyperplasia to fulminant lymphoma [13]. The incidence of PTLD in transplant recipients depends on the immunosuppressive regimen, the allografted organs, and the age of the patients [64–66]. Tumor onset is believed to originate from massively reduced numbers of EBV-specific cytotoxic T cells, which allows outgrowth of EBV-infected B cells and presumably uncontrolled virus replication and dispersal. The reduced immune competence results from substantial iatrogenic immunosuppression required after of transplantation to avert possible graft rejection. The observation that a strong cytotoxic T-cell response can prevent PTLD is supported by several observations: (i) the first cases of PTLD emerged after the introduction of cyclosporine A, a compound, which dampens T-cell responses [67]. (ii) In bone marrow transplantations in which the transplant is T cell-depleted the patient's risk of developing PTLD increases from less than 1% to 24% [68]. (iii) Transplant recipients carry a substantially increased risk of developing

PTLD if they are sero-negative for EBV at the time of transplantation [69]. This empiric data has been accumulated from observations during first-line treatment: reduction of immunosuppression in solid organ transplants leads to tumor regression in about 20–50% of patients [64, 70, 71]. This strategy almost completely fails in bone marrow recipients because immune reconstitution by donor marrow takes a long time during which any effective anti-tumor response is compromised.

13.4.2
Adoptive T-cell Therapy

The grade of immunosuppression is directly correlated with the risk of developing an EBV-associated tumor and provides a rationale for the first steps towards anti-viral adoptive T-cell immunotherapy. The aim of such a therapeutical approach is the selective reconstitution of anti-EBV immunity by administration of EBV-specific CTLs. Viral epitopes presented by HLA class I molecules on infected or growth transformed B-lymphocytes will be detected by the T-cell receptor of administered CTLs. Two effector pathways will destroy the infected tumor cell upon T-cell recognition [72]. Upon antigen recognition CTLs release perforin- and serine esterases- (granzymes) containing granules, which lead to apoptosis of target cells. Cell death can be induced by CD95-ligand expression on CTLs, leading to CD95- (Fas) mediated apoptosis. Additionally, cytokines like TNF-α, TNF-β, IFN-α, -β, and -γ will exert strong anti-viral and anti-tumor effects.

Detection of antigen and likewise of malignant cells underlies a strong HLA restriction, which is mediated by positive and negative selection of reactive T cells in the thymus. Antigen-selected T-cells detect epitopes bound to HLA molecules of their own allelic type but ignore antigenic epitopes in combination with non-self HLA molecules. This restriction has a severe impact on the outcome of adoptive T-cell transfer strategies, depending on the different type of transplantation. In bone marrow recipients, PTLD originates almost exclusively from engrafted donor cells and can therefore be treated with the transfer of donor-derived T cells. They recognize the PTLD tumor cells, since virus-derived antigenic epitopes are presented on HLA molecules which are of donors composition and therefore regarded as self-HLA. In contrast, solid organ transplantations most commonly trigger a recipient-derived tumor, which can be selectively treated with autologous CTLs (or allogenic CTLs with a partial HLA mismatch).

The treatment of PTLD with adoptive T-cell therapy carries a certain risk of developing graft-versus-host-disease (GvHD) [73, 74] which is caused by allospecific T-cells contaminating the curative infusion and can lead to multiple organ damage including the immune system itself. Graft-versus-host T cells produce an excess of cytokines, including TNF-α and IFN-γ. To circumvent severe GvHD effects by alloreactive T cells, most studies use *in vitro* generated and well-defined polyclonal EBV-specific CTL lines [75]. The technical procedure to establish such CTL lines comprises of two parts: (i) *in vitro* expanded T cells stem from the EBV-specific memory T-cell pool of EBV sero-positive donors and (ii) antigen-presenting stimulatory cells such as lymphoblastoid cell lines (LCLs). They originate from autologous peripheral B cells as a result of their *in vitro* infection with EBV, a process called growth transformation [1]. LCLs display latency type III and therefore present all latent antigens, among them the immunodominant EBNA3s and LMP2. To avoid carry-over of growth-transformed LCLs, these cells are irradiated sub-lethally prior to incubation with T cells in the presence of IL-2, which will lead to the proliferation and differentiation of EBV-specific CTLs. Immunophenotyping reveals their $CD8^+$ nature including several activation markers such as CD28, CD38, CD45-RO, CD69, CD150, and HLA-DR [75, 76]. In accordance with the *in vivo* situation, this procedure yields an oligoclonal T-cell population characterized by a high reactivity against epitopes from EBNA3 and LMP2 along with a small proportion of epitopes from lytic gene products [77].

Several clinical studies demonstrate the successful adoptive transfer of CTL lines generated by this means. Among others, Rooney *et al.* administered EBV-specific T cells to bone marrow transplant patients presenting with progressive PTLD. All patients showed a complete and persistent remission and drop in viral load. None developed PTLD [78–80]. The successful treatment of PTLD was shown not only in bone marrow recipients but also among recipients of solid organ transplants. Khanna *et al.* infused autologous EBV-specific CTLs to treat a patient for whom the reduction of immunosuppression did not result in tumor regression [81]. Two infusions of CTLs were needed for complete remission but secondary PTLD in this patient was unresponsive to further treatment (the reason for this will be discussed later).

Regardless of whether solid organ or bone marrow transplant recipients are treated prophylactically or after tumor onset, the use of autologous CTLs faces two major obstacles. The first relates to technical restrictions since the generation of autologous EBV-specific T cells is very time consuming. It takes up to 3 months to obtain a sufficient number of CTLs for adoptive therapy. Secondly, patients running the highest risk of developing PTLD are EBV sero-negative recipients, who lack an EBV-specific memory T-cell pool and hence rely on naïve EBV-specific T cells. There have been major difficulties in establishing autologous CTLs from blood of this high-risk group. Therefore various research has endeavored to circumvent the need for autologous CTLs. Allogenic T-cell transfer is based on the administration of partially HLA-matched CTLs derived from the blood of healthy EBV sero-positive donors. Haque *et al.* generated EBV-specific T-cell clones from approximately 100 donors and created a cell bank, so that the cells were ready to thaw and administer to patients in need [82]. After careful HLA matching and *in vitro* cytotoxicity assays, eight patients for whom any standard treatment had failed were infused with these partially HLA-matched CTLs. Three patients experienced complete remission, two showed no effect and one patient had only a partial response. Two patients died before any anti-tumor effect could be expected. All four successfully treated patients showed a decrease in EBV-DNA in their blood and constant high levels of EBV-specific T cells for up to 7 weeks. Importantly, no infusion-related adverse reaction was observed in any of these cases. This is a major advantage as compared to infusion of non-separated PBMCs and is encouraging for further development of prophylaxis and treatment.

Several questions remain unanswered. One concerns the possible expansion and long-term survival of administered EBV-specific T cells. Studies addressing this potential problem reveal that infused autologous cells were detected in extreme cases for up to several years whereas the median is about 11 weeks [78, 80, 83]. This finding indicates the presence of a memory T-cell population in transfused patients. Another question refers to the role of $CD4^+$ T cells in adoptive T-cell transfers. Animal models showed that infusion of $CD4^+$-depleted T cells led to diminished anti-tumor effects [84, 85]. Co-administered $CD4^+$ T cells may regulate $CD8^+$ T cells by secreting cytokines such as IL-2, IL-12 or Il-15 [86–88]. Despite the fact that $CD4^+$ T-cell responses to latent EBV epitopes are weaker compared to $CD8^+$ T-cell response, their effect may be very important. Epitopes of the viral protein EBNA1 are readily detected by $CD4^+$ but not by $CD8^+$ T cells. Therefore $CD4^+$ T cells might not only support CTL function but may also close the gap that $CD8^+$ T cells leave open in the detection of EBV epitopes. Hence, elucidation of EBV reactivities in the $CD4^+$ T-cell pool is of considerable interest [63].

13.4.3
Burkitt's Lymphoma and Hodgkin's Lymphoma

The situation in PTLD is relatively easy to cover because the tumor cells express latency program III in which all viral latent proteins are expressed. As a consequence, CTLs specific for several viral antigens can execute their anti-tumor effect. In other prominent EBV-associated malignancies the situation is far more restricted. Hodgkin lymphoma (HL) is characterized by latency II program which includes the expression of EBNA1, LMP1 and LMP2A, only [7]. The viral epitopes, which give rise to the strongest and most robust cytotoxic T-cell response stem from members of the EBNA3 family, which are not expressed. In Burkitt's lymphoma (BL) the pattern is even more restricted to the exclusive expression of EBNA1 [16].

In order to gain access to these lymphomas as well as to PTLD cases with resistance to infused CTLs alternative strategies must be developed. Resistance to cell therapy with adoptively transplanted CTLs can arise due to several tumor evasion mechanisms. The administration of polyclonal CTLs may result in the selection of tumor cells with restricted viral expression, which escape the CTL response; a situation, which might explain the emerging resistance to secondary PTLD in one case reported by Khanna et al. [81]. Downregulation of HLA molecules by several mechanisms [57, 59] can also lead to a failure of adoptively transferred CTLs. Alternatively, selection pressure can cause the dominance of mutated EBV proteins as seen in the study by Gottschalk and co-workers, in which a deletion in the EBNA3B gene resulted in immune evasion [32]. In HL and BL, tumors bear an intrinsic evasion mechanism which consists of a restricted viral expression pattern and/or the exclusive expression of viral antigens, which are only marginally immunogenic *in vivo*. In the case of HL and BL, it would be of great value to identify CTLs specific for EBNA1, LMP1, and LMP2. Several protocols have been established and tested in clinical studies to generate CTLs which specifically detect epitopes that stem from the latent membrane proteins. One method would be the selective expansion of CTLs by stimulation with antigen-presenting cells exogenously loaded with synthetic EBV peptides, transfected with viral RNA, or infected with *vaccinia* or adenoviral vectors expressing the desired EBV proteins [89–91]. Alternative methods include the transfer of T cell receptor (TCR) genes into T cells [92]. These bispecific T cells adopt the inserted TCR together with its novel and defined specificity. The use of chimeric TCRs consisting of immunoglobulin-derived antigen specificity fused to the TCR signaling domain circumvents the possible immune evasion caused by HLA downregulation [93, 94].

It is obvious that adoptive T-cell therapy has been successfully applied in several clinical entities. It is equally obvious that much more has to be learned in order to introduce this therapeutic option into the clinic as a standard regimen.

References

1 Kieff, E. and Rickinson, A.B. (2007) Epstein-Barr Virus and its replication, in *Fields' Virology*, 5th edn (eds D.M. Knipe and P.M. Howley Wolters Kluver, Lippincott - Williams & Wilkins, Philadelphia.

2 Thompson, M.P. and Kurzrock, R. (2004) Epstein-Barr virus and cancer. *Clinical Cancer Research*, **10**, 803–821.

3 Epstein, M.A., Achong, B.G. and Barr, Y.M. (1964) Virus particles in cultured lymphoblasts from Burkitt's lymphoma. *Lancet*, **15**, 702–703.

4 Anon (1997) *International Agency for Research on Cancer: Epstein-Barr virus and Kaposi's sarcoma herpesvirus/human herpesvirus 8*, **70**, IARC, Lyon.

5 Kuppers, R. (2003) B cells under influence: transformation of B cells by Epstein-Barr virus. *Nature Reviews. Immunology*, **3**, 801–812.

6 Feederle, R., Neuhierl, B., Bannert, H., Geletneky, K., Shannon-Lowe, C. and Delecluse, H.J. (2007) Epstein-Barr virus B95.8 produced in 293 cells shows marked tropism for differentiated primary epithelial cells and reveals interindividual variation in susceptibility to viral infection. *International Journal of Cancer*, **121**, 588–594.

7 Shannon-Lowe, C.D., Neuhierl, B., Baldwin, G., Rickinson, A.B. and Delecluse, H.J. (2006) Resting B cells as a transfer vehicle for Epstein-Barr virus infection of epithelial cells. *Proceedings of the National Academy of Sciences of the United States of America*, **103**, 7065–7070.

8 Nemerow, G.R., Mold, C., Schwend, V.K., Tollefson, V. and Cooper, N.R. (1987) Identification of gp350 as the viral glycoprotein mediating attachment of Epstein-Barr virus (EBV) to the EBV/C3d receptor of B cells: sequence homology of gp350 and C3 complement fragment C3d. *Journal of Virology*, **61**, 1416–1420.

9 Borza, C.M. and Hutt-Fletcher, L.M. (2002) Alternate replication in B cells and epithelial cells switches tropism of Epstein-Barr virus. *Nature Medicine*, **8**, 594–599.

10 Thorley-Lawson, D.A. (2001) Epstein-Barr virus: exploiting the immune system. *Nature Reviews. Immunology*, **1**, 75–82.

11 Hislop, A.D., Taylor, G.S., Sauce, D. and Rickinson, A.B. (2007) Cellular responses to viral infection in humans: lessons from Epstein-Barr virus. *Annual Review of Immunology*, **25**, 587–617.

12 Rezk, S.A. and Weiss, L.M. (2007) Epstein-Barr virus-associated lymphoproliferative disorders. *Human Pathology*, **38**, 1293–1304.

13 Dharnidharka, V.R. and Araya, C.E., (2007) Post-transplant lymphoproliferative disease. *Pediatric Nephrology (Berlin, Germany)*.

14 Pallesen, G., Sandvej, K., Hamilton-Dutoit, S.J., Rowe, M. and Young, L.S. (1991) Activation of Epstein-Barr virus

replication in Hodgkin and Reed-Sternberg cells. *Blood*, **78**, 1162–1165.
15 Laichalk, L.L. and Thorley-Lawson, D.A. (2005) Terminal differentiation into plasma cells initiates the replicative cycle of Epstein-Barr virus in vivo. *Journal of Virology*, **79**, 1296–1307.
16 Rickinson, A.B. and Kieff, E. (2007) Epstein-Barr virus, in *Fields' Virology* 5th edn, (eds D.M. Knipe and P.M. Howley), Wolters Kluver, Lippincott - Williams & Wilkins, Philadelphia.
17 Wang, J. and Sugden, B. (2005) Origins of bidirectional replication of Epstein-Barr virus: models for understanding mammalian origins of DNA synthesis. *Journal of Cellular Biochemistry*, **94**, 247–256.
18 Nanbo, A., Sugden, A. and Sugden, B. (2007) The coupling of synthesis and partitioning of EBV's plasmid replicon is revealed in live cells. *The EMBO Journal*, **26**, 4252–4262.
19 Humme, S., Reisbach, G., Feederle, R., Delecluse, H..-J., Bousset, K., Hammerschmidt, W. and Schepers, A., (2003) The EBV nuclear antigen 1 (EBNA1) enhances B-cell immortalization several thousand-fold. *Proceedings of the National Academy of Sciences of the United States of America*, **100**, 10989–10994.
20 Thomae, A.W., Pich, D., Brocher, J., Spindler, M.P., Berens, C., Hock, R., Hammerschmidt, W. and Schepers, A. (2008) Interaction between HMGA1a and the origin recognition complex creates site-specific replication origins. *Proceedings of the National Academy of Sciences of the United States of America*, **105**, 1692–1697.
21 Schepers, A., Ritzi, M., Bousset, K., Kremmer, E., Yates, J.L., Harwood, J., Diffley, J.F. and Hammerschmidt, W. (2001) Human origin recognition complex binds to the region of the latent origin of DNA replication of Epstein-Barr virus. *The EMBO Journal*, **20**, 4588–4602.

22 Altmann, M., Pich, D., Ruiss, R., Wang, J., Sugden, B. and Hammerschmidt, W. (2006) Transcriptional activation by EBV nuclear antigen 1 is essential for the expression of EBV's transforming genes. *Proceedings of the National Academy of Sciences of the United States of America*, **103**, 14188–14193.
23 Kang, M.S., Lu, H., Yasui, T., Sharpe, A., Warren, H., Cahir-McFarland, E., Bronson, R., Hung, S.C. and Kieff, E. (2005) Epstein-Barr virus nuclear antigen 1 does not induce lymphoma in transgenic FVB mice. *Proceedings of the National Academy of Sciences of the United States of America*, **102**, 820–825.
24 Wilson, J.B., Bell, J.L. and Levine, A.J. (1996) Expression of Epstein-Barr virus nuclear antigen-1 induces B cell neoplasia in transgenic mice. *The EMBO Journal*, **15**, 3117–3126.
25 Ling, P.D., Peng, R.S., Nakajima, A., Yu, J.H., Tan, J., Moses, S.M., Yang, W.H., Zhao, B., Kieff, E., Bloch, K.D. and Bloch, D.B. (2005) Mediation of Epstein-Barr virus EBNA-LP transcriptional coactivation by Sp100. *The EMBO Journal*, **24**, 3565–3575.
26 Sinclair, A.J., Palmero, I., Peters, G. and Farrell, P.J. (1994) EBNA-2 and EBNA-LP cooperate to cause G0 to G1 transition during immortalization of resting human B lymphocytes by Epstein-Barr virus. *The EMBO Journal*, **13**, 3321–3328.
27 Selkoe, D.J. and Wolfe, M.S. (2007) Presenilin: running with scissors in the membrane. *Cell*, **131**, 215–221.
28 Zimber-Strobl, U. and Strobl, L.J. (2001) EBNA2 and Notch signalling in Epstein-Barr virus mediated immortalization of B lymphocytes. *Seminars in Cancer Biology*, **11**, 423–434.
29 Zhao, B., Marshall, D.R. and Sample, C.E. (1996) A conserved domain of the Epstein-Barr virus nuclear antigens 3A and 3C binds to a discrete domain of Jkappa. *Journal of Virology*, **70**, 4228–4236.

30 Robertson, E.S., Lin, J. and Kieff, E. (1996) The amino-terminal domains of Epstein-Barr virus nuclear proteins 3A, 3B, and 3C interact with RBPJ(kappa). *Journal of Virology*, **70**, 3068–3074.

31 Tomkinson, B., Robertson, E. and Kieff, E. (1993) Epstein-Barr virus nuclear proteins EBNA-3A and EBNA-3C are essential for B-lymphocyte growth transformation. *Journal of Virology*, **67**, 2014–2025.

32 Gottschalk, S., Ng, C.Y., Perez, M., Smith, C.A., Sample, C., Brenner, M.K., Heslop, H.E. and Rooney, C.M. (2001) An Epstein-Barr virus deletion mutant associated with fatal lymphoproliferative disease unresponsive to therapy with virus-specific CTLs. *Blood*, **97**, 835–843.

33 Knight, J.S., Sharma, N. and Robertson, E.S. (2005) Epstein-Barr virus latent antigen 3C can mediate the degradation of the retinoblastoma protein through an SCF cellular ubiquitin ligase. *Proceedings of the National Academy of Sciences of the United States of America*, **102**, 18562–18566.

34 Bajaj, B.G., Murakami, M., Cai, Q., Verma, S.C., Lan, K. and Robertson, E.S. (2008) Epstein-Barr virus nuclear antigen 3C interacts with and enhances the stability of the c-Myc oncoprotein. *Journal of Virology*, **82**, 4082–4090.

35 Kieser, A. (2007) Signal transduction by the Epstein-Barr virus oncogene latent membrane protein 1 (LMP1). *Signal Transduction*, **7**, 20–33.

36 Kieser, A., Kilger, E., Gires, O., Ueffing, M., Kolch, W. and Hammerschmidt, W. (1997) Epstein-Barr virus latent membrane protein-1 triggers AP-1 activity via the c-Jun N-terminal kinase cascade. *The EMBO Journal*, **16**, 6478–6485.

37 Kieser, A., Kaiser, C. and Hammerschmidt, W. (1999) LMP1 signal transduction differs substantially from TNF receptor 1 signaling in the molecular functions of TRADD and TRAF2. *The EMBO Journal*, **18**, 2511–2521.

38 Gires, O., Zimber-Strobl, U., Gonnella, R., Ueffing, M., Marschall, G., Zeidler, R., Pich, D. and Hammerschmidt, W. (1997) Latent membrane protein 1 of Epstein-Barr virus mimics a constitutively active receptor molecule. *The EMBO Journal*, **16**, 6131–6140.

39 Gires, O., Kohlhuber, F., Kilger, E., Baumann, M., Kieser, A., Kaiser, C., Zeidler, R., Scheffer, B., Ueffing, M. and Hammerschmidt, W. (1999) Latent membrane protein 1 of Epstein-Barr virus interacts with JAK3 and activates STAT proteins. *The EMBO Journal*, **18**, 3064–3073.

40 Wang, D., Liebowitz, D. and Kieff, E. (1985) An EBV membrane protein expressed in immortalized lymphocytes transforms established rodent cells. *Cell*, **43**, 831–840.

41 Baichwal, V.R. and Sugden, B. (1988) Transformation of Balb 3T3 cells by the BNLF-1 gene of Epstein-Barr virus. *Oncogene*, **2**, 461–467.

42 Moorthy, R.K. and Thorley-Lawson, D.A. (1993) All three domains of the Epstein-Barr virus-encoded latent membrane protein LMP-1 are required for transformation of rat-1 fibroblasts. *Journal of Virology*, **67**, 1638–1646.

43 Uchida, J., Yasui, T., Takaoka-Shichijo, Y., Muraoka, M., Kulwichit, W., Raab-Traub, N. and Kikutani, H. (1999) Mimicry of CD40 signals by Epstein-Barr virus LMP1 in B lymphocyte responses. *Science*, **286**, 300–303.

44 Kulwichit, W., Edwards, R.H., Davenport, E.M., Baskar, J.F., Godfrey, V. and Raab-Traub, N. (1998) Expression of the Epstein-Barr virus latent membrane protein 1 induces B cell lymphoma in transgenic mice. *Proceedings of the National Academy of Sciences of the United States of America*, **95**, 11963–11968.

45 Kilger, E., Kieser, A., Baumann, M. and Hammerschmidt, W. (1998) Epstein-Barr virus-mediated B-cell proliferation is dependent upon latent membrane

protein 1, which simulates an activated CD40 receptor. *The EMBO Journal*, **17**, 1700–1709.

46 Mancao, C. and Hammerschmidt, W. (2007) Epstein-Barr virus latent membrane protein 2A is a B-cell receptor mimic and essential for B-cell survival. *Blood*, **110**, 3715–3721.

47 Chen, F., Zou, J.Z., di Renzo, L., Winberg, G., Hu, L.F., Klein, E., Klein, G. and Ernberg, I. (1995) A subpopulation of normal B cells latently infected with Epstein-Barr virus resembles Burkitt lymphoma cells in expressing EBNA-1 but not EBNA-2 or LMP1. *Journal of Virology*, **69**, 3752–3758.

48 Longnecker, R. (2000) Epstein-Barr virus latency: LMP2, a regulator or means for Epstein-Barr virus persistence? *Advances in Cancer Research*, **79**, 175–200.

49 Ikeda, M., Fukuda, M. and Longnecker, R. (2005) *Function of Latent Membrane Protein 2A*, Caister Academic Press.

50 Caldwell, R.G., Wilson, J.B., Anderson, S.J. and Longnecker, R. (1998) Epstein-Barr virus LMP2A drives B cell development and survival in the absence of normal B cell receptor signals. *Immunity*, **9**, 405–411.

51 Casola, S., Otipoby, K.L., Alimzhanov, M., Humme, S., Uyttersprot, N., Kutok, J.L., Carroll, M.C. and Rajewsky, K. (2004) B cell receptor signal strength determines B cell fate. *Nature Immunology*, **5**, 317–327.

52 Mancao, C., Altmann, M., Jungnickel, B. and Hammerschmidt, W., (2005) Rescue of 'crippled' germinal center B cells from apoptosis by Epstein-Barr virus. *Blood*, **106**, 4339–4344.

53 Chaganti, S., Bell, A.I., Pastor, N.B., Milner, A.E., Drayson, M., Gordon, J. and Rickinson, A.B. (2005) Epstein-Barr virus infection *in vitro* can rescue germinal center B cells with inactivated immunoglobulin genes. *Blood*, **106**, 4249–4252.

54 Bechtel, D., Kurth, J., Unkel, C. and Kuppers, R. (2005) Transformation of BCR-deficient germinal-center B cells by EBV supports a major role of the virus in the pathogenesis of Hodgkin and posttransplantation lymphomas. *Blood*, **106**, 4345–4350.

55 Wen, W., Iwakiri, D., Yamamoto, K., Maruo, S., Kanda, T. and Takada, K. (2007) Epstein-Barr virus BZLF1 gene, a switch from latency to lytic infection, is expressed as an immediate-early gene after primary infection of B lymphocytes. *Journal of Virology*, **81**, 1037–1042.

56 Altmann, M. and Hammerschmidt, W. (2005) Epstein-Barr virus provides a new paradigm: a requirement for the immediate inhibition of apoptosis. *PLoS Biology*, **3**, e404.

57 Zeidler, R., Eissner, G., Meissner, P., Uebel, S., Tampe, R., Lazis, S. and Hammerschmidt, W. (1997) Downregulation of TAP1 in B lymphocytes by cellular and Epstein-Barr virus-encoded interleukin-10. *Blood*, **90**, 2390–2397.

58 Pudney, V.A., Leese, A.M., Rickinson, A.B. and Hislop, A.D. (2005) CD8+ immunodominance among Epstein-Barr virus lytic cycle antigens directly reflects the efficiency of antigen presentation in lytically infected cells. *The Journal of Experimental Medicine*, **201**, 349–360.

59 Hislop, A.D., Ressing, M.E., van Leeuwen, D., Pudney, V.A., Horst, D., Koppers-Lalic, D., Croft, N.P., Neefjes, J.J., Rickinson, A.B. and Wiertz, E.J. (2007) A CD8+T cell immune evasion protein specific to Epstein-Barr virus and its close relatives in Old World primates. *The Journal of Experimental Medicine*, **204**, 1863–1873.

60 Woodberry, T., Suscovich, T.J., Henry, L.M., Davis, J.K., Frahm, N., Walker, B.D., Scadden, D.T., Wang, F. and Brander, C. (2005) Differential targeting and shifts in the immunodominance of Epstein-Barr virus--specific CD8 and CD4 T cell responses during acute and persistent infection. *The Journal of Infectious Diseases*, **192**, 1513–1524.

61 Precopio, M.L., Sullivan, J.L., Willard, C., Somasundaran, M. and Luzuriaga, K. (2003) Differential kinetics and specificity of EBV-specific CD4+ and CD8+T cells

during primary infection. *Journal of Immunology (Baltimore, Md: 1950)*, **170**, 2590–2598.

62 Adhikary, D., Behrends, U., Moosmann, A., Witter, K., Bornkamm, G.W. and Mautner, J. (2006) Control of Epstein-Barr virus infection *in vitro* by T helper cells specific for virion glycoproteins. *The Journal of Experimental Medicine*, **203**, 995–1006.

63 Adhikary, D., Behrends, U., Boerschmann, H., Pfunder, A., Burdach, S., Moosmann, A., Witter, K., Bornkamm, G.W. and Mautner, J. (2007) Immunodominance of lytic cycle antigens in Epstein-Barr virus-specific CD4+T cell preparations for therapy. *PLoS ONE*, **2**, e583.

64 Burns, D.M. and Crawford, D.H. (2004) Epstein-Barr virus-specific cytotoxic T-lymphocytes for adoptive immunotherapy of post-transplant lymphoproliferative disease. *Blood Reviews*, **18**, 193–209.

65 Ghelani, D., Saliba, R. and Lima, M. (2005) Secondary malignancies after hematopoietic stem cell transplantation. *Critical Reviews in Oncology/Hematology*, **56**, 115–126.

66 Taylor, A.L., Marcus, R. and Bradley, J.A. (2005) Post-transplant lymphoproliferative disorders (PTLD) after solid organ transplantation. *Critical Reviews in Oncology/Hematology*, **56**, 155–167.

67 Brumbaugh, J., Baldwin, J.C., Stinson, E.B., Oyer, P.E., Jamieson, S.W., Bieber, C.P., Henle, W. and Shumway, N.E. (1985) Quantitative analysis of immunosuppression in cyclosporine-treated heart transplant patients with lymphoma. *The Journal of Heart Transplantation*, **4**, 307–311.

68 Shapiro, R.S., McClain, K., Frizzera, G., Gajl-Peczalska, K.J., Kersey, J.H., Blazar, B.R., Arthur, D.C., Patton, D.F., Greenberg, J.S., Burke, B. *et al.* (1988) Epstein-Barr virus associated B cell lymphoproliferative disorders following bone marrow transplantation. *Blood*, **71**, 1234–1243.

69 Ho, M., Jaffe, R., Miller, G., Breinig, M.K., Dummer, J.S., Makowka, L., Atchison, R.W., Karrer, F., Nalesnik, M.A. and Starzl, T.E. (1988) The frequency of Epstein-Barr virus infection and associated lymphoproliferative syndrome after transplantation and its manifestations in children. *Transplantation*, **45**, 719–727.

70 Masmoudi, A., Toumi, N., Khanfir, A., Kallel-Slimi, L., Daoud, J., Karray, H. and Frikha, M. (2007) Epstein-Barr virus-targeted immunotherapy for nasopharyngeal carcinoma. *Cancer Treatment Reviews*, **33**, 499–505.

71 Paya, C.V., Fung, J.J., Nalesnik, M.A., Kieff, E., Green, M., Gores, G., Habermann, T.M., Wiesner, P.H., Swinnen, J.L., Woodle, E.S. and Bromberg, J.S. (1999) Epstein-Barr virus-induced posttransplant lymphoproliferative disorders. ASTS/ASTP EBV-PTLD Task Force and The Mayo Clinic Organized International Consensus Development Meeting. *Transplantation*, **68**, 1517–1525.

72 Barry, M. and Bleackley, R.C. (2002) Cytotoxic T lymphocytes: all roads lead to death. *Nature Reviews. Immunology*, **2**, 401–409.

73 Papadopoulos, E.B., Ladanyi, M., Emanuel, D., Mackinnon, S., Boulad, F., Carabasi, M.H., Castro-Malaspina, H., Childs, B.H., Gillio, A.P., Small, T.N. *et al.* (1994) Infusions of donor leukocytes to treat Epstein-Barr virus-associated lymphoproliferative disorders after allogeneic bone marrow transplantation. *The New England Journal of Medicine*, **330**, 1185–1191.

74 Nalesnik, M.A., Rao, A.S., Zeevi, A., Fung, J.J., Pham, S., Furukawa, H., Gritsch, A., Klein, G. and Starzl, T.E. (1997) Autologous lymphokine-activated killer cell therapy of lymphoproliferative disorders arising in organ transplant recipients. *Transplantation Proceedings*, **29**, 1905–1906.

75 Haque, T., Amlot, P.L., Helling, N., Thomas, J.A., Sweny, P., Rolles, K.,

Burroughs, A.K., Prentice, H.G. and Crawford, D.H. (1998) Reconstitution of EBV-specific T cell immunity in solid organ transplant recipients. *Journal of Immunology (Baltimore, Md: 1950)*, **160**, 6204–6209.

76 Wilkie, G.M., Taylor, C., Jones, M.M., Burns, D.M., Turner, M., Kilpatrick, D., Amlot, P.L., Crawford, D.H. and Haque, T. (2004) Establishment and characterization of a bank of cytotoxic T lymphocytes for immunotherapy of Epstein-Barr virus-associated diseases. *Journal of Immunotherapy, (1997)*, **27**, 309–316.

77 Ibisch, C., Saulquin, X., Gallot, G., Vivien, R., Ferrand, C., Tiberghien, P., Houssaint, E. and Vie, H. (2000) The T cell repertoire selected in vitro against EBV: diversity, specificity, and improved purification through early IL-2 receptor alpha-chain (CD25)-positive selection. *Journal of Immunology (Baltimore, Md: 1950)*, **164**, 4924–4932.

78 Rooney, C.M., Smith, C.A., Ng, C.Y., Loftin, S.K., Sixbey, J.W., Gan, Y., Srivastava, D.K., Bowman, L.C., Krance, R.A., Brenner, M.K. and Heslop, H.E. (1998) Infusion of cytotoxic T cells for the prevention and treatment of Epstein-Barr virus-induced lymphoma in allogeneic transplant recipients. *Blood*, **92**, 1549–1555.

79 Rooney, C.M., Smith, C.A., Ng, C.Y., Loftin, S., Li, C., Krance, R.A., Brenner, M.K. and Heslop, H.E. (1995) Use of gene-modified virus-specific T lymphocytes to control Epstein-Barr-virus-related lymphoproliferation. *Lancet*, **345**, 9–13.

80 Straathof, K.C., Savoldo, B., Heslop, H.E. and Rooney, C.M. (2002) Immunotherapy for post-transplant lymphoproliferative disease. *British Journal of Haematology*, **118**, 728–740.

81 Khanna, R., Bell, S., Sherritt, M., Galbraith, A., Burrows, S.R., Rafter, L., Clarke, B., Slaughter, R., Falk, M.C., Douglass, J., Williams, T., Elliott, S.L. and Moss, D.J. (1999) Activation and adoptive transfer of Epstein-Barr virus-specific cytotoxic T cells in solid organ transplant patients with posttransplant lymphoproliferative disease. *Proceedings of the National Academy of Sciences of the United States of America*, **96**, 10391–10396.

82 Haque, T., Wilkie, G.M., Taylor, C., Amlot, P.L., Murad, P., Iley, A., Dombagoda, D., Britton, K.M., Swerdlow, A.J. and Crawford, D.H. (2002) Treatment of Epstein-Barr-virus-positive post-transplantation lymphoproliferative disease with partly HLA-matched allogeneic cytotoxic T cells. *Lancet*, **360**, 436–442.

83 Heslop, H.E., Ng, C.Y., Li, C., Smith, C.A., Loftin, S.K., Krance, R.A., Brenner, M.K. and Rooney, C.M. (1996) Long-term restoration of immunity against Epstein-Barr virus infection by adoptive transfer of gene-modified virus-specific T lymphocytes. *Nature Medicine*, **2**, 551–555.

84 Matloubian, M., Concepcion, R.J. and Ahmed, R. (1994) CD4+ T cells are required to sustain CD8+ cytotoxic T-cell responses during chronic viral infection. *Journal of Virology*, **68**, 8056–8063.

85 Zajac, A.J., Blattman, J.N., Murali-Krishna, K., Sourdive, D.J., Suresh, M., Altman, J.D. and Ahmed, R. (1998) Viral immune evasion due to persistence of activated T cells without effector function. *The Journal of Experimental Medicine*, **188**, 2205–2213.

86 Yee, C., Thompson, J.A., Byrd, D., Riddell, S.R., Roche, P., Celis, E. and Greenberg, P.D. (2002) Adoptive T cell therapy using antigen-specific CD8+ T cell clones for the treatment of patients with metastatic melanoma: in vivo persistence, migration, and antitumour effect of transferred T cells. *Proceedings of the National Academy of Sciences of the United States of America*, **99**, 16168–16173.

87 Kieper, W.C., Prlic, M., Schmidt, C.S., Mescher, M.F. and Jameson, S.C. (2001) Il-12 enhances CD8 T cell homeostatic expansion. *Journal of Immunology (Baltimore, Md: 1950)*, **166**, 5515–5521.

88 Mueller, Y.M., Bojczuk, P.M., Halstead, E.S., Kim, A.H., Witek, J., Altman, J.D. and Katsikis, P.D. (2003) IL-15 enhances survival and function of HIV-specific CD8+ T cells. *Blood*, **101**, 1024–1029.

89 Khanna, R., Burrows, S.R., Nicholls, J. and Poulsen, L.M. (1998) Identification of cytotoxic T cell epitopes within Epstein-Barr virus (EBV) oncogene latent membrane protein 1 (LMP1): evidence for HLA A2 supertype-restricted immune recognition of EBV-infected cells by LMP1-specific cytotoxic T lymphocytes. *European Journal of Immunology*, **28**, 451–458.

90 Sing, A.P., Ambinder, R.F., Hong, D.J., Jensen, M., Batten, W., Petersdorf, E. and Greenberg, P.D. (1997) Isolation of Epstein-Barr virus (EBV)-specific cytotoxic T lymphocytes that lyse Reed-Sternberg cells: implications for immune-mediated therapy of EBV+ Hodgkin's disease. *Blood*, **89**, 1978–1986.

91 Gottschalk, S., Edwards, O.L., Sili, U., Huls, M.H., Goltsova, T., Davis, A.R., Heslop, H.E. and Rooney, C.M. (2003) Generating CTLs against the subdominant Epstein-Barr virus LMP1 antigen for the adoptive immunotherapy of EBV-associated malignancies. *Blood*, **101**, 1905–1912.

92 Orentas, R.J., Roskopf, S.J., Nolan, G.P. and Nishimura, M.I. (2001) Retroviral transduction of a T cell receptor specific for an Epstein-Barr virus-encoded peptide. *Clinical Immunology (Orlando, Fla)*, **98**, 220–228.

93 Eshhar, Z., Waks, T., Gross, G. and Schindler, D.G. (1993) Specific activation and targeting of cytotoxic lymphocytes through chimeric single chains consisting of antibody-binding domains and the gamma or zeta subunits of the immunoglobulin and T-cell receptors. *Proceedings of the National Academy of Sciences of the United States of America*, **90**, 720–724.

94 Eshhar, Z., Waks, T., Bendavid, A. and Schindler, D.G. (2001) Functional expression of chimeric receptor genes in human T cells. *Journal of Immunological Methods*, **248**, 67–76.

14
Human Papillomavirus (HPV) Tumor-associated Antigens
Andreas M. Kaufmann

14.1
Introduction

The Human Papillomaviruses (HPVs) are a family of double-stranded DNA viruses with over 100 different genotypes. HPV genotypes are divided into low-risk and high-risk for their potential to transform cells. Fifteen types of HPV are classified as high-risk types (16, 18, 31, 33, 35, 39, 45, 51, 52, 56, 58, 59, 68, 73 and 82), three are classified as probable high-risk types (26, 53, and 66) and 18 are classified as low-risk types (6, 11, 34, 40, 42–44, 54, 55, 57, 61, 67, 70–72, 81, 83, and 84) [1]. While low-risk types induce benign genital condylomata and low-grade squamous intraepithelial lesions the high-risk types are associated with anogenital cancers and can be detected in more than 99% of cervical cancers, with type HPV16 found in about 50% of cases [2, 3]. In general HPV infection *per se* causes no problem for the host. Latent infection may become obvious when the infected epithelium reacts and hyperproliferation of the infected cells leads to wart formation or dysplasia. Low grade lesions (cervical intraepithelial neoplasia I, CIN I) are regarded as stages when the productive life cycle of the virus is initiated and infectious particles are produced. Such lesions either regress spontaneously to latent infection or resolve completely, or can progress to high grade lesions (CIN II/III). In persistent high grade lesions the HPV-infected epithelial cells no longer differentiate and virus production is generally shut off. These pre-malignant lesions are regarded as obligatory precursors for HPV-induced cancer.

Cervical cancer is the second leading cause of cancer deaths in women worldwide. Since the disease and its pre-malignant stages are generally associated with HPV infection, this presents a unique opportunity to prevent and treat cervical cancer through anti-viral vaccination using viral proteins as tumor associated antigens (TAA). It became obvious that other malignancies, albeit with lower rates of association, are also induced by HPV infection. This is true for other anogenital squamous cell carcinoma such as vulvar and vaginal cancer, anal cancer, and penile cancer. Infection by the high-risk types of HPV is also not confined to the anogenital

Tumor-Associated Antigens. Edited by Olivier Gires and Barbara Seliger
Copyright © 2009 WILEY-VCH Verlag GmbH & Co. KGaA, Weinheim
ISBN: 978-3-527-32084-4

area, since 18.3% of cancers of the oropharynx contain DNA from these types [4]. However, the low-risk types of HPV can cause other diseases such as recurrent respiratory papillomatosis. In general, about 75% of the sexually active population acquires at least one genital HPV type during their lifetime [5]. Likewise, virtually everyone is infected by dermal HPV types which potentially cause common warts like *verrucae vulgares* on the hand and fingers and *verrucae plantares* on the feet [6]. Most individuals remain asymptomatic and clear HPV infections spontaneously. This is most probably due to an active immune response. A small percentage of patients develop clinically or histologically recognizable pre-cancers that can persist and may develop over time into invasive cancers, like certain types of non-melanoma skin cancer. This is especially problematic in the immune suppressed such as HIV-positive patients or iatrogenically immune suppressed transplant recipients [7].

14.2
HPV-encoded TAA

14.2.1
The Proteins of HPV

Papillomaviruses are old from an evolutionary point of view and are well-adapted DNA viruses. A particular characteristic of papillomaviruses is that the partly overlapping open reading frames are arranged on only one DNA strand of only 8 kb. The relative arrangement of the eight to 10 open reading frames within the genome is the same in all papillomavirus types. The genome can be divided into three functional regions (i) the long control region (LCR), (ii) the region of 'early proteins' (E1-E8) expressed during the non-productive part of the viral life cycle, and (iii) the region of the late proteins (L1, L2). A schematic representation of the genome structure is shown in Figure 14.1. A common event during transformation is the random integration of the HPV genome into the host cell genome. Since the circular genome has to open, this can result in the destruction of open reading frames. However, the viral oncogenes E6 and E7 have, to date, never been found to be destroyed.

Since all viral proteins are potential TAAs their characteristics are listed in Table 14.1.

The long control region has a length of 500–1000 bp and contains the origin of replication and regulation elements of gene expression. It does not code for any proteins. The early proteins are numbered consecutively according to their sequence in the genome. E1 is a 68–85-kDa protein with a helicase function. It is essential for viral replication and control of gene transcription. The E2 protein has a size of 48 kDa and is a viral transcription factor that is essential for replication, control of gene transcription, genome segregation, and encapsidation. The function of E3 is not known and it is only present in a few HPV types. E4 is formed by a splicing event of E1^E4 sequences and has a size of 10–44 kDa. Expression of E4 is confined to suprabasal differentiating cells in the epithelium. It binds to cytoskeletal proteins and may have a function during virus release. E5 has a M_r of 14 kDa and has been shown

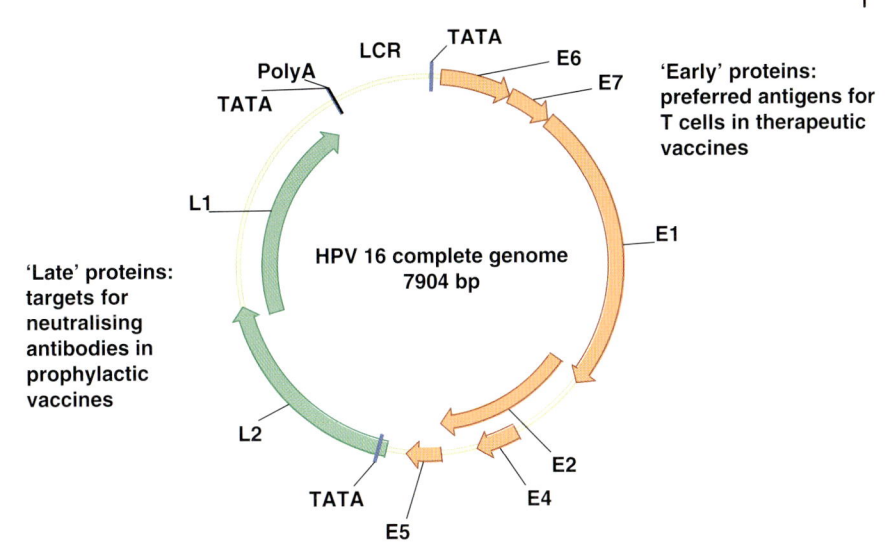

Figure 14.1 The HPV 16 genome structure as a prototype for the canonical PV genome. 'E' designates proteins expressed early during the viral life cycle while 'L' indicates the late expressed structural proteins of the viral capsid. Known transcription start sites (TATA) and polyadenylation sites (poly A) are indicated. The long control region (LCR) harbors sequences that interact with cellular transcription factors and control HPV gene expression and replication. All open reading frames (arrows) are transcribed in the same direction and are located on a single DNA coding strand.

to interact with EGF/PDGF-receptors, thereby stimulating proliferation. The two most important early proteins with respect to transformation and carcinogenesis are the viral oncogenes E6 and E7. E6 has a M_r of 16–18 kDa and has a scaffolding function. It interacts with several cellular proteins. Its most prominent effects are degradation of p53 and activation of telomerase. These are critical events during immortalization of cells: the inactivation of apoptosis inducing p53 and the activation of telomerase leading to chromosome instability.

E7 is a 10-kDa protein and also has several cellular target proteins. Its interaction with pRb leads to release of the transcription factor E2F and thus to trans-activation of E2F-dependent promoters, many of which are associated with proliferation. Finally the E8^-E2C splicing event results in the long-distance transcription and generation of replication a repressor protein that has its function during latency of the virus in basal epithelial cells.

All these sequences and proteins – except for E6 and E7 – can be destroyed or deleted upon integration of the virus and are therefore not reliable TAAs. The viral oncogenes E6 and E7 are mandatorily expressed in all HPV-transformed cells and knock-down of either of the two drives infected cells into apoptosis. Therefore, they are regarded as the most interesting viral tumor-associated antigens.

The viral capsid is built of only two proteins and has no envelope. The major capsid protein is L1, a 57-kDa protein comprising the core structure of the viral shell. The

Table 14.1 Proteins of the HPV.

Viral protein	Molecular weight (kDa)	Function	Targeted as TAA
Early proteins			
E1	68–85	Helicase, essential for viral replication and gene transcription, similarity among types	Not used
E2	48	Viral transcription factor, essential for viral replication and control of gene transcription, genome segregation and encapsidation	In some approaches
E3	Unknown	Function not known and present in few HPV types	Not used
E1^E4	10–44	Binding to cytoskeletal protein	Not used
E5	14	Interaction with EGF/PDGF receptors	Not used
E6	16–18	Interacting with several cellular proteins, degradation of p53 and activation of telomerase	Predominantly used
E7	~10	Interacting with several cellular proteins, interaction with pRB and transactivation of E2F-dependent transcription	Predominantly used
E8	20	Long distance transcription and replication repressor protein	Not used
Late proteins			
L1	57	Major capsid protein	Prophylactic vaccines, carrier in combination vaccines
L2	43–53	Minor capsid protein	Prophylactic/therapeutic vaccines

minor capsid protein L2 (43–53 kDa) is directed inwards and connects the shell to the encapsidated DNA [8].

The virus uses these few multifunctional proteins to fulfil a life cycle that is highly adapted to the epithelial target cell and strictly host species-specific. Infection targets the basal cell layer of the epithelium where the virus can persist for prolonged periods of time, often for years. These cells have stem cell-like properties rejuvenating the basal cell layer and giving rise to the cell compartment consisting of proliferating cells which build up the different layers of the epithelium. The early proteins with exception of E4 are exclusively expressed here. The virus is propagated directly in descendent cells and virus copy number is controlled by an unknown mechanism. When the cell migrates upwards in the epithelium and starts to differentiate the expression pattern changes and E4 and L1, L2 are expressed, viral genomes are produced, and encapsidated. Release of the virus occurs passively by cell shaling, which is an important peculiarity of HPV. The virus avoids any non-physiologic cell death or alterations and does not deliver any danger signals to which the immune system can react.

Human papillomaviruses are extremely well adapted to evade the host immune system. They infect epithelial cells exclusively and fulfil their lifecycle within the epithelium without liberating antigens into the parenteral space thus avoiding exposure to the lymphoid system. They have developed strategies to suppress Langerhans cell activation and interferon signaling. The multifunctional viral oncogenes E6 and E7 alter expression of interferon response genes, NF kappa B stimulated genes, and cell cycle regulation genes [9, 10]. This strategy guarantees persistent infection for a mean duration of 8–18 months and are important features with respect to the evasion of the immune system.

14.2.2
Vaccine Development

Prophylactic and therapeutic vaccine development has focused on HPV proteins as target antigens. Prophylactic vaccines to HPV have been proven highly efficacious. They were introduced in 2006 and have the potential to reduce the burden of cervical cancer by approximately 70% [11–13]. They have, however, no therapeutic potential due to their mechanism of action via antibodies targeting the L1 capsid antigen which is not expressed in the persistently HPV-infected basal epithelial cells [14]. The capsid proteins L1 and L2 are also not tumor-associated viral antigens because they are not expressed in high-grade lesions (CIN II/III) or in cervical cancer cells. Therefore, therapeutic vaccines based on other HPV-encoded proteins still offer benefits for the treatment of high-grade lesions (or worse), which will still develop despite the availability of prophylactic vaccines because (i) prophylactic vaccines do not protect against all carcinogenic HPV types, (ii) women already infected today might still progress to cancer if left untreated, (iii) when protective immunity fades breakthrough infections may arise, and (iv) not everyone will be vaccinated.

14.2.3
Therapeutic Vaccine Strategies

Therapeutic vaccines target antigens that are present inside the infected cell. In the case of HPV-induced cancers, the antigens which are targeted predominantly are the viral oncogenes E6 and E7, because their sustained expression is required for the maintenance of the cancerous phenotype [8]. Preclinical research has demonstrated excellent efficacy of immunization with these antigens by different vaccination approaches in tumor protection and regression experiments in animal tumor models [15].

Several therapeutic vaccines have been tested in clinical trials over the past 15 years, which target the E2, E5, E6 and/or E7 proteins by many different strategies [16]. These vaccines activate the patient's cellular immune response to recognize and kill exclusively cells that express HPV proteins. Vaccine safety was demonstrated in initial trials in patients with advanced cervical cancer, who tend to have decreased

Table 14.2 Clinical trials with therapeutic HPV vaccines

Vaccine type	Target Disease	Reference
Peptide	CxCa, CIN, VIN	[16]
		[26]
		[27]
Recombinant protein	CxCa, CIN, VIN, anal HSIL,	[16]
CVLP	CIN	[28]
DNA	CIN, cervical/anal HSIL	[16]
Recombinant Vaccinia Virus, or Modified Virus Ankara (MVA)	CxCa, CIN VIN,	[16]
		[29]
		[30]
		[31]
Dendritic cells	CxCa	[16]
		[32]

immune function. Patients with earlier invasive or premalignant disease were included in more recent and ongoing immunotherapy trials (Table 14.2).

In addition, the choice of antigens is changing: initially and for pragmatic reasons the HPV E7 protein was regarded as the ideal target. E7 is a smaller protein than E6 and displays higher solubility under physiologic conditions. Recently, it was shown that the larger E6 and E2 proteins might be more immunogenic and hence more effective targets [17–19]. It has been demonstrated that in the healthy population immunity to the E6 and E2 antigens prevail, indicating protective/ therapeutic efficacy while immunity to E7 dominates in diseased individuals [18]. Therefore, a re-design of vaccines to include or combine several HPV antigens is underway.

Also new basic immunology findings such as the presence and characterization of immunosuppressive regulatory T cells have influenced the strategies of immunotherapy [20]. To date, however, there has been no convincing evidence that therapeutic vaccination prompt an immune response that correlates with clinical outcome. Since the local tumor environment is immunosuppressive and not immunostimulatory, and cervical tumors may be directly accessible, a combination of vaccination with topical immune stimulatory agents such as Imiquimod or other pro-inflammatory mediators like interferons might improve the performance of systemic immunization with HPV vaccines [21].

Examples of therapeutic vaccination trials with experimental vaccines targeting HPV antigens are shown in Table 14.2. The initial phase I trials that were carried out in end-stage cervical cancer patients showed no overt clinical response. Results were generally frustrating. However, with most vaccination strategies safety and immunogenicity even in advanced disease was demonstrated. As experience was gained and vaccine toxicity reduced, studies were carried out in early invasive and also in pre-malignant stages of the disease. Patients with lesions induced by mucosal or cutaneous HPVs such as cervical and vaginal intraepithelial lesions, anal intraepithelial lesions, condylomas and anogenital warts, head and neck squamous cell carcinomas and recurrent respiratory papillomatosis were treated. These trials were placebo-controlled and blinded, and used higher patient numbers and fewer immunosuppressed individuals. It was hoped that this would facilitate interpretation of antigen-specific immunity and its correlation to clinical efficacy. Specific immune responses can regularly be observed at least in a subset of patients. However, clinical responses are rare and more frequently observed in cases of premalignant disease that also show spontaneous resolution. This can be enhanced to some extent but reliable cure rates have not been achieved. Also correlation of induced immune responses with clinical responses has not been convincingly shown so far [16].

These vaccines targeted HPV antigens that had been shown to be immunogenic and effective in animal model systems, or were recognized by tumor-infiltrating lymphocytes. Also information on the chosen antigen was available e.g. HLA-restricted peptide epitopes, which could be identified [22].

The lack of clinical responses even in the situation of vaccine-induced strong T-cell responses produced a dilemma. Initially, due to the presence of highly immunogenic viral tumor-associated antigens, cervical cancer was thought to be an ideal entity for immunotherapy. Most HPV infections resolve spontaneously while HPV-specific T cells were observed in regressing warts and in tumor-infiltrating lymphocytes [23]. Only after a long period of persistent infection and premalignant disease, lasting for around 10 years, does invasive cancer develop. During this time period the immune system tackles the tumor cells thus selecting those cells which are capable of evading immune attack. In addition by the action of the apoptosis-abrogating effect of E6 and the mitotic spindle-disturbing activity of E7, oncogene products, mutations, and chromosomal aberrations accumulate. As a consequence highly selected cells that have evolved by immune evasion will constitute the HPV-induced tumor. These might display high resistance towards any further attack by T cells. It has been shown in cervical cancer that the loss of antigenicity can occur at all levels of antigen processing and presentation [24]. Since the antigens E6 and E7 are both obligatory for the maintenance of the immortal and transformed phenotype, their expression is never lost in HPV-transformed cells. Therefore, antigen-loss variants as are observed in other malignancies are never found for E6 and E7. Immune evasion is achieved by different means including progressive loss of MHC alleles or gene locus, compromised peptide processing and transport to the endoplasmatic reticulum, and loss of stability of MHC by lack of beta2-microglobulin, among others.

Characterization of tumor-infiltrating lymphocytes has demonstrated high numbers of immunosuppressive regulatory T cells and features of T-cell exhaustion in the antigen-specific T-cell compartment (our own unpublished observations). Taken together it seems unlikely that end-stage cervical cancer can be effectively cured by immunotherapy alone. Recently, clinical trials focused on pre-invasive lesions and on infection by HPV to prevent baleful events early enough before the accumulation of malign alterations. Moreover, there is no therapy for HPV infection and women diagnosed with an infection can not be offered any treatment options. In these trials, there were indications for the first time that immunotherapy enhances regression of lesions. This has been observed in conjunction with several antigens such as HPV16

E6/E7, E7 alone, and even bovine papillomavirus E2 that putatively cross react with HPV E2 (16). There is hope that the process of immune evasion can be abrogated if an immune response to HPV antigens and cervical cancer TAA is induced early and strongly enough before resistant variants develop.

14.2.4
HPV-induced Cellular TAA

Although HPV-encoded TAA seem to be ideal targets for immunotherapy they have the disadvantage of being HPV type-specific. Only very restricted cross reactivity has been found for E6 and E7 antigens of different HPVs and it is probable that individual therapeutic vaccines will have to be offered for each individual HPV type. About 70% of cervical cancer is induced by only two HPV types, HPV16 and HPV18. Fifteen different HPV types account for the remaining 30% of cancers. To simplify vaccine design and production, a generic TAA for HPV-induced cancer is desirable.

There are candidates under investigation for this purpose. HPV infection and progression towards malignancy lead to the induction and over-expression of cellular proteins that could serve as generic 'HPV-associated' antigens. Examples of such self-antigens are $p16^{INK4a}$, a cell cycle regulating protein that is greatly over-expressed in high-risk HPV-infected cells due to a positive feedback loop [8]. This protein is normally expressed only in senescent non-proliferating cells and would not be expected to show tolerance or autoimmunity. Another candidate is NET-1/C4.8. This protein which belongs to the tetraspanin family of scaffolding proteins has been shown to be over-expressed in a subset of CIN III lesions (that presumably progress to invasive lesions) and in most invasive cervical cancers [25]. This protein is not widely expressed in normal tissues. These potential TAA for cervical cancer or other HPV-induced malignancies await further investigation.

14.3
Conclusions and Future Perspectives

HPV-related antigens and HPV-induced cancer can be regarded as an ideal model system for investigating tumor-directed immune responses and immune evasion mechanisms. Despite this promising setting initial clinical trials were disappointing with respect to treatment efficacy. The use of vaccines in the earlier stages of disease has produced better results. Premalignant dysplasia and HPV infection will be the diseases to aim for in the future in order to impede immune evasion by tumor cells. Since HPV has evolved as a highly adapted and immunosuppressive virus the choice of the most effective antigen and powerful adjuvant will be of paramount importance. If therapy of invasive cancer is the goal, means will need to be developed to reverse immune escape by the tumor cells and to reactivate exhausted T cells.

In conclusion a combination of several strategies will be the aim of future preclinical and clinical studies in order to identify the best antigen and most effective delivery together with an adjuvant that supports the right immune response and a local treatment to sensitize tumor cells to immune attack.

References

1 Munoz, N., Bosch, F.X., de Sanjose, S., Herrero, R., Castellsague, X., Shah, K.V., Snijders, P.J. and Meijer, C.J. (2003) Epidemiologic classification of human papillomavirus types associated with cervical cancer. *The New England Journal of Medicine*, **348**, 518–527.

2 Walboomers, J.M., Jacobs, M.V., Manos, M.M., Bosch, F.X., Kummer, J.A., Shah, K.V., Snijders, P.J., Peto, J., Meijer, C.J. and Munoz, N. (1999) Human papillomavirus is a necessary cause of invasive cervical cancer worldwide. *The Journal of Pathology*, **189**, 12–19.

3 Bosch, F.X., Manos, M.M., Munoz, N., Sherman, M., Jansen, A.M., Peto, J., Schiffman, M.H., Moreno, V., Kurman, R. and Shah, K.V. (1995) Prevalence of human papillomavirus in cervical cancer: a worldwide perspective. International biological study on cervical cancer (IBSCC) Study Group. *Journal of the National Cancer Institute*, **87**, 796–802.

4 Herrero, R., Castellsague, X., Pawlita, M., Lissowska, J., Kee, F., Balaram, P., Rajkumar, T., Sridhar, H., Rose, B., Pintos, J., Fernandez, L., Idris, A. *et al.* (2003) Human papillomavirus and oral cancer: the International Agency for Research on Cancer multicenter study. *Journal of the National Cancer Institute*, **95**, 1772–1783.

5 Koutsky, L. (1997) Epidemiology of genital human papillomavirus infection. *The American Journal of Medicine*, **102**, 3–8.

6 Nindl, I., Gottschling, M. and Stockfleth, E. (2007) Human papillomaviruses and non-melanoma skin cancer: Basic virology and clinical manifestations. *Disease Markers*, **23**, 247–259.

7 Grulich, A.E., van Leeuwen, M.T., Falster, M.O. and Vjadic, C.M. (2007) Incidence of cancers in people with HIV/AIDS compared with immunosuppressed transplant recipients: a metaanalysis. *Lancet*, **370**, 59–67.

8 zur Hausen, H. (2002) Papillomaviruses and cancer: from basic studies to clinical application. *Nature Reviews. Cancer*, **2**, 342–350.

9 Chang, Y.E. and Laimins, L.A. (2000) Microarray analysis identifies interferon-inducible genes and Stat-1 as major transcriptional targtes of human papillomavirus type 31. *Journal of Virology*, **74**, 4174–4182.

10 Nees, M., Geoghegan, J.M., Hyman, T., Frank, S., Miller, L. and Woodworth, C.D. (2001) Papillmavirus type 16 oncogenes downregulate expression of interferon-responsive genes and upregulate proliferation-associated and NF-kappaB-responsive genes in cervical keratinocytes. *Journal of Virology*, **75**, 4283–4296.

11 Future, II., Study Group. (2007) Quadrivalent vaccine against human papillomavirus to prevent high-grade cervical lesions. *The New England Journal of Medicine*, **356**, 1915–1927.

12 Harper, D.M., Franco, E.L., Wheeler, C.M., Moscicki, A.B., Romanowski, B., Roteli-Martins, C.M., Jenkins, D., Schuind, A., Costa Clemens, S.A. and Dubin, G. (2006) Sustained efficacy up to 4.5 years of a bivalent L1 virus-like particle vaccine against human papillomavirus types 16 and 18: follow-up from a randomised control trial. *Lancet*, **367**, 1247–1255.

13 Paavonen, J., Jenkins, D., Bosch, F.X., Naud, P., Salmeron, J., Wheeler, C.M., Chow, S.N., Apter, D.L., Kitchener, H.C., Castellsague, X., de Carvalho, N.S., Skinner, S.R. *et al.* (2007) Efficacy of a prophylactic adjuvanted bivalent L1 virus-

like-particle vaccine against infection with human papillomavirus types 16 and 18 in young women: an interim analysis of a phase III double-blind, randomised controlled trial. *Lancet*, **369**, 2161–2170.

14 Hildesheim, A., Herrero, R., Wacholder, S., Rodriguez, A.C., Solomon, D., Bratti, M.C., Schiller, J.T., Gonzalez, P., Dubin, G., Porras, C., Jimenez, S.E., Lowy, D.R. *et al.* (2007) Effect of human papillomavirus 16/18 L1 viruslike particle vaccine among young women with preexisting infection: a randomized trial. *The Journal of the American Medical Association*, **298**, 743–753.

15 Brinkman, J.A., Hughes, S.H., Stone, P., Caffrey, A.S., Muderspach, L.I., Roman, L.D., Weber, J.S. and Kast, W.M. (2007) Therapeutic vaccination for HPV induced cervical cancers. *Disease Markers*, **23**, 337–352.

16 Schreckenberger, C. and Kaufmann, A.M. (2004) Vaccination strategies for the treatment and prevention of cervical cancer. *Current Opinion in Oncology*, **16**, 485–491.

17 de Jong, A., van der Burg, S.H., Kwappenberg, K.M., van der Hulst, J.M., Franken, K.L., Geluk, A., van Meijgaarden, K.E., Drijfhout, J.W., Kenter, G., Vermeij, P., Melief, C.J. and Offringa, R. Frequent detection of human papillomavirus 16 E2-specific T-helper immunity in healthy subjects. *Cancer Research*, **62**, 472–479.

18 Welters, M.J., de Jong, A., van den Eeden, S.J., van der Hulst, J.M., Kwappenberg, K.M., Hassane, S., Franken, K.L., Drijfhout, J.W., Fleuren, G.J., Kenter, G., Melief, C.J., Offringa, R. *et al.* (2003) Frequent display of human papillomavirus type 16 E6-specific memory T-Helper cells in the healthy population as witness of previous viral encounter. *Cancer Research*, **63**, 636–641.

19 Garcia-Hernandez, E., Gonzalez-Sanchez, J.L., Andrade-Manzano, A., Contreras, M.L., Padilla, S., Guzman, C.C., Jimenez, R., Reyes, L., Morosoli, G., Verde, M.L. and Rosales, R. (2006) Regression of papilloma high-grade lesions (CIN 2 and CIN 3) is stimulated by therapeutic vaccination with MVA E2 recombinant vaccine. *Cancer Gene Therapy*, **13**, 592–597.

20 Molling, J.W., de Gruijl, T.D., Glim, J., Moreno, M., Rozendaal, L., Meijer, C.J., van den Eertwegh, A.J., Scheper, R.J., von Blomberg, M.E. and Bontkes, H.J. (2007) CD4(+)CD25(hi) regulatory T-cell frequency correlates with persistence of human papillomavirus type 16 and T helper cell responses in patients with cervical intraepithelial neoplasia. *International Journal of Cancer*, **121**, 1749–1755.

21 Stanley, M.A. (2002) Imiquimod and the imidazoquinolones: mechanism of action and therapeutic potential. *Clinical and Experimental Dermatology*, **27**, 571–577.

22 Ressing, M.E., Sette, A., Brandt, R.M., Ruppert, J., Wentworth, P.A., Hartman, M., Oseroff, C., Grey, H.M., Melief, C.J. and Kast, W.M. (1995) Human CTL epitopes encoded by human papillomavirus type 16 E6 and E7 identified through *in vivo* and *in vitro* immunogenicity studies of HLA-A*0201-binding peptides. *Journal of Immunology (Baltimore, Md: 1950)*, **154**, 5934–5943.

23 Coleman, N., Birley, H.D., Renton, A.M., Hanna, N.F., Ryait, B.K., Byrne, M., Taylor-Robinson, D. and Stanley, M.A. (1994) Immunological events in regressing genital warts. *American Journal of Clinical Pathology*, **102**, 768–774.

24 Ritz, U., Momburg, F., Pilch, H., Huber, C., Maeurer, M.J. and Seliger, B. (2001) Deficient expression of components of the MHC class I antigen processing machinery in human cervical carcinoma. *International Journal of Oncology*, **19**, 1211–1220.

25 Wollscheid, V., Kühne-Heid, R., Stein, I., Jansen, L., Köllner, S., Schneider, A. and Dürst, M. (2002) Identification of a new proliferation-associated protein NET-1/C4.8 characteristic for a subset of high-grade cervical intraepithelial neoplasia and cervical carcinomas. *International Journal of Cancer*, **99**, 771–775.

26 Welters, M.J., Kenter, G.G., Piersma, S.J., Vloon, A.P., Löwik, M.J., Berends-van der Meer, D.M., Drijfhout, J.W., Valentijn, A.R., Wafelman, A.R., Oostendorp, J., Fleuren, G.J., Offringa, R., Melief, C.J. and van der Burg, S.H. (2008) Induction of tumour-specific CD4+ and CD8+ T-cell immunity in cervical cancer patients by a human papillomavirus type 16 E6 and E7 long peptides vaccine. *Clinical Cancer Research*, **14**, 178–187.

27 Kenter, G.G., Welters, M.J., Valentijn, A.R., Lowik, M.J., Berends-van der Meer, D.M., Vloon, A.P., Drijfhout, J.W., Wafelman, A.R., Oostendorp, J., Fleuren, G.J., Offringa, R., van der Burg, S.H. and Melief, C.J. (2008) Phase I immunotherapeutic trial with long peptides spanning the E6 and E7 sequences of high-risk human papillomavirus 16 in end-stage cervical cancer patients shows low toxicity and robust immunogenicity. *Clinical Cancer Research*, **14**, 169–177.

28 Kaufmann, A.M., Nieland, J.D., Jochmus, I., Baur, S., Friese, K., Gabelsberger, J., Gieseking, F., Gissmann, L., Glasschröder, B., Grubert, T., Hillemanns, P., Höpfl, R., Ikenberg, H., Schwarz, J., Karrasch, M., Knoll, A., Küppers, V., Lechmann, M., Lelle, R.J., Meissner, H., Müller, R.T., Pawlita, M., Petry, K.U., Pilch, H., Walek, E. and Schneider, A. (2007) Vaccination trial with HPV16 L1E7 chimeric virus-like particles in women suffering from high grade cervical intraepithelial neoplasia (CIN 2/3). *International Journal of Cancer*, **121**, 2794–2800.

29 Corona Gutierrez, C.M., Tinoco, A., Navarro, T., Contreras, M.L., Cortes, R.R., Calzado, P., Reyes, L., Posternak, R., Morosoli, G., Verde, M.L. and Rosales, R. (2004) Therapeutic vaccination with MVA E2 can eliminate precancerous lesions (CIN 1, CIN 2, and CIN 3) associated with infection by oncogenic human papillomavirus. *Human Gene Therapy*, **15**, 421–431.

30 García-Hernández, E., González-Sánchez, J.L., Andrade-Manzano, A., Contreras, M.L., Padilla, S., Guzmán, C.C., Jiménez, R., Reyes, L., Morosoli, G., Verde, M.L. and Rosales, R. (2006) Regression of papilloma high-grade lesions (CIN 2 and CIN 3) is stimulated by therapeutic vaccination with MVA E2 recombinant vaccine. *Cancer Gene Therapy*, **13**, 592–597.

31 Albarran, Y., Carvajal, A., de la Garza, A., Cruz Quiroz, B.J., Vazquez Zea, E., Díaz Estrada, I., Mendez Fuentez, E., López Contreras, M., Andrade-Manzano, A., Padilla, S., Varela, A.R. and Rosales, R. (2007) MVA E2 recombinant vaccine in the treatment of human papillomavirus infection in men presenting intraurethral flat condyloma: a phase I/II study. *BioDrugs: Clinical Immunotherapeutics, Biopharmaceuticals and Gene Therapy*, **21**, 47–59.

32 Santin, A.D., Bellone, S., Palmieri, M., Zanolini, A., Ravaggi, A., Siegel, E.R., Roman, J.J., Pecorelli, S. and Cannon, M.J. (2008) Human papillomavirus type 16 and 18 E7-pulsed dendritic cell vaccination of stage IB or IIA cervical cancer patients: a phase I escalating-dose trial. *Journal of Virology*, **82**, 1968–1979.

15
Circulating TAAs: Biomarkers for Cancer Diagnosis, CA125

Angela Coliva, Ettore Seregni, and Emilio Bombardieri

15.1
Introduction

Mucins comprise a class of molecules expressed by various epithelial cell types. Their principal role is to maintain homeostasis and to prompt cell survival in harsh environments, but mucins have other functions such as signal transduction and also serve as receptors. Mucins are implicated in the pathogenesis of cancer, as tumors often grow in hypoxic, acidic and proteases-rich sites. MUC16 is a mucin of high molecular weight, which is heavily O-glycosylated and membrane associated. The molecular structure is quite complex and is composed of an extracellular, a transmembrane, and a cytoplasmic region. The molecule can undergo proteolytic cleavage in the extracellular domain and the component released (ektodomain) is known as CA125. This secreted antigen is easily detectable in serum. CA125 is shed in more than 80% of epithelial ovarian cancers, but also in normal tissues or in other non-ovarian carcinomas. The determination of normal values was carried out in the early 1980s when an assay method was developed. The clinical use of CA125 as a tumor marker is hampered by the fact that many factors affect the serum levels of healthy individuals. CA125 could have been an ideal screening agent because of its relatively low cost and non-invasive nature, but the low positive predictive value in detecting ovarian cancer in an asymptomatic population has limited its use. It is more useful to use CA125 as a tool for assessing prognosis, monitoring therapy response, and for follow-up.

15.2
Definition and Classification of Mucins

The mucin family of proteins represents a component of mucus, the hydrophilic mixture of the epithelial layers which has a protective role against pathogens or stresses, and are responsible for its physical properties. Mucins can be classified into two subfamilies according to their location relative to the cell surface.

Tumor-Associated Antigens. Edited by Olivier Gires and Barbara Seliger
Copyright © 2009 WILEY-VCH Verlag GmbH & Co. KGaA, Weinheim
ISBN: 978-3-527-32084-4

The gel-forming (secreted) mucins, comprising five molecules, MUC6, MUC2, MUC5AC, MUC5B and MUC19, are entirely extracellular and form oligomeric structures [1].

In normal physiology, the secreted mucins, in particular the polymeric mucins MUC5AC and MUC5B, provide the physical barrier that protects the epithelial cells lining the respiratory and gastrointestinal tracts and the ductal surfaces of organs such as the liver, pancreas and kidney, and are major contributors to the rheological properties of the barrier. Overproduction of mucins is an important factor in the morbidity and mortality of chronic airway diseases. The secreted mucins are encoded by a cluster of genes at the chromosomal locus 11p15 and exhibit amino acid homology [2]. A tumor suppressor function has been hypothesized based on the finding that genetic inactivation of the MUC2 gene induces the formation of intestinal tumors [3] while they are not known to contribute to tumor development.

The second subfamily comprises the membrane-tethered mucins. As secreted mucins, cell surface mucins can also be involved in the formation of the protective mucus gel through ektodomains that extend from the apical cell surface.

Mucins are large proteins characterized by a variable number of tandem repeats (VNTR) and heavy glycosylation of the extracellular domain. Like secreted mucins, membrane-bound mucins contain at least one mucin-like domain, characterized by a serine- and threonine-rich composition. Serine and threonine hydroxyl groups provide sites for the extensive O-glycosylation found on mature mucins in which up to 80% of the mass is O-glycans, with galactose, N-acetylgalactosamine, N-acetylglucosamine, and sialic acid as the main sugars with minor amounts of fucose, mannose, and glucose. The large number of O-glycans present in the ektodomain, together with the sheer size of the mucin tandem repeats domains, imparts to mucins a particular 'bottle brush' tertiary conformation [4] which protrudes from the cell surface at a remarkable distance (up to 500 nm).

Characteristically, an extracellular structural motif designated 'SEA' (sea urchin sperm protein, enterokinase and agrin) can be identified in mucins [5]. Proteins containing this module are generally highly O-linked glycosylated, adhesive proteins that are associated with the cell membrane. These proteins include single-pass transmembrane proteins, GPI-linked membrane proteins (the 63-kDa GPI anchored sea urchin sperm protein), and the protein enterokinase, which harbors a hydrophobic anchor sequence near its N-terminus. As proteins carrying the SEA module contain a considerable number of O-glycosidic-linked carbohydrates, it has been proposed that this common module might function in binding to neighboring carbohydrate residues. Moreover, in these proteins, the SEA module may function as a proteolytic cleavage site [6].

The membrane-bound mucins may be further subdivided in two distinct groups: small mucins and large mucins. Large mucins are both located at the cell surface and shed into the mucus gel, as their genes may encode splicing variants for secreted proteins or proteolysis to cleave and release the heavily O-glycosylated extracellular portion. The large released portion may contribute to the physical properties of mucus gels in contrast to small mucin molecules [7, 8].

15.3 Structure of MUC 16

MUC16 is a transmembrane mucin that corresponds to the CA125 antigen, a marker for ovarian cancer. MUC16 is a large protein with a molecular weight of 2.5 MDa in its unglycosylated form and a potential molecular weight of approximately 20 MDa when fully glycosylated.

The amino acid sequence of CA125/MUC16 resembles other mucins in having serine, threonine, and proline as major amino acids. However, its high content of leucine is characteristic as is the unusual length of the repeat units (156 amino acids). The presence of numerous lysine and arginine residues that are remote from the O-glycosylation sites is also interesting. Searching for conserved domains revealed the presence of many SEA domains in the protein structure.

15.3.1 Gene and Protein Structure

The *muc16* gene encodes a transmembrane-bound molecule named CA125. CA125 is a mucin-type O-linked glycoprotein whose gene was recently cloned by two independent groups and the studies published in the same year [9, 10]. In the genome, *muc16* is localized in the 19p13.2 chromosome and is a large (usually >10 kb) genomic DNA that contains repeated coding sequences in tandem which are typically subject to a length polymorphism due to Variation in the Number of Tandem Repeats (VNTR).

Structurally, three distinct domains can be identified: the Amino Terminal Domain (Domain I), the Tandem Repeat region (Domain II) and the Carboxy Terminal Domain (Domain III) (Figure 15.1). The largest part of the molecule is extracellular and only a short tail is cytoplasmic.

15.3.1.1 Domain I
The N-terminal domain comprises more than 1600 amino acids, is rich in serine/threonine residues, and accounts for the major O-glycosylation of MUC16. The amino terminal domain is encoded by five genomic exons covering about 13·250 bp. Exon 2 is extraordinarily large (amino acids 34–1593), while the other four exons are small (exon 1 1–33 amino acids (aa); exon 3 1594–1605 aa; exon 4 1606–1617 aa; exon 5 1618–1637 aa) [8].

15.3.1.2 Domain II
The MUC16 protein back bone is dominated by a tandem repeat region which contains more than 60 repeat SEA domains, according to the variant, each composed of 156 amino acids. Though all the individual repeat units are different, most of them occur more than once in the sequence. The amino acid sequences in the TRs are not perfectly conserved, although 81 positions contain conserved amino acids and certain motifs, e.g. GPLYSCRLTLLR, ELGPYTL, FTLNFTIXNL, and PGSRKFNXT are found in all or most of the TRs. Two closely spaced cysteine residues (20 amino

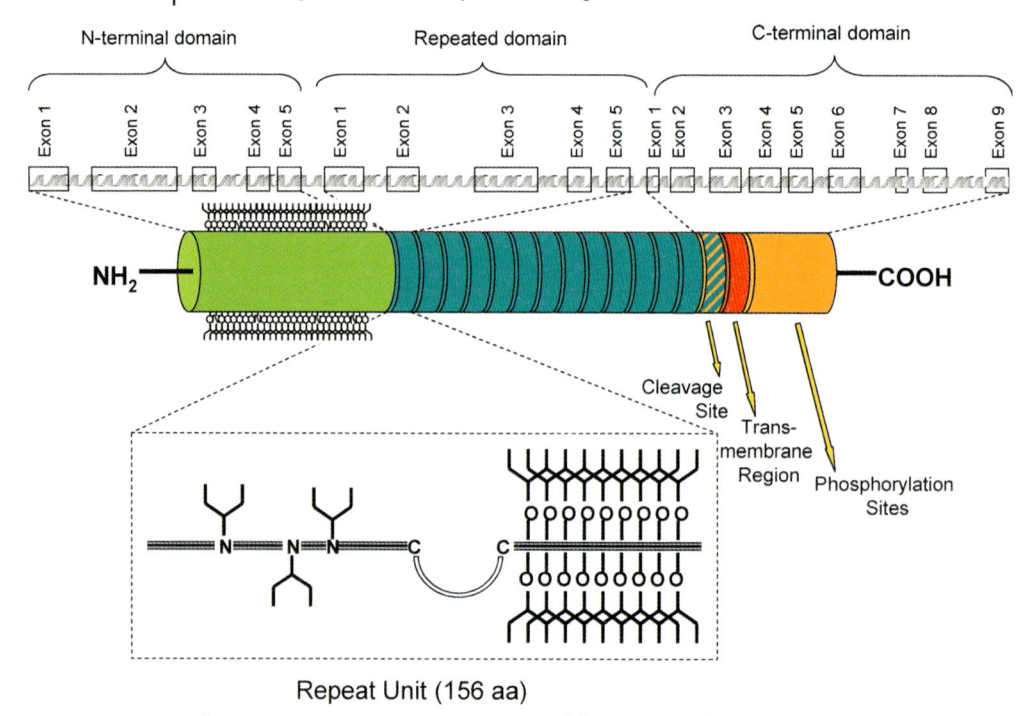

Figure 15.1 Diagrammatic representation of the structure of MUC16.

acids apart), which could form interchain disulfide bonded loops in the structure, are also perfectly conserved. Serine and threonine residues, representing potential O-glycosylation sites, are scattered throughout the sequence but blocks of clustered serine and threonine residues are evident in the TR region. These regions have adjacent or nearby proline residues, a motif that is frequently found in O-glycosylation sites. Numerous potential N-glycosylation sites (Asn-X-Ser/Thr, where X is any amino acid except Pro) are also found in the sequence, including two that are perfectly conserved in the TR region. It is unlikely, however, that the content of N-linked glycan chains in purified CA125 is very low [10].

Every unit of 156 amino acids encompasses N-glycosylation sites at the amino end, the cysteine-enclosed loop with the epitope binding sites and a region with highly O-glycosylated residues at the carboxy-end of the repeat. SEA domain forms a unique α/β sandwich fold composed of two α helices and four antiparallel β strands with the N and C termini on the same side of the molecule and has a characteristic turn known as the TY-turn between α1 and α2. The two conserved residues of the TY-turn, Thr-27 and Tyr-30, and the linker region between them, play an important role in maintaining the structure. The internal mobility is low throughout the domain. The residues that form the hydrophobic core and the TY-turn are fully conserved in all SEA domain sequences, indicating that the fold is common in the family. Two cysteine residues (Cys-57 and Cys-78) are located next to each other on the β2 and β3 strands and are close enough to form a disulfide bond. The peptide loop between the cysteine residues includes the nonoverlapping epitopes for anti-MUC16 antibodies [6].

Each repeat is composed of five exons covering approximately 1900 bases of genomic DNA. Exon 1 consists of 42 amino acids (1–42), exon 2, 23 amino acids (43–65), exon 3, 58 amino acids (66–123), exon 4, 12 amino acids (124–135) and exon 5, 21 amino acids (136–156).The antibody binding sites are inserted in the C-loop region encoded by exons 2 and 3. The functional nature of this region is supported by the fact that it does not present any N- or O-glycosylation sites which are extensively represented in the other exons [9].

15.3.1.3 Domain III

The carboxy terminal domain is a 284-amino acids sequence downstream from the repeat domain. As in the remainder of the protein, one short serine/threonine-rich sequence (PTSSSST) and numerous lysines are likewise found in the C-terminal region. Three major subdomains can be differentiated: an extracellular SEA domain (which does not have any homology to other known domains), a transmembrane sequence and cytoplasmic tail. The extracellular part of the carboxy-terminal domain has many N-glycosylation sites and some O-glycosylation sites. Several analyses suggested the presence of a cleavage site in the extracellular part at about 50 amino acids upstream from the transmembrane domain that may facilitate proteolytic cleavage and release of CA125. Nevertheless, the mechanism of CA125 secretion from primary ovarian tumors and permanent carcinoma cell lines is unclear. In fact, the other possibility is that alternatively spliced mRNAs are generated which lack the transmembrane region and as such code for secreted CA125. The sequence of hydrophobic amino acids located at positions 230–252 may represent the membrane-spanning region, as the same transmembrane motifs have been found in five other mucins (MUC-1, -3, -4, -12, and 13). The short cytoplasmic tail (31 amino acids) is characterized by a highly basic sequence adjacent to the membrane and contains a putative tyrosine phosphorylation site (RRKKEGEY). If CA125 is released after proteolytic cleavage at the extracellular SEA domain, the release is subsequent to phosphorylation. Phosphorylation can occur at serine/threonine sites and at tyrosine sites, all of which are represented in the cytoplasmic tail. Whether or not the tyrosine residue is phosphorylated in CA125 antigen is not known. Fendrick *et al.* [11] reported the presence of phosphate in CA125 from WISH cells by labeling with $^{32}PO_4^{3-}$ and immunoprecipitation analysis, but concluded that the phosphorylation sites are on serines or threonines. Hence, the possibility that CA125/MUC16 is phosphorylated on tyrosines and is involved in intracellular signaling needs further investigation [10]. The addition of the phosphate group creates a conformational change: CA125 is phosphorylated when it is in the cell (molecule in its membrane-bound state) and dephosphorylated prior to its release from the cell (secreted molecule) [11].

The carboxy-terminal domain covers more than 14 000 genomic bp and it is encoded by nine exons [9].

15.3.2
Evolutionary Considerations

Human MUC16 is unique among the membrane-bound mucins as it contains many SEA domains rather than a single domain. Chicken (*Gallus gallus*) has four consecutive SEA domains resembling MUC16. A phylogenetic tree was constructed with alignment of the SEA domain sequences of chicken and MUC16. The tree showed that the chicken SEA domains 1, 2 and 4 are closely related to the MUC16 SEA domains, indicating that human MUC16 and the related chicken gene evolved from a common ancestor. A phylogenetic tree was also constructed using the different MUC16 domain sequences. The resulting tree indicated that the multiple MUC16 SEA domains arose from repeated duplication events. The most ancient SEA domain that was duplicated and shuffled is located at the C-terminal in relation to the other SEA domains. Moreover, in contrast to other mucins such as MUC1, 3, 12, 13 and 17, MUC16 evolved separately from agrin [12].

15.4
Distribution of MUC16

The tissue distribution of MUC16/CA125 has been investigated by immunohistochemistry studies exploiting the availability of OC125, a murine monoclonal antibody specific for CA125. The CA125 antigen is associated with more than 80% of epithelial ovarian cancers of serous, endometrial, clear cell and undifferentiated types. Nevertheless, CA125 is also present in a number of normal adult and fetal tissues, as well as in other neoplasms of non-ovarian origin. In fetal tissues, MUC16 has been detected in amnion and in derivatives of the coelomic epithelium, such as the Müllerian epithelium and the lining cells of the peritoneum, pleura, and pericardium. Within adult tissues, the antigen is found in the epithelium of fallopian tubes, endometrium, and endocervix. MUC16 is expressed also in cells of mesothelial origin such as pleural, pericardial, and peritoneal cells, particularly in areas of inflammation and adhesion [13]. CA125 was detected in epithelia of kidney, lung, stomach, gall bladder, pancreas, and colon [14]. In neoplastic tissues of non-ovarian origin MUC16 is expressed in adenocarcinomas of the endocervix, endometrium, and fallopian tube and in mesotheliomas.

15.5
MUC16 Function in Normal Tissues

Membrane-associated mucins, such as MUC16, are multifunctional molecules combining classical mucin functions with a signaling function. O-glycosylation is crucial not only for structural aspects, but also to determine the function of the molecule. Mucin-type oligosaccharides are involved in specific ligand–receptor interactions, confer hydroscopic properties, and might bind various small molecules and proteins, as sialyl Lewis A and sialyl Lewis X in MUC1 which interact with different lectin-like molecules influencing the general properties of cell adhesion [15, 16].

Like all mucins, MUC16 is a component of the mucus layers and as such is involved in the maintenance of hydration and lubrication of the epithelial surface, protection from proteases, defence against pathogens, and formation of a disadhesive barrier. The reported extensive glycosylation of MUC16 provides a hydrophilic environment that is ideal for hydration and lubrification of epithelia. At the same time, the large protein core and dense sugar chains prevent access to the epithelium underneath. The sugar chains branch out from the axis of the molecule much like bristles on a brush creating an interlocking mesh. A network of mucins would create steric hindrance and thus a barrier to larger particles and provide a strongly negatively charged milieu around airway epithelia and thus a repulsive force to aid in the expulsion of bacteria. Mucins may have both anti- and pro-adhesive capacities. The extended and bulky structure of mucins disrupts close contact between cells while the peptide core and carbohydrate modifications provide epitopes for a variety of adhesion molecules. A pro-adhesive function has been attributed to MUC16 as a consequence of its ability to bind the tumor marker mesothelin [17].

Experimental studies support these findings. Using a human telomerase immortalized human corneal epithelial cell line in which MUC16 expression was stably knocked down using an siRNA approach, adherence of *Staphylococcus aureus* to epithelial cells increased, indicating that MUC16 originally prevented adhesion of pathogens. In addition, this mucin serves as a barrier against the potential penetration of agents into epithelial cells, since knockdown of MUC16 increases penetration of rose Bengal dye [18]. Another function is associated with the polybasic sequence of amino acids that bind to the ezrin/radixin/moesin (ERM) family of proteins. This association anchors the mucin to the actin cytoskeleton within microplicae on the surface of epithelial cells. The function of this association is as yet unclear, but it probably facilitates surface cell membrane folding [19].

The role of mucins in cancer is supposedly not very different from their role in normal cells. They protect cells from adverse growth conditions and control the local molecular microenvironment. Furthermore, interactions of the extracellular domain of MUC16 are implicated in facilitating metastases formation. In fact, ovarian cancer cells expressing MUC16 bind to cells transfected with mesothelin, which provides a mechanism for the heterocellular adhesion characteristic of metastatic disease [17].

15.6
Release of MUC16: CA125 Tumor Marker

15.6.1
History

The CA125 antigen was firstly described by Bast and colleagues in 1981 [20] who developed a murine monoclonal antibody (OC125) against the OVCA 433 cell line that was derived from the ascitic fluid of patients with serous papillary cystadenocarcinoma of the ovary. The antibody reacted with all epithelial ovarian carcinoma cell lines tested and with 60% of primary tumor tissue from ovarian cancer patients. By contrast, the antibody did not bind to a variety of non-malignant tissues, including adult and fetal ovary. OC125 reacts with only one of 14 cell lines derived from non-ovarian neoplasms and failed to react with cryostat sections from 12 non-ovarian carcinomas. Another antibody, M11, has been described to recognize a different epitope of the CA125 antigen in a separate antigenic domain.

15.6.2
Assay Methods

The CA125 antigen is located at the luminal cell surface, is actively secreted into the lumen, and then into the serum where it can be detected. The availability of monoclonal antibodies that recognize specifically the antigen and the proven overexpression of the antigen by ovarian carcinomas have motivated researchers to develop an adequate and sensitive assay method. Nevertheless, the use of CA125 for diagnostic purposes was complicated by the fact that the antigen is also significantly expressed in normal tissue, in benign disease such as endometriosis or pelvic inflammatory diseases and in amniotic fluid during gestation [21]. Furthermore, the monoclonal antibody OC125 appeared to react with both the normal and tumor-derived antigen. Using OC125 as both capture and labeled antibody, Bast et al. [20] developed an immunoradiometric assay for the quantification of CA125 levels in serum. In a typical assay, the unlabeled antibody is bound to a solid surface and the sample to be tested brought into contact with it. After a suitable time period of incubation, which is typically sufficient to allow formation of an antibody–antigen binary complex, a second antibody, labeled with a reporter molecule capable of producing a detectable signal, is then added. After washing steps the presence of the antigen is determined by observation of a signal produced by the reporter molecule. 'Reporter molecule' refers to a chemical compound that provides an analytical indicator and allows for the detection of antigen-bound antibody. The most commonly used reporter molecules are enzymes, fluorophores or radionuclide-containing molecules. When using an enzyme-based system the substrates to be used are generally chosen for their production of a detectable color change upon hydrolysis. It is also possible to employ fluorogenic substrates, which yield a fluorescent product. Alternately, fluorescent compounds may be chemically coupled to antibodies and the signal observed is fluorescence. Other reporter molecules include radioisotope, chemiluminescent, or bioluminescent molecules. The results may either be qualitative, by simple observation of the visible signal, or may be quantified by comparison with a control sample containing known amounts of antigen. Variations on the assay include a simultaneous assay, in which both sample and labeled antibody are added simultaneously to the bound antibody, or a reverse assay in which the labeled antibody and sample to be tested are first combined, incubated and then added to the unlabeled surface-bound antibody.

Recently, the CA125 assay was modified, i.e. a new antibody known as M11 replaced OC125 as the capture antibody. This modified assay, which became known as CA125 II, is now used for the routine determination of CA125.

15.6.3
Serum CA125 Levels in Healthy Subjects

With the assay developed by Bast and colleagues [22], 1% of 888 apparently healthy subjects (537 males and 351 females) were found to have serum levels of the marker > 35 kU/L. Since then, a threshold value of 35 kU/L has been widely adopted as the cut-off point for CA125. Nevertheless, many factors are known to influence serum CA125 levels in apparently healthy women [23] as listed in Table 15.1. These include the following.

15.6.3.1 Age
Multiple studies have shown that healthy pre-menopausal women have higher serum CA125 levels than post-menopausal women. This was evidenced by non-linearity in CA125 values with respect to age: relatively constant values up to the age of 45 years followed by a decrease up to age 55 years and a further stabilization after 55 years of age. The application of age-adjusted cut-off concentrations may further improve the identification of subgroups with a high risk for epithelial ovarian carcinoma [24].

Table 15.1 Main causes of elevated CA125.

Gynecologic origin	Non-gynecologic origin
Malignant conditions	
Ovaria	Lung
Endometrium	Pancreas
Endocervix	Breast
Fallopian tube	Colon
	Rectum
Non-malignant conditions	
Fibromas	Cirrhosis
Endometriosis	Hepatitis
Uterine myomas	Pancreatitis
Salpingitis	Tubercolosis
Menstruation	Renal failure
Pregnancy	Peritonitis
Pelvic inflammatory disease	Heart failure
	Previous surgery
	Peritoneas dialysis
	Collagen vascular diseases with pleural effusion

15.6.3.2 Menstrual cycle

In some women, serum levels of CA125 fluctuate throughout the menstrual cycle with higher values being observed at the time of menstruation than at other stages of the cycle [25].

15.6.3.3 Pregnancy

CA125 levels can also increase during pregnancy, especially during the first trimester, likely derived from the decidualized endometrium [14].

15.6.3.4 Race

Significantly higher levels were found in post-menopausal Caucasian women (median value, 14.2 kU/L) compared to either Asian (median value, 13 kU/L) or African women (median value, 9.0 kU/L) [25].

15.6.3.5 Other Factors that may affect CA125 Levels

Some studies showed that smoking and caffeine consumption decreased CA125 levels. Parity, hormone replacement therapy, use of oral contraceptives, and unilateral oophorectomy, however, do not appear to affect CA125 levels. In contrast to unilateral oophorectomy, hysterectomy significantly decreased CA125 levels [26]. Other factors that may alter serum CA125 levels include recent surgery, administration of mouse antibodies (e.g. radiolabeled antibodies against CA125), and the presence of endogenously produced antibodies [27].

15.6.4 Serum CA125 Levels in Patients with Benign Diseases

Multiple benign diseases both gynecological and non-gynecological can give rise to elevated serum levels of CA125. Benign gynecological disorders that may be associated with increased levels include endometriosis, fibromas, uterine myomas, acute and chronic salpingitis, pelvic inflammatory disease, and Meig's syndrome. Non-gynecological benign diseases shown to increase CA125 levels include liver cirrhosis, chronic active hepatitis, acute and chronic pancreatitis, and lung and pleural disease. Serum CA125 levels may also be elevated by ascites of benign origin or by any disorder that inflames the peritoneum, pericardium, or pleura [28].

15.6.5 Serum CA125 Levels in Non-ovarian Cancer

Elevated levels of CA125 can be observed in most types of adenocarcinoma, especially if distant metastases are present, including those derived from breast, colorectum, pancreas, lung, endometrium, cervix, and fallopian tube [28].

15.6.6 Serum CA125 Levels in Ovarian Cancers

Bast and colleagues [22] found elevated levels of CA125 (>35 kU/L) in 83 of 101 (82%) patients with ovarian cancer. Subsequent studies, however, reported that both the proportion of patients with elevated levels and the extent of elevation depended primarily on disease stage and histology type [27]. Jacobs and Bast [14] combined the data from 15 different studies and showed that CA125 levels were increased in 49 of 96 (50%) patients with FIGO stage I disease, 55 of 61 (90%) in stage II, 199 of 216 (92%) with stage III, and 77 of 82 (94%) in stage IV disease. In 14 out of 15 of these studies, the cut-off point used for CA125 was 35 kU/L, while in the remaining study it was 25 kU/L. Furthermore, combining data from 12 separate studies, it appeared that elevated levels were detected in 254 of 317 (80%) patients with tumors of serous type, 35/51(69%) with mucinous type, 39/52 (75%) with endometrial type, 28/36 (78%) with clear cell type, and 56/64 (88%) with undifferentiated type. Some but not all studies reported that serum CA125 levels were higher in patients with undifferentiated rather than differentiated ovarian cancers. As a result, the frequency of positivity is greater in patients with non-mucinous rather than mucinous tumors and in patients with high stage (FIGO II, III, or IV) rather than stage I disease.

15.7
Clinical Applications

15.7.1
Screening

Screening procedures for ovarian cancer include trans-abdominal ultrasound, trans-vaginal ultrasound with or without color Doppler and assessment of serum CA125 levels. Among these, assaying CA125 levels appeared to be the more attractive variant as it is non-invasive, comparably cheap, and widely available. Use of CA125 alone, however, has several limitations that compromise its utility. As previously mentioned, it has a low sensitivity for early or stage I disease and it lacks specificity especially in pre-menopausal women. This lack of sensitivity and specificity when combined with the low prevalence of ovarian cancer in the general population means that CA125 alone has a low positive predictive value (PPV) in detecting ovarian cancer in an asymptomatic population. An ovarian cancer screening strategy should achieve a minimum PPV of 10% and, to reach this value in a post-menopausal population screening, a diagnostic test alone or in combination with imaging techniques will need to show a minimum specificity of 99.6% [29].

Different strategies have been suggested to enhance the clinical utility of CA125 as a screening test [23].These include the assay of other markers in addition to CA125, sequential assays of CA125, and a combination of CA125 with ultrasound (i.e. multimodal screening). In a study carried out on 22·000 women aimed at the detection of ovarian cancer, Jacobs *et al.* reported a PPV for CA125 followed by ultrasound of 26% [30]. Another promising approach involves the elaboration of an algorithm incorporating the subject's age, rate of change in CA125 level, and absolute levels of CA125. This algorithm was based on the observation that CA125 levels increase in women with ovarian malignancy while constant or declining levels have been determined in other diseases. The algorithm calculates the best-fit line and higher risk of ovarian cancer is associated with a greater slope or intercept. In a study including more than 5000 subjects, a sensitivity of 83%, a specificity of 99.7%, and a PPV of 16% in predicting ovarian cancer in the year following the last screening was achieved [31]. The third approach involves the introduction of complementary markers such as tumor-associated trypsin inhibitor, CA19-9, CA72.4, and inhibin or others. Petricoin and colleagues [32] described the use of proteomics for the diagnosis of ovarian cancer. This methodology yielded a sensitivity of 100%, specificity of 95%, and a PPV of 94%. For comparison, the PPV for CA125 in the same masked patient group was 34%, but in this study, the population was artificially constructed to have a prevalence of ovarian cancer close to 50%. In a further study on proteomics, Zhang *et al.* [33] identified three markers that were altered in ovarian cancer and for which an immunoassay was available. The markers were apolipoprotein A1 (down-regulated), a truncated form of transthyretin (down-regulated), and a cleavage fragment of inter-alpha-trypsin inhibitor chain H4 (up-regulated). The sensitivity of the three markers combined with CA125 was 74% versus 65% for CA125 alone at a matched specificity of 97%. Although screening with CA125 can detect ovarian cancer in some asymptomatic women, there is currently no evidence as to whether screening reduces mortality. The EGTM Panel therefore recommends that screening for ovarian cancer in asymptomatic women without a family history of the disease using either CA125 alone or in combination with other modalities should not be carried out except in clinical trials.

15.7.2
Evaluating Pelvic Masses

In a multicenter prospective trial, 228 post-menopausal women with pelvic masses were evaluated with CA125, trans-vaginal ultrasound, and pelvic examination [34]. The accuracy in differentiating between benign and malignant lesions was almost identical with all the methods applied, 77% with CA125 (cut-off level, 35 kU/L), 76% with pelvic examination and 74% with ultrasound. The regression analysis of data indicated that the stronger predictor of disease diagnosis was pelvic examination followed by CA125 and ultrasound. These results established a role for CA125 in distinguishing malignant from benign pelvic masses especially in postmenopausal subjects. In 1990 Jacobs *et al.* [35] developed a risk of malignancy index incorporating CA125, sonography, and menopausal status: this index has demonstrated a sensitivity of 85% and a specificity of 97% in estimating the probability of malignant potential for a pelvic mass. The interpretation of these tests will improve the management of patients and facilitate referral to the best-qualified surgeon for definitive surgery.

15.7.3
Assessing Prognosis

The prognostic value of CA125 in ovarian carcinoma was evaluated by the Gynaecology Working Party of the Medical Research Council in the United Kingdom combining the results from 11 British institutions to give a total of 248 patients [36].

Three different models were followed to investigate the impact of CA125 levels on patient outcome during primary chemotherapy. These models were:

a. CA125 levels pre- and after one course of chemotherapy. The parameter considered was the CA125 half-life: cut-off, 20 days.
b. CA125 absolute value after two courses of chemotherapy: cut-off point, 35 kU/L.
c. CA125 comparison between pre-and after one course of chemotherapy values: cut-off, sevenfold decline.

The analysis of data showed that the absolute value of CA125 levels after two courses of chemotherapy was the best factor for predicting progression at 12 months; nevertheless, it gave a false positive rate of 19%. CA125 alone was not sufficiently accurate to be beneficial with respect to the management of such patients, while it should provide prognostic information during the early phases of first-line chemotherapy. In another study, the prognostic value of CA125 was found to be relevant only for the first year after surgery, but opinions are still conflicting [37].

15.7.4
Assessing Response to Therapy

The standard treatment for patients with advanced ovarian cancer is cyto-reductive surgery followed by chemotherapy with a taxoid and a platinum compound [38]. After this type of treatment many patients have low-volume disease that may not be palpable or detectable by radiologic procedures and assessing response can be difficult. In this situation the availability of a serum marker such as CA125 may be useful as serum levels are elevated in over 80% of patients with advanced ovarian cancer. Various studies have shown that CA125 levels increase with tumor progression and decrease with regression [28]. Nonetheless, a univocal definition of the alteration in marker levels that reliably correlates with response has not been yet reached. Rustin [39] proposed the following definitions: response occurred if there was either a 50% or 75% reduction in CA125 levels. The introduction of two different levels is due to the fact that sometimes only one criterion can be achieved in the assessment of response. In order to establish a definite response as characterized by a 50% reduction, four separate samples are required, i.e. two initial samples showing an elevation with two subsequent samples showing a decrease in levels of CA125. For a 75% response, only three samples are necessary, which must exhibit a serial decrease of at least 75% in levels of CA125. In both response conditions, the final sample must be taken at least 28 days after the previous sample. Patients with initial concentrations of CA125 less than 40 kU/L cannot be evaluated using these response definitions. The Gynecologic Cancer Intergroup (GCIG) has recently recommended a simpler CA125-based response definition to be used in evaluating therapies for relapsed ovarian cancer. The response is defined as a 50% reduction in CA125 levels with respect to the pre-treatment value which must be maintained for at least 20 days. Patients can be evaluated only if the pre-treatment sample is twice the normal upper limit and is taken within 2 weeks prior to the start of treatment. In assessing response to initial therapy, which is surgery and chemotherapy, both treatments lead to changes in CA125 levels and the effects caused by each therapy cannot be distinguished [40]. Nevertheless, EGTM recommends assaying CA125 levels for monitoring chemotherapy in order to aid patients' management. Levels or trends suggesting treatment failure should result in discontinuation of ineffective therapy, switch to an alternative therapy, or randomization in trials evaluating novel treatments. On the other hand, trends indicating response should encourage the continuation of potentially toxic chemotherapy.

15.7.5
Follow-up after Completion of Initial Therapy

Monitoring CA125 levels can be used to follow up patients who have been diagnosed with a malignancy and are apparently cured, as Tuxen [28] showed that the lead-time between increases in CA125 levels and clinical progression of disease varied from 1 to 15 months, the median lead-time being about 3–4 months. The possibility of pre-clinical detection of recurrent/metastatic disease may lead to the early administration of salvage chemotherapy, with the assumption that this will improve the patient's outcome. Multiple definitions of tumor progression with regard to CA125 levels have been described. In a study involving 255 patients, Rustin *et al.* [41] first reported that an increase in CA125 serum values to levels more than double the upper limit of normal would predict recurrence (sensitivity of 84% and a false-positive rate of < 2%). In another study of 88 patients, Rustin *et al.* confirmed that a doubling of CA125 levels from its nadir predicted progression with a sensitivity of 94% and specificity of almost 100% [42]. Although these definitions require further validation, they have been adopted by the GCIG for defining progression following first-line chemotherapy.

Although it is established that serial measurements of CA125 support early detection of relapsed ovarian cancer, the clinical value of this early warning is less clear, as recurrent ovarian cancer is usually incurable with existing treatments. Thus, the oncologist has to balance the potential advantage derived from the early administration of palliative chemotherapy, with the side-effects and subsequent decreased quality of life for the patient. A multivariate analysis carried out on 704 platinum pre-treated patients, showed that tumor size and number of disease sites (as well as serosal histology) were independent predictors of response to subsequent chemotherapy. A randomized trial is currently ongoing to ascertain whether administering chemotherapy based on CA125 elevation enhances outcome compared to the administration of chemotherapy based on clinical evidence of recurrence [43].

15.8
Novel Markers

Marker expression associated with ovarian cancer is rather heterogeneous among different patients. For this reason it is unlikely that a single marker would

Table 15.2 The most important new markers in ovarian carcinoma.

Marker	Reference
MUC1	[44]
HE4	[45]
Osteopontin	[46]
M-CSF	[47]
Lysophosphatidic acid	[48]
Kallikrein 6	[49]
Kallikrein 10	[50]
Mesothelin	[51]
Prostasin	[52]
CLDN3	[53]
VEGF	[44]

lead to the detection of different subtypes and stages with high sensitivity and specificity. The use of a combination of a panel of biomarkers may provide greater potential for the early detection of ovarian cancer. Gene expression in normal ovarian epithelial cells and in ovarian cancer cells was profiled and found to differ in relation to stage, grade and histotype thus leading to the identification of candidate markers. New candidates including mesothelin, M-CSF, lysophosphatidic acid (a lipid that is elevated in serum and ascites fluid), HE4, different kallikreins, prostasin, osteopontin, VEGF, and IL8, have all been identified using various technologies [56]. The most important markers are listed in Table 15.2.

For a subset of these markers specific antibodies were available, allowing analysis of protein expression in tissue arrays. An analysis of 158 ovarian cancer tissues showed that in each sample at least one of four protein markers (CA125, CLDN3, MUC1, and VEGF) was identified, and hence that a combination of a limited number of markers may suffice to identify >99% of epithelial ovarian cancers [44].

The multiplexed measurements of a combination of biomarkers may improve the sensitivity and specificity of the diagnosis of early-stage ovarian cancer. The analysis of a combination of markers requires novel statistical techniques such as the artificial neural network developed by Zhang et al. for early stage disease. At a fixed specificity of 98%, an artificial neural network including CA125, CA72-4, CA15-3, and M-CSF exhibited a sensitivity of 72%, whereas CA125 produced a sensitivity of 48% [54]. Skates et al. obtained similar results using mixtures of multivariate normal distributions [55].

Proteomic analysis also shows great promise, although its utility in early-stage disease remains uncertain. Two strategies can be applied in this field, the first aimed at the identification of a distinctive pattern of peptide expression in serum or urine while the second is aimed at the isolation of specific peptides and the development of individualized assays to analyze marker combinations.

15.9
Conclusions

CA125 is a very useful tool for clinicians as it is easily and reliably measurable in blood samples. Unfortunately, the values obtained are subject to various conditions, both benign and malignant, that hamper the interpretation of the results. These difficulties reduce the reliability of CA125 levels as a screening procedure, at least as a single test. Researchers are attempting to couple a tumor marker assay with other instrumental techniques in order to improve the sensitivity and specificity and have at their disposal a useful screening agent. On the other hand, measurement of CA125 levels is a very functional and simple procedure for assessing response to therapy. The utility of the method is conditional on the availability of a basal value that is higher than the normal upper limit, but provides a non-invasive method for obtaining important information to guide patient management. The CA125 assay offers the possibility of pre-clinical detection of recurrent/metastatic disease even though it has not been demonstrated that the early administration of chemotherapy will enhance outcome. Research is ongoing with regard to finding novel tumor markers associated with ovarian cancer, but it is unlikely that a single marker would be able to detect different subtypes and stages with high sensitivity and specificity. A combination of different biomarkers may provide greater potential for the early detection of ovarian cancer in the future.

References

1 Thornton, D.J., Rousseau, K. and McGuckin, M.A. (2008) Structure and function of the polymeric mucins in airways mucus. *Annual Review of Physiology*, **70**, 459–486.

2 Desseyn, J.L., Buisine, M.P., Porchet, N., Aubert, J.P., Degand, P. and Laine, A. (1998) Evolutionary history of the 11p15 human mucin gene family. *Journal of Molecular Evolution*, **46**, 102–106.

3 Velcich, A., Yang, W., Heyer, J., Fragale, A., Nicholas, C., Viani, S., Kucherlapati, R., Lipkin, M., Yang, K., Augenlicht, L. (2002) Colorectal cancer in mice genetically deficient in the mucin MUC2. *Science*, **295**, 1726–1729.

4 Gipson, I.K., Hori, Y. and Argueso, P. (2004) Character of ocular surface mucins and their alteration in dry eye disease. *Ocul Surf*, **2**, 131–148.

5 Bork, P. and Patthy, L. (1995) The SEA module: a new extracellular domain associated with O-glycosilation. *Protein Science: A Publication of the Protein Society*, **4**, 1421–1425.

6 Maeda, T., Inoue, M., Koshiba, S., Yabuki, T. and Aoki, M. (2004) Solution structure of the SEA domains from the murine homologue of ovarian cancer antigen CA125 (MUC16). *The Journal of Biological Chemistry*, **279**, 13174–13182.

7 Desseyn, J.L., Tetaert, D. and Gouyer, V. (2008) Architecture of the large membrane-bound mucins. *Gene*, **410**, 215–222.

8 Duraisamy, S., Ramasamy, S., Kharbanda, S. and Kufe, D. (2006) Distinct evolution of the human carcinoma-associated transmamebrane mucins, MUC1, MUC4 and MUC16. *Gene*, **373**, 28–34.

9 O'Brien, T.J., Beard, J.B., Underwood, L.J., Dennis, R.A., Santin, A.D. and York, L. (2001) The CA125 gene: an extracellular superstructure dominated

by repeat sequences. *Tumor Biology*, **22**, 348–366.

10. Yin, T.W.T. and Lloyd, K.O. (2001) Molecular cloning of the CA125 ovarian cancer antigen Identification as a new mucin (MUC16). *The Journal of Biological Chemistry*, **276**, 27371–27375.

11. Fendrick, J.L., Konishi, I., Geary, S.M., Parmley, T.H., Quirk, J.G. and O'Brien, T.J. (1997) CA125 phosphorilation is associated with its secretion from the WISH human amnion cell line. *Tumor Biology*, **18**, 278–289.

12. Duraisamy, S., Ramasamy, S., Kharbanda, S. and Kufe, D. (2006) Distinct evolution of the human carcinoma-associated transmembrane mucins, MUC1, MUC4 AND MUC16. *Gene*, **373**, 28–34.

13. Kabawat, S.E., Bast, R.C., Bhan, A.K., Welch, W.R., Knapp, R.C. and Colvin, R.B. (1983) Tissue distribution of a coelomic-epithelium-related antigen recognized by the monoclonal antibody OC125. *International Journal of Gynecological Pathology*, **2**, 275–285.

14. Jacobs, I. and Bast, R.C. (1989) The CA125 tumour-associated antigen: A review of the literature. *Human Reproduction (Oxford, England)*, **4**, 1–12.

15. Hollingsworth, M.A. and Swanson, B.J. (2004) Mucins in cancer: protection and control of the cell surface. *Nature Reviews. Cancer*, **4**, 45–60.

16. McDermott, K.M., Crocker, P.R., Harris, A., Burdick, M.D., Hinoda, Y., Hayashi, T., Imai, K. and Hollingsworth, M.A. (2001) Overexpression of MUC1 reconfigures the binding properties of tumor cells. *International Journal of Cancer*, **94** (6), 783–791.

17. Hattrup, C.L. and Gendler, S.J. (2008) Structure and function of the cell surface (tethered) mucins. *Annual Review of Physiology*, **70**, 431–457.

18. Blalock, T., Spurr-Michaud, S., Tisdale, A., Heimer, S., Gilmore, M., Ramesh, V. and Gipson, I. (2007) Functions of MUC16 in corneal epithelial cells. *Investig Ophtalmol Vis*, **48**, 4509–4518.

19. Perez, B.H. and Gipson, I.K. (2008) Focus on molecules: human mucin MUC16. *Experimental Eye Research*, doi:10.1016/j.exer.2007.12.008

20. Bast, R.C., Feeney, M., Lazarus, H., Nadler, L.M., Colvin, R.C. and Knapp, R.C. (1981) Reactivity of a monoclonal antibody with human ovarian carcinoma. *The Journal of Clinical Investigation*, **68**, 1331–1337.

21. Hardardottir, H., Parmley, T.H., 2nd, Quirk, J.G., Jr, Sanders, M.M., Miller, F.C. and O'Brien, T.J. (1990) Distribution of CA 125 in embryonic tissues and adult derivatives of the fetal periderm. *American Journal of Obstetrics and Gynecology*, **163**, 1925–1931.

22. Bast, R.C., Klug, T.L., St John, E., Jenison, E., Niloff, J.M., Lazarus, H., Berkowitz, R.S., Leavitt, T., Griffiths, C.T. and Parker, L. (1983) A radioimmunoassay using a monoclonal antibody to monitor the course of epithelial ovarian cancer. *The New England Journal of Medicine*, **309**, 883–887.

23. Duffy, M.J., Bonfrer, J.M., Kulpa, J., Rustin, G.J.S., Soletormos, G., Torre, G.C., Tuxen, M.K. and Zwirnery, M. (2005) CA125 in ovarian cancer: European Group on Tumor Markers guidelines for clinical use. *International Journal of Gynecological Cancer*, **15**, 679–691.

24. Bonfrer, J.M.G., Korse, C.M., Verstraeten, R.A. *et al.* (1997) Clinical evaluation of the Byk LIA-mat CA125 II assay: discussion of a reference value. *Clinical Chemistry*, **43**, 491–497.

25. Grover, S., Koh, H., Weideman, P. and Quinn, M.A. (1992) The effect of menstrual cycle on serum CA125 levels: a population study. *American Journal of Obstetrics and Gynecology*, **167**, 1379–1381.

26. Pauler, D.K., Menon, U., McIntosh, M. *et al.* (2001) Factors influencing serum CA125II levels in healthy postmenopausal women. *Cancer Epidemiology, Biomarkers & Prevention: A Publication of the American Association for Cancer Research, Cosponsored*

by the *American Society of Preventive Oncology*, **10**, 489–493.

27 Turpeinen, U., Lehtovirta, P., Afthan, H. and Stenman, U.H. (1990) Interference by human anti-mouse antibodies in CA125 assay after immunoscintigraphy: anti-idiotypic antibodies not neutralised by mouse IgG but removed by chromatography. *Clinical Chemistry*, **36**, 1333–1338.

28 Tuxen, M.K. (2001) Tumor marker CA125 in ovarian cancer. *Journal of Tumor Marker Oncology*, **16**, 49–68.

29 Gagnon, A. and Ye, B. (2008) Discovery and application of protein biomarkers for ovarian cancer. *Current Opinion in Obstetrics & Gynecology*, **20**, 9–13.

30 Jacobs, I., Davies, A.P., Bridges, J. et al. (1993) Prevalence screening for ovarian cancer in postmenopausal women by CA125 measurements and ultrasonography. *British Medical Journal*, **306**, 1030–1034.

31 Skates, S.J., Xy, F..-J., Yu, Y..-H. et al. (1995) Towards an optimal algorithm for ovarian cancer screening with longitudinal tumor markers. *Cancer*, **76**, 2004–2010.

32 Petricoin, E.F., III, Ardekani, A.M., Hitt, B.A., Levine, P.J., Fusaro, V.A. and Steinberg, S.M. (2002) Use of proteomic patterns in serum to identify ovarian cancer. *Lancet*, **359**, 572–577.

33 Zhang, Z., Bast, R.C., Jr Yu, Y. et al. (2004) Three biomarkers identified from serum proteomic analysis for the detection of early stage ovarian cancer. *Cancer Research*, **64**, 5882–5890.

34 Schutter, E.M.J., Kenemans, P., Sohn, C. et al. (1994) Diagnostic value of pelvic examination, ultrasound and serum CA125 in post-menopausal women with a pelvic mass. *Cancer*, **74**, 1398–1406.

35 Jacobs, I., Oram, D., Fairbanks, J., Turner, J., Frost, C. and Grudzinskas, J.G. (1990) A risk of malignancy index incorporating CA 125, ultrasound and menopausal status for the accurate preoperative diagnosis of ovarian cancer. *British Journal of Obstetrics and Gynaecology*, **97**, 922–929.

36 Fayers, P.M., Rustin, G., Wood, R. et al. (1993) The prognostic value of serum CA125 in patients with advanced ovarian carcinoma: an analysis of 573 patients by the Medical Research Council Working Party on Gynaecological Cancer. *International Journal of Gynecological Cancer*, **3**, 285–292.

37 Clark, T.G., Stewart, M.E., Altman, D.G., Gabra, H. and Smyth, J.F. (2001) A prognostic model for ovarian cancer. *British Journal of Cancer*, **85**, 944–952.

38 Berek, J.S., Bertelsen, K., du Bois, A. et al. (1999) Advanced epithelial ovarian cancer: 1998 consensus statements. *Annals of Oncology*, **10**, (Suppl. 1) S87–92.

39 Rustin, G.J.S. (2003) Use of CA125 to assess response to new agents in ovarian cancer trials. *Journal of Clinical Oncology*, **21**, 187s–s193.

40 Rustin, G.J.S., Quinn, M., Thigpen, T. et al. (2004) Re: new guidelines to evaluate the response to treatment in solid tumors (ovarian cancer). *Journal of the National Cancer Institute*, **96**, 487–488.

41 Rustin, G.J.S., Nelstrop, A.E., McClean, P. et al. (1996) Defining response of ovarian carcinoma to initial chemotherapy according to serum CA125. *Journal of Clinical Oncology*, **14**, 1545–1551.

42 Rustin, G.J.S., Marples, M., Nelstrop, A.E. et al. (2001) Use of CA125 to define progression of ovarian cancer in patients with persistently elevated levels. *Journal of Clinical Oncology*, **10**, 4054–4057.

43 Eisenhauer, E.A., Vermorken, J.B. and van Glabbeke, M. (1997) Predictors of response to subsequent chemotherapy in platinum pretreated ovarian cancer: a multivariate analysis of 704 patients. *Annals of Oncology*, **8**, 963–968.

44 Lu, K.H., Patterson, A.P., Wang, L., Marquez, R.T., Atkinson, E.N., Baggerly, K.A. et al. (2004) Selection of Potential Markers for Epithelial Ovarian Cancer with Gene Expression Arrays and Recursive Descent Partition Analysis. *Clinical Cancer Research*, **10**, 3291–3300.

45 Schummer, M., Ng, W.V., Bumgarner, R.E. et al. (1999) Comparative hybridization of an array of 21,500 ovarian cDNAs for the discovery of genes overexpressed in ovarian carcinomas. *Gene*, **238**, 375–385.

46 Kim, J.H., Skates, S.J., Uede, T. et al. (2002) Osteopontin as a potential diagnostic marker for ovarian cancer. *The Journal of the American Medical Association*, **287**, 1671–1679.

47 Xu, F.J., Ramakrishnan, S., Daly, L. et al. (1991) Increased serum levels of macrophage colony-stimulating factor in ovarian cancer. *American Journal of Obstetrics and Gynecology*, **165**, 1356–1362.

48 Xu, Y., Shen, Z., Wiper, D. et al. (1998) Lysophosphatidic acid as a potential biomarker for ovarian and other gynaecological cancers. *The Journal of the American Medical Association*, **280**, 719–723.

49 Diamandis, E.P., Yousef, G.M., Soosaipillai, A.R. and Bunting, P. (2000) Human kallikrein 6 (zyme/protease M/neurosin): a new serum biomarker of ovarian carcinoma. *Clinical Biochemistry*, **33**, 579–583.

50 Luo, L.Y., Bunting, P., Scorilas, A. and Diamandis, E.P. (2001) Human kallikrein 10: a novel tumor marker for ovarian carcinoma? *Clinica Chimica Acta; International Journal of Clinical Chemistry*, **306**, 111–118.

51 Scholler, N., Fu, N., Yang, Y. et al. (1999) Soluble member(s) of the mesothelin/megakaryocyte potentiating factor family are detectable in sera from patients with ovarian carcinoma. *Proceedings of the National Academy of Sciences of the United States of America*, **96**, 11531–11536.

52 Mok, S., Chao, J., Skates, S. et al. (2001) Prostasin, a potential serum marker for ovarian cancer: identification through microarray technology. *Journal of the National Cancer Institute*, **93**, 1458–1464.

53 Rangel, L.B., Agarwal, R., D'Souza, T., Pizer, E.S., Alò, P.L., Lancaster, W.D., Gregoire, L., Schwartz, D.R., Cho, K.R. and Morin, P.J. (2003) Tight junction proteins claudin-3 and claudin-4 are frequently overexpressed in ovarian cancer but not in ovarian cystadenomas. *Clinical Cancer Research*, **9**, 2567–2575.

54 Zhang, Z., Yu, Y., Xu, F., Berchuck, A., van Haaften-Day, C., Havrilesky, L.J., de Bruijn, H.W., van der Zee, A.G., Woolas, R.P., Jacobs, I.J., Skates, S., Chan, D.W. and Bast, R.C. Jr. (2007) Combining multiple serum tumor markers improves detection of stage I epithelial ovarian cancer. *Gynecologic Oncology*, **107**, 526–531.

55 Skates, S.J., Horick, N., Yu, Y. et al. (2004) Pre-operative sensitivity and specificity for early stage ovarian cancer when combining CA 125, CA 15.3, CA 72.4 and M-CSF using mixtures of multivariate normal distributions. *Journal of Clinical Oncology*, **22**, 4059–4066.

56 Bast, R.C., Badgwell, D., Lu, Z., Marquez, R., Rosen, D., Liu, J., Baggerly, K.A., Atkinson, E.N., Skates, S., Zhang, Z., Lokshin, A., Menon, U., Jacobs, I. and Lu, K., (2005) New tumor markers: CA125 and beyond. *International Journal of Gynecological Cancer*, **15** (Suppl. 3), 274–281.

Part Four
Clinical Applications of TAAs

16
Overview of Cancer Vaccines
John Copier and Angus Dalgleish

16.1
Introduction

It is now well established that the immune system can be induced to reject tumors. Vaccines can elicit immune responses that target tumor-associated antigens (TAAs) and thus act as therapeutic modalities for the treatment of existing cancers. The effect of stimulating immune responses against cancer was first observed by William Coley, a surgeon working in New York in the early 1900s, who recognized that infection of a recurrent sarcoma led to its subsequent regression [1]. Coley developed the first treatment that might be recognized as a cancer vaccine when he treated sarcoma patients with a mixture of *Streptococcus pyogenes* and *Serrelia marcescens*. Subsequently Coley observed the importance of inducing a fever, indicating an active immune response. Coley's vaccine appears to have induced an anti-tumor response which was more likely to be mediated by a non-tumor-specific activation. It is probable that many bacterial preparations, such as those used in adjuvants (e.g. mycobacterial preparations), are stimulating immune cells through Pattern Recognition Receptors (PRRs) like Toll-like receptors (TLRs), which recognize chemical characteristics of molecules uniquely expressed in bacterial and viral pathogens. However, over the past two decades much research has been carried out to characterize tumor-specific antigens that may be targeted by the immune system (much of this has been described in earlier chapters) and to establish principles for optimum delivery of cancer vaccines. In this chapter we describe the technologies, which have been developed for the effective stimulation of tumor-specific immune responses and discuss the advances made towards the clinical use of these technologies.

16.2
Cellular Immune Responses in Tumor Rejection

Broadly speaking the specific immune response is divided into the humoral (antibody) and cellular (T lymphocyte) responses. Although early cancer vaccines

Tumor-Associated Antigens. Edited by Olivier Gires and Barbara Seliger
Copyright © 2009 WILEY-VCH Verlag GmbH & Co. KGaA, Weinheim
ISBN: 978-3-527-32084-4

specifically targeted antibody responses, mostly through the injection of recombinant TAAs, these approaches fell into decline as few studies demonstrated efficacy of humoral responses in humans. However, advances in the production of recombinant, humanized monoclonal antibodies mAb have renewed interest in this modality. These have shown clinical efficacy and FDA approved antibodies are available for the treatment of a variety of tumors including breast cancer (Trastuzumab), colorectal cancer (Cetuximab, Bevacizumab) and B cell lymphoma (Rituximab) [2]. These drugs, although clearly working through immunological mechanisms, are not strictly vaccines despite recent evidence to suggest that they may trigger anti-idiotype responses.

16.3
Selection of TAAs for the Development of Cancer Vaccines

Ideally TAAs selected for use in vaccination are specific to the tumor and not to the host (i.e. those that are not subject to central tolerance). Thus antigens that are uniquely expressed within the tumor environment (e.g. developmental antigens that are not expressed post-natally, antigens that carry unique mutations, for example mutations of p53, or those that have very limited tissue expression such as the cancer-testis antigens) represent the most likely candidates. Another class of antigens that may be considered are those that are over-expressed by the tumor (e.g. HER-2 neu) or those that are post-translationally modified resulting in the generation of new epitopes (e.g. MUC-1).

Although there is rare evidence that the immune response can spontaneously reject tumors, in the majority of individuals there exist mechanisms employed by the tumor to evade or suppress the immune response. Several of these are important in terms of the selection of specific TAAs for the development of a vaccine. The selection of antigen has become an important determinant in the subsequent long-term efficacy of the vaccine since tumors are capable of altering the expression profile of their antigens depending on factors such as stage and metastatic status. Substantial differences can be demonstrated between primary and metastatic sites and even between metastases [3, 4]. Thus, in designing vaccines for the treatment of a tumor it is necessary to ascertain the optimum, most stably expressed TAA or spectrum of TAAs for formulation of the vaccine. This becomes an issue when targeting single TAAs, for example those expressing a mutation in a single antigen. The strategies used to overcome these issues will be discussed below in the context of different vaccination strategies.

As has been described in earlier chapters the spectrum of TAAs varies significantly between different tumor types. However, biomarker research is now determining the essential differences and similarities in antigen expression between different tumors [5]. It is interesting to speculate that future vaccines may be developed for families of tumors expressing similar antigen profiles.

In order to promote cellular responses TAA must have epitopes that can be presented through the patient's HLA class I and II molecules. Thus vaccine design must take into account the availability of HLA-restricted epitopes in a TAA and the variability of HLA types within a population. It is for this last reason that most vaccines concentrate on epitopes restricted to HLA-A2, since this is the most common HLA type in the Caucasian population. However, knowledge of the patient's HLA type and the determination of relevant HLA-restricted epitopes in the vaccine are important.

As with TAAs, tumors are also able to downregulate the expression of HLA and thus escape the cellular immune response [6]. In fact one possible consequence of monovalent (single antigen/epitope) vaccine approaches is the positive selection of antigen-escape or HLA-escape phenotypes. This will be discussed further below in the context of individual vaccine approaches.

16.4
Types of Cancer Vaccine

Vaccines for cancer fall into two reasonably distinct categories; those with defined TAAs and those relying on undefined TAAs. Here the current technologies for the delivery of cancer vaccines and the relative merits of using defined and undefined antigens will be discussed. Vaccines with demonstrable clinical benefit will be described.

16.4.1
Defined Antigen Approaches

Approaches using defined antigens potentially suffer from the problems of antigen escape and HLA class I surface downregulation described above.

16.4.1.1 Peptides
There has been considerable preclinical research, which demonstrates the efficacy of peptide vaccination in raising immune responses and in eliciting tumor rejection. Despite this, a number of technical challenges exist in bringing these to the clinic. Peptides designed to stimulate cytotoxic T-cell responses must have a defined HLA class I restriction. Furthermore, peptides are largely non-immunogenic and vaccination with peptide requires the use of adjuvant to stimulate cellular responses. Most clinical trials using peptides have concentrated on standard immunological adjuvants. However, in recent years trials have demonstrated the efficacy of other combinations which enhance peptide immunogenicity. Thus in melanoma patients T-cell responses to peptides increase when peptide is given with incomplete Freund's adjuvant and in the presence of either GM-CSF or CpG 7909 (a TLR agonist that has been shown to enhance immune responses to hepatitis B vaccine) [7, 8]. New generations of adjuvants based on cytokines or TLR agonists may lead to more effective responses to peptides in future generations of vaccine.

Despite their lack of immunogenicity peptide vaccines have demonstrated clinical responses in a number of cases. For example in melanoma vaccines a degree of clinical efficacy has been demonstrated using p53 or K-ras peptides [9], a mixture of gp-100 and tyrosinase peptides with different HLA restrictions [10], and a mixture of HLA class I- and class II-restricted NY-ESO peptides [11].

Peptides may also be used in combination with dendritic cells (DCs). Studies in which DCs are loaded with class I binding epitopes have demonstrated immunological responses in several cancer types including colorectal cancer and melanoma [12, 13].

16.4.1.2 Recombinant Protein

Unlike peptides recombinant proteins can be made for selected TAAs, which contain a broader spectrum of epitopes suitable for a wider range of HLA-class I molecules. It is also likely that recombinant proteins provide class II epitopes which provide $CD4^+$ T cell help. HLA-binding epitopes have been described preclinically for many TAAs including MUC1, HER-2, NY-ESO, and MAGE-3 and it has further been demonstrated that these TAAs can elicit cellular and humoral responses [14–16]. The assumption underlying such responses is that DCs at the injection site take up and process recombinant antigen for cross presentation on HLA class I molecules. In fact, the immunogenicity of recombinant proteins is usually low and requires the addition of adjuvant (see discussion of MAGE-3 vaccine below).

Recombinant proteins have been shown to have clinical efficacy in melanoma. In one study recombinant NY-ESO, a cancer-testis antigen, was used in combination with a saponin-based adjuvant (ISCOMATRIX) in a phase I clinical trial in melanoma patients. No objective responses were reported in this trial but it appears that patients in a placebo group or recombinant alone had a poorer clinical outcome than those treated with recombinant plus adjuvant [14].

Recombinant proteins offer the possibility of modification and manipulation such that immune responses can be improved. GlaxoSmithKline (GSK) is developing a vaccine based on a recombinant fusion protein consisting of MAGE-A3 and lipidated protein D (a protein derived from *H. influenzae*, which is incorporated into the fusion protein to provide an enhanced immunogenicity and helper epitopes). A MAGE-3 recombinant vaccine in combination with AS02B adjuvant (an oil-in-water emulsion of monophosphoryl A and QS21) has been shown to elicit cellular and humoral responses in metastatic melanoma patients [16]. Among 33 patients in a phase I/II trial receiving vaccination with recombinant MAGE-3 plus adjuvant, five patients had objective responses (two partial responses, two mixed responses and one stable disease) [17]. GSK have expanded these studies to examine the efficacy of the MAGE fusion protein injected intradermally and subcutaneously in the absence of adjuvant [18]. These studies demonstrated the importance of including an adjuvant since none of the patients raised an IgG response to MAGE-3. $CD4^+$ T cell responses were also limited. Despite this a number of patients had clinical responses; one partial response and four mixed responses. Moreover, one patient who was exposed to a 'maintenance regime' (a second cycle of MAGE-3 fusion protein vaccinations interspersed with MAGE-2, HLA-A2-restricted peptide vaccine) demonstrated a complete regression. Whether this reflects the use of combined vaccination protocols or simply persistent, long-term vaccination remains unclear.

A modified protein is also being generated for HER-2 vaccination. Thus, although previous clinical responses to HER-2 recombinant vaccines have been poor, new approaches are being used to develop more immunologically active HER-2 vaccines. Pharmexa are making recombinant HER-2 protein into which helper epitopes from tetanus toxoid have been engineered [19].

As has been described for peptides, recombinant protein can be loaded *ex-vivo* onto autologous DCs for use as a cellular vaccine. In this context it is noteworthy that the use of recombinant protein to pulse DCs has led to the most advanced clinical trial to date for a cancer vaccine. Dendreon Corporation has recently completed phase III trials for sipuleucel-T for hormone-refractory prostate cancer. Sipuleucel-T is a vaccine in which autologous DCs have been loaded with a recombinant fusion protein composed of prostatic acid phosphatase (PAP) and GM-CSF. Thus while the PAP acts as a TAA, GM-CSF targets the recombinant protein to DCs. In a randomized, double-blind phase III trial the primary endpoint (time-to-progression) failed to achieve statistical significance [20]. However, evaluation of the overall survival strongly suggested that there might be a significant clinical benefit from this vaccine. Based on these data Dendreon has begun a new phase III trial with overall survival (OS) as a primary endpoint. Data are expected in 2010. It is worth noting that all these patients had a taxotere-based chemotherapy on progression and hence the vaccine may be priming for a better chemotherapy outcome.

16.4.1.3 DNA

DNA vaccines are based on the cloning and expression of the cDNA for TAAs into plasmid vectors. The mechanism of efficacy of this vaccination strategy is unclear. Plasmids are delivered into the tissue (by injection or by 'gene-gun' type methods) where they are either directly taken up and expressed by APCs or are expressed in tissues and subsequently cross-presented by DCs. DNA vaccines have a number of key advantages: they are easy to produce in bulk and relatively cheap to purify to Good Manufacturing Practice (GMP) standards. Furthermore, due to their bacterial origin, they contain unmethylated CpGs, which act to stimulate TLRs and it is postulated that this enhances immune responses raised by DNA vaccination. Many of the arguments used for the implementation of recombinant proteins are also applicable here since DNA vaccines can be used to encode whole TAAs but can also be used to deliver multiple epitopes.

It is well established that immune responses can be raised in cancer patients to a number of plasmid-encoded TAAs including PSA [21, 22] and CEA [23]. However, not all DNA cancer vaccines have shown such promise with some studies demonstrating a complete lack of immune responses to MART-1 [24] and gp-100 [25]. The essential differences between these studies remain unclear. Although very effective in murine models, the activity of DNA vaccines is very elusive in humans. It is unclear whether this is a dose-related issue or whether there is a requirement for better delivery, for example through microneedle-based injection systems, 'gene gun' type, or electroporation-based mechanisms. Routes of vaccination and methodology for vaccination are under investigation in an attempt to improve delivery of

DNA vaccines to APCs [26]. Moreover, DNA vaccines are still in the developmental stage and the best combinations of vector and adjuvant have yet to be clarified. DNA vaccines have been shown to raise both humoral and cellular responses, but so far none of the trials has been able to demonstrate substantial clinical responses [23, 25, 27].

16.4.1.4 Viral Vectors

TAAs may be cloned into viral vectors for expression in human tissues. Thus, the virus can deliver TAA genes to tissues or APCs *in vivo* where they are expressed and presented on MHC class I molecules. A number of different vectors have been developed for this purpose, most of which are derived from Pox viridae family of viruses including ALVAC (canary pox), vectors based on fowl pox and MVA (modified vaccinia Ankara). Preclinical experiments have demonstrated the ability of viral vectors to induce immune responses against TAAs suggesting that these may form the basis of an effective cancer vaccine [28, 29].

Viral vaccines have several noteworthy advantages. As described for recombinant vectors (above) viruses can carry genes for the expression of whole proteins. In some cases several genes have been engineered into viral vectors for coordinate expression (e.g. IL-2 and MUC-1 [30]). Viruses also bear moieties which stimulate through TLRs (e.g. double stranded RNA) and theoretically help to elicit stronger immune responses.

Clinical trials have been carried out in a variety of cancer backgrounds including colorectal cancer, melanoma, and breast carcinoma. In the majority of these trials viral vaccines induced either cellular or humoral responses to the target antigen and a number of trials have demonstrated clinical responses. Although a complete review of the field is beyond the scope of this chapter (for alternative reviews see [26, 31]) a few trials are worth mentioning to exemplify the progress in this field.

An MVA virus is under development by Oxford Biomedica (TroVax) for the treatment of metastatic colorectal cancer. TroVax encodes the 5T4 antigen which is found in colorectal, ovarian, gastric and renal cancers. In a phase I/II trial this vector induced cellular responses to 5T4 in 16 of 17 evaluable patients and antibody response in 14/17 patients [32]. Although not a primary endpoint in this study, a retrospective analysis of the time-to-progression demonstrated a potential correlation between strong antibody responses to 5T4 and clinical outcome.

One major disadvantage of viral vectors is their tendency to elicit immune responses, and particularly neutralizing antibodies, against the virus itself. In response to this problem the concept of using mixed vaccines for priming and boosting has emerged. One example of this approach is that of van Baren *et al.* who demonstrated immunological and clinical responses to an ALVAC vector encoding MAGE-1 and MAGE-3 epitopes [33]. In this vaccination strategy the first four vaccinations used the ALVAC-MAGE vector and this was followed by three vaccinations with MAGE-1 and -3 peptides. This strategy gave rise to cellular responses against MAGE-3 amongst a majority of patients. Of 30 evaluable melanoma patients one showed a partial response, two had stable disease of greater than 6 months and four patients had mixed responses.

A fowl pox vector for the treatment of metastatic melanoma was developed by Therion Biologics Corporation. A variety of gp-100 constructs have been tested in clinical trials either with or without addition of IL-2 [34]. A vaccination regime using a fowl pox vector, containing a minigene construct of gp-100 in which HLA-A2 epitopes were modified to increase binding to MHC, followed by high dose IL-2, induced regressions in six out of 12 patients treated, with three of these being complete responses.

Arlen et al. carried out trials in prostate cancer using a PSA-based vaccine in a prime–boost sequence using human- and fowl-based pox vectors given before endocrine therapy [35]. Results have been encouraging with progression free survival (PFS) being significantly longer than that observed in those patients randomized to receive standard endocrine treatment followed by vaccine on progression.

Perhaps the most advanced study of viral vectors is that of the Therion vaccine PANVAC-VF, which encodes the TAAs MUC-1 and CEA, but also the costimulatory molecules ICAM-1, B7.1, and LFA-3 (which were also used in the above-mentioned Arlen et al. study [35]). The initial promise of phase I/II trials, which suggested clinical benefits in advanced stage pancreatic cancer, was not borne out in subsequent phase III trials (reviewed in Finke et al. [36]). Following this somewhat expected result Therion closed its research programs and has gone into administration. BNpharma have taken over this project.

16.4.2
Undefined Antigens

There are a number of approaches involving vaccination with whole tumor cells, lysates of tumors, or using DCs pulsed with tumor lysate that have demonstrated great success in the clinic [20, 37, 38]. Such vaccines rely on the presence of relevant tumor antigens in the source material. However, despite an incomplete understanding of constituent antigens the presence of defined TAAs in the preparations is important in the selection of cells, the determination of their stability, and the measurement of the immunological outcome of vaccination (see below).

Such 'polyvalent', cellular vaccines contain many antigens, both of known and unknown origin, which are probably identical to the patient's tumor (autologous preparations) or have an overlapping spectrum of antigens (allogeneic preparations). Several advantages are inherent in these approaches. Since a large number of tumor-specific antigens are present, it is unlikely that antigen-escape phenotypes will give any advantage to the tumor. Furthermore, cellular vaccines can incorporate modifications, such as the ability to secrete cytokines or to express relevant MHC or costimulatory molecules [39].

16.4.2.1 Autologous Tumor Cell Vaccines
In preclinical studies autologous vaccines have been shown to generate anti-tumor immune responses and to induce regression of syngeneic tumors [40, 41]. However, a number of problems arise in translating these studies to the clinic. Enough cells must be acquired from resection to prepare sufficient vaccine and the

number of immunizations is always limited by this factor. Where the concentration of cells is not sufficient, an *in-vitro* culture step can be undertaken to increase the number of cells available. However, there is a high failure rate for primary culture of tumor cells with the result that this can be an unreliable technology. Furthermore the delay necessary to scale up the number of cells may be critical since patients' diseases potentially progress during this period. Despite these problems two vaccines have reached phase III trials and show promise in colorectal and renal carcinomas [37, 38] (see also Table 16.1).

16.4.2.2 Allogeneic Tumor Cell Vaccines

An alternative to the use of autologous tumor (which is essentially a bespoke vaccine) is the development of an off-the-shelf vaccine using allogeneic tumor cells, in which optimal cell lines are selected for their stability in culture and a range of standard, known TAAs. Allogeneic cell lines not only express similar shared antigens but also induce a 'danger signal' due to their foreign allo-antigens.

A number of allogeneic cellular vaccines have been tested in phase III clinical trials with variable success (see Table 16.1). Critically these vaccines have used the expression profile of known TAAs in order to determine the cell lines to be used. For example Canvaxin (an allogeneic cellular vaccine composed of three melanoma cell lines: M14, M24 and M101) was initially selected for the expression of melanoma TAAs of immunological importance: gangliosides (GD2, GM2 and O-acetyl GD3), a lipoprotein (M-TAA), and two glycoproteins (M-fetal antigen and O-urinary antigen) [43]. Although the selection of such antigens at initial stages in vaccine development does not guarantee their effectiveness in eliciting immune responses, it does allow the cells to be monitored for their stability (i.e. any change in expression profile in culture) and for a prediction of their relevance to the tumor background to be obtained.

Another advantage of using cell lines is that they can be selected for their capacity to be modified. GVAX vaccines, in which cell lines have been modified to express GM-CSF, have been shown to have clinical efficacy and have progressed to phase III studies (see Table 16.1). GM-CSF is known to induce DC migration and maturation and it is believed that its secretion at the tumor site enhances antitumor responses by increasing uptake of vaccine by DCs at the site of injection. GVAX vaccines are currently being trialled in prostate and pancreatic cancer patients (see Table 16.1). However, the prostate trial has now been halted due to failure to meet primary endpoints.

16.4.2.3 Loading DCs with Tumor Preparations

Autologous DCs may be used for preparation of vaccines. The development of DC vaccines has been focused largely on the study of the DC maturation cycle (described in [50]). Immature DCs can take up tumor material in a number of different forms including apoptotic tumor cells, tumor protein in the form of a mechanical lysate, and tumor mRNA. These last two methods of loading tumor material have been extensively studied and have been shown to have clinical efficacy in melanoma [13], prostate cancer [20], and glioma [51] amongst others. The issues related to DC vaccines will be discussed in further detail in following chapters.

16.4 Types of Cancer Vaccine | 291

Table 16.1 Noteworthy clinical trials using cellular vaccines.

Vaccine/Company	Autologous/allogeneic	Tumor type	Clinical results	Reference
OncoVAX®	Autologous whole cell	Colorectal stage II	Improvement in 5-year recurrence-free survival 37.7% (compared to control, no vaccine arm 21.3%). $p = 0.009$	[37]
Liponova	Autologous lysate	Renal Cell Carcinoma, stage III	Improved 5-year OS, 27.3% (compared to control arm, 17.5%). $p = 0.01$ Improved 5-year PFS, 67.5% (compared to surgery alone, 49.7%). $p = 0.039$	[38]
Canvaxin	Allogeneic whole cell	Metastatic melanoma	Phase I and II studies showed promise with • Increased median survival • Patients with low volume disease had complete remission Significant survival benefit in stage III melanoma patients post-resection Phase III studies were halted when primary endpoints (OS and DFS[a]) failed to diverge from control arm	[42, 43]
GVAX	Allogeneic whole cell modified to secrete GM-CSF	Metastatic prostate cancer	Phase II study showed • Trend towards increased 2-year survival • Longer median survival compared to standard chemotherapy Phase III trial failed	[44, 45]
GVAX	Allogeneic whole cell modified to secrete GM-CSF	Stage I, II and III pancreatic adenocarcinoma	1- and 2-year survival greater in treatment groups (88 and 76%) than in historical controls (63 and 42%)	[46]
Onyvax-p	Allogeneic whole cell vaccine	Hormone resistant prostate cancer	Phase I data • Increased time-to-progression in treatment group (58 weeks) compared to historical controls (26 weeks) • 11/26 patients had prolonged decrease in PSA velocity Phase IIB trial is now recruiting	[47]
Melacine	Allogeneic lysate	Primary cutaneous melanoma	Phase I study showed a 6.1% response rate Phase III study failed to show differences between treatment and control arms. Retrospective analysis identified an HLA profile that may be used to stratify patients (HLA-A2, -A28, -B44, -B45 and -C3). A new Phase III trial is ongoing to test this hypothesis	[48]

[a]OS, overall survival; DFS, disease free survival. These trials are reviewed more extensively in Copier et al. [49].

16.5
Immune Biomarkers

It has become increasingly important to include assays for the measurement of cellular and humoral responses in patients undergoing treatment with cancer vaccines in order to determine the extent and dynamics of the immune response. The relationship between clinical and immunological responses remains unclear with few trials showing clear correlations. Although controversial, some trials have shown correlations between clinical outcome in vaccinated melanoma patients and the extent of delayed type hypersensitivity (DTH) induration [52–54]. Furthermore, the determination of T_H1/T_H2 bias has been correlated with clinical outcome when combined with biochemical response (i.e. PSA slope) in an analysis of a whole cell vaccine trial for hormone-resistant prostate cancer [47]. However, there are very few demonstrable correlates between antigen-specific responses and clinical outcome. Many groups assess the antigen specificity of responses through tetramer analysis of circulating antigen-specific lymphocytes, but few if any trials to date have demonstrated clinical correlates with the antigen-specific response. It is therefore noteworthy that clinical correlates have been successfully demonstrated by determining the presence of tetramer-positive lymphocyte populations in either DTH sites [55] or in tumors [54]. This is likely to be a measure of the ability of a vaccine to promote cellular responses that induce lymphocyte homing.

Often conventional endpoints in the final assessment of clinical trials are not appropriate for cancer vaccines. Currently the gold standard endpoint for many clinical trials of cancer vaccines is overall survival, although in some cancer backgrounds such as hormone-resistant prostate cancer (HRPC) progression free survival is a good candidate endpoint. However, understanding the dynamics of immune responses, the degree of T-cell infiltration into tumors, and the importance of antigen specificity in promoting a clinical outcome will help to determine more useful surrogate endpoints. These may ultimately be used as surrogate markers for early assessment of clinical efficacy. To date no such biomarker exists that may be used as a primary endpoint. Nevertheless there are a number of serum tumor antigens, which indicate response to treatment such as a reduction in the rate of rise of PSA for prostate cancer and reduction and stabilization of CA19.9, CA-125 and CEA in prostate, ovarian and colorectal cancers, respectively.

16.6
Problems with the Assessment of Clinical Trials for Cancer Vaccines

There are many reasons why drugs may be active in single center phase II trials and yet fail in multicenter phase III trials. These include the obvious ones such as, the drug did not really work and the phase II was a chance or placebo effect. In addition the patient population may vary subtly between phase II and phase III as well as protocol changes and other treatments becoming available, which even given sequentially may affect the outcome. All of these problems apply even more to vaccine trials. Within the melanoma field there are a number of high profile vaccines, effective in phase II studies that have dramatically failed to show any clinical advantage in randomized studies. This could underline a specific problem with melanoma.

A ganglioside-based vaccine pioneered by Livingstone and colleagues showed a benefit when used with Bacillus Calmette Guerin (BCG) as an adjuvant, which was only significant if the immune response to the vaccine was taken into consideration [56]. A commercial preparation using the more powerful QS-21-based adjuvant not only failed to induce a better clinical outcome, but was worse than high-dose IFNα in a randomized phase III study and was worse than observations in a randomized phase II study [57]. The most obvious difference here was the vaccine adjuvant with the more powerful QS-21 inducing a strong humoral response which may be detrimental. Indeed, many good vaccine outcomes correlate with a cell-mediated, T_H1 response and not a humoral T_H2 response.

The other high-profile failure is that of CancerVAX, a polyantigen cell-based vaccine, which was associated with very impressive 5-year survival results, when used in post-resected stage III and stage IV melanoma patients at the John Wayne Cancer Center in California. A randomized phase III including hundreds of patients in many different countries comparing BCG plus vaccine to BCG plus placebo initially looked in favor of the vaccine arm with regard to time-to-progression, but the trial had to be stopped early because of lack of significant separation. Now several years after the first patients were enrolled there appears to be a survival benefit for the BCG plus placebo arm. However, both arms did better than predicted so it is unlikely that the vaccine was detrimental. There are many reasons why this trial failed. One of the most important concerned the need for first class surgery and complete resection. It was already clear that this was not practised on stage IV patients in many centers outside of the John Wayne. Another aspect may well have been the selection, as highly motivated patients self-refer to the John Wayne, whereas once the trial is open in many centers 'all comers' are admitted providing that they fit the inclusion criteria. However, how can the survival benefit of BCG be explained? The dose and schedule was determined from years of experience using a quantifiable response to purified protein derivative (PPD). This may have led to an optimal cell-mediated stimulation including non-antigen specific responses such as those mediated by NK cells. The improvement in survival of the BCG arm over BCG plus polyantigens is hard to explain unless it is related to dominant antigen down-regulation as described for peptide-based vaccines where the targeted epitope can completely disappear from residual tumors in a couple of months. It remains possible that this process is occurring amongst a few dominant epitopes even in a multi-antigen, cell-based vaccine.

Another reason that trials may fail is that they are poorly conducted when they go into many new centers that are inexperienced in the use of a completely new modality. *Mycobacterium vaccae* (also known as SLR-172) gave a survival benefit when added to chemotherapy in patients with lung cancer in randomized single center trials [58]. The developers had determined that optimal response required intradermal (i.d.) and not subcutaneous (s.c.) administration and that a regular booster should be given.

In spite of a protocol prescribing at least eight vaccines, less than half received three or more doses and the majority was given s.c. when the vaccine entered a multicenter, randomized study [59]! Despite this, symptomatic control in lung cancer has been demonstrated through combined SLR-172 and chemotherapy [60].

16.6.1
Biomarkers

It is clear that cancers are heterologous and that although some immunological responses are dramatic most patients do not respond to vaccines in randomized trials. Therefore valuable biomarkers, to determine which patient will benefit from a vaccine, are urgently needed. A good example is HER-2/neu as a biomarker for Herceptin, which although very successful would not have reached clinical significance if tried in all advanced breast cancers. Comparable markers for other tumors may make such a difference in vaccine development, and hence are urgently required. Nevertheless the ability to respond in a non-specific, cell-mediated, T_H1 dominant manner (as opposed to T_H2) may be just as important. A recent vaccine based on cells, and hence specific antigens, induced a T_H1 dominant response and a reduction in the rate of rise (velocity) in PSA in patients responding clinically with prolonged time-to-disease-progression [47] (see also Table 16.1). This was only clear after using Artificial Neural Networking (ANN) analysis. These results are in keeping with the Megavax vaccine, which underperformed when given with an adjuvant that boosts the humoral or T_H2 immune response component.

16.6.2
Schedule and Protocol

Many vaccination trials have been very optimistic setting the targets very high, often out of necessity, in particular patient populations defined by safety and ethical issues. For instance all the available evidence to date suggests that vaccine-based therapies should be given before or with other modalities such as radiotherapy, chemotherapy or hormonal therapies. Few studies to date have addressed these issues. Indeed the priming effect of vaccines for other treatments may be the major clinical benefit. This may be the case for the Dendreon studies where there is a survival benefit for vaccine plus chemotherapy versus chemotherapy alone in spite of no benefit being noted prior to additional treatments being given at further progression [20].

16.7
Combined Therapies; Synergy with Conventional Approaches

During the development of cancer vaccines it has always been assumed that standard chemo- and radiation therapies will interfere with the patient's ability to mount an immune response. Indeed high dose chemotherapy regimes that induce bone marrow ablation clearly have detrimental effects on the immune response unless immune reconstitution via transplantation occurs. There has been mounting interest in the possibility that cancer vaccines can act synergistically with standard medicines. It is known that some low dose irradiation can increase the potential immunogenicity of tumor cells through changes in antigen profile and upregulation of MHC, costimulatory markers, and Fas [61–63]. In addition chemotherapeutics can induce changes that may be immunologically beneficial. Doxorubicin, a chemotherapeutic agent that induces apoptosis, has been shown to render tumor cells more immunogenic in preclinical models [64]. Recent evidence indicates that anthracyclines like doxorubicin induce a specific apoptotic phenotype that allow tumor uptake by DCs providing a potential mechanism for the induction of immunogenicity [65].

As more data accumulates it is becoming clear that many different drugs have unexpected effects that may induce better immune responses. In our laboratory the use of IMiDs (which have a known anti-angiogenic effect) has been demonstrated to increase costimulatory activity [66].

There is a growing body of evidence to suggest that chemotherapy may have clinical benefit when used in combination with cancer vaccines. In recent clinical trials the use of low dose 'metronomic' cyclophosphamide was shown to substantially reduce the numbers of circulating T regulatory cells (Tregs) [67]. Furthermore there is tentative evidence that the use of other drugs, such as ONTAK (a fusion protein consisting of diphtheria toxin and IL-2), which reduce numbers of Tregs can enhance the efficacy of cancer vaccines [68, 69]. A number of trials have reported the synergistic action of standard chemotherapy or radiotherapy with cancer vaccines [35, 70]. The priming effect of "vaccines" or non-specific stimulation such as by TLR agonists or cytokines has been reported to enhance radiotherapy responses to pre-armed radio-resistant tumors e.g. melanoma, since the 1960s. Indeed the number of reports even recently raises the question as to why this is not yet standard treatment. The benefit of chemotherapy and immune stimulation is confined to low or moderate dose regimes.

The use of low dose cyclophosphamide exemplifies how reduced doses of standard chemotherapeutics may be used with vaccines indicating a possible window of therapeutic activity which may reduce the toxicity associated with such drugs. Much work has yet to be done to define the impact of different drugs on the immune system and on tumor cells before these interactions can be fully understood.

16.8
Conclusion

Cancer vaccines offer a very real possibility for use in the clinic within 10 years. These treatments are attractive since they are well tolerated in the majority of cases with very few side effects. The evidence that vaccines may work in the adjuvant setting for surgery and chemotherapy, and may even synergize with conventional chemo- and radiotherapies suggests that vaccines may ultimately act as a powerful tool for suppression of tumor growth that complements the currently available repertoire of treatment modalities.

It is interesting to speculate on the future direction of cancer vaccines, the recent history of which has been the constant striving to produce antigen- or tumor-specific responses. A century on from Coley there is a burgeoning interest in the development of non-specific 'vaccines' and drugs that stimulate innate (NK and $\gamma\delta$ T-cell) immune responses for the treatment of cancer, which may constitute the next generation of cancer treatments or work alongside conventional vaccinetherapies.

Acknowledgment

J. Copier is supported by the Cancer Vaccine Institute.

References

1 Coley, W.B., (1991) The treatment of malignant tumors by repeated inoculations of erysipelas. With a report of ten original cases. *Clinical Orthopaedics and Related Research*, **262**, 3–11.

2 http://www.fda.gov/cder/cancer/druglistframe.htm Food and Drug Administration website.

3 Varambally, S., Yu, J., Laxman, B., Rhodes, D.R., Mehra, R., Tomlins, S.A., Shah, R.B., Chandran, U., Monzon, F.A., Becich, M.J., Wei, J.T., Pienta, K.J., Ghosh, D., Rubin, M.A. and Chinnaiyan, A.M. (2005) Integrative genomic and proteomic analysis of prostate cancer reveals signatures of metastatic progression. *Cancer Cell*, **8**, 393–406.

4 Shah, R.B., Mehra, R., Chinnaiyan, A.M., Shen, R., Ghosh, D., Zhou, M., Macvicar, G.R., Varambally, S., Harwood, J., Bismar, T.A., Kim, R., Rubin, M.A. and Pienta, K.J. (2004) Androgen-independent prostate cancer is a heterogeneous group of diseases: lessons from a rapid autopsy program. *Cancer Research*, **64**, 9209–9216.

5 Copier, J., Whelan, M. and Dalgleish, A. (2006) Biomarkers for the development of cancer vaccines: current status. *Molecular Diagnosis and Therapy*, **10**, 337–343.

6 Rodriguez, T., Mendez, R., Del Campo, A., Jimenez, P., Aptsiauri, N., Garrido, F. and Ruiz-Cabello, F. (2007) Distinct mechanisms of loss of IFN-gamma mediated HLA class I inducibility in two melanoma cell lines. *BMC Cancer*, 7.

7 Speiser, D.E., Lienard, D., Rufer, N., Rubio-Godoy, V., Rimoldi, D., Lejeune, F., Krieg, A.M., Cerottini, J.C. and Romero, P. (2005) Rapid and strong human CD8+ T cell responses to vaccination with peptide, IFA, and CpG oligodeoxynucleotide 7909. *The Journal of Clinical Investigation*, **115**, 739–746.

8 Weber, J., Sondak, V.K., Scotland, R., Phillip, R., Wang, F., Rubio, V., Stuge, T.B., Groshen, S.G., Gee, C., Jeffery, G.G., Sian, S. and Lee, P.P. (2003) Granulocyte-macrophage-colony-stimulating factor added to a multipeptide vaccine for resected Stage II melanoma. *Cancer*, **97**, 186–200.

9 Carbone, D.P., Ciernik, I.F., Kelley, M.J., Smith, M.C., Nadaf, S., Kavanaugh, D., Maher, V.E., Stipanov, M., Contois, D., Johnson, B.E., Pendleton, C.D., Seifert, B., Carter, C., Read, E.J., Greenblatt, J., Top, L.E., Kelsey, M.I., Minna, J.D. and Berzofsky, J.A. (2005) Immunization with mutant p53- and K-ras-derived peptides in cancer patients: immune response and clinical outcome. *Journal of Clinical Oncology*, **23**, 5099–5107.

10 Slingluff, C.L., Jr. Petroni, G.R., Yamshchikov, G.V., Hibbitts, S.,

Grosh, W.W., Chianese-Bullock, K.A., Bissonette, E.A., Barnd, D.L., Deacon, D.H., Patterson, J.W., Parekh, J., Neese, P.Y., Woodson, E.M., Wiernasz, C.J. and Merrill, P. (2004) Immunologic and clinical outcomes of vaccination with a multiepitope melanoma peptide vaccine plus low-dose interleukin-2 administered either concurrently or on a delayed schedule. *Journal of Clinical Oncology*, **22**, 4474–4485.

11 Khong, H.T., Yang, J.C., Topalian, S.L., Sherry, R.M., Mavroukakis, S.A., White, D.E. and Rosenberg, S.A. (2004) Immunization of HLA-A*0201 and/or HLA-DPbeta1*04 patients with metastatic melanoma using epitopes from the NY-ESO-1 antigen. *Journal of Immunotherapie (1997)*, **27**, 472–477.

12 Kavanagh, B., Ko, A., Venook, A., Margolin, K., Zeh, H., Lotze, M., Schillinger, B., Liu, W., Lu, Y., Mitsky, P., Schilling, M., Bercovici, N., Loudovaris, M., Guillermo, R., Lee, S.M., Bender, J., Mills, B. and Fong, L. (2007) Vaccination of metastatic colorectal cancer patients with matured dendritic cells loaded with multiple major histocompatibility complex class I peptides. *Journal of Immunotherapie (1997)*, **30**, 762–772.

13 Banchereau, J., Ueno, H., Dhodapkar, M., Connolly, J., Finholt, J.P., Klechevsky, E., Blanck, J.P., Johnston, D.A., Palucka, A.K. and Fay, J. (2005) Immune and clinical outcomes in patients with stage IV melanoma vaccinated with peptide-pulsed dendritic cells derived from CD34+ progenitors and activated with type I interferon. *Journal of Immunotherapie (1997)*, **28**, 505–516.

14 Davis, I.D., Chen, W., Jackson, H., Parente, P., Shackleton, M., Hopkins, W., Chen, Q., Dimopoulos, N., Luke, T., Murphy, R., Scott, A.M., Maraskovsky, E., McArthur, G., MacGregor, D., Sturrock, S., Tai, T.Y., Green, S., Cuthbertson, A., Maher, D., Miloradovic, L., Mitchell, S.V., Ritter, G., Jungbluth, A.A., Chen, Y.T., Gnjatic, S., Hoffman, E.W., Old, L.J. and Cebon, J.S. (2004) Recombinant NY-ESO-1 protein with ISCOMATRIX adjuvant induces broad integrated antibody and CD4(+) and CD8(+) T cell responses in humans. *Proceedings of the National Academy of Sciences of the United States of America* **101**, 10697–10702.

15 Disis, M.L., Schiffman, K., Guthrie, K., Salazar, L.G., Knutson, K.L., Goodell, V., dela Rosa, C. and Cheever, M.A. (2004) Effect of dose on immune response in patients vaccinated with an her-2/neu intracellular domain protein--based vaccine. *Journal of Clinical Oncology*, **22**, 1916–1925.

16 Vantomme, V., Dantinne, C., Amrani, N., Permanne, P., Gheysen, D., Bruck, C., Stoter, G., Britten, C.M., Keilholz, U., Lamers, C.H., Marchand, M., Delire, M. and Gueguen, M. (2004) Immunologic analysis of a phase I/II study of vaccination with MAGE-3 protein combined with the AS02B adjuvant in patients with MAGE-3-positive tumors. *Journal of Immunotherapie (1997)*, **27**, 124–135.

17 Marchand, M., Punt, C.J., Aamdal, S., Escudier, B., Kruit, W.H., Keilholz, U., Hakansson, L., van Baren, N., Humblet, Y., Mulders, P., Avril, M.F., Eggermont, A.M., Scheibenbogen, C., Uiters, J., Wanders, J., Delire, M., Boon, T. and Stoter, G. (2003) Immunisation of metastatic cancer patients with MAGE-3 protein combined with adjuvant SBAS-2: a clinical report. *European Journal of Cancer (Oxford, England: 1990)*, **39**, 70–77.

18 Kruit, W.H., van Ojik, H.H., Brichard, V.G., Escudier, B., Dorval, T., Dreno, B., Patel, P., van Baren, N., Avril, M.F., Piperno, S., Khammari, A., Stas, M., Ritter, G., Lethe, B., Godelaine, D., Brasseur, F., Zhang, Y., van der Bruggen, P., Boon, T., Eggermont, A.M. and Marchand, M. (2005) Phase 1/2 study of subcutaneous and intradermal immunization with a recombinant MAGE-3 protein in patients with detectable metastatic melanoma. *International Journal of Cancer*, **117**, 596–604.

19 Renard, V. and Leach, D.R. (2007) Perspectives on the development of a therapeutic HER-2 cancer vaccine. *Vaccine*, **25** (Suppl 2), B17–B23.

20 Small, E.J., Schellhammer, P.F., Higano, C.S., Redfern, C.H., Nemunaitis, J.J., Valone, F.H., Verjee, S.S., Jones, L.A. and Hershberg, R.M. (2006) Placebo-controlled phase III trial of immunologic therapy with sipuleucel-T (APC8015) in patients with metastatic, asymptomatic hormone refractory prostate cancer. *Journal of Clinical Oncology*, **24**, 3089–3094.

21 Pavlenko, M., Roos, A.K., Lundqvist, A., Palmborg, A., Miller, A.M., Ozenci, V., Bergman, B., Egevad, L., Hellstrom, M., Kiessling, R., Masucci, G., Wersall, P., Nilsson, S. and Pisa, P. (2004) A phase I trial of DNA vaccination with a plasmid expressing prostate-specific antigen in patients with hormone-refractory prostate cancer. *British Journal of Cancer*, **91**, 688–694.

22 Miller, A.M., Ozenci, V., Kiessling, R. and Pisa, P. (2005) Immune monitoring in a phase 1 trial of a PSA DNA vaccine in patients with hormone-refractory prostate cancer. *Journal of Immunotherapy*, **28**, 389–395.

23 Conry, R.M., Curiel, D.T., Strong, T.V., Moore, S.E., Allen, K.O., Barlow, D.L., Shaw, D.R. and LoBuglio, A.F. (2002) Safety and immunogenicity of a DNA vaccine encoding carcinoembryonic antigen and hepatitis B surface antigen in colorectal carcinoma patients. *Clinical Cancer Research*, **8**, 2782–2787.

24 Triozzi, P.L., Aldrich, W., Allen, K.O., Carlisle, R.R., LoBuglio, A.F. and Conry, R.M. (2005) Phase I study of a plasmid DNA vaccine encoding MART-1 in patients with resected melanoma at risk for relapse. *Journal of Immunotherapie (1997)*, **28**, 382–388.

25 Rosenberg, S.A., Yang, J.C., Sherry, R.M., Hwu, P., Topalian, S.L., Schwartzentruber, D.J., Restifo, N.P., Haworth, L.R., Seipp, C.A., Freezer, L.J., Morton, K.E., Mavroukakis, S.A. and White, D.E. (2003) Inability to immunize patients with metastatic melanoma using plasmid DNA encoding the gp100 melanoma-melanocyte antigen. *Human Gene Therapy*, **14**, 709–714.

26 Anderson, R.J. and Schneider, J. (2007) Plasmid DNA and viral vector-based vaccines for the treatment of cancer. *Vaccine*, **25** (Suppl 2), B24–B34.

27 Tagawa, S.T., Lee, P., Snively, J., Boswell, W., Ounpraseuth, S., Lee, S., Hickingbottom, B., Smith, J., Johnson, D. and Weber, J.S. (2003) Phase I study of intranodal delivery of a plasmid DNA vaccine for patients with Stage IV melanoma. *Cancer*, **98**, 144–154.

28 Palmowski, M.J., Choi, E.M., Hermans, I.F., Gilbert, S.C., Chen, J.L., Gileadi, U., Salio, M., Van Pel, A., Man, S., Bonin, E., Liljestrom, P., Dunbar, P.R. and Cerundolo, V. (2002) Competition between CTL narrows the immune response induced by prime-boost vaccination protocols. *Journal of Immunology (Baltimore, Md: 1950)*, **168**, 4391–4398.

29 Amara, R.R., Villinger, F., Staprans, S.I., Altman, J.D., Montefiori, D.C., Kozyr, N.L., Xu, Y., Wyatt, L.S., Earl, P.L., Herndon, J.G., McClure, H.M., Moss, B. and Robinson, H.L. (2002) Different patterns of immune responses but similar control of a simian-human immunodeficiency virus 89.6P mucosal challenge by modified vaccinia virus Ankara (MVA) and DNA/MVA vaccines. *Journal of Virology*, **76**, 7625–7631.

30 Scholl, S.M., Balloul, J.M., Le Goc, G., Bizouarne, N., Schatz, C., Kieny, M.P., von Mensdorff-Pouilly, S., Vincent-Salomon, A., Deneux, L., Tartour, E., Fridman, W., Pouillart, P. and Acres, B. (2000) Recombinant vaccinia virus encoding human MUC1 and IL2 as immunotherapy in patients with breast cancer. *Journal of Immunotherapie (1997)*, **23**, 570–580.

31 Harrop, R., John, J. and Carroll, M.W. (2006) Recombinant viral vectors: cancer vaccines. *Advanced Drug Delivery Reviews*, **58**, 931–947.

32 Harrop, R., Connolly, N., Redchenko, I., Valle, J., Saunders, M., Ryan, M.G., Myers, K.A., Drury, N., Kingsman, S.M., Hawkins, R.E. and Carroll, M.W. (2006) Vaccination of colorectal cancer patients with modified vaccinia Ankara delivering the tumor antigen 5T4 (TroVax) induces immune responses which correlate with disease control: a phase I/II trial. *Clinical Cancer Research*, **12**, 3416–3424.

33 van Baren, N., Bonnet, M.C., Dreno, B., Khammari, A., Dorval, T., Piperno-Neumann, S., Lienard, D., Speiser, D., Marchand, M., Brichard, V.G., Escudier, B., Negrier, S., Dietrich, P.Y., Maraninchi, D., Osanto, S., Meyer, R.G., Ritter, G., Moingeon, P., Tartaglia, J., van der Bruggen, P., Coulie, P.G. and Boon, T. (2005) Tumoral and immunologic response after vaccination of melanoma patients with an ALVAC virus encoding MAGE antigens recognized by T cells. *Journal of Clinical Oncology*, **23**, 9008–9021.

34 Rosenberg, S.A., Yang, J.C., Schwartzentruber, D.J., Hwu, P., Topalian, S.L., Sherry, R.M., Restifo, N.P., Wunderlich, J.R., Seipp, C.A., Rogers-Freezer, L., Morton, K.E., Mavroukakis, S.A., Gritz, L., Panicali, D.L. and White, D.E. (2003) Recombinant fowlpox viruses encoding the anchor-modified gp100 melanoma antigen can generate antitumor immune responses in patients with metastatic melanoma. *Clinical Cancer Research*, **9**, 2973–2980.

35 Arlen, P.M., Gulley, J.L., Todd, N., Lieberman, R., Steinberg, S.M., Morin, S., Bastian, A., Marte, J., Tsang, K.Y., Beetham, P., Grosenbach, D.W., Schlom, J. and Dahut, W. (2005) Antiandrogen, vaccine and combination therapy in patients with nonmetastatic hormone refractory prostate cancer. *The Journal of Urology*, **174**, 539–546.

36 Finke, L.H., Wentworth, K., Blumenstein, B., Rudolph, N.S., Levitsky, H. and Hoos, A. (2007) Lessons from randomized phase III studies with active cancer immunotherapies-outcomes from the 2006 meeting of the Cancer Vaccine Consortium (CVC). *Vaccine*, **25** (Suppl 2), B97–B109.

37 Uyl-de Groot, C.A., Vermorken, J.B., Hanna, M.G., Jr. Verboom, P., Groot, M.T., Bonsel, G.J., Meijer, C.J. and Pinedo, H.M. (2005) Immunotherapy with autologous tumor cell-BCG vaccine in patients with colon cancer: a prospective study of medical and economic benefits. *Vaccine*, **23**, 2379–2387.

38 Jocham, D., Richter, A., Hoffmann, L., Iwig, K., Fahlenkamp, D., Zakrzewski, G., Schmitt, E., Dannenberg, T., Lehmacher, W., von Wietersheim, J. and Doehn, C. (2004) Adjuvant autologous renal tumour cell vaccine and risk of tumour progression in patients with renal-cell carcinoma after radical nephrectomy: phase III, randomised controlled trial. *Lancet*, **363**, 594–599.

39 Copier, J. and Dalgleish, A. (2006) Overview of tumor cell-based vaccines. *International Reviews of Immunology*, **25**, 297–319.

40 Knight, B.C., Souberbielle, B.E., Rizzardi, G.P., Ball, S.E. and Dalgleish, A.G. (1996) Allogeneic murine melanoma cell vaccine: a model for the development of human allogeneic cancer vaccine. *Melanoma Research*, **6**, 299–306.

41 Souberbielle, B.E., Westby, M., Ganz, S., Kayaga, J., Mendes, R., Morrow, W.J. and Dalgleish, A.G. (1998) Comparison of four strategies for tumour vaccination in the B16-F10 melanoma model. *Gene Therapy*, **5**, 1447–1454.

42 Hsueh, E.C., Nathanson, L., Foshag, L.J., Essner, R., Nizze, J.A., Stern, S.L. and Morton, D.L. (1999) Active specific immunotherapy with polyvalent melanoma cell vaccine for patients with in-transit melanoma metastases. *Cancer*, **85**, 2160–2169.

43 Morton, D.L., Hsueh, E.C., Essner, R., Foshag, L.J., O'Day, S.J., Bilchik, A., Gupta, R.K., Hoon, D.S., Ravindranath, M., Nizze, J.A., Gammon, G., Wanek, L.A., Wang, H.J. and Elashoff, R.M. (2002)

Prolonged survival of patients receiving active immunotherapy with Canvaxin therapeutic polyvalent vaccine after complete resection of melanoma metastatic to regional lymph nodes. *Annals of Surgery*, **236**, 438–448.

44 Simons, J.W. and Sacks, N. (2006) Granulocyte-macrophage colony-stimulating factor-transduced allogeneic cancer cellular immunotherapy: the GVAX vaccine for prostate cancer. *Urologic Oncology*, **24**, 419–424.

45 Small, E., Higano, D., Smith, D., Corman, J., Centeno, A., Steidle, C., Gittelman, M., Hudes, G.R., Sacks, N. and Simons, J. (2004) A phase II study of an allogeneic GM-CSF gene-transduced prostate cancer cell line vaccine in patients with metastatic hormone-refractory prostate cancer (HRPC). *Proceedings of the American Society of Clinical Oncology*, Abstract no 4565.

46 Jaffee, E.M., Hruban, R.H., Biedrzycki, B., Laheru, D., Schepers, K., Sauter, P.R., Goemann, M., Coleman, J., Grochow, L., Donehower, R.C., Lillemoe, K.D., O'Reilly, S., Abrams, R.A., Pardoll, D.M., Cameron, J.L. and Yeo, C.J. (2001) Novel allogeneic granulocyte-macrophage colony-stimulating factor-secreting tumor vaccine for pancreatic cancer: a phase I trial of safety and immune activation. *Journal of Clinical Oncology*, **19**, 145–156.

47 Michael, A., Ball, G., Quatan, N., Wushishi, F., Russell, N., Whelan, J., Chakraborty, P., Leader, D., Whelan, M. and Pandha, H. (2005) Delayed disease progression after allogeneic cell vaccination in hormone-resistant prostate cancer and correlation with immunologic variables. *Clinical Cancer Research*, **11**, 4469–4478.

48 Sondak, V.K. and Sosman, J.A. (2003) Results of clinical trials with an allogenic melanoma tumor cell lysate vaccine: Melacine. *Seminars in Cancer Biology*, **13**, 409–415.

49 Copier, J., Ward, S. and Dalgleish, A. (2007) Tumour cell cancer vaccines: regulatory and commercial development. *Vaccine*, (in press).

50 Banchereau, J. and Palucka, A.K. (2005) Dendritic cells as therapeutic vaccines against cancer. *Nature Reviews Immunology*, **5**, 296–306.

51 Rutkowski, S., De Vleeschouwer, S., Kaempgen, E., Wolff, J.E., Kuhl, J., Demaerel, P., Warmuth-Metz, M., Flamen, P., Van Calenbergh, F., Plets, C., Sorensen, N., Opitz, A. and Van Gool, S.W. (2004) Surgery and adjuvant dendritic cell-based tumour vaccination for patients with relapsed malignant glioma, a feasibility study. *British Journal of Cancer*, **91**, 1656–1662.

52 Escobar, A., Lopez, M., Serrano, A., Ramirez, M., Perez, C., Aguirre, A., Gonzalez, R., Alfaro, J., Larrondo, M., Fodor, M., Ferrada, C. and Salazar-Onfray, F. (2005) Dendritic cell immunizations alone or combined with low doses of interleukin-2 induce specific immune responses in melanoma patients. *Clinical and Experimental Immunology*, **142**, 555–568.

53 Hersey, P., Menzies, S.W., Halliday, G.M., Nguyen, T., Farrelly, M.L., DeSilva, C. and Lett, M. (2004) Phase I/II study of treatment with dendritic cell vaccines in patients with disseminated melanoma. *Cancer Immunology, Immunotherapy*, **53**, 125–134.

54 Haanen, J.B., Baars, A., Gomez, R., Weder, P., Smits, M., de Gruijl, T.D., von Blomberg, B.M., Bloemena, E., Scheper, R.J., van Ham, S.M., Pinedo, H.M. and van den Eertwegh, A.J. (2006) Melanoma-specific tumor-infiltrating lymphocytes but not circulating melanoma-specific T cells may predict survival in resected advanced-stage melanoma patients. *Cancer Immunology, Immunotherapy*, **55**, 451–458.

55 de Vries, I.J., Bernsen, M.R., Lesterhuis, W.J., Scharenborg, N.M., Strijk, S.P., Gerritsen, M.J., Ruiter, D.J., Figdor, C.G., Punt, C.J. and Adema, G.J. (2005) Immunomonitoring tumor-specific T cells in delayed-type hypersensitivity skin biopsies after

dendritic cell vaccination correlates with clinical outcome. *Journal of Clinical Oncology*, **23**, 5779–5787.

56 Livingston, P.O., Wong, G.Y., Adluri, S., Tao, Y., Padavan, M., Parente, R., Hanlon, C., Calves, M.J., Helling, F., Ritter, G. *et al.* (1994) Improved survival in stage III melanoma patients with GM2 antibodies: a randomized trial of adjuvant vaccination with GM2 ganglioside. *Journal of Clinical Oncology*, **12**, 1036–1044.

57 Kirkwood, J.M., Ibrahim, J.G., Sosman, J.A., Sondak, V.K., Agarwala, S.S., Ernstoff, M.S. and Rao, U. (2001) High-dose interferon alfa-2b significantly prolongs relapse-free and overall survival compared with the GM2-KLH/QS-21 vaccine in patients with resected stage IIB-III melanoma: results of intergroup trial E1694/S9512/C509801. *Journal of Clinical Oncology*, **19**, 2370–2380.

58 Assersohn, L., Souberbielle, B.E., O'Brien, M.E., Archer, C.D., Mendes, R., Bass, R., Bromelow, K.V., Palmer, R.D., Bouilloux, E., Kennard, D.A. and Smith, I.E. (2002) A randomized pilot study of SRL172 (Mycobacterium vaccae) in patients with small cell lung cancer (SCLC) treated with chemotherapy. *Clinical Oncology (Royal College of Radiologists (Great Britain)*, **14**, 23–27.

59 O'Brien, M.E., Anderson, H., Kaukel, E., O'Byrne, K., Pawlicki, M., Von Pawel, J. and Reck, M. (2004) SRL172 (killed Mycobacterium vaccae) in addition to standard chemotherapy improves quality of life without affecting survival, in patients with advanced non-small-cell lung cancer: phase III results. *Annals of Oncology*, **15**, 906–914.

60 Harper-Wynne, C.L., Sumpter, K., Ryan, C., Priest, K., Norton, A., Ross, P., Ford, H.E., Johnson, P. and O'Brien, M.E. (2005) Addition of SRL 172 to standard chemotherapy in small cell lung cancer (SCLC) improves symptom control. *Lung Cancer (Amsterdam, Netherlands)*, **47**, 289–290.

61 Ogawa, Y., Nishioka, A., Hamada, N., Terashima, M., Inomata, T., Yoshida, S., Seguchi, H. and Kishimoto, S. (1997) Expression of fas (CD95/APO-1) antigen induced by radiation therapy for diffuse B-cell lymphoma: immunohistochemical study. *Clinical Cancer Research*, **3**, 2211–2216.

62 Klein, B., Loven, D., Lurie, H., Rakowsky, E., Nyska, A., Levin, I. and Klein, T. (1994) The effect of irradiation on expression of HLA class I antigens in human brain tumors in culture. *Journal of Neurosurgery*, **80**, 1074–1077.

63 Garnett, C.T., Palena, C., Chakraborty, M., Tsang, K.Y., Schlom, J. and Hodge, J.W. (2004) Sublethal irradiation of human tumor cells modulates phenotype resulting in enhanced killing by cytotoxic T lymphocytes. *Cancer Research*, **64**, 7985–7994.

64 Casares, N., Pequignot, M.O., Tesniere, A., Ghiringhelli, F., Roux, S., Chaput, N., Schmitt, E., Hamai, A., Hervas-Stubbs, S., Obeid, M., Coutant, F., Metivier, D., Pichard, E., Aucouturier, P., Pierron, G., Garrido, C., Zitvogel, L. and Kroemer, G. (2005) Caspase-dependent immunogenicity of doxorubicin-induced tumor cell death. *The Journal of Experimental Medicine*, **202**, 1691–1701.

65 Obeid, M., Tesniere, A., Ghiringhelli, F., Fimia, G.M., Apetoh, L., Perfettini, J.L., Castedo, M., Mignot, G., Panaretakis, T., Casares, N., Metivier, D., Larochette, N., van Endert, P., Ciccosanti, F., Piacentini, M., Zitvogel, L. and Kroemer, G. (2007) Calreticulin exposure dictates the immunogenicity of cancer cell death. *Nature Medicine*, **13**, 54–61.

66 Dredge, K., Marriott, J.B., Todryk, S.M., Muller, G.W., Chen, R., Stirling, D.I. and Dalgleish, A.G. (2002) Protective antitumor immunity induced by a costimulatory thalidomide analog in conjunction with whole tumor cell vaccination is mediated by increased Th1-type immunity. *Journal of Immunology (Baltimore, Md: 1950)*, **168**, 4914–4919.

67 Ghiringhelli, F., Menard, C., Puig, P.E., Ladoire, S., Roux, S., Martin, F., Solary, E.,

Le Cesne, A., Zitvogel, L. and Chauffert, B. (2007) Metronomic cyclophosphamide regimen selectively depletes CD4+CD25+ regulatory T cells and restores T and NK effector functions in end stage cancer patients. *Cancer Immunology, Immunotherapy*, **56**, 641–648.

68 Dannull, J., Su, Z., Rizzieri, D., Yang, B.K., Coleman, D., Yancey, D., Zhang, A., Dahm, P., Chao, N., Gilboa, E. and Vieweg, J. (2005) Enhancement of vaccine-mediated antitumor immunity in cancer patients after depletion of regulatory T cells. *The Journal of Clinical Investigation*, **115**, 3623–3633.

69 Mahnke, K., Schonfeld, K., Fondel, S., Ring, S., Karakhanova, S., Wiedemeyer, K., Bedke, T., Johnson, T.S., Storn, V., Schallenberg, S. and Enk, A.H. (2007) Depletion of CD4+CD25+ human regulatory T cells *in vivo*: kinetics of Treg depletion and alterations in immune functions *in vivo* and *in vitro*. *International Journal of Cancer*, **120**, 2723–2733.

70 Antonia, S.J., Mirza, N., Fricke, I., Chiappori, A., Thompson, P., Williams, N., Bepler, G., Simon, G., Janssen, W., Lee, J.H., Menander, K., Chada, S. and Gabrilovich, D.I. (2006) Combination of p53 cancer vaccine with chemotherapy in patients with extensive stage small cell lung cancer. *Clinical Cancer Research*, **12**, 878–887.

17
Tumor-associated Antigenic Peptides as Vaccine Candidates

Dhiraj Hans, Paul R. Young, and David P. Fairlie

17.1
Introduction

An attractive strategy for 21st century prevention and treatment of cancer is the development of vaccines. Polypeptides and short peptides have been widely studied as prospective vaccines for a wide range of diseases [1–3]. By virtue of their natural amino acid composition, successful production in high quantity and with high purity via protein expression or peptide synthesis, peptides have some clear advantages as vaccines over live and attenuated viruses and most cellular vaccines which suffer from quality control issues and biological contaminants that can cause side-effects. On the other hand, expensive manufacture, susceptibility to proteolytic degradation, and conformational instability in water are some of the problems that have limited their development.

Vaccination to prevent or treat cancer relies upon the administration of tumor-associated antigens (TAAs) to instill an immunological memory and to prime the host immune response to carcinogenesis. TAAs are amino acid sequences within proteins that are expressed by cancer cells during neoplasia and they have become extremely important mechanistic probes in immunotherapy for cancer. A landmark study by van der Bruggen *et al.* identified one of the first human tumor antigens, cloned from the MAGE-1 (melanoma antigen) gene for melanoma, to be recognized by T cells [4]. Since the 1990s a large number of new TAAs have been identified by various computational or experimental methods. The antigenic sequences are usually peptide segments of 8–10 amino acids within proteins, and these sequences could potentially be developed into peptide vaccines to prevent or combat cancer. Ideally, an anticancer peptide vaccine should induce a strong cytotoxic T-cell (CTL) response that is cancer cell specific, and that can also be tailored to a particular patient or patient population. One goal for such vaccines is to effect the elimination of cancer cells that do not express MHC I molecules (MHC I$^-$).

More than 1000 tumor epitopes have been described in the literature to date, but less than 150 such peptides have reached phase I clinical trials, less than 30 have progressed to phase II, and there are only a few peptide vaccines that have entered phase III clinical trials. Despite this lack of clinical success to date, immunotherapy and vaccination strategies based on the use of peptidic tumor epitopes do appear to be promising approaches for stimulating anti-tumor immune responses that can increase the survival rates of cancer patients. Ideally TAAs are defined and classified based on their antigenic restriction pattern and cellular distribution, and peptides chosen for trials tend to be epitopes based on HLA association, either class I HLA (A, B and C) or class II (DP, DQ, and DR) or in some cases both class I and II. An ideal TAA peptide should be predominantly expressed in cancer cells and, most importantly, the antigen should be immunogenic but not cytotoxic. When expressed on normal cells as well as cancer cells, there is an increased risk of cytotoxicity, which is of course less desirable. TAAs are usually classified by their relative expression in tumor versus normal cells (http://www.cancerimmunity.org/) and this will be used below to categorize TAAs.

Immunization has involved administering the peptide alone, with various adjuvants, in combination with other peptides, attached to dendritic cells, and in combination with chemotherapy or surgery. Recent combinations of chemotherapy/cytokine therapy with immunization using peptide vaccines have attracted attention due to the effective killing of MHC I$^-$ cells. Use of peptide vaccines in conjunction with chemotherapeutic agents can be tricky as the latter are often dose-dependent immunosuppressants that need to be carefully dose-controlled as they can counter the immunological responses desired from a vaccine. There is currently no peptide vaccine, based on a tumor-associated antigen, available for human use.

This chapter summarizes some tumor-associated antigenic peptides (8–15 residues), categorized here based on their relative expression levels in tumor versus normal cells, that have shown early promise in the clinical trials as anticancer vaccines (Table 17.1).

17.2
Different Types of Antigens

17.2.1
Mutation Antigens

These antigens generally arise due to point mutation in the coding region of a ubiquitously expressed gene. These mutations could be found in tumors of just one or a few patients and are not necessarily shared by other tumors. Any directed patient immunotherapy based on such specific peptide mutations are obviously heavily personalized and are thus not useful generally. Some of the important peptides belonging to this category are from bcr-abl (b3a2), SYT-SSX, K-ras, N-ras, SIRT2, ARTC1, α-actinin-4, B-RAF, caspases, β-catenin, cdc27, p53, FLT3-ITD (www.cancerimmunity.org).

Bcr-abl (break point cluster region-abelson) is a chimeric gene product of the translocation of c-abl oncogene on chromosome 9 to the bcr gene on chromosome 22, producing the Philadelphia chromosome. The resulting gene fusion translates into a protein that generates typically abnormal tyrosine kinase activity during disease.

Table 17.1 Examples of peptide vaccines in the clinical trials from each tumor associated antigen class.

Antigen class	Cancer	Peptide sequence	Protein source	Adjuvant and route
Mutation antigens	Pancreatic adenocarcinoma	^{5}KLVVVGAGGVGK SALTI^{21}	P21 ras	GM-CSF, i.d.
	Myeloid leukaemia	ATGFKQSSK+ KQSSKALQR+ HSATGFKQSSK+ GFKQSSKAL+ IVHSATGFKQSS KALQRPVASDFEP	A11 binding A3 peptide A3 and A11 binding B8 binding B3a2-CML (2–25) peptide	QS-21+ GM-CSF. s.c.
	Pancreatic and lung cancer	^{217}VVPCEPPEV^{225}	p53 peptide	PBMC pulsed with peptide as infusion
	Metastatic cancer (advanced)	^{5}KLVVVGA*V*GVGKS ^{17} KLVVVGA*D*GVGKS KLVVVGA*C*GVGKS	ras peptides	
Differentiation antigens	Melanoma	^{240}DAEKSDICTDEY^{251S} ^{369}YMDGTMSQV^{377D} ^{280}YLEPGPVTA^{288} ^{17}ALLAVGATK^{25} AQYIKANSKFIGITEL ^{26}ELAGIGILTV^{35}	Tyrosinase + tyrosinase gp100+gp100 tetanus helper peptide Melan-A/ Mart-1 peptide	GM-CSF + Montanide ISA-51/IL-2, s.c.+i.d. Montanide AS02, s.c.
Shared tumor specific antigens	Melanoma	MAGE-A1$_{96-104}$ MAGE-10A$_{254-262}$ gp100$_{614-622}$	Peptides	Montanide ISA 51 and GM-CSF
	Melanoma and hepatocellular carcinoma	^{157}SLLMWITQCFL^{167} ^{157}SLLMWITQC^{165} ^{163}QLSLLMWIT^{155}	NY-ESO-1 peptide tetramers	±GM-CSF, s.c.
Over-expressed antigens	Breast cancer, lung cancer, and leukemia	^{235}CYTWNQMNL^{243}	WT1 gene peptide	Montanide ISA51. i.d.
	Breast and ovarian cancer	ECD vaccine (p42–56+ p98–114+p328–345) ICD vaccine (p776–790+ p927–941+ p1166–p1180), +helper peptides (p369–384+ p688–p703+p971–984)	HER-2/neu derived peptides	GM-CSF, i.d.
	Lung cancer, colorectal carcinoma, and prostate cancer (except ART4$^{13-20}$)	^{690}EYRGFTQDF^{699} ^{93}DYSARWNEI^{101} ^{161}AYDLFYNYL^{169} ^{899}SYTRLFLIL^{907} ^{109}VYDYNCHVDL^{118}	SART1 SART2 SART2 SART2 SART3	± Estramustine phosphate, i.d.

(Continued)

Table 17.1 (Continued)

Antigen class	Cancer	Peptide sequence	Protein source	Adjuvant and route
		^{315}AYIDFEMKI^{323}	SART3	
		^{84}KFHRVIKDF^{92}	CyB	
		^{91}DFMIQGGDF^{99}	CyB	
		^{208}HYTNASDGL^{216}	Lck	
		^{486}TFDYLRSVL^{494}	Lck	
		^{488}DYLRSVLEDF^{497}	Lck	
		^{170}EYCLKFTKL^{178}	ART1	
		^{13}AFLRHAAL^{20}	ART4	
		^{75}DYPSLSATDI^{84}	ART4	
Viral antigens	Advanced cervical carcinoma	^{11}YMLDLQPETT^{2}		
		^{86}TLGIVCPI^{93}	HPV16 E7	Montanide
			HPV16 E7	ISA 51, s.c.

Bcr-abl has a unique amino acid sequence at the fusion point, which is immunogenic in nature. Peptides based on a nine-residue component of Bcr-abl were used as vaccines in a small phase I and II trial using the lipid adjuvant QS-21 but did not reduce tumor burden [5, 6]. A subsequent, statistically more relevant, phase II trial in combination with QS-21 and granulocyte macrophage colony stimulating factor (GM-CSF) as adjuvants was undertaken, with one group of chronic myeloid leukemia (CML) patients given imatinib and a second group treated with interferon α prior to vaccination. All patients were then immunized with vaccine containing 100 µg of each peptide plus GM-CSF and QS-21. All but one patient in each group experienced cytogenetic remission (reduction in detectable bcr-abl transcripts). Patients vaccinated after chemotherapy were found to have better immunogenic responses than patients taking the vaccine alone [7]. Although it was not clear how much effect the co-adjuvant GM-CSF had, it was clear that immunotherapy after chemotherapy gave better anti-tumor responses in CML patients.

In another phase I study, *p53* and *K-ras* derived peptides were given as PBMC loaded cellular infusions to lung/pancreatic carcinoma patients without noticeable toxicity and encouraging CTL responses. However, recruitment of only 38 patients out of 276 underlines the difficulties associated with targeting populations with specific mutation antigens [8]. In another trial a *SYT-SSX* junction peptide was safely administered (0.1 mg or 1.0 mg) to eight synovial sarcoma patients [9]. In each of these cases, CTL induction was poor but these are examples where application of new technologies might dramatically improve results.

17.2.2
Shared Tumor-specific Antigens

Unlike mutation antigens, these antigens are shared among tumors and are expressed at much lower levels by normal cells and tissues. Important antigens in this category are MAGE, NY-ESO, BAGE, GAGE, LAGE, SAGE, SSX-2/4, GnTv, TRAG-3 and mucin (www.cancerimmunity.org).

MAGE-A1-A4, A6, A10, A12 are HLA-I restricted antigens that are very attractive candidates due to their $\geq$1000-fold greater expression in tumor cells. MAGE peptides were initially examined as vaccines together with dendritic cells (DCs). In a phase I trial, stage IV melanoma patients were immunized five times with mature DCs pulsed with MAGE-A3$_{167-175}$ peptide and tetanus toxin or tuberculin. The first three vaccinations were administered subcutaneously at two sites 1.5×10^6 DCs each and intradermally at 10 sites 3×10^5 DCs each, followed by two intravenous injections of 6×10^6 and 12×10^6 antigen-pulsed DCs every 14 days [10]. Toxicity and safety were assessed for all patients, with 11/11 showing grade I or II side-effects, 8/11 with significant CD8+ T cell precursor induction that reduced after the intravenous injections. Regression of skin, lymph node, lung, and liver metastasis was also observed in 6/11 patients. Interestingly, when these responses were compared with those following intracutaneous immunization, it was found that intravenous injections were comparatively less effective. Recent phase III data, from 190 stage IV melanoma patients in a multi-center open label randomized trial, compared MAGE peptides from class I and II HLAs using pulsed DC vaccine therapy versus dacarbazine monochemotherapy [11]. The vaccine was initially injected subcutaneously every 2 weeks for the first five injections and later every 4 weeks. In both dacarbazine and DC vaccination groups, only 5.5% and 3.8% respectively of patients responded to treatment, but HLA-A2+ and HLA-B44– patients survived longer than patients with any other HLA type or those treated with dacarbazine alone. It may be interesting in the future to examine only HLA-A2+ and HLA-B44– patients using DC vaccination. MAGE peptides (not DC pulsed) are also in upcoming phase II or III clinical trials, along with other antigens such as NY-ESO-1, Melan, GnTV, survivin and gp100. In those clinical trials, several other adjuvants will also be examined, including rhIL-10, KLH, QS-21, and Montanide ISA51 (www.clinicaltrials.gov).

NY-ESO (New York esophageous) was initially identified by serological analysis of recombinant cDNA library expression (SEREX). It is expressed in melanoma, breast cancer, lung cancer, and synovial sarcoma. Immunogenicity of the NY-ESO derived peptides has been studied in both melanoma and hepatocellular carcinoma models and shown to elicit both cellular and humoral immune responses. NY-ESO$_{157-165}$ was investigated in a phase I trial using GM-CSF as adjuvant [12] and was tested along with NY-ESO$_{157-167}$ and NY-ESO$_{155-163}$. Twelve HLA-A2 positive patients were immunized intradermally with 100 μg of peptide dissolved in 33% DMSO and PBS once a week for 4 weeks. GM-CSF was administered subcutaneously at 75 μg/day (from 3 days prior to the start of vaccination until 2 days after vaccination) to the patients with no disease progression on day 50. Both cellular (measured by ELISPOT) and humoral (measured by ELISA) responses were measured. Delayed type hypersensitivity (DTH) responses were also measured 48 h after injection. All NY-ESO peptides had either no or low reactivity. In three patients strong inflammatory responses were elicited after 3–4 injections with 5/7 NY-ESO antibody negative patients experiencing either stabilization or regression of metastasis. One melanoma patient did not show any increase in lymph node tumor for 8 months post vaccination, but a new lymph node metastasis was detected at day 258. A different melanoma patient, who had minor metastatic regression, developed necrosis in one lymph node for which the biopsy showed high intratumoral infiltration of CD4+ and CD8+ T cells. In both patients the study was discontinued. Also in NY-ESO antibody-positive patients, disease was stabilized in three of five patients while the remaining two patients experienced disease progression. All five patients developed lymph node metastasis and brain metastasis after three cycles of immunization, forcing discontinuation of treatment for all five patients [12].

17.2.3
Differentiation Antigens

Whereas shared tumor specific antigens are expressed on cancer cells only, differentiation antigens are expressed on normal tissue/cells as well as cancer cells. Some of the important antigens are CEA, gp100, PSA, Melan-A/MART-1, and NY-BR-1 (www.cancerimmunity.org).

Peptide sequences corresponding to small epitopes of the protein tyrosine kinases and gp100 have been among the most extensively studied for immunotherapy. In various phase I clinical trials, these peptides have been investigated with different adjuvants (including Montanide ISA 51, estramustine phosphate, IFA and progenipoietin, GM-CSF, QS-21, IL-2, Montanide ISA 720 and Flt-3 ligand) in conjunction with DCs [1]. $Gp100_{280-288}$ has also been safely delivered with tetanus peptide, Montanide ISA 51 or QS-21 [13, 14]. In a phase II trial, gp100 and tyrosinase peptides (alone and in conjugation with DCs) were administered together with Montanide ISA 51 or GM-CSF. IFN-γ responses were compared from PBLs (peripheral-blood lymphocytes) and SIN (sentinel immunized node) by ELISPOT assay. The T-cell responses to these treatments were different depending upon the analysis method, with higher responses when SINs (lymph nodes draining a vaccination site) rather than PBLs were monitored (80% patient responses versus 42%). There were 11/13 patients with progressive tumor in response to treatment with DC pulsed peptides, 8/12 patients had progressive tumor following treatment with peptide and GM-CSF, while one patient who was immunized with peptide adjuvanted with Montanide ISA 51 had progressive tumor [15]. In an important study, $MART-1_{26-35}$, $gp100_{209-217\,(210M)}$, and $tyrosinase_{368-376}$ were given to 15 patients along with progenipoietin (chemotherapeutic agonist of GM-CSF and FLT-3 receptors) and incomplete Freund's adjuvant (IFA) as adjuvant. Almost all patients were safely immunized with only grade I or II toxicity, except one patient who developed antibody-mediated leucopenia. This is one of the more promising peptide combinations, restricted by both class I and II HLA types [16], and is being evaluated in phase III trials along with sargramostim/NA17A.

Melan-A\MART-1_{27-35} has been used as immunogen in various phase I and II clinical trials in melanoma patients along with Montanide AS02/IL-12 alone or with other $gp100_{209-217}$, and $tyrosinase_{368-376}$ peptides with KLH (carrier) and GM-CSF (adjuvant) [17–20]. In one trial HLA-A2$^+$ patients who showed Melan-A expression were injected every 3 weeks with Melan-A\MART-1_{27-35} loaded on PBMC, using rhIL-12 (4 µg) as adjuvant [19]. This vaccine was safely injected with grade 1 or 2 toxicity in all patients, the most common side-effects being fever and fatigue. Only 2/20 patients had a complete response, 9/20 had progressive disease, 5/20 had mixed or minor responses, and 4/20 had stabilization of disease. Survival of patients was correlated with lactate dehydrogenase (LDH) levels, as LDH is a known negative prognostic factor. Direct ELISPOT assay was carried out to assess T-cell responses. Median survival of patients was just over 12 months rather than the expected 6–9 months. However, no correlation between elevated IFN-γ producing CD8+ T-cells and tumor volume was observed in patients with progressive disease. Further investigations are required to understand how cancer cells escape elimination from T-cells *in vivo*.

17.2.4
Over-Expressed Antigens

These antigens are expressed in normal cells as wells as over-expressed in various tumors. Many over-expressed antigens also have alternatively spliced variants that can react with normal patient sera as well as infected patient sera. Some of the important over-expressed antigens are HER-2/neu, hAFP, LAGE-1, XAGE-1, SSX, cyclin D1, MCSP, mdm-2, MMP-2, p53, PSMA, SOX10, survivin, telomerase, and WT1 (www.cancerimmunity.org). Since these antigens are over-expressed, a threshold response needs to be maintained to prevent autoimmunity. A number of identified peptides have been investigated by Itoh *et al.* over the last 5 years in various clinical trials as personalized peptide vaccines [21–23].

Personalized peptide vaccines (PPVs) can be designed from autologous tumors or their components, from peptides expressed by autologous tumors, and from peptides that induce higher CTL precursor frequencies. The first two approaches have not been very successful clinically for several reasons. Studying expression profiles in a tumor can be tricky depending on the difficulties in obtaining a sample from chronically ill patients. The latter approach has been widely investigated in various clinical trials, including those for lung, stomach, skin, brain, uterine, pancreas, and colorectal tumors. In a phase I trial for colorectal carcinoma, HLA-A24-positive patients were treated with the HLA-A24 binding peptides, $SART1_{690}$, $SART2_{93/161/899}$, $SART3_{109/315}$, $CypB_{84/91}$, $lck_{208/486/488}$, $ART1_{170}$, and $ART4_{13/75}$. PBMCs were first isolated from patients to study peptide-specific CTL precursor frequencies [21]. Patients were then vaccinated subcutaneously three times every 14 days with selected peptides (3 mg/ml) using Montanide ISA-51 as adjuvant. All injected peptides generated either very low or no toxicity. Half the patients (5/10) developed peptide-specific immune responses, while 7/10 patients had anti-peptide IgG responses post vaccination. One patient did not respond to vaccination at all and another partially responded.

Similarly, another study was carried out in HLA-A2/HLA-A24-positive patients with gynecological cancers following a similar protocol [22]. In this study, four HLA–A24$^+$ patients were administered with peptide vaccines chosen based on CTLs that were reactive to squamous cell carcinoma and adenocarcinoma. Another six HLA–A24$^+$ and four HLA–A2$^+$ patients were given different peptides based on CTL precursors identified before vaccination. The former group did not generate significant immune responses, while the latter group had high levels of both cellular and humoral responses in 7/10 patients. Similarly, PPVs have been tested in phase I trials for hormone-refractory prostate and gastric cancer patients with encouraging results. PPVs have also been investigated with a 5-fluorouracil derivative, TS-1, in advanced gastric or colorectal carcinoma patients. CTL responses were not greatly enhanced with doses of TS-1 at 20 mg/m^2/day, but 40–80 mg/m^2/day resulted in increased IgG and CTL responses [23]. There might be some safety concerns with this combination as 1/11 patients developed anemia and neutropenia. Overall there were some encouraging signs for these personalized peptide vaccines.

HER-2/*neu* (Human Epidermal Receptor-2/neurological) belongs to the tyrosine kinase family of growth receptors over-expressed in breast, ovary, and colorectal carcinoma. HER-2/*neu* is restricted by both class I and II HLA types. HER-2/*neu*$_{369-377}$ was initially identified in ovarian cancer as a TAA [24]. Since then this peptide has been examined using GM-CSF and IFA as adjuvant. It has also been investigated when coupled to KLH as carrier, using QS-21 or detox as adjuvant. Initially examined with IFA as adjuvant, it was found that CTLs generated by the peptide in HLA-A2$^+$ patients did not react with specific tumors [25]. When this peptide was injected with GM-CSF (125 µg), 3/4 patients showed peptide-specific IFN-γ production, however all patients developed a peptide specific DTH response [26]. In another study, 100 µg of a combination of three peptides from the HER-2/*neu* intracellular domain (p776–790, p927–941, p1166–1180) or extracellular domain (p42–56, p98–114, 328–345) was injected with GM-CSF (250 µg) as adjuvant in breast and ovarian cancer patients. T-cell responses were generated in 8/8 patients but a large-scale clinical analysis of these vaccine candidates will need to be undertaken to provide a thorough understanding of efficacy. T-cell responses observed for extracellular and intracellular domain vaccine were specific to the peptide as well HER-2/*neu* protein. Interestingly vaccines also elicited immune response to protein domains not included in the vaccine. One patient developed lymphocytic infiltration, urticaria and generalized puritis to the extracellular domain vaccine, particularly in response to the p98–114 peptide. The DTH response which developed after vaccination requires further investigation [27].

WT1 (Wilm's tumor) is another promising TAA over-expressed in leukemia, breast cancer, and glioblastoma and presented by both class I and II HLA types. In a recently published phase I and II trial, safety and toxicity of WT1$_{235-243}$ was examined [28]. Of 10 patients who received vaccinations weekly for 12 weeks with Montanide ISA 51 adjuvant, nine were reported to show grade 1 or 2 toxicity, five showed stabilization of disease and four patients had progressive disease. All but one patient had previously had surgery and chemotherapy/radiotherapy. No patient had had any therapy at least 28 days prior to vaccination and there were no other toxicities observed. This study confirmed that WT1 peptide vaccination is safe with mild local inflammatory responses, but peptide efficacy really needs to be examined now in a larger number of patients. In a further study, evidence was provided that the WT1 peptide was able to induce immune responses even at a low dose of 5 µg injected biweekly. In another phase I study, the WT1$_{126-134}$ peptide was investigated in an AML patient injected subcutaneously with 0.2 mg of peptide admixed with 1 mg of KLH and 75 µg/day

GM-CSF was injected daily from -2 to $+1$ day. After four cycles of vaccination over a year the patient was reported to have WT1 mRNA levels that were reduced by approximately 10-fold, comparable to the levels of WT1 in the bone marrow of a healthy individual [29]. A phase II study is yet to begin for this candidate vaccine. In other phase II trials that are yet to start, the $WT1_{126-134}$ peptide will also be investigated with IFA, sargramostim and $PR1_{169-177}$. Apart from these, other important peptides in clinical trials include MUC1 (phase II) and hTERT (phase I), which are now in progress (www.clinicaltrials.gov).

17.2.5
Viral Antigens

Viruses are known to be responsible for at least 20% of the global cancer burden and so viral proteins are important antigens in connection with cancer immunotherapeutics. Persistent viral infections that appear to be linked with the development of cancer include human papilloma virus HPV (cervical cancer), Epstein-Barr virus EBV (lymphoma and Hodgkin's disease, nasopharyngeal carcinoma), human endogenous retroviruses or HERV (germ cell tumors, prostate, renal, and breast cancers), human T-cell leukemia virus HTLV (T-cell leukemia), hepatitis B and C viruses (liver cancer). In relation to HPV, there are now two VLP-based prophylactic vaccines that can be successfully used to prevent cervical cancer caused by specific high-risk strains of HPV [30]. Towards the development of therapeutic vaccines, several short HPV peptide epitopes, mostly based on the E6 and E7 viral proteins, have shown some promise against HPV-induced cervical cancer. For example, HPV 16 $E7_{12-20}$ has been studied in phase I trials in cervical and vulvar cancer. In a phase I and II trial, HLA-A0201$^+$ patients with HPV16-associated advanced cervical carcinoma were immunized with HPV 16 $E7_{12-20}$ + HPV 16 $E7_{86-93}$ using Montanide ISA 51 as adjuvant [31]. Patients were given four subcutaneous vaccinations 3 weeks apart. As patients treated with 100 μg or 300 μg of each peptide experienced no side-effects a further nine patients received 1000 μg of each peptide. All patients had previously received chemotherapy, radiotherapy, surgery or some combination. Two patients, who had had radiotherapy, then the vaccination schedule, followed by chemotherapy showed regression of lesions. A patient, who had received chemotherapy, then immunization with the vaccine, had 6/7 lesions which were refractory to subsequent chemotherapy. One patient, who had chemotherapy for the first time after immunization, showed tumor regression but later died of the disease.

Immunotherapy for virus-induced cancer is greatly affected by the complexity associated with viruses, their numerous serotypes, and incomplete understanding of CTL escape. Peptide vaccines for virally-induced cancers are few at the present time. They have been heavily studied for HPV due to pioneering work in the field of cervical cancer particularly by Frazer *et al.* [32, 33], but for most other viruses peptide vaccines are still in early development. For Epstein-Barr virus, a peptide vaccine has recently been reported to have been safely administered to EBV seronegative patients in a phase I trial. Only mild side-effects accompanied otherwise encouraging T-cell responses [34].

17.3
Conclusions

Although cancer therapy is still very much a combination of chemo- and immunotherapy, surgery and radiation treatment, the 21st century promises the use of safer, less invasive, and more cancer-specific therapies. The discovery of tumor-specific antigens has led to the notion that peptides of 8–13 amino acid residues might provide new starting points for the development of both prophylactic and therapeutic vaccines against cancers. Presently, such small peptides suffer from poor immunogenicity, which may be due to a combination of their conformational instability as well as their susceptibility to proteolytic degradation. Current research is in progress to increase the immune response to short peptides by chemically stabilizing their structures, using various presentation strategies such as on dendritic cells and with different adjuvants [3]. Peptides pulsed on dendritic cells are being evaluated in several clinical trials and have shown encouraging results. Terminal modifications, C-terminal amidation or polyethylene glycolation and/or N-terminal acetylation, to peptides to increase their half-lives have also been shown to be possible without losing immunogenic integrity *in vitro* [35]. Technologies that incorporate strategies for recognizing and better defining T-cell and B-cell epitopes are also only now beginning to evolve. While such studies continue to progress, some encouraging proof-of-concept investigations on the viability of unmodified peptides as vaccines based on TAAs have been presented in this chapter.

Numerous clinical trials have reported on the promise of new peptide vaccine candidates in patients. Among these have been hTERT peptide (GV1001) in phase II/III, bcr-abl in phase II, MAGE in phase II, gastrin-related peptides in phase III, gp100 and tyrosinase in phase III, and MUC1 in phase II trials (www.clinicaltrials.gov). Nevertheless, there is still a need to investigate and develop new technologies for more effective vaccine delivery and outcomes, and to develop more effective adjuvants with no or less toxicity.

Immunization with *mutant peptides* is mostly patient specific and the time and costs involved in identifying specific mutations for each individual should preclude the standard use of personalized approaches for some time. *Shared tumor specific antigens* are the safest alternative for immunization given their significantly low or lack of expression in normal cells. These antigens seem to be progressing well in combination with *differentiation antigens*. The progress of the gp100 + tyrosinase + NY-ESO combination with a chemotherapeutic agent into a phase III clinical trial is encouraging, but it is important that dose optimization studies are done on a larger scale to statistically evaluate metastasis in normal tissues and other side-effects because differentiation antigens are also expressed on normal cells/tissues. *Overexpressed antigens* have been investigated for safety as personalized peptide vaccines in phase I trials but their efficacy is yet to be established in phase II and III trials.

In summary, peptide vaccines can be used safely and effectively for immunization, but there is still a long way to go to identify the most appropriate tumor-associated antigens and their peptide derivatives, to enhance peptide stability and immunogenicity and to identify more effective and safer adjuvants that can provide appropriate

T-cell help. The appropriate composition of prophylactic versus therapeutic vaccines will also require considerable additional study. On average a new vaccine takes about 15–20 years to reach the marketplace, so it may take another decade before we can judge whether peptide vaccines have realized their early promise, especially for the prevention and/or treatment of cancers.

Acknowledgments

We thank the Australian Research Council and the National Health and Medical Research Council for financial support.

References

1 Hans, D., Young, P.R. and Fairlie, D.P. (2006) Current status of short synthetic peptides as vaccines. *Medicinal Chemistry*, 2, 627–646.

2 Naz, R.K. and Dabir, P. (2007) Peptide vaccines against cancer, infectious diseases, and conception. *Frontiers in Bioscience: a Journal and Virtual Library*, 12, 1833–1844.

3 Purcell, A.W., McCluskey, J. and Rossjohn, J. (2007) More than one reason to rethink the use of peptides in vaccine design. *Nature Reviews. Drug Discovery*, 6, 404–414.

4 van der Bruggen, P., Traversari, C., Chomez, P., Lurquin, C., De Plaen, E., Van den Eynde, B., Knuth, A. and Boon, T. (1991) A gene encoding an antigen recognized by cytolytic T lymphocytes on a human melanoma. *Science*, 254, 1643–1647.

5 Cathcart, K., Pinilla-Ibarz, J., Korontsvit, T., Schwartz, J., Zakhaleva, V., Papadopoulos, E.B. and Scheinberg, D.A. (2004) A multivalent bcr-abl fusion peptide vaccination trial in patients with chronic myeloid leukemia. *Blood*, 103, 1037–1042.

6 Pinilla-Ibarz, J., Cathcart, K., Korontsvit, T., Soignet, S., Bocchia, M., Caggiano, J., Lai, L., Jimenez, J., Kolitz, J. and Scheinberg, D.A. (2000) Vaccination of patients with chronic myelogenous leukemia with bcr-abl oncogene breakpoint fusion peptides generates specific immune responses. *Blood*, 95, 1781–1787.

7 Bocchia, M., Gentili, S., Abruzzese, E., Fanelli, A., Iuliano, F., Tabilio, A., Amabile, M., Forconi, F., Gozzetti, A., Raspadori, D., Amadori, S. and Lauria, F. (2005) Effect of a p210 multipeptide vaccine associated with imatinib or interferon in patients with chronic myeloid leukaemia and persistent residual disease: a multicentre observational trial. *Lancet*, 365, 657–662.

8 Carbone, D.P., Ciernik, I.F., Kelley, M.J., Smith, M.C., Nadaf, S., Kavanaugh, D., Maher, V.E., Stipanov, M., Contois, D., Johnson, B.E., Pendleton, C.D., Seifert, B., Carter, C., Read, E.J., Greenblatt, J., Top, L.E., Kelsey, M.I., Minna, J.D. and Berzofsky, J.A. (2005) Immunization with mutant p53- and K-ras-derived peptides in cancer patients: immune response and clinical outcome. *Journal of Clinical Oncology*, 23, 5099–5107.

9 Kawaguchi, S., Wada, T., Ida, K., Sato, Y., Nagoya, S., Tsukahara, T., Kimura, S., Sahara, H., Ikeda, H., Shimozawa, K., Asanuma, H., Torigoe, T., Hiraga, H., Ishii, T., Tatezaki, S.I., Sato, N. and Yamashita, T. (2005) Phase I vaccination trial of SYT-SSX junction peptide in patients with disseminated synovial sarcoma. *Journal of Translational Medicine*, 3, 1.

10 Thurner, B., Haendle, I., Roder, C., Dieckmann, D., Keikavoussi, P., Jonuleit, H., Bender, A., Maczek, C., Schreiner, D., von den Driesch, P., Brocker, E.B., Steinman, R.M., Enk, A., Kampgen, E. and Schuler, G. (1999) Vaccination with mage-3A1 peptide-pulsed mature, monocyte-derived dendritic cells expands specific cytotoxic T cells and induces regression of some metastases in advanced stage IV melanoma. *The Journal of Experimental Medicine*, **190**, 1669–1678.

11 Schadendorf, D., Ugurel, S., Schuler-Thurner, B., Nestle, F.O., Enk, A., Brocker, E.B., Grabbe, S., Rittgen, W., Edler, L., Sucker, A., Zimpfer-Rechner, C., Berger, T., Kamarashev, J., Burg, G., Jonuleit, H., Tuttenberg, A., Becker, J.C., Keikavoussi, P., Kampgen, E. and Schuler, G. (2006) Dacarbazine (DTIC) versus vaccination with autologous peptide-pulsed dendritic cells (DC) in first-line treatment of patients with metastatic melanoma: a randomized phase III trial of the DC study group of the DeCOG. *Annals of Oncology*, **17**, 563–570.

12 Jager, E., Gnjatic, S., Nagata, Y., Stockert, E., Jager, D., Karbach, J., Neumann, A., Rieckenberg, J., Chen, Y.T., Ritter, G., Hoffman, E., Arand, M., Old, L.J. and Knuth, A. (2000) Induction of primary NY-ESO-1 immunity: CD8+ T lymphocyte and antibody responses in peptide-vaccinated patients with NY-ESO-1+ cancers. *Proceedings of the National Academy of Sciences of the United States of America*, **97**, 12198–12203.

13 Slingluff, C.L., Jr. Yamshchikov, G., Neese, P., Galavotti, H., Eastham, S., Engelhard, V.H., Kittlesen, D., Deacon, D., Hibbitts, S., Grosh, W.W., Petroni, G., Cohen, R., Wiernasz, C., Patterson, J.W., Conway, B.P. and Ross, W.G. (2001) Phase I trial of a melanoma vaccine with gp100 (280–288) peptide and tetanus helper peptide in adjuvant: immunologic and clinical outcomes. *Clinical Cancer Research*, **7**, 3012–3024.

14 Yamshchikov, G.V., Barnd, D.L., Eastham, S., Galavotti, H., Patterson, J.W., Deacon, D.H., Teates, D., Neese, P., Grosh, W.W., Petroni, G., Engelhard, V.H. and Slingluff, C.L. Jr. (2001) Evaluation of peptide vaccine immunogenicity in draining lymph nodes and peripheral blood of melanoma patients. *International Journal of Cancer*, **92**, 703–711.

15 Slingluff, C.L., Jr. Petroni, G.R., Yamshchikov, G.V., Barnd, D.L., Eastham, S., Galavotti, H., Patterson, J.W., Deacon, D.H., Hibbitts, S., Teates, D., Neese, P.Y., Grosh, W.W., Chianese-Bullock, K.A., Woodson, E.M., Wiernasz, C.J., Merrill, P., Gibson, J., Ross, M. and Engelhard, V.H. (2003) Clinical and immunologic results of a randomized phase II trial of vaccination using four melanoma peptides either administered in granulocyte-macrophage colony-stimulating factor in adjuvant or pulsed on dendritic cells. *Journal of Clinical Oncology*, **21**, 4016–4026.

16 Pullarkat, V., Lee, P.P., Scotland, R., Rubio, V., Groshen, S., Gee, C., Lau, R., Snively, J., Sian, S., Woulfe, S.L., Wolfe, R.A. and Weber, J.S. (2003) A phase I trial of SD-9427 (progenipoietin) with a multipeptide vaccine for resected metastatic melanoma. *Clinical Cancer Research*, **9**, 1301–1312.

17 Atzpodien, J., Fluck, M. and Reitz, M. (2004) Individualized synthetic peptide vaccines with GM-CSF in locally advanced melanoma patients. *Cancer Biotherapy & Radiopharmaceuticals*, **19**, 758–763.

18 Jager, E., Ringhoffer, M., Dienes, H.P., Arand, M., Karbach, J., Jager, D., Ilsemann, C., Hagedorn, M., Oesch, F. and Knuth, A. (1996) Granulocyte-macrophage-colony-stimulating factor enhances immune responses to melanoma-associated peptides *in vivo*. *International Journal of Cancer*, **67**, 54–62.

19 Peterson, A.C., Harlin, H. and Gajewski, T.F. (2003) Immunization with Melan-A peptide-pulsed peripheral blood mononuclear cells plus recombinant human interleukin-12 induces clinical activity and T-cell responses in advanced

melanoma. *Journal of Clinical Oncology*, **21**, 2342–2348.

20 Speiser, D.E., Lienard, D., Pittet, M.J., Batard, P., Rimoldi, D., Guillaume, P., Cerottini, J.C. and Romero, P. (2002) In vivo activation of melanoma-specific CD8(+) T cells by endogenous tumor antigen and peptide vaccines. A comparison to virus-specific T cells. *European Journal of Immunology*, **32**, 731–741.

21 Sato, Y., Maeda, Y., Shomura, H., Sasatomi, T., Takahashi, M., Une, Y., Kondo, M., Shinohara, T., Hida, N., Katagiri, K., Sato, K., Sato, M., Yamada, A., Yamana, H., Harada, M., Itoh, K. and Todo, S. (2004) A phase I trial of cytotoxic T-lymphocyte precursor-oriented peptide vaccines for colorectal carcinoma patients. *British Journal of Cancer*, **90**, 1334–1342.

22 Tsuda, N., Mochizuki, K., Harada, M., Sukehiro, A., Kawano, K., Yamada, A., Ushijima, K., Sugiyama, T., Nishida, T., Yamana, H., Itoh, K. and Kamura, T. (2004) Vaccination with predesignated or evidence-based peptides for patients with recurrent gynecologic cancers. *Journal of Immunotherapy*, **27**, 60–72.

23 Sato, Y., Fujiwara, T., Mine, T., Shomura, H., Homma, S., Maeda, Y., Tokunaga, N., Ikeda, Y., Ishihara, Y., Yamada, A., Tanaka, N., Itoh, K., Harada, M. and Todo, S. (2007) Immunological evaluation of personalized peptide vaccination in combination with a 5-fluorouracil derivative (TS-1) for advanced gastric or colorectal carcinoma patients. *Cancer Science,* **98**, 1113–1119.

24 Fisk, B., Chesak, B., Pollack, M.S., Wharton, J.T. and Ioannides, C.G. (1994) Oligopeptide induction of a cytotoxic T lymphocyte response to HER-2/neu protooncogene in vitro. *Cellular Immunology*, **157**, 415–427.

25 Zaks, T.Z. and Rosenberg, S.A. (1998) Immunization with a peptide epitope (p369–377) from HER-2/neu leads to peptide-specific cytotoxic T lymphocytes that fail to recognize HER-2/neu+ tumors. *Cancer Research*, **58**, 4902–4908.

26 Murray, J.L., Gillogly, M.E., Przepiorka, D., Brewer, H., Ibrahim, N.K., Booser, D.J., Hortobagyi, G.N., Kudelka, A.P., Grabstein, K.H., Cheever, M.A. and Ioannides, C.G. (2002) Toxicity, immunogenicity, and induction of E75-specific tumor-lytic CTLs by HER-2 peptide E75 (369–377) combined with granulocyte macrophage colony-stimulating factor in HLA-A2+ patients with metastatic breast and ovarian cancer. *Clinical Cancer Research*, **8**, 3407–3418.

27 Disis, M.L., Grabstein, K.H., Sleath, P.R. and Cheever, M.A. (1999) Generation of immunity to the HER-2/neu oncogenic protein in patients with breast and ovarian cancer using a peptide-based vaccine. *Clinical Cancer Research*, **5**, 1289–1297.

28 Tsuboi, A., Oka, Y., Osaki, T., Kumagai, T., Tachibana, I., Hayashi, S., Murakami, M., Nakajima, H., Elisseeva, O.A., Fei, W., Masuda, T., Yasukawa, M., Oji, Y., Kawakami, M., Hosen, N., Ikegame, K., Yoshihara, S., Udaka, K., Nakatsuka, S., Aozasa, K., Kawase, I. and Sugiyama, H. (2004) WT1 peptide-based immunotherapy for patients with lung cancer: report of two cases. *Microbiology and Immunology*, **48**, 175–184.

29 Mailander, V., Scheibenbogen, C., Thiel, E., Letsch, A., Blau, I.W. and Keilholz, U. (2004) Complete remission in a patient with recurrent acute myeloid leukemia induced by vaccination with WT1 peptide in the absence of hematological or renal toxicity. *Leukemia*, **18**, 165–166.

30 Boccardo, E. and Villa, L.L. (2007) Viral origins of human cancer. *Current Medicinal Chemistry*, **14**, 2526–2539.

31 van Driel, W.J., Ressing, M.E., Kenter, G.G., Brandt, R.M., Krul, E.J., van Rossum, A.B., Schuuring, E., Offringa, R., Bauknecht, T., Tamm-Hermelink, A., van Dam, P.A., Fleuren, G.J., Kast, W.M., Melief, C.J. and Trimbos, J.B. (1999) Vaccination with HPV16 peptides of patients with advanced cervical carcinoma: clinical evaluation of a phase I-II trial.

European Journal of Cancer (Oxford, England: 1990), **35**, 946–952.

32 Liu, X.S., Abdul-Jabbar, I., Qi, Y.M., Frazer, I.H. and Zhou, J. (1998) Mucosal immunisation with papillomavirus virus-like particles elicits systemic and mucosal immunity in mice. *Virology*, **252**, 39–45.

33 Zhao, K.N., Frazer, I.H., Jun Liu, W., Williams, M. and Zhou, J. (1999) Nucleotides 1506–1625 of bovine papillomavirus type 1 genome can enhance DNA packaging by L1/L2 capsids. *Virology*, **259**, 211–218.

34 Elliott, S.L., Suhrbier, A., Miles, J.J., Lawrence, G., Pye, S.J., Le, T.T., Rosenstengel, A., Nguyen, T., Allworth, A., Burrows, S.R., Cox, J., Pye, D., Moss, D.J. and Bharadwaj, M. (2008) Phase I trial of a CD8+ T-cell peptide epitope-based vaccine for infectious mononucleosis. *Journal of Virology*, **82**, 1448–1457.

35 Brinckerhoff, L.H., Kalashnikov, V.V., Thompson, L.W., Yamshchikov, G.V., Pierce, R.A., Galavotti, H.S., Engelhard, V.H. and Slingluff, C.L. Jr. (1999) Terminal modifications inhibit proteolytic degradation of an immunogenic MART-1 (27–35) peptide: implications for peptide vaccines. *International Journal of Cancer*, **83**, 326–334.

18
DNA Vaccines for the Human Papilloma Virus

Barbara Ma, Chien-Fu Hung, and T.-C. Wu

18.1
Introduction

HPV-associated malignancies represent one of the best models for the development of antigen-specific immunotherapy. Contrary to many tumor antigens, HPV-encoded antigens represent true tumor-specific antigens since they are uniquely expressed in the HPV-associated tumors but not in HPV-infected non-tumor cells. It is now evident that HPV represents the etiological factor for development of cervical cancer. Thus, HPV-encoded antigens serve as important molecular targets for therapeutic intervention. As a result, cancer immunotherapy targeting HPV antigens has significant clinical applications.

18.1.1
HPV and Cervical Cancer

Cervical cancer is the second leading cause of cancer deaths in women worldwide with approximately 500 000 new cases diagnosed and an estimated 250 000 deaths per year [1]. More than 99% of cervical cancers contain human papillomavirus (HPV) DNA. It is now clear that persistent infection with HPV is a necessary trigger for the development of premalignant squamous intraepithelial lesions (SIL) as well as cervical cancer [2, 3]. Over two hundred HPV genotypes have been identified and are classified into low- or high-risk types. While 'low-risk' types, including HPV-6 and HPV-11, can cause low-grade benign lesions, high-risk types are linked to the development of high-grade lesions and malignant tumors [4]. Among the high-risk types, HPV 16, 18, 31, and 45 account for about 80% of cervical cancer. HPV-16 is the most prevalent, present in about half of all cervical cancers [5]. Thus, it is the focus of many recent HPV vaccine developments. The establishment of a close link between HPV infection and cervical cancer suggests that effective vaccines against HPV infection are warranted as important measures in the control of HPV-associated malignancies.

18.1.2
Events in the Progression from HPV Infection to Cervical Cancer

Understanding the molecular virology of HPV is important for the design of a successful vaccine. Several studies have shown a generally conserved HPV genomic organization. HPVs are double-stranded, non-enveloped, circular DNA viruses of the family *Papillomaviridae* whose genome encodes early proteins and late proteins. The early proteins facilitate viral DNA replication (E1, E2), cytoskeleton reorganization (E4) and cell transformation (E5, E6, E7), while the late proteins (L1, L2) comprise the structural components of the viral capsid and initiate the packaging of new virions. Though HPV DNA usually replicates in episomal form, it may integrate into host DNA, which often results in the deletion of several early (E2, E4, E5) and late (L1, L2) genes, leaving E6 and E7 as the principal proteins expressed within the infected cell. The E6 and E7 proteins interact with p53 and Rb, respectively, and render these cellular proteins dysfunctional, resulting in the disruption of cell cycle regulation and genomic instability. Hence, integration of high-risk HPV into the genome of infected cells contributes to the progression of cancer (for a review, see [6]).

18.1.3
HPV Antigens for Vaccine Development

Based on this understanding, the choice of target antigen is important in vaccine design. Preventive vaccination could be implemented to prevent infection by generating neutralizing antibodies to block HPV viral infection. The structural components of infectious viral capsids, L1 and L2 proteins, have been extensively used as targets for the development of preventive vaccines by generating neutralizing antibodies against HPV infection. Since L1 and L2 capsid proteins are expressed on the viral surface, they represent ideal targets for HPV preventive vaccines. However, L1 and L2 are not suitable targets for the development of HPV therapeutic vaccines. Development of a therapeutic HPV vaccine requires the generation of T-cell mediated immune responses, which is important for the control of established HPV infection and HPV-associated lesions. Most HPV infection is initiated in the basal cell area. L1 and L2 are not expressed in basal cells infected with HPV, whereas the HPV early proteins are expressed early in viral infection at all levels of the infected epithelium and help regulate the progression of the disease. Thus, HPV early proteins, such as E6 and E7, represent more suitable targets for the development of therapeutic HPV vaccines.

Antigen-specific immunotherapy targeting HPV E6 and E7 represents one of the most promising approaches for the control of HPV-associated malignancies. As mentioned above, E6 and E7 are critical for the induction and maintenance of cellular transformation in HPV-associated malignancies. Therefore, it is unlikely that tumor cells can escape immune attack through loss of E6 and E7 antigen. Furthermore, since E6 and E7 are totally foreign proteins, immunization against HPV-associated tumors circumvents some common cancer vaccine-associated problems such as tolerance and autoimmune response. Thus, E6 and E7 in cervical cancer represent ideal targets in the development of therapeutic vaccines.

The development of therapeutic HPV vaccines targeting HPV E6 and E7 has been an active area of research. Various forms of vaccines, such as vector-based vaccines, cell-based vaccines, protein/peptide-based and DNA-based vaccines have been described in experimental systems targeting HPV-16 E6 and/or E7 proteins (for reviews, see [7, 8]). Among these strategies, DNA vaccines have emerged as an attractive therapeutic HPV vaccine approach because of their safety, stability, ease of storage and preparation, ability to bypass MHC restriction and potential for sustained expression of antigen on MHC-peptide complexes compared to peptide or protein vaccines. Thus, this chapter will mainly focus on the discussion of HPV DNA vaccination targeting HPV-encoded antigens as an attractive approach for antigen-specific cancer immunotherapy.

18.2
DNA Vaccines for HPV

DNA vaccines are made up of an antigen gene of interest cloned into a mammalian expression vector engineered for optimal expression in eukaryotic cells. Vaccination with DNA encoding the antigen of interest may induce cellular and humoral immune response against the encoded antigen. DNA has several advantages compared to conventional viral vectors. Not only do DNA vaccines have an outstanding safety profile, but they are also easy to prepare on a large scale with high purity and high stability. Additionally, DNA vaccines have the capacity for repeated administration. Thus, DNA vaccines have emerged as an attractive approach for antigen-specific immunotherapy. HPV DNA vaccines can be conceptually classified into two major categories: (a) preventive and (b) therapeutic HPV vaccines. Preventive vaccines aim to induce neutralizing antibodies to protect against new HPV infection while therapeutic vaccines aim to eliminate existing HPV-associated lesions.

18.2.1
Preventive HPV DNA Vaccines

Prophylactic HPV vaccines generate protection via the induction of a protective humoral immune response against HPV infection. Current strategies for the development of safe and effective preventive vaccines are based on the induction of neutralizing antibodies against the capsid proteins, L1 and/or L2, in the form of virus-like particles (VLPs) or capsomeres. Two HPV L1 VLP vaccines have been developed, Cervarix^{TM}, a bivalent HPV16/18 VLP vaccine produced by GlaxoSmithKline, and Gardasil^{TM}, a quadrivalent HPV16/18/6/11 VLP vaccine manufactured by Merck.

L1 DNA vaccines may serve as an alternative approach to VLP vaccines for the induction of protective immunity. L1 DNA vaccines have been shown to generate L1-specific humoral immune responses [9–13] and protective effects [9, 14, 15] in preclinical models. Furthermore, L1 DNA vaccines have been shown to induce

L1-specific cell-mediated immunity, contributing to the protective immunity against papillomavirus infection [16].

Several strategies have been employed to enhance the immunogenicity of L1 DNA vaccines, including codon optimization [17–19] and genetic fusion to a chemokine and secretory signal peptide sequences. For example, Kim et al. demonstrated that HPV L1 DNA vaccine could be enhanced by genetic fusion to a secretory signal peptide sequence and RANTES, a chemokine known to mediate endocytosis important for intracellular delivery and processing of secreted chemokine-fused antigen. Vaccination with the chimeric DNA vaccine led to enhanced L1-specific antibody response [20]. However, it is not clear whether these antibodies contain neutralizing antibodies capable of generating protective immunity.

L1 DNA vaccines face the same limitation as VLP vaccines (i.e. type specificity). One way to resolve this issue is to include DNA encoding L1 from different types of HPV. Recently, Gasparic et al. showed that combination of different papillomavirus L1 DNA vaccines could induce L1-specific antibody responses against the different papillomavirus types included in the DNA vaccine [21]. However, there was strong interference when plasmids encoding different L1 genes were used together. For example, the co-administration of DNA vaccines encoding L1 DNA derived from HPV 11 and/or bovine papillomavirus (BPV) 1 suppressed HPV 16 L1-specific immune responses generated by HPV 16 L1 DNA vaccine. Gasparic et al. further demonstrated that this repression could be overcome by separation of immunization either by vaccination with different L1 DNA vaccines at different time points or at different locations [21].

An alternative approach to circumvent the limitation of type specificity is the employment of the minor capsid protein L2 in DNA vaccine development. L2 is a conserved antigen among different types of HPV. Thus, neutralizing antibodies against L2 of a particular type of HPV will potentially cross-react to other types of HPV, resulting in broader cross-protection. This implies that L2-based vaccines could potentially protect against a large spectrum of high-risk HPV types, as well as low-risk HPV types. While many studies have been carried out on protein-based vaccines targeting L2, there are limited studies on L2 DNA vaccines. One major limitation of L2-based vaccines compared to L1-based vaccines, however, is the limited immunogenicity of the L2 protein. Therefore, it is important to consider strategies to enhance DNA vaccine potency targeting L2. For example, employment of codon optimization represents a potential strategy to improve immunogenicity of L2 DNA vaccines [17].

There are several challenges for the future development of HPV preventive vaccines. First, it is very important to consider safety in the development of preventive vaccines. HPV DNA preventive vaccines have to present either similar, or better safety profiles compared to the current commercially available HPV preventive vaccines, Gardasil™ and Cervarix™. Second, the immunogenicity generated by the HPV DNA preventive vaccines should be comparable or better compared to VLP vaccines. Third, preventive HPV DNA vaccines should generate longer duration of protection relative to VLP vaccines. Additional factors to consider include costs, frequency of vaccination and route of administration. All these issues remain to be addressed before preventive HPV DNA vaccines can be considered as a feasible alternative to existing VLP vaccines.

18.2.2
Therapeutic HPV DNA Vaccines

It is also important to consider the development of therapeutic HPV vaccines for several reasons. First, it is estimated that it would take approximately 20 years from the implementation of mass vaccination for the effective preventive vaccines to impact the cervical cancer rates due to the prevalence of a significant population with existing HPV infections and HPV-associated lesions as well as the long latency from HPV infection to the development of invasive cervical cancer. Secondly, although preventive HPV vaccines can produce neutralizing antibodies against HPV capsid antigens, they are unlikely to eliminate established HPV infection and HPV-associated lesions. As mentioned above, HPV infection tends to initiate at the basal epithelial cells. Since infected basal cells do not express detectable levels of HPV capsid antigen, preventive HPV vaccines targeting L1 and/or L2 would be unlikely to kill the infected basal cells for therapeutic control of HPV infections. Thus, it is important to develop therapeutic HPV vaccines targeting the appropriate HPV antigens to alleviate the burden of established HPV infections and the associated lesions, and to accelerate the effect of vaccination on the incidence of cervical cancer and other HPV-associated malignancies.

A successful therapeutic vaccine should target HPV antigens that are continuously expressed in the infected cells and/or cancer cells. As described in the Introduction, early viral proteins such as E6 and E7 are potentially ideal target antigens since they are expressed early in viral infection and are essential for transformation. Furthermore, E6 and E7 are co-expressed in the majority of HPV-associated precursor lesions as well as in malignant lesions, but not in HPV-infected non-transformed cells. Thus, many therapeutic HPV vaccine approaches have targeted HPV E6 and/or E7 antigens. These approaches include various forms of vaccines, such as vector-based, protein/peptide, dendritic cell-based, tumor-based, RNA and DNA vaccines. The advantages and disadvantages of the various forms of therapeutic HPV vaccines have been summarized in Table 18.1.

Among these systems, DNA vaccination has emerged as a particularly attractive approach. DNA vaccination is a safe, simple and stable form of immunotherapy capable of generating antigen-specific immunity. In addition, compared to viral vector-based vaccines, DNA vaccines have the capacity for repeated administration without losing the ability to boost the immune response since they do not elicit neutralizing antibodies against the vector. Furthermore, DNA vaccines are easy to prepare on a large scale with high purity and high stability and can be engineered to express tumor antigenic peptides or proteins. However, DNA vaccines suffer from low immunogenicity since DNA lacks the intrinsic ability to amplify or spread from transfected cells to surrounding cells *in vivo*. The potency of DNA vaccines may be improved by targeting DNA or its encoded antigen directly to professional antigen presenting cells (APCs), and by modifying the properties

Table 18.1 Characteristics of therapeutic HPV approaches.

Vaccine approach	Advantages	Disadvantages
Live vector-based	• High immunogenicity • Wide variety of vectors available; customizable	• Risk of toxicity • Potential pre-existing immunity may influence potency • Inhibited repeat immunization
Peptide-based	• Stable • Easy to produce • Safe	• Weak immunogenicity • May not be universally effective • HLA restricted
Protein-based	• Safe • No HLA restriction	• Weak immunogenicity • Better induction of antibody response than CTL response
Dendritic cell-based	• High immunogenicity • Methods available to generate large numbers of DCs	• Labor intensive, costly • Requires individualized production
Tumor cell-based	• Useful if tumor antigen unknown • Potency can be enhanced by gene transduction or cytokine treatment • Likely to express relevant tumor antigens	• Safety concerns about injecting tumor cells into patients • Labor-intensive preparation • Weak antigen presentation by tumor cells • Requires availability of tumor cell lines or autologous tumor cells
RNA	• Non-infectious, transient • Capable of multiple immunizations • RNA replicons replicate in the cell and enhance Ag expression	• Unable to store and handle • Labor-intensive preparation • Difficult to prepare large amounts
DNA	• Safe • Easy to produce, store and transport • Capable of repeat immunizations	• Intrinsically weak immunogenicity

of APCs to enhance the antigen-specific immune response elicited by DNA vaccines.

18.2.2.1 Importance of DCs in Enhancing DNA Vaccine Potency

It is well known that professional APCs, such as dendritic cells (DCs), are the primary initiators of anti-tumor immunity. APCs are specialized cells that can present antigens to naïve $CD8^+$ and $CD4^+$ T cells and activate them to become armed effector T cells. $CD8^+$ and $CD4^+$ T cells play an important role in both cell-mediated and humoral immunity. Cytotoxic $CD8^+$ T cells allow for the direct killing of infected cells while $CD4^+$ T-helper cells can facilitate the generation of humoral immunity and augment $CD8^+$ immune responses. Dendritic cells are equipped to capture antigens, process them into antigenic peptides and load them onto major histocompatibility complex (MHC) class I and II molecules in order to efficiently present the antigens on the cell surface. In the presence of maturation-inducing stimuli such as inflammatory cytokines or stimulation via CD40, DCs upregulate co-stimulatory molecules (i.e. B7), MHC-I and MHC-II molecules, and adhesion molecules, such as intercellular cell adhesion molecule-1 (ICAM-1) [22]. These activated DCs migrate to the lymphoid organs, where they activate naïve T cells to become antigen-specific effector T cells. Therefore, many therapeutic vaccine strategies have focused on targeting antigen to professional APCs and enhancing antigen processing and presentation in DCs in order to augment T cell-mediated immune responses (for review, see [22–24]).

Since DCs play an important role in DNA vaccine-mediated immune responses, direct targeting of DNA or its encoded antigen to DCs and modification of the properties of DCs represent promising approaches to improve DNA vaccine potency. These strategies include (i) increasing the number of antigen-expressing DCs, (ii) improving antigen expression, processing, and presentation in DCs, and (iii) enhancing DC interaction with T cells during T cell priming to augment T cell-mediated immune responses (for review, see [25, 26]). These strategies are summarized in Table 18.2.

Strategies for Increasing the Number of Antigen-expressing DCs

Intradermal Administration of DNA Vaccines as an Efficient Route for Delivery of DNA to DCs The opportunity to efficiently deliver DNA into tissues containing significant number of DCs *in vivo* will increase the likelihood of transfecting large numbers of local DCs resulting in strong $CD8^+$ T-cell immune responses. Among the different routes of DNA administration, intradermal vaccination via gene gun has been to shown to be one of the most efficient methods of delivery of DNA directly into DCs. The gene gun delivers DNA-coated gold particles into the skin using pressurized helium and efficiently transfects intradermal DCs (Langerhans cells) that can mature and migrate to the lymphoid organs for T-cell priming [27]. In a head-to-head comparison of different routes of administration of E7-expressing DNA vaccines, gene gun administration has been shown to be the most efficient in generating E7-specific $CD8^+$ T cell immune responses [28]. Furthermore, gene gun has been shown to be significantly more dose-efficient than intramuscular or subcutaneous injection in generating a comparable level of antigen-specific immune responses [29]. Additionally, intradermal vaccination with DNA vaccines facilitates direct priming, where antigen expressed in DCs is directly processed within the cell and presented on MHC class I molecules to $CD8^+$ T cells. Therefore, intradermal administration of DNA vaccines via gene gun represents a convenient and effective method for the *in vivo* delivery of naked DNA into DCs. Furthermore, intradermal administration of DNA vaccines represents an ideal system for testing strategies that require the direct delivery of DNA into DCs (for a review, see [25, 26]).

Linkage of Antigen to Molecules Capable of Binding to DCs as an Efficient Strategy to Target Antigen to DCs DNA vaccines encoding antigen linked to molecules that can

Table 18.2 Strategies to enhance HPV DNA vaccine potency by modification of properties of APCs.

Strategies	Methods	Study
Strategies to increase the number of antigen-expressing DCs	Intradermal administration of DNA vaccines via gene gun	[29]
	Linkage of antigen to ligands specific for DCs	
	• Flt3 ligand	[30]
	• Heat shock proteins	[28, 31, 32]
	Intercellular spreading of encoded antigen	
	• VP22	[41, 42]
	• MVP22	[44]
Strategies to improve antigen expression, processing, and presentation in DCs	Codon optimization	[19, 48–50]
	Using intracellular targeting strategies	
	• Enhancing MHC class I antigen presentation	[30, 51–54]
	• Enhancing MHC class II antigen presentation	[55–58]
	Circumventing antigen processing via MHC class I SCT	[60]
Strategies to enhance DC and T cell interaction	Prolonging DC survival via anti-apoptotic proteins	[61, 62]
Combination strategies	Prolong DC life + circumventing antigen processing via MHC class I SCT	[66]
	Prolong DC life + enhance intracellular targeting	[63–65, 67]
	Prolong DC life + enhance intracellular targeting + recruit CD4 + T cell help	[68]

DC, dendritic cell; MHC, major histocompatibility complex; SCT, single chain trimer.

bind to DCs represents another strategy to enhance the antigen-specific immune response. It is important to consider molecules that are highly expressed on the surface of DCs, allowing HPV antigens to be targeted to DCs. For example, HPV-16 E7 antigen has been linked to flt3 ligand, which binds with flt3 receptors on DCs for DNA vaccine development [30]. In addition, DNA vaccines encoding heat shock proteins (Hsp), which bind with scavenger receptors on DCs such as CD91 have been employed to enhance HPV E7-specific immune responses [28, 31–33]. Other potential molecules that may be used for enhancing therapeutic HPV DNA vaccine potency include cytotoxic T lymphocyte antigen 4 (CTLA-4), which binds to B7-expressing DCs [34], and the Fc portion of immunoglobulin G, which can target antigen to Fc receptors on the surface of DCs [35].

Intercellular Spreading of Encoded Antigen as a Strategy to Increase Antigen-Expressing DC Populations Intercellular spreading of encoded antigen is an approach that allows DNA vaccines to potentially overcome the inability of naked DNA to naturally spread among cells *in vivo*. A promising technique for broadening the cellular localization of encoded antigen is to fuse it with herpes simplex virus type 1 VP22 (HSV-1), a viral protein that has been shown to distribute proteins to the nuclei of many surrounding cells [36–40]. HSV-1 VP22 has been employed in DNA vaccines to enhance vaccine-elicited immune responses against HPV-16 E6 [41] and HPV-16 E7 [42, 43]. In addition, vaccination with a DNA vaccine encoding a gene consisting of Marek's disease virus VP22 (MVP22) linked to HPV-16 E7 has also demonstrated enhanced E7-specific CD8$^+$ T cell responses and anti-tumor effects against E7-expression tumors in mice [44]. Although the intercellular trafficking ability of VP22 remains controversial [45], DNA vaccines encoding antigen linked to VP22 have been shown to generate enhanced immune responses against the encoded antigen [46, 47].

Strategies for Enhancing Antigen Expression, Processing, and Presentation in DCs

Codon Optimization for Enhancement of Antigen Expression in DCs Codon optimization is one method for improving antigen expression of HPV DNA vaccines in DCs. Antigenic gene sequences are modified by replacing codons that are rarely recognized by cellular protein synthesis machinery with more commonly recognized codons in order to enhance translation of a DNA vaccine in DCs. It has also been shown to be effective in increasing the HPV-16 E6/E7-specific cytotoxic T lymphocyte (CTL) response induced by DNA vaccines [19, 48–50].

Using Intracellular Targeting Strategies to Enhance MHC Class I and Class II Antigen Presentation in DCs Increasing understanding of the antigen processing and presentation pathways has created an opportunity for improving DNA vaccine potency. Enhancing MHC class I and II antigen processing and presentation in DCs may lead to the activation and expansion of antigen-specific T cells. Figure 18.1 summarizes the strategies to enhance DNA vaccine potency via targeting MHC class I and II antigen presentation pathways. For example, several studies have shown that DNA vaccines encoding E7 linked to molecules capable of enhancing MHC class I processing and presentation, have increased E7-specific CD8+ T cell immune responses, resulting in significant anti-tumor effects. These molecules include *M. tuberculosis* hsp70 [51], calreticulin (CRT) [52], γ-tubulin [53], the extracellular domain of Flt3 (Fms-like tyrosine kinase 3)-ligand [30], and the translocation domain of *Pseudomonas aeruginosa* exotoxin A [54].

Strategies to enhance MHC class II presentation of HPV antigen encoded by DNA vaccines have also been tested in preclinical models. Fusion of antigen to MHC class II-targeting molecules can redirect an antigen to the class II pathway and generate greater CD4$^+$ T-cell responses that augment CTL responses and generate memory cells (for a review, see [25, 26]). For example, it has been previously shown that linkage of E7 antigen to the sorting signal of lysosome-associated

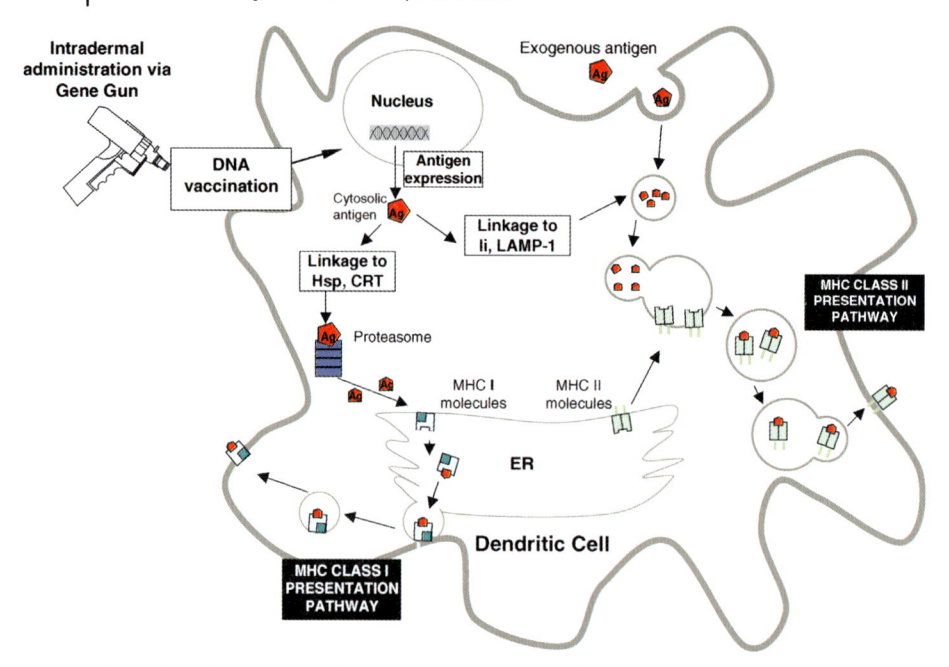

Figure 18.1 Strategies to enhance DNA vaccine potency. Delivering DNA vaccines via a gene gun directly into the nucleus of DCs increases the likelihood of transfecting large numbers of local DCs, increasing the number of antigen-expressing DCs and leading to stronger CD8+ T cell immune responses. Fusion of an antigen with molecules that target the antigen to the endoplasmic reticulum (ER) can enhance MHC class I presentation, while fusion of the antigen with endosomal/lysosomal-targeting molecules improves class II presentation. Hsp, heat shock protein; CRT, calreticulin; ER, endoplasmic reticulum; LAMP-1, lysosome-associated membrane protein type 1; Ii, MHC-II associated invariant chain.

membrane protein 1 (LAMP-1) can enhance MHC class II antigen presentation and generate significant E7-specific CD4+ T-cell responses [55]. This fusion gene, termed Sig/E7/LAMP-1, was tested in the context of a DNA vaccine and produced greater numbers of E7-specific CD4+ T cells and also higher E7-specific CTL activity in vaccinated mice compared to wild-type E7 DNA alone [56].

Another strategy to improve HPV DNA vaccine potency is the utilization of MHC class II invariant chain (Ii). In the endoplasmic reticulum, Ii binds with MHC class II molecules and the class II-associated Ii peptide (CLIP) region of the Ii chain occupies the class II peptide-binding groove, preventing premature binding of antigenic peptides into the groove. In the endosomal compartments, CLIP is replaced by an antigenic peptide in the peptide-binding groove of the MHC class II molecule, allowing for the presentation of the antigenic peptide on the surface of the DC. Hung et al. have demonstrated that DNA vaccines encoding Ii chain replacing the CLIP region with the pan-human leukocyte antigen (HLA)-DR-binding epitope (PADRE) (termed Ii-PADRE) were capable of eliciting strong PADRE-specific CD4+ T cell immune responses. Furthermore, co-administration of DNA encoding E7 and DNA encoding Ii-PADRE in mice was shown to elicit potent E7-specific CD8$^+$ T cell responses compared to co-administration of DNA encoding E7 and DNA encoding unmodified Ii [57]. Furthermore, a recent study demonstrated that mice vaccinated with DNA encoding E7 fused to invariant chain (IiE7) were capable of generating E7-specific CD8$^+$ as well as CD4$^+$ T cell immune response, resulting in protective anti-tumor immunity against E7-expressing tumors [58].

The intracellular targeting strategy has also been combined with other strategies to generate an enhanced anti-tumor effect. For example, Kim et al. demonstrated that E7 DNA vaccines combining codon optimization and LAMP-1 targeting strategy resulted in dramatic enhancement in E7-specific T cell-mediated immune response and anti-tumor protection, compared to E7 DNA vaccine employing codon optimization or DNA vaccine encoding lysosome-targeted E7 alone [59].

Circumventing Antigen Processing as a Method for Generating Stable Antigen Presentation in DCs Bypassing antigen processing and presentation in DCs altogether may allow for the generation of stable MHC class I presentation of a peptide encoded by a DNA vaccine. One approach is the use of MHC class I single-chain trimer (SCT) technology, which involves linking an antigenic peptide to β2-microglobulin and MHC class I heavy chain. This technology yields a single-chain construct encoding an MHC class I

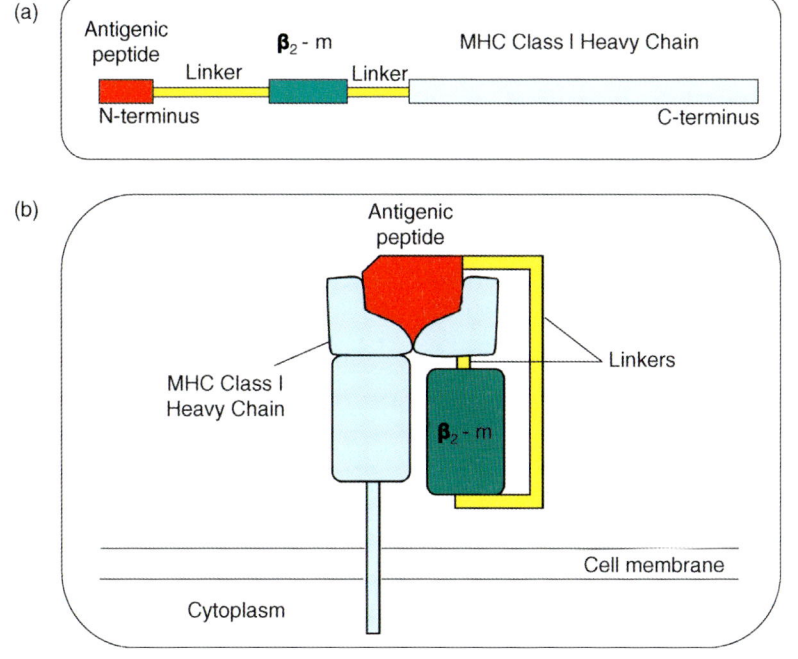

Figure 18.2 (a) Gene construct of SCT. (b) SCT comprising antigenic peptide, β2-m, and MHC I heavy chain on the cell surface. β2-m, β2-microgobulin; MHC, Major histocompatibility complex; SCT, single-chain trimer.

molecule fused to the peptide antigen. Figure 18.2 illustrates the SCT gene construct and its expression on the cell surface. Cells transfected with DNA encoding SCT of an immunodominant CTL epitope of HPV-16 E6, β2-microglobulin, and H-2K^b MHC class I heavy chain (termed pIRES-E6- β2m-K^b) resulted in the stable expression of a chimeric molecule, capable of activating E6-specific CD8$^+$ T cells. Furthermore, mice vaccinated with this DNA vaccine showed increased E6 peptide-specific CD8$^+$ T-cell responses and significant anti-tumor effects against E6-expressing tumors, compared to mice vaccinated with DNA encoding wild-type E6 [60].

Strategies for Enhancing DC and T Cell Interaction

Prolonging DC Survival to Enhance T Cell Interaction One important strategy to enhance the interaction between T cells and DCs is to prolong the life of DCs. DCs become susceptible to T cell-mediated apoptosis after T-cell priming. Thus, strategies that are capable of prolonging the life of DCs may improve the number of antigen-expressing DCs and enhance the ability of DCs to prime more T cells, resulting in augmented T-cell responses. Based on this notion, Kim *et al.* have previously shown that co-delivery of DNA encoding HPV-16 E7 antigen with DNA encoding inhibitors of apoptosis such as Bcl-xL, Bcl-2, X-linked inhibitor of apoptosis protein, and dominant-negative caspases, enhanced E7-specific CD8$^+$ T-cell responses in mice [61]. However, introduction of DNA encoding anti-apoptotic proteins into cells raises concerns of oncogenicity. An alternative approach, such as inhibition of pro-apoptotic proteins using RNA interference (RNAi), may potentially alleviate these problems. Kim *et al.* have demonstrated that a co-administration of DNA vaccines encoding E7 with short interfering RNA (siRNA) targeting the key pro-apoptotic proteins Bak and Bax was able to effectively improve DC resistance to apoptotic cell death and enhance anti-tumor CD8$^+$ T-cell responses in mice [62]. It would be of interest to further explore co-administration of DNA vaccines with siRNA targeting other key pro-apoptotic proteins, such as caspase-8, caspase-9 and/or caspase-3, to enhance DNA vaccine potency.

DNA vaccines employing strategies to prolong DC life have also been used in conjunction with other strategies to further improve DNA vaccine potency [63–68]. Several studies employed this notion, combining DNA vaccines using strategies to prolong DC life, such as DNA encoding Bcl-xL or serine protease inhibitor, with DNA vaccine encoding Sig/E7/LAMP-1, which enhances intracellular E7 antigen processing through the MHC class II pathway. This combination resulted in increased numbers of E7 CD8$^+$ T cells, leading to improved anti-tumor effects when compared to mice vaccinated without the combination strategies [63–65]. Prolonging DC survival has also been used in conjunction with strategies to enhance antigen processing and presentation in DCs. Huang *et al.* showed that DNA vaccines combining an anti-apoptotic strategy with SCT technology (see 'Circumventing antigen processing' section above) also enhanced DNA vaccine potency by augmenting antigen-specific CD8$^+$ T cell immune responses and anti-tumor effects. Mice vaccinated with SCT-E6 DNA combined with Bcl-xL DNA generated enhanced E6-specific CD8$^+$ T cell immune responses compared to mice vaccinated with SCT-E6 DNA and non-functional mutant Bcl-xL DNA [66]. Additionally, it was recently shown that the DNA vaccine encoding E7 linked to IL-6 was capable of significantly increasing E7-specific T-cell immunity, anti-E7 antibody responses and anti-tumor effects against E7-expressing tumors in vaccinated mice, presumably through the mechanisms of prolonging DC survival and enhancing antigen processing via MHC Class I pathways [67].

It is also possible to combine more than two strategies for improved efficacy of DNA vaccines. Employing strategies to prolong DC life with intracellular targeting strategies and strategies to improve $CD4^+$ T cell help to augment CTL responses, has been effectively used to improve HPV DNA vaccine potency. For example, it was recently shown that E7 DNA vaccines using intracellular targeting strategies (Sig/E7/LAMP-1 DNA or CRT/E7 DNA) combined with $CD4^+$ T cell help strategy and a strategy to prolong DC survival (Bcl-xL DNA) was further enhanced by co-administration of DNA vaccines with DNA encoding Ii-PADRE. This combination led to significant enhancement of E7-specific $CD8^+$ T cells as well as improved therapeutic effects against established E7-expressing tumors in tumor-challenged mice [68]. All these strategies employed in DNA vaccine development modified the properties of dendritic cells through different mechanisms. It is conceivable that with our continued understanding of the biology of the dendritic cell, additional innovative strategies to enhance DNA vaccine potency will be identified.

18.2.2.2 Clinical Progress in HPV DNA Vaccine Development

The impressive results of therapeutic HPV DNA vaccines in preclinical models have led to several therapeutic HPV DNA vaccine trials in patients. These are summarized in Table 18.3. For example, a microencapsulated DNA vaccine encoding multiple HLA-A2-restricted E7-derived epitopes, termed ZYC-101 (ZYCOS, Inc., now acquired by MGI Pharma), has been tested in patients with CIN-2/3 lesions [69] and in patients with high-grade anal intra-epithelial lesions [70]. The vaccine was well tolerated in both trials. Five out of 15 women with CIN-2/3 had complete histological responses and 11 of this group showed HPV-specific T-cell responses [69]. In addition, out of the 12 individuals with anal dysplasia, 10 had increased immune response to the peptide epitopes encoded within the DNA vaccine [70]. A new version of the vaccine, termed ZYC-101a (ZYCOS, Inc., now acquired by MGI Pharma), encodes HPV-16 and HPV-18 E6- and E7-derived epitopes and was shown to resolve CIN-2/3 lesions in a subset of young women enrolled in the trial [71].

Another phase I trial using a DNA vaccine encoding modified HPV-16 E7 DNA (with a mutation that abolished the Rb binding site) linked to *M. tuberculosis* hsp70 (pNGVL4a-Sig/E7 (detox/hsp70)) was undertaken in patients with CIN-2/3 lesions at the Johns Hopkins Hospital. The vaccine was well tolerated by all patients, and some of the patients who received the maximum dose of DNA vaccine (4 mg/vaccination) generated stronger E7-specific $CD8^+$ T cell immune responses in PBMCs compared to those in patients who received lower doses. Three of nine patients receiving the maximum dose of DNA vaccine showed disease regression post-vaccination. The analysis regarding the correlation of the immune response with lesion regression is currently ongoing. The same DNA vaccine has also been

Table 18.3 Selected therapeutic HPV DNA vaccine clinical trials.

HPV Type	Ag	Sponsor	Patient Population	Ref
HPV-16	E7 epitope (aa 83 to 95)	MGI Pharma Biologics (formerly ZYCOS Inc)	Phase I trial in patients with anal HSIL	[70]
			Phase II/III trial in patients with CIN2/3	[69]
	pNGVL4a-Sig/E7 (detox)/hsp70 plasmid	NCI	Clinical trials have begun for patients with CIN2/3	a
			Clinical trials to begin soon for patients with advanced HNSCC	a
	CRT-E7 detox plasmid		Clinical trials to begin soon for patients with stage 1B1 cervical cancer using gene gun	b
HPV-16 and -18	E6 and E7 epitopes	MGI Pharma Biologics (formerly ZYCOS Inc)	Phase II/III trial in patients with CIN2/3	[71]

aa, amino acids; CIN, cervical intraepithelial neoplasia; CRT, calreticulin; HNSCC, head and neck squamous cell carcinoma; HPV, human papillomavirus; HSIL, high grade squamous intraepithelial lesion.
[a] HPV DNA vaccine clinical trials at the Johns Hopkins Hospital.
[b] HPV DNA vaccine trials being planned at University of Alabama at Birmingham in collaboration with Johns Hopkins University.

tested and shown to produce appreciable E7-specific immune responses in a subset of HPV-16-positive patients with head and neck squamous cell carcinoma at the Johns Hopkins University.

Another candidate DNA vaccine that is currently being prepared for clinical trials conducted at the University of Alabama at Birmingham in collaboration with Johns Hopkins Hospital is a DNA vaccine encoding calreticulin (CRT) fused to HPV-16 E7 (E7detox). This trial involves cluster (short-interval) intradermal CRT/E7 DNA vaccination using the PowderMed/Pfizer proprietary gene gun device ND-10, which is an individualized gene gun device suitable for clinical trials in HPV-16-positive patients with stage 1B1 cervical cancer. Cluster intradermal CRT/E7 DNA vaccination has been shown to rapidly generate significant E7 antigen-specific immune responses in preclinical models. The proposed trial using the cluster vaccination regimen will potentially allow the completion of the vaccination regimen before tumor resection thus facilitating assessment of the influence of the DNA vaccination on the tumor microenvironment without compromising the standard care.

18.2.3
Combined Approaches

18.2.3.1 Combined Preventive and Therapeutic HPV DNA Vaccines

An ideal HPV vaccine should be safe, simple and cost-effective, and have the ability to prevent new HPV infections as well as treat established HPV infections and HPV-associated lesions. This has led to research involving a combination of preventive and therapeutic vaccines that target both early and late HPV antigens, aiming to induce neutralizing antibodies as well as early protein-specific cellular immunity. For example, recent preclinical studies have shown the efficiency of HPV 16 L1/E7 DNA immunization in inducing L1-specific antibodies as well as E7-specific CTL responses in vaccinated mice [72]. Furthermore, Kim et al. have generated a potential preventive and therapeutic HPV DNA vaccine using human calreticulin (CRT) linked to HPV16 early proteins, E6 and E7 and the late protein L2 (hCRTE6E7L2). Vaccination with hCRTE6E7L2 DNA vaccine induced a potent E6/E7-specific $CD8^+$ T cell immune response, resulting in a significant therapeutic effect against E6/E7-expressing tumor cells as well as significant L2-specific neutralizing antibody responses, protecting against pseudovirion infection (T. W. Kim, personal communication).

18.2.3.2 HPV DNA Vaccines in Combination with Other Therapies

HPV DNA vaccines can also be combined with other forms of therapy in order to enhance their anti-tumor effects. Because DNA vaccines often generate relatively weak CTL responses, prime-boost strategies have been used to enhance vaccine potency. For example, in preclinical models, priming the immune system with a HPV E7 DNA vaccine and then boosting with a live *vaccinia* E7 vaccine has been shown to elicit stronger E7-specific $CD8^+$ T cell immune responses in vaccinated mice [73]. Based on this preclinical data, a clinical trial using pNGVL4a/Sig/E7(detox)/Hsp70 DNA prime followed by HPV-16/18 E6/E7 *vaccinia* (TA-HPV) boost is currently being planned at Johns Hopkins University in patients with CIN2/3 lesions. TA-HPV is a recombinant *vaccinia* virus encoding HPV-16/18 E6 and E7 mutated to inactivate p53 and Rb binding.

HPV DNA vaccines have also been used in conjunction with other therapeutic agents to improve therapeutic effects. For example, Kang et al. recently showed that the chemotherapeutic agent epigallocatechin-3-Gallate (EGCG), a chemical derived from green tea, could induce tumor cellular apoptosis and enhance the tumor antigen-specific T cell immune responses elicited by DNA vaccination in a therapeutic context [74]. This has led to the planning of a phase I clinical trial at Johns Hopkins University involving the combination of oral EGCG administration with intradermal administration of CRT/E7 DNA vaccination via gene gun in patients with advanced HPV-associated head and neck squamous cell carcinomas (HPV-HNSCC). It may be of interest to explore whether other chemotherapeutic agents and adjuvants could exhibit similar synergistic effects when combined with HPV DNA vaccines.

18.3
Conclusion

Cancer vaccines targeting HPV-encoded antigens represent one of the best models for the development of antigen-specific cancer immunotherapy. HPV-encoded antigens are important tumor-associated antigens in that they are expressed uniquely and universally in HPV-associated malignancies. Furthermore, they are totally foreign to humans, thus circumventing the issue of immune tolerance. In the development of preventive HPV vaccines, investigators have focused on targeting L1 and L2 viral capsid proteins to generate neutralizing antibodies to block HPV infection. Though the development of commercially available prophylactic HPV vaccines represents a great step forward in controlling HPV through preventing new infections, the large existing burden of established HPV infections and HPV-associated lesions demonstrates the need for continued efforts to develop therapeutic HPV vaccines. For the development of therapeutic HPV vaccines, most investigators have focused on HPV E6 and E7 antigens, which are crucial for the initiation and maintenance of malignant transformation in HPV-infected cells. The promising results from preclinical models have led to several ongoing clinical trials with therapeutic HPV DNA vaccines in patients with HPV-associated cervical lesions and a subset of HPV-associated head and neck squamous cell carcinoma. These clinical studies allow us to identify the relevant characteristics and mechanisms of the immune responses that lead to the greatest enhancements in therapeutic HPV DNA vaccine potency. Such immunological parameters will be useful for characterizing effective immune mechanisms in the development of better HPV DNA vaccines. The innovative strategies for therapeutic HPV DNA vaccines may potentially be applicable to other cancers with well-defined tumor-associated antigens.

Acknowledgments

This review is not intended to be encyclopedic, and the authors apologize to those not cited. We gratefully acknowledge Ms. Archana Monie for her assistance in preparation of the manuscript. This work was supported by the National Cancer Institute SPORE in Cervical Cancer P50 CA098252 and the 1 RO1 CA114425-01.

References

1 Parkin, D.M., Bray, F., Ferlay, J. and Pisani, P. (2005) Global cancer statistics, 2002. CA: A Cancer Journal for Clinicians, 55, 74–108.
2 Walboomers, J.M., Jacobs, M.V., Manos, M.M. et al. (1999) Human papillomavirus is a necessary cause of invasive cervical cancer worldwide. The Journal of Pathology, 189, 12–19.
3 Ferenczy, A. and Franco, E. (2002) Persistent human papillomavirus infection and cervical neoplasia. The Lancet Oncology, 3, 11–16.

4 de Villiers, E.M., Fauquet, C., Broker, T.R., Bernard, H.U. and zur Hausen, H. (2004) Classification of papillomaviruses. *Virology*, **324**, 17–27.

5 Bosch, F.X., Manos, M.M., Munoz, N. *et al.* (1995) Prevalence of human papillomavirus in cervical cancer: a worldwide perspective. International biological study on cervical cancer (IBSCC) Study Group. *Journal of the National Cancer Institute*, **87**, 796–802.

6 zur Hausen, H. (2002) Papillomaviruses and cancer: from basic studies to clinical application. *Nature Reviews. Cancer*, **2**, 342–350.

7 Lin, Y.Y., Alphs, H., Hung, C.F., Roden, R.B. and Wu, T.C. (2007) Vaccines against human papillomavirus. *Frontiers in Bioscience: a Journal and Virtual Library*, **12**, 246–264.

8 Wu, T.C. (2007) Therapeutic human papillomavirus DNA vaccination strategies to control cervical cancer. *European Journal of Immunology*, **37**, 310–314.

9 Donnelly, J.J., Martinez, D., Jansen, K.U., Ellis, R.W., Montgomery, D.L. and Liu, M.A. (1996) Protection against papillomavirus with a polynucleotide vaccine. *The Journal of Infectious Diseases*, **173**, 314–320.

10 Schreckenberger, C., Sethupathi, P., Kanjanahaluethai, A. *et al.* (2000) Induction of an HPV 6bL1-specific mucosal IgA response by DNA immunization. *Vaccine*, **19**, 227–233.

11 Sun, X., Si, L., Cao, Z., Wang, Y., Liu, T. and Guo, J. (2000) Immune response to plasmid DNA encoding HPV16-L1 protein. *Chinese Medical Journal*, **113**, 277–280.

12 Matsumoto, K., Kawana, K., Yoshikawa, H., Taketani, Y., Yoshiike, K. and Kanda, T. (2000) DNA vaccination of mice with plasmid expressing human papillomavirus 6 major capsid protein L1 elicits type-specific antibodies neutralizing pseudovirions constructed *in vitro*. *Journal of Medical Virology*, **60**, 200–204.

13 Rocha-Zavaleta, L., Alejandre, J.E. and Garcia-Carranca, A. (2002) Parenteral and oral immunization with a plasmid DNA expressing the human papillomavirus 16-L1 gene induces systemic and mucosal antibodies and cytotoxic T lymphocyte responses. *Journal of Medical Virology*, **66**, 86–95.

14 Stanley, M.A., Moore, R.A., Nicholls, P.K. *et al.* (2001) Intra-epithelial vaccination with COPV L1 DNA by particle-mediated DNA delivery protects against mucosal challenge with infectious COPV in beagle dogs. *Vaccine*, **19**, 2783–2792.

15 Maeda, H., Kubo, K., Sugita, Y. *et al.* (2005) DNA vaccine against hamster oral papillomavirus-associated oral cancer. *The Journal of International Medical Research*, **33**, 647–653.

16 Hu, J., Cladel, N.M., Budgeon, L.R., Reed, C.A., Pickel, M.D. and Christensen, N.D. (2006) Protective cell-mediated immunity by DNA vaccination against Papillomavirus L1 capsid protein in the Cottontail Rabbit Papillomavirus model. *Viral Immunology*, **19**, 492–507.

17 Leder, C., Kleinschmidt, J.A., Wiethe, C. and Muller, M. (2001) Enhancement of capsid gene expression: preparing the human papillomavirus type 16 major structural gene L1 for DNA vaccination purposes. *Journal of Virology*, **75**, 9201–9209.

18 Liu, W.J., Zhao, K.N., Gao, F.G., Leggatt, G.R., Fernando, G.J. and Frazer, I.H. (2001) Polynucleotide viral vaccines: codon optimisation and ubiquitin conjugation enhances prophylactic and therapeutic efficacy. *Vaccine*, **20**, 862–869.

19 Cheung, Y.K., Cheng, S.C., Sin, F.W. and Xie, Y. (2004) Plasmid encoding papillomavirus Type 16 (HPV16) DNA constructed with codon optimization improved the immunogenicity against HPV infection. *Vaccine*, **23**, 629–638.

20 Kim, S.J., Lee, C., Lee, S.Y. *et al.* (2003) Enhanced immunogenicity of human papillomavirus 16 L1 genetic vaccines fused to an ER-targeting secretory

signal peptide and RANTES. *Gene Therapy*, **10**, 1268–1273.

21 Gasparic, M., Rubio, I., Thones, N., Gissmann, L. and Muller, M. (2007) Prophylactic DNA immunization against multiple papillomavirus types. *Vaccine*, **25**, 4540–4553.

22 Guermonprez, P., Valladeau, J., Zitvogel, L., Thery, C. and Amigorena, S. (2002) Antigen presentation and T cell stimulation by dendritic cells. *Annual Review of Immunology*, **20**, 621–667.

23 Steinman, R.M. (1991) The dendritic cell system and its role in immunogenicity. *Annual Review of Immunology*, **9**, 271–296.

24 Banchereau, J., Briere, F., Caux, C. et al. (2000) Immunobiology of dendritic cells. *Annual Review of Immunology*, **18**, 767–811.

25 Hung, C.F. and Wu, T.C. (2003) Improving DNA vaccine potency via modification of professional antigen presenting cells. *Current Opinion in Molecular Therapeutics*, **5**, 20–24.

26 Tsen, S.W., Paik, A.H., Hung, C.F. and Wu, T.C. (2007) Enhancing DNA vaccine potency by modifying the properties of antigen-presenting cells. *Expert Review of Vaccines*, **6**, 227–239.

27 Porgador, A., Irvine, K.R., Iwasaki, A., Barber, B.H., Restifo, N.P. and Germain, R.N. (1998) Predominant role for directly transfected dendritic cells in antigen presentation to CD8+ T cells after gene gun immunization. *The Journal of Experimental Medicine*, **188**, 1075–1082.

28 Trimble, C., Lin, C.T., Hung, C.F. et al. (2003) Comparison of the CD8+ T cell responses and antitumor effects generated by DNA vaccine administered through gene gun, biojector, and syringe. *Vaccine*, **21**, 4036–4042.

29 Gurunathan, S., Klinman, D.M. and Seder, R.A. (2000) DNA vaccines: immunology, application, and optimization. *Annual Review of Immunology*, **18**, 927–974.

30 Hung, C.-F., Hsu, K.-F., Cheng, W.-F. et al. (2001) Enhancement of DNA vaccine potency by linkage of antigen gene to a gene encoding the extracellular domain of Flt3-ligand. *Cancer Research*. **61**, 1080–1088.

31 Hauser, H. and Chen, S.Y. (2003) Augmentation of DNA vaccine potency through secretory heat shock protein-mediated antigen targeting. *Methods*, **31**, 225–231.

32 Hauser, H., Shen, L., Gu, Q.L., Krueger, S. and Chen, S.Y. (2004) Secretory heat-shock protein as a dendritic cell-targeting molecule: a new strategy to enhance the potency of genetic vaccines. *Gene Therapy*, **11**, 924–932.

33 Huang, C.Y., Chen, C.A., Lee, C.N. et al. (2007) DNA vaccine encoding heat shock protein 60 co-linked to HPV16 E6 and E7 tumor antigens generates more potent immunotherapeutic effects than respective E6 or E7 tumor antigens. *Gynecologic Oncology*, **107**, 404–412.

34 Boyle, J.S., Brady, J.L. and Lew, A.M. (1998) Enhanced responses to a DNA vaccine encoding a fusion antigen that is directed to sites of immune induction. *Nature*, **392**, 408–411.

35 You, Z., Huang, X., Hester, J., Toh, H.C. and Chen, S.Y. (2001) Targeting dendritic cells to enhance DNA vaccine potency. *Cancer Research*, **61**, 3704–3711.

36 Elliott, G. and O'Hare, P. (1997) Intercellular trafficking and protein delivery by a herpesvirus structural protein. *Cell*, **88**, 223–233.

37 Phelan, A., Elliott, G. and O'Hare, P. (1998) Intercellular delivery of functional p53 by the herpesvirus protein VP22. *Nature Biotechnology*, **16**, 440–443.

38 Dilber, M.S., Phelan, A., Aints, A. et al. (1999) Intercellular delivery of thymidine kinase prodrug activating enzyme by the herpes simplex virus protein, VP22. *Gene Therapy*, **6**, 12–21.

39 Wybranietz, W.A., Gross, C.D., Phelan, A. et al. (2001) Enhanced suicide gene effect by adenoviral transduction of a VP22-cytosine deaminase (CD) fusion gene. *Gene Therapy*, **8**, 1654–1664.

40 Michel, N., Osen, W., Gissmann, L., Schumacher, T.N., Zentgraf, H. and

Muller, M. (2002) Enhanced immunogenicity of HPV 16 E7 fusion proteins in DNA vaccination. *Virology*, **294**, 47–59.

41 Peng, S., Trimble, C., Ji, H. et al. (2005) Characterization of HPV-16 E6 DNA vaccines employing intracellular targeting and intercellular spreading strategies. *Journal of Biomedical Science*, **12**, 689–700.

42 Hung, C.F., Cheng, W.F., Chai, C.Y. et al. (2001) Improving vaccine potency through intercellular spreading and enhanced MHC class I presentation of antigen. *Journal of Immunology (Baltimore, Md: 1950)*, **166**, 5733–5740.

43 Kim, T.W., Hung, C.F., Kim, J.W. et al. (2004) Vaccination with a DNA vaccine encoding herpes simplex virus type 1 VP22 linked to antigen generates long-term antigen-specific CD8-positive memory T cells and protective immunity. *Human Gene Therapy*, **15**, 167–177.

44 Hung, C.F., He, L., Juang, J., Lin, T.J., Ling, M. and Wu, T.C. (2002) Improving DNA Vaccine Potency by Linking Marek's Disease Virus Type 1 VP22 to an Antigen. *Journal of Virology*, **76**, 2676–2682.

45 Perkins, S.D., Hartley, M.G., Lukaszewski, R.A., Phillpotts, R.J., Stevenson, F.K. and Bennett, A.M. (2005) VP22 enhances antibody responses from DNA vaccines but not by intercellular spread. *Vaccine*, **23**, 1931–1940.

46 Oliveira, S.C., Harms, J.S., Afonso, R.R. and Splitter, G.A. (2001) A genetic immunization adjuvant system based on BVP22-antigen fusion. *Human Gene Therapy*, **12**, 1353–1359.

47 Sciortino, M.T., Taddeo, B., Poon, A.P., Mastino, A. and Roizman, B. (2002) Of the three tegument proteins that package mRNA in herpes simplex virions, one (VP22) transports the mRNA to uninfected cells for expression prior to viral infection. *Proceedings of the National Academy of Sciences of the United States of America*, **99**, 8318–8323.

48 Liu, W.J., Gao, F., Zhao, K.N. et al. (2002) Codon modified human papillomavirus type 16 E7 DNA vaccine enhances cytotoxic T-lymphocyte induction and anti-tumour activity. *Virology*, **301**, 43–52.

49 Steinberg, T., Ohlschlager, P., Sehr, P., Osen, W. and Gissmann, L. (2005) Modification of HPV 16 E7 genes: correlation between the level of protein expression and CTL response after immunization of C57BL/6 mice. *Vaccine*, **23**, 1149–1157.

50 Lin, C.T., Tsai, Y.C., He, L. et al. (2006) A DNA Vaccine Encoding a Codon-Optimized Human Papillomavirus Type 16 E6 Gene Enhances CTL Response and Anti-tumor Activity. *Journal of Biomedical Science*, **13**, 481–488.

51 Chen, C.-H., Wang, T.-L., Hung, C.-F. et al. (2000) Enhancement of DNA vaccine potency by linkage of antigen gene to an HSP70 gene. *Cancer Research*, **60**, 1035–1042.

52 Cheng, W.F., Hung, C.F., Chai, C.Y. et al. (2001) Tumor-specific immunity and antiangiogenesis generated by a DNA vaccine encoding calreticulin linked to a tumor antigen. *The Journal of Clinical Investigation*, **108**, 669–678.

53 Hung, C.F., Cheng, W.F., He, L. et al. (2003) Enhancing major histocompatibility complex class I antigen presentation by targeting antigen to centrosomes. *Cancer Research*, **63**, 2393–2398.

54 Hung, C.-F., Cheng, W.-F., Hsu, K.-F. et al. (2001) Cancer immunotherapy using a DNA vaccine encoding the translocation domain of a bacterial toxin linked to a tumor antigen. *Cancer Research*, **61**, 3698–3703.

55 Wu, T.C., Guarnieri, F.G., Staveley-O'Carroll, K.F. et al. (1995) Engineering an intracellular pathway for major histocompatibility complex class II presentation of antigens. *Proceedings of the National Academy of Sciences of the United States of America*, **92**, 11671–11675.

56 Ji, H., Wang, T.L., Chen, C.H. et al. (1999) Targeting human papillomavirus type 16 E7 to the endosomal/lysosomal compart-

ment enhances the antitumor immunity of DNA vaccines against murine human papillomavirus type 16 E7-expressing tumors. *Human Gene Therapy*, **10**, 2727–2740.

57 Hung, C.F., Tsai, Y.C., He, L. and Wu, T.C. (2007) DNA Vaccines Encoding Ii-PADRE Generates Potent PADRE-specific CD4(+) T cell Immune Responses and Enhances Vaccine Potency. *Molecular Therapy: The Journal of the American Society of Gene Therapy*, **15**, 1211–1219.

58 Brulet, J.M., Maudoux, F., Thomas, S. *et al.* (2007) DNA vaccine encoding endosome-targeted human papillomavirus type 16 E7 protein generates CD4+ T cell-dependent protection. *European Journal of Immunology*, **37**, 376–384.

59 Kim, M.S. and Sin, J.I. (2005) Both antigen optimization and lysosomal targeting are required for enhanced anti-tumour protective immunity in a human papillomavirus E7-expressing animal tumour model. *Immunology*, **116**, 255–266.

60 Huang, C.H., Peng, S., He, L. *et al.* (2005) Cancer immunotherapy using a DNA vaccine encoding a single-chain trimer of MHC class I linked to an HPV-16 E6 immunodominant CTL epitope. *Gene Therapy*, **12**, 1180–1186.

61 Kim, T.W., Hung, C.F., Ling, M. *et al.* (2003) Enhancing DNA vaccine potency by coadministration of DNA encoding antiapoptotic proteins. *The Journal of Clinical Investigation*, **112**, 109–117.

62 Kim, T.W., Lee, J.H., He, L. *et al.* (2005) Modification of professional antigen-presenting cells with small interfering RNA *in vivo* to enhance cancer vaccine potency. *Cancer Research*, **65**, 309–316.

63 Kim, T.W., Hung, C.F., Boyd, D. *et al.* (2003) Enhancing DNA vaccine potency by combining a strategy to prolong dendritic cell life with intracellular targeting strategies. *Journal of Immunology (Baltimore, Md: 1950)*, **171**, 2970–2976.

64 Kim, T.W., Hung, C.F., Boyd, D.A. *et al.* (2004) Enhancement of DNA vaccine potency by coadministration of a tumor antigen gene and DNA encoding serine protease inhibitor-6. *Cancer Research*, **64**, 400–405.

65 Kim, T.W., Hung, C.F., Zheng, M. *et al.* (2004) A DNA vaccine co-expressing antigen and an anti-apoptotic molecule further enhances the antigen-specific CD8+ T cell immune response. *Journal of Biomedical Science*, **11**, 493–499.

66 Huang, B., Mao, C.P., Peng, S., He, L., Hung, C.F. and Wu, T.C. (2007) Intradermal administration of DNA vaccines combining a strategy to bypass antigen processing with a strategy to prolong dendritic cell survival enhances DNA vaccine potency. *Vaccine*, **25**, 7824–7831.

67 Hsieh, C.Y., Chen, C.A., Huang, C.Y. *et al.* (2007) IL-6-Encoding tumor antigen generates potent cancer immunotherapy through antigen processing and anti-apoptotic pathways. *Molecular Therapy*, **15**, 1890–1897.

68 Kim, D., Hoory, T., Wu, T.C. and Hung, C.F. (2007) Enhancing DNA vaccine potency by combining a strategy to prolong dendritic cell life and intracellular targeting strategies with a strategy to boost CD4(+) T cell. *Human Gene Therapy*, **18**, 1129–1140.

69 Sheets, E.E., Urban, R.G., Crum, C.P. *et al.* (2003) Immunotherapy of human cervical high-grade cervical intraepithelial neoplasia with microparticle-delivered human papillomavirus 16 E7 plasmid DNA. *American Journal of Obstetrics and Gynecology*, **188**, 916–926.

70 Klencke, B., Matijevic, M., Urban, R.G. *et al.* (2002) Encapsulated plasmid DNA treatment for human papillomavirus 16-associated anal dysplasia: a Phase I study of ZYC101. *Clinical Cancer Research*, **8**, 1028–1037.

71 Garcia, F., Petry, K.U., Muderspach, L. *et al.* (2004) ZYC101a for treatment of high-grade cervical intraepithelial neoplasia: a randomized controlled trial. *Obstetrics and Gynecology*, **103**, 317–326.

72 Kuck, D., Leder, C., Kern, A. *et al.* (2006) Efficiency of HPV 16 L1/E7 DNA immunization: influence of cellular localization and capsid assembly. *Vaccine*, **24**, 2952–2965.

73 Chen, C.H., Wang, T.L., Hung, C.F., Pardoll, D.M. and Wu, T.C. (2000) Boosting with recombinant vaccinia increases HPV-16 E7-specific T cell precursor frequencies of HPV-16 E7-expressing DNA vaccines. *Vaccine*, **18**, 2015–2022.

74 Kang, T.H., Lee, J.H., Song, C.K. *et al.* (2007) Epigallocatechin-3-gallate enhances CD8+ T cell-mediated antitumor immunity induced by DNA vaccination. *Cancer Research*, **67**, 802–811.

19
Adoptive T-cell Transfer in Cancer Treatment
Andreas Moosmann and Angela Krackhardt

19.1
Introduction

Adoptive transfer of antigen-specific T cells has the potential to prevent and cure cancer. Epstein-Barr virus-associated lymphomas in immunosuppressed individuals illustrate the severe effects of an interrupted surveillance by specific T cells and the benefits of restoring surveillance by T cell transfer. For malignant diseases not associated with viral infections, the protective role of adaptive immune responses is more controversially discussed [1, 2]. However, encouraging initial clinical results and important methodical innovations will help make T-cell therapy more universally applicable. In the present chapter, recent developments and perspectives for adoptive T-cell therapy of malignant disease will be discussed.

19.2
T Cell-based Therapies

19.2.1
Therapy of Virus-associated Cancers with Antigen-specific T Cells

Since the 1990s, EBV-specific therapy has established the principle that human immunogenic tumors can be prevented or successfully treated by the adoptive transfer of antigen-specific T cells [3, 4]. This success was favored by the unusual immunogenicity of transplantation-associated EBV-induced cancers, which present many viral antigens that can be recognized by T cells, and by the clinical context in which such cancers arise (donor availability and patient lymphopenia) [5]. However, EBV is associated with a wider range of malignant diseases, mainly of B-cell or epithelial origin [6]. Each of these types of cancer expresses a typical, often restricted selection of EBV proteins, and the different types of B cell-derived tumors have markedly different B cell phenotypes and immunogenicities. Therefore, strategies of EBV-specific adoptive T-cell therapy have to be adapted to the characteristics of the malignancy in question.

Conditions for T-cell therapy are especially favorable in EBV-associated lymphomas arising in immunodeficient individuals after hematopoietic stem cell transplantation (HSCT) [5]. The malignant B cells in EBV-associated post-transplant lymphoproliferative disease (EBV-PTLD) express a broad range of immunogenic EBV proteins. This includes the full set of EBV latent antigens, most of which are required for cell proliferation and transformation. Proteins of the EBV lytic cycle, required for productive virus replication, are generally co-expressed. Both latent and lytic proteins are efficiently presented to specific $CD8^+$ or $CD4^+$ T cells by the malignant cell population [7]. In EBV-PTLD after HSCT, the transplant donor is usually EBV-positive, the PTLD cells are of donor origin, and it is logical to assume that the transfer of EBV-infected B cells from donor to patient is the decisive etiologic factor. In healthy EBV carriers, EBV is controlled by a varied and expanded EBV-specific T-cell repertoire [7]; this implies that these donor's EBV-specific T cells represent an ideal means to treat the disease. Engraftment of transferred T cells in EBV-PTLD patients after HSCT is facilitated by the patient's lymphopenia.

A convenient *in vitro* model of EBV-PTLD with similar or equal biological characteristics exists in the form of lymphoblastoid cell lines (LCL), which can easily be generated by infection of peripheral B cells from patients or donors with EBV *in vitro* [8]. LCL are excellent tools to reactivate, analyze, and monitor EBV-specific T-cell reactivities [9]. Most importantly, stimulation of PBMCs with donor-autologous LCL remains the most important method to generate EBV-specific T cells for therapy purposes, due to the convenience of the procedure and the broad spectrum of EBV-specific T cells that are reactivated by LCL stimulation *in vitro*. It appears that LCL stimulate most of the various antigen specificities that are present in the T-cell memory of healthy EBV carriers. Mirroring T-cell frequencies in healthy carriers, LCL-stimulated cultures tend to be dominated by $CD8^+$ T cells specific for latent and early lytic cycle proteins, which efficiently recognize and kill EBV-infected B cells *in vitro* [7]. In addition, such cultures usually contain a smaller proportion of $CD4^+$ T cells [4]. Although it was postulated that the presence of these $CD4^+$ T cells importantly contributes to the success of adoptive transfer therapy [10], it has not been easy to detail the principle behind that observation. LCL stimulation may induce $CD4^+$ T cells specific for EBV latent antigens, but often these T cells fail to mobilize effector functions against EBV-infected targets [11]. However, recent results indicate that T cells against a broad range of late antigens, many of them EBV structural proteins, are present in the $CD4^+$ T-cell repertoire in healthy EBV carriers, and dominate in the $CD4^+$ component of LCL-stimulated T-cell preparations [12]. Because such T cells efficiently attack EBV antigen-presenting targets, their contribution to successful therapy is probably important [13].

Rooney, Heslop, and colleagues established the field of EBV-specific adoptive T-cell therapy by studying prophylaxis and treatment of EBV-PTLD after hematopoietic stem cell transplantation (HSCT) with polyclonal EBV-specific T-cell preparations obtained by stimulation of donor peripheral blood cells with the donor

LCL [3, 4, 14]. Prophylaxis was successful, with no side effects, in all of 39 children who were considered at risk of EBV lymphoma after receiving T cell-depleted allogeneic HSCT [4], while 11% of historical control patients developed EBV-PTLD. Six patients with manifest EBV lymphomas were also successfully cured, with the remarkable exception of one patient whose EBV-PTLD escaped T-cell control by deletion of sequences from the EBV genome that coded for epitopes recognized by the T cells used for therapy [15]. Additional patients have been successfully treated by this group [16], and other researchers have confirmed that the infusion of EBV-specific T cells efficiently and safely protects from EBV lymphoma – except in rare cases when T-cell lines lacking EBV antigen-specific reactivity, prepared from EBV-negative donors, were used [17]. In spite of these results, today very few high-risk HSCT patients are in a situation to receive EBV-specific T cells as a curative or prophylactic treatment. The relative complexity of current T-cell preparation protocols, increasing regulatory demands and associated costs to provide the necessary infrastructure, and the scarcity of specialized expertise might all contribute to this unfortunate situation. To make EBV-specific T-cell therapy more broadly available, efforts are being made to accelerate and facilitate the preparation of EBV-specific T cells or to widen their applicability and scope. For example, direct isolation of EBV-specific T cells and formalin fixation of the LCL used as stimulator can be combined to increase the speed of preparation and to formally exclude the admixture of live virus-infected cells [18]. Superinfection of LCL with a recombinant adenovirus carrying a cytomegalovirus antigen produces stimulator cells that can be used to generate cultures containing T cells specific for each of the three viruses, and such T cells can control infection with any of the three viruses in patients after HSCT [19].

EBV-PTLD in patients after solid organ transplantation (SOT) would appear to pose a greater obstacle to specific T-cell therapy [20]. These lymphomas, generally of patient origin, arise in the presence of sustained pharmacological immunosuppression, but in the absence of a generalized lymphopenia, factors that might not favor T-cell therapy. It was a concern that EBV-specific T cells may be difficult to obtain, especially if the patient was EBV-negative at the time of transplantation. Still, patient-derived EBV-specific T cells could be successfully generated in several larger sets of patients and were therapeutically infused for PTLD pre-emption upon EBV reactivation or for treatment of established disease [21, 22]. T-cell transfer was well tolerated, and PTLD did not emerge in any of the patients (seven patients in the first, 11 patients in the second study, which also included a case of manifest PTLD which underwent partial remission). Therefore, the transferred EBV-specific T cells appeared to exert a protective function, although T-cell expansion and maintenance *in vivo* was less pronounced than in HSCT recipients [22]. In an alternative approach, Haque *et al.* have pioneered the therapeutic use of EBV-specific T cells prepared by LCL stimulation from HLA-matched third party donors [23]. In a phase II study, they investigated the treatment of cases of established PTLD that had resisted standard treatment, and observed complete remissions in 14 of 33 patients after T-cell transfer. Thus, EBV-specific T cells appear to have considerable potential in the more difficult situation of PTLD therapy after SOT.

The spectrum of viral antigens expressed and presented by EBV-associated nasopharyngeal carcinoma (NPC) and Hodgkin's disease (HD) is much more restricted, possibly reflecting some selective pressure exerted by the patient's functional immune system. In NPC, the viral antigens EBNA1, LMP2, and occasionally LMP1 are found; in the malignant Reed-Sternberg cells of HD, all three antigens are expressed. As with PTLD, therapeutic studies on specific T-cell transfer have mainly used polyclonal EBV-specific T cells generated by stimulation with autologous LCL. Such T-cell preparations tend to contain much fewer T cells specific for LMP2, LMP1 or EBNA1 than for the immunodominant EBNA3 proteins and immediate-early proteins. Nonetheless, some encouraging clinical results were achieved. In studies on therapy of advanced NPC, control of disease progression in a majority [24] and even complete remissions in a minority of patients [25] were reported. In a therapeutic study on HD, there were complete responses in two patients with less advanced disease from a cohort of 11 patients treated [26]. Using an improved protocol to generate T cells, raising the proportion of LMP2-specific T cells by overexpressing this viral antigen in the LCL used for T-cell stimulation, clinical efficacy could be improved, and complete responses (sustained for at least 9 months) were now observed in four of six patients [27], including patients with bulky disease.

There are human cancers associated with viruses other than EBV, which are possible targets of virus-specific adoptive T-cell therapy. Much effort has been dedicated to the development of such therapies and the expansion and analysis of relevant T cells, especially against human papillomavirus (HPV) type 16/18 [28] which is associated with cervical carcinomas. Patients tend to carry HPV-specific T cells [29] that can recognize cervical carcinoma cells which weakly but consistently express the target antigens, the oncoproteins E6 and E7 [30]. However, the generation of HPV-specific T cells remains challenging; technological advances [31] might facilitate their future clinical use. Adoptive therapy of hepatocellular carcinoma with hepatitis B or C virus-specific T cells is also under consideration; however, liver toxicity of virus-specific T cells is a concern here [32]. Adult T-cell leukemias infected with human T-cell leukemia virus type 1 (HTLV-1) may retain expression of Tax, a major target of specific T cells [33]. Though the frequency of Tax-specific T cells in patients is correlated with relapse prevention after allo-HSCT [34], HTLV-1-specific T cells have not been used in human therapy so far.

Taken together, the therapeutic efficiency and safety of EBV-specific T-cell transfer is well established for EBV-PTLD after HSCT, and very promising results have also been achieved for several other EBV-associated malignancies. Therefore, it is desirable to make this form of therapy much more widely available to patients with EBV-associated cancers. There are grounds for optimism that other virus-associated malignancies might also be successfully treated by specific T-cell transfer in the future.

19.2.2
T Cells with Specificity for Self-antigens

Whereas viral antigens expressed in human tumors represent a favorable target antigen which can be well recognized by T cells, targeting of tumor-associated self-antigens in cancer proved much more difficult. Most anti-tumor responses have been reported against melanoma cells and there is considerable evidence that melanocyte differentiation antigens including cancer testis antigens presented to the immune system of melanoma patients can overcome tolerance and induce T cell-based immunity [35, 36]. Transfer of autologous tumor-reactive T cells has therefore been mostly applied in melanoma patients. PBMC-derived CTL were used to treat patients with refractory metastatic melanoma, about 50% of treated patients developed objective clinical responses [37, 38]. Although current approaches using autologous CTL or TIL show promising results, the long-term and overall response has been disappointing in cancer patients. It has been demonstrated in different models that T cells seem to have limited function *in vivo*, although infiltrating T cells with potent tumor reactivity can be readily detected and isolated from the tumor site [37, 39–41]. This might be explained by the induction of tolerance by the tumor environment or ignorance of tumors by immune cells due to a immunologically privileged site of malignancies [2, 40]. Moreover, a rapid tumor progression is also likely to play an important role in tumor escape [41, 42]. Numerous mechanisms may additionally influence anti-tumor reactivity *in vitro* and *in vivo* including expression of surface molecules such as MHC, co-stimulatory and pro-apoptotic molecules, production of inhibitory molecules such as TGF-β and IL-10, and the recruitment of regulatory T cells (Treg) [43–48]. Thus, although autologous T cells may specifically recognize abundantly detectable antigen on the surface of tumor cells even including ubiquitously expressed self-antigens [49], they often cannot elicit an effective immune response *in vivo*. The different mechanisms involved need to be intensively addressed in order to develop effective adoptive T-cell therapies in regard to functional efficiency as well as long-term survival of transferred T cells. Different approaches are currently being investigated to cope with these problems. Minimal residual disease or limited tumor burden is regarded as the best setting in which to start with adoptive T-cell therapies. In addition, it has previously been observed that transferred T cells only persisted if the infused T-cell product contained CD4+ T helper cells and that transfused clonal or oligoclonal T-cell lines induced T-cell escape mutants [15, 50]. Thus, therapeutic T-cell vaccines should target multiple antigens and should contain both CD8- and CD4-positive T cells.

19.2.3
T Cells with Specificity for Alloantigens

In the allogeneic system, donor leukocyte infusions have been shown to be highly effective and possess a curative potential in patients with hematological malignancies by inducing a graft versus leukemia and lymphoma effect (GvL) [51]. Although there are also encouraging data using reduced conditioning regimens followed by allogeneic stem cell transplantation in several solid malignancies, this approach has been much less successful in these diseases [52, 53]. This may be partially due to a special immunological privilege at the site of solid tumors, a problem likewise encountered in the autologous system. However, allogeneic stem cell transplantation in hematological as well as solid malignancies is often associated with severe graft versus host disease (GvHD) as allogeneic T cells may have high avidity T-cell receptors (TCR) towards their antigen due to the lack of these foreign MHC molecules during thymic development. An important future goal is therefore the improvement of specificity of anti-tumor immune responses by reduction of GvHD and increase of graft versus tumor (GvT) effects in order to reach a broader patient population. One approach to separate GvL/GvT from GvHD is the generation of T cells specifically recognizing minor histocompatibility antigens with restricted tissue expression in a matched major histocompatibility (MHC) context [54, 55]. However, this approach is primarily limited by the difficulty of finding matched donors. Another approach is the generation of allorestricted or xenorestricted peptide-specific T cells, which are specific for a tumor-associated antigen (TAA) presented by an allogeneic or xenogenic MHC molecule [56, 57]. In this case, an elaborate donor search is not necessary as any suitable antigen identified by any methodology can be selected for the generation of these specific T cells [58]. However, despite aberrant expression of TAA in tumor cells, many of the TAA are also expressed to some degree in non-malignant peripheral adult tissues [59] and high-avidity allorestricted T cells may harbor an enhanced risk for cross-reactivity compared to T cells restricted to autologous MHC molecules [58, 60]. Thus, the delicate balance between tumor immunity and alloreactivity may impose limitations on the use of allorestricted T cells in anti-cancer immunotherapy.

19.2.4
Engineered T Cells

The routine isolation and expansion of tumor-specific T cells from cancer patients and healthy donors have proven to be difficult. This limitation can be circumvented by cloning the genes that encode the α and β chains from specified auto- or alloreactive tumor-specific T cells and introduce them into recipient T cells to transfer the specificity of the donor T cell [61]. This procedure allows the rapid production of antigen-specific T cells after TCR transfer and has been successfully applied to different targets such as melanoma antigens, minor histocompatibility antigens and common oncoproteins [62–68]. The feasibility of this exciting approach in the clinic has been demonstrated in the treatment of metastatic melanoma [69]. However, response rates were lower compared to conventional adoptive T-cell therapy with expanded tumor-infiltrating lymphocytes [69, 70]. Limited avidity of the transferred TCR and/or inefficient TCR expression may be responsible for this reduced outcome.

The inefficient expression of introduced TCR α and β-chains in T lymphocytes can be one of the rate-limiting steps for TCR gene therapy [71, 72]. As TCR surface expression requires association of CD3 γ, δ, ε, and ζ-chains, the introduced TCR competes with endogenous TCR for a limited number of CD3 molecules. In addition, the introduced TCR chains may mispair with endogenous chains, thus further reducing the expression of relevant TCR αβ heterodimers on the surface of transduced T cells. Different approaches to improve the expression and function of TCR-transduced T cells are conceivable. The use of synthetic genes with an optimized codon can improve *in vitro* TCR expression and TCR function by avoiding *cis*-acting sequence motifs [73, 74]. Moreover, the introduction of an additional disulfide bond and mutations at the constant alpha/beta chain interface may facilitate TCR pairing and expression in human T cells [75–77]. In addition, hybrid TCR chains in which the human constant region is exchanged for murine sequences have been shown to display improved TCR pairing and enhanced association with the CD3 molecules in human T cells [78]. However, the functional benefit these T-cell modifications may represent *in vivo* needs to be further analyzed [79]. Further attempts are being made to increase the functional avidity and affinity of specific TCR by TCR engineering in order to improve tumor reactivity. Using a yeast or bacterial phage display approach high affinity TCR can be generated which however, also showed increased cross-reactivity [60, 80, 81].

In addition to the optimization of the specific TCR candidate, the optimal target population used for TCR transfer needs to be defined. Different target populations are currently under investigation including different T-cell subpopulations and progenitor cells [82–86].

An alternative approach to modifying the specificity of T cells involves the use of genes that encode monoclonal antibody chains specific for TAA [87]. Chimeric antigen receptors (CAR) are composed of an extracellular domain, which is responsible for antigen recognition containing a scFv with the heavy and light variable chains of a monoclonal antibody joined by a flexible linker. The scFv is then linked to an intracellular signaling domain, which usually consists of the TCRζ chain (CD3-ζ) or IgE high affinity receptors. CARs recognize their target antigen independently of MHC and may also be specific for carbohydrates and glycolipids. Preclinical studies in murine models have demonstrated that T cells modified to express CARs can efficiently eliminate tumors *in vivo* [88], and phase I clinical trials of adoptively transferred CAR-modified T cells have been initiated [89]. However, T-cell activation and signaling needs to be further optimized [90, 91].

19.2.5
Safety Concerns and Suicide Gene Transfer

Adoptive T-cell therapy harbors different risks, which need to be carefully addressed. TCRs and CARs may be transferred either by retroviral or lentiviral gene transfer [92, 93]. The use of retroviral vectors in gene therapy incorporates a genotoxic risk due to uncontrolled insertion into the human genome [94]. The risk for malignant transformation may be dependent on the virus, vector, and the disease as well as the cell type that is transduced, however, the consequences to the biology and function of transplanted T cells seems to be limited in most applications [94–97]. Another concern is the induction of autoimmunity as well as the potential of autonomous cell growth of genetically manipulated transferred T cells. Incorporation of suicide genes or targeting of TCR-integrated tags may therefore be necessary in order to offer more safety for clinical trials [98–100].

19.3
Conclusion

Adoptive T-cell therapies have proved to be powerful tools in the treatment of PTLD as well as in the allogeneic transplant setting for the treatment of hematological malignancies. In both applications, effector T cells from immunocompetent individuals are transferred to the highly immunocompromised patient facilitating the engraftment and efficacy of the transferred T cells. Adoptive T-cell therapies in the immunocompetent host and in solid tumors are still in their infancy. There are several major challenges which need to be intensively addressed in order to establish adoptive T-cell therapies more broadly: (1) adoptive T-cell therapies need to be well balanced between tolerance and autoimmunity in order to be effective against the tumor without causing major harm to normal tissue; (2) adoptive T-cell therapies need to be effective even at immunologically privileged sites such as in solid tumors, need to be able to treat fast-growing tumors, and need to include memory for long-lasting effects, and last but not least (3) costs and complexity of adoptive T-cell therapies need to be limited in order to reach a broader patient population. The highly promising novel developments and great efforts currently being undertaken in this field will hopefully overcome these hurdles and will soon lead to effective adoptive T-cell therapies in the treatment of a broad spectrum of malignant diseases.

References

1 Galon, J., Costes, A., Sanchez-Cabo, F., Kirilovsky, A., Mlecnik, B., Lagorce-Pages, C., Tosolini, M., Camus, M., Berger, A., Wind, P., Zinzindohoue, F., Bruneval, P., Cugnenc, P.H., Trajanoski, Z., Fridman, W.H. and Pages, F. (2006) Type, density, and location of immune cells within human colorectal tumors predict clinical outcome. *Science*, **313**, 1960.

2 Willimsky, G. and Blankenstein, T. (1995) Sporadic immunogenic tumours avoid destruction by inducing T-cell tolerance. *Nature*, **437**, 141.

3 Rooney, C.M., Smith, C.A., Ng, C.Y., Loftin, S., Li, C., Krance, R.A., Brenner, M.K. and Heslop, H.E. (1995) Use of gene-modified virus-specific T lymphocytes to control Epstein-Barr-virus-related lymphoproliferation. *Lancet*, **345**, 9.

4 Rooney, C.M., Smith, C.A., Ng, C.Y., Loftin, S.K., Sixbey, J.W., Gan, Y., Srivastava, D.K., Bowman, L.C., Krance, R.A., Brenner, M.K. and Heslop, H.E. (1998) Infusion of cytotoxic T cells for the prevention and treatment of Epstein-Barr virus-induced lymphoma in allogeneic transplant recipients. *Blood*, **92**, 1549.

5 Moss, P. and Rickinson, A. (2005) Cellular immunotherapy for viral infection after HSC transplantation. *Nature Reviews. Immunology*, **5**, 9.

6 Kuppers, R. (2003) B cells under influence: transformation of B cells by Epstein-Barr virus. *Nature Reviews. Immunology*, **3**, 801.

7 Hislop, A.D., Taylor, G.S., Sauce, D. and Rickinson, A.B. (2007) Cellular responses to viral infection in humans: lessons from Epstein-Barr virus. *Annual Review of Immunology*, **25**, 587.

8. Pope, J.H., Horne, M.K. and Scott, W. (1968) Transformation of foetal human leukocytes *in vitro* by filtrates of a human leukaemic cell line containing herpes-like virus. *International Journal of Cancer*, **3**, 857.
9. Rickinson, A.B., Moss, D.J., Allen, D.J., Wallace, L.E., Rowe, M. and Epstein, M.A. (1981) Reactivation of Epstein-Barr virus-specific cytotoxic T cells by in vitro stimulation with the autologous lymphoblastoid cell line. *International Journal of Cancer*, **27**, 593.
10. Haque, T., Wilkie, G.M., Jones, M.M., Higgins, C.D., Urquhart, G., Wingate, P., Burns, D., McAulay, K., Turner, M., Bellamy, C., Amlot, P.L., Kelly, D., MacGilchrist, A., Gandhi, M.K., Swerdlow, A.J. and Crawford, D.H. (2007) Allogeneic cytotoxic T-cell therapy for EBV-positive posttransplantation lymphoproliferative disease: results of a phase 2 multicenter clinical trial. *Blood*, **110**, 1123.
11. Long, H.M., Haigh, T.A., Gudgeon, N.H., Leen, A.M., Tsang, C.W., Brooks, J., Landais, E., Houssaint, E., Lee, S.P., Rickinson, A.B. and Taylor, G.S. (2005) CD4+T-cell responses to Epstein-Barr virus (EBV) latent-cycle antigens and the recognition of EBV-transformed lymphoblastoid cell lines. *Journal of Virology*, **79**, 4896.
12. Adhikary, D., Behrends, U., Boerschmann, H., Pfunder, A., Burdach, S., Moosmann, A., Witter, K., Bornkamm, G.W. and Mautner, J. (2007) Immunodominance of lytic cycle antigens in Epstein-Barr virus-specific CD4+T cell preparations for therapy. *PLoS ONE*, **2**, e583.
13. Adhikary, D., Behrends, U., Moosmann, A., Witter, K., Bornkamm, G.W. and Mautner, J. (2006) Control of Epstein-Barr virus infection *in vitro* by T helper cells specific for virion glycoproteins. *The Journal of Experimental Medicine*, **203**, 995.
14. Heslop, H.E., Ng, C.Y., Li, C., Smith, C.A., Loftin, S.K., Krance, R.A., Brenner, M.K. and Rooney, C.M. (1996) Long-term restoration of immunity against Epstein-Barr virus infection by adoptive transfer of gene-modified virus-specific T lymphocytes. *Nature Medicine*, **2**, 551.
15. Gottschalk, S., Ng, C.Y., Perez, M., Smith, C.A., Sample, C., Brenner, M.K., Heslop, H.E. and Rooney, C.M. (2001) An Epstein-Barr virus deletion mutant associated with fatal lymphoproliferative disease unresponsive to therapy with virus-specific CTLs. *Blood*, **97**, 835.
16. Gottschalk, S., Rooney, C.M. and Heslop, H.E. (2005) Post-transplant lymphoproliferative disorders. *Annual Review of Medicine*, **56**, 29.
17. Gustafsson, A., Levitsky, V., Zou, J.Z., Frisan, T., Dalianis, T., Ljungman, P., Ringden, O., Winiarski, J., Ernberg, I. and Masucci, M.G. (2000) Epstein-Barr virus (EBV) load in bone marrow transplant recipients at risk to develop posttransplant lymphoproliferative disease: prophylactic infusion of EBV-specific cytotoxic T cells. *Blood*, **95**, 807.
18. Hammer, M.H., Brestrich, G., Mittenzweig, A., Roemhild, A., Zwinger, S., Subklewe, M., Beier, C., Kurtz, A., Babel, N., Volk, H.D. and Reinke, P. (2007) Generation of EBV-specific T cells for adoptive immunotherapy: a novel protocol using formalin-fixed stimulator cells to increase biosafety. *Journal of Immunotherapy (1997)*, **30**, 817.
19. Leen, A.M., Myers, G.D., Sili, U., Huls, M.H., Weiss, H., Leung, K.S., Carrum, G., Krance, R.A., Chang, C.C., Molldrem, J.J., Gee, A.P., Brenner, M.K., Heslop, H.E., Rooney, C.M. and Bollard, C.M. (2006) Monoculture-derived T lymphocytes specific for multiple viruses expand and produce clinically relevant effects in immunocompromised individuals. *Nature Medicine*, **12**, 1160.
20. Williams, H. and Crawford, D.H. (2006) Epstein-Barr virus: the impact of scientific

advances on clinical practice. *Blood*, **107**, 862.

21 Comoli, P., Labirio, M., Basso, S., Baldanti, F., Grossi, P., Furione, M., Vigano, M., Fiocchi, R., Rossi, G., Ginevri, F., Gridelli, B., Moretta, A., Montagna, D., Locatelli, F., Gerna, G. and Maccario, R. (2002) Infusion of autologous Epstein-Barr virus (EBV)-specific cytotoxic T cells for prevention of EBV-related lymphoproliferative disorder in solid organ transplant recipients with evidence of active virus replication. *Blood*, **99**, 2592.

22 Savoldo, B., Goss, J.A., Hammer, M.M., Zhang, L., Lopez, T., Gee, A.P., Lin, Y.F., Quiros-Tejeira, R.E., Reinke, P., Schubert, S., Gottschalk, S., Finegold, M.J., Brenner, M.K., Rooney, C.M. and Heslop, H.E. (2006) Treatment of solid organ transplant recipients with autologous Epstein Barr virus-specific cytotoxic T lymphocytes (CTLs). *Blood*, **108**, 2942.

23 Haque, T., Wilkie, G.M., Taylor, C., Amlot, P.L., Murad, P., Iley, A., Dombagoda, D., Britton, K.M., Swerdlow, A.J. and Crawford, D.H. (2002) Treatment of Epstein-Barr-virus-positive post-transplantation lymphoproliferative disease with partly HLA-matched allogeneic cytotoxic T cells. *Lancet*, **360**, 436.

24 Comoli, P., Pedrazzoli, P., Maccario, R., Basso, S., Carminati, O., Labirio, M., Schiavo, R., Secondino, S., Frasson, C., Perotti, C., Moroni, M., Locatelli, F. and Siena, S. (2005) Cell therapy of stage IV nasopharyngeal carcinoma with autologous Epstein-Barr virus-targeted cytotoxic T lymphocytes. *Journal of Clinical Oncology*, **23**, 8942.

25 Straathof, K.C., Bollard, C.M., Popat, U., Huls, M.H., Lopez, T., Morriss, M.C., Gresik, M.V., Gee, A.P., Russell, H.V., Brenner, M.K., Rooney, C.M. and Heslop, H.E. (2005) Treatment of nasopharyngeal carcinoma with Epstein-Barr virus–specific T lymphocytes. *Blood*, **105**, 1898.

26 Bollard, C.M., Aguilar, L., Straathof, K.C., Gahn, B., Huls, M.H., Rousseau, A., Sixbey, J., Gresik, M.V., Carrum, G., Hudson, M., Dilloo, D., Gee, A., Brenner, M.K., Rooney, C.M. and Heslop, H.E. (2004) Cytotoxic T lymphocyte therapy for Epstein-Barr virus+ Hodgkin's disease. *The Journal of Experimental Medicine*, **200**, 1623.

27 Bollard, C.M., Gottschalk, S., Leen, A.M., Weiss, H., Straathof, K.C., Carrum, G., Khalil, M., Wu, M.F., Huls, M.H., Chang, C.C., Gresik, M.V., Gee, A.P., Brenner, M.K., Rooney, C.M. and Heslop, H.E. (2007) Complete responses of relapsed lymphoma following genetic modification of tumor-antigen presenting cells and T-lymphocyte transfer. *Blood*, **110**, 2838.

28 Ressing, M.E., Sette, A., Brandt, R.M., Ruppert, J., Wentworth, P.A., Hartman, M., Oseroff, C., Grey, H.M., Melief, C.J. and Kast, W.M. (1995) Human CTL epitopes encoded by human papillomavirus type 16 E6 and E7 identified through *in vivo* and *in vitro* immunogenicity studies of HLA-A*0201-binding peptides. *Journal of Immunology (Baltimore, Md: 1950)*, **154**, 5934.

29 Youde, S.J., Dunbar, P.R., Evans, E.M., Fiander, A.N., Borysiewicz, L.K., Cerundolo, V. and Man, S. (2000) Use of fluorogenic histocompatibility leukocyte antigen-A*0201/HPV 16 E7 peptide complexes to isolate rare human cytotoxic T-lymphocyte-recognizing endogenous human papillomavirus antigens. *Cancer Research*, **60**, 365.

30 Crook, T., Morgenstern, J.P., Crawford, L. and Banks, L. (1989) Continued expression of HPV-16 E7 protein is required for maintenance of the transformed phenotype of cells co-transformed by HPV-16 plus EJ-ras. *The EMBO Journal*, **8**, 513.

31 Zentz, C., Wiesner, M., Man, S., Frankenberger, B., Wollenberg, B., Hillemanns, P., Zeidler, R., Hammerschmidt, W. and Moosmann, A.

(2007) Activated B cells mediate efficient expansion of rare antigen-specific T cells. *Human Immunology*, **68**, 75.

32 Butterfield, L.H. (2007) Recent advances in immunotherapy for hepatocellular cancer. *Swiss Medical Weekly*, **137**, 83.

33 Kannagi, M. (2007) Immunologic control of human T-cell leukemia virus type I and adult T-cell leukemia. *International Journal of Hematology*, **86**, 113.

34 Harashima, N., Kurihara, K., Utsunomiya, A., Tanosaki, R., Hanabuchi, S., Masuda, M., Ohashi, T., Fukui, F., Hasegawa, A., Masuda, T., Takaue, Y., Okamura, J. and Kannagi, M. (2004) Graft-versus-Tax response in adult T-cell leukemia patients after hematopoietic stem cell transplantation. *Cancer Research*, **64**, 391.

35 Coulie, P.G., Brichard, V., Van Pel, A., Wolfel, T., Schneider, J., Traversari, C., Mattei, S., De Plaen, E., Lurquin, C., Szikora, J.P., Renauld, J.C. and Boon, T. (1994) A new gene coding for a differentiation antigen recognized by autologous cytolytic T lymphocytes on HLA-A2 melanomas. *The Journal of Experimental Medicine*, **180**, 35.

36 Kawakami, Y., Eliyahu, S., Sakaguchi, K., Robbins, P.F., Rivoltini, L., Yannelli, J.R., Appella, E. and Rosenberg, S.A. (1994) Identification of the immunodominant peptides of the MART-1 human melanoma antigen recognized by the majority of HLA-A2-restricted tumor infiltrating lymphocytes. *The Journal of Experimental Medicine*, **180**, 347.

37 Dudley, M.E., Wunderlich, J.R., Robbins, P.F., Yang, J.C., Hwu, P., Schwartzentruber, D.J., Topalian, S.L., Sherry, R., Restifo, N.P., Hubicki, A.M., Robinson, M.R., Raffeld, M., Duray, P., Seipp, C.A., Rogers-Freezer, L., Morton, K.E., Mavroukakis, S.A., White, D.E. and Rosenberg, S.A. (2002) Cancer regression and autoimmunity in patients after clonal repopulation with antitumor lymphocytes. *Science*, **298** 850.

38 Dudley, M.E., Wunderlich, J.R., Yang, J.C., Sherry, R.M., Topalian, S.L., Restifo, N.P., Royal, R.E., Kammula, U., White, D.E., Mavroukakis, S.A., Rogers, L.J., Gracia, G.J., Jones, S.A., Mangiameli, D.P., Pelletier, M.M., Gea-Banacloche, J., Robinson, M.R., Berman, D.M., Filie, A.C., Abati, A. and Rosenberg, S.A. (2005) Adoptive cell transfer therapy following non-myeloablative but lymphodepleting chemotherapy for the treatment of patients with refractory metastatic melanoma. *Journal of Clinical Oncology*, **23**, 2346.

39 Goedegebuure, P.S., Douville, L.M., Li, H., Richmond, G.C., Schoof, D.D., Scavone, M. and Eberlein, T.J. (1995) Adoptive immunotherapy with tumor-infiltrating lymphocytes and interleukin-2 in patients with metastatic malignant melanoma and renal cell carcinoma: a pilot study. *Journal of Clinical Oncology*, **13**, 1939.

40 Ochsenbein, A.F., Klenerman, P., Karrer, U., Ludewig, B., Pericin, M., Hengartner, H. and Zinkernagel, R.M. (1999) Immune surveillance against a solid tumor fails because of immunological ignorance. *Proceedings of the National Academy of Sciences of the United States of America*, **96**, 2233.

41 Speiser, D.E., Miranda, R., Zakarian, A., Bachmann, M.F., McKall-Faienza, K., Odermatt, B., Hanahan, D., Zinkernagel, R.M. and Ohashi, P.S. (1997) Self antigens expressed by solid tumors Do not efficiently stimulate naive or activated T cells: implications for immunotherapy. *The Journal of Experimental Medicine*, **186**, 645.

42 Wick, M., Dubey, P., Koeppen, H., Siegel, C.T., Fields, P.E., Chen, L., Bluestone, J.A. and Schreiber, H. (1997) Antigenic cancer cells grow progressively in immune hosts without evidence for T cell exhaustion or systemic anergy. *The Journal of Experimental Medicine*, **186**, 229.

43 Curiel, T.J., Coukos, G., Zou, L., Alvarez, X., Cheng, P., Mottram, P., Evdemon-Hogan, M., Conejo-Garcia, J.R., Zhang, L., Burow, M., Zhu, Y., Wei, S., Kryczek, I., Daniel, B., Gordon, A., Myers, L., Lackner, A., Disis, M.L., Knutson, K.L., Chen, L. and Zou, W. (2004) Specific recruitment of regulatory T cells in ovarian carcinoma fosters immune privilege and predicts reduced survival. *Nature Medicine*, **10** 942.

44 Dong, H., Strome, S.E., Salomao, D.R., Tamura, H., Hirano, F., Flies, D.B., Roche, P.C., Lu, J., Zhu, G., Tamada, K., Lennon, V.A., Celis, E. and Chen, L. (2002) Tumor-associated B7-H1 promotes T-cell apoptosis: a potential mechanism of immune evasion. *Nature Medicine*, **8**, 793.

45 Hahne, M., Rimoldi, D., Schroter, M., Romero, P., Schreier, M., French, L.E., Schneider, P., Bornand, T., Fontana, A., Lienard, D., Cerottini, J. and Tschopp, J. (1996) Melanoma cell expression of Fas(Apo-1/CD95) ligand: implications for tumor immune escape. *Science*, **274**, 1363.

46 Seliger, B., Hohne, A., Knuth, A., Bernhard, H., Meyer, T., Tampe, R., Momburg, F. and Huber, C. (1996) Analysis of the major histocompatibility complex class I antigen presentation machinery in normal and malignant renal cells: evidence for deficiencies associated with transformation and progression. *Cancer Research*, **56**, 1756.

47 Suzuki, T., Tahara, H., Narula, S., Moore, K.W., Robbins, P.D. and Lotze, M.T. (1995) Viral interleukin 10 (IL-10), the human herpes virus 4 cellular IL-10 homologue, induces local anergy to allogeneic and syngeneic tumors. *The Journal of Experimental Medicine*, **182**, 477.

48 Thomas, D.A. and Massague, J. (2005) TGF-beta directly targets cytotoxic T cell functions during tumor evasion of immune surveillance. *Cancer Cell*, **8**, 369.

49 Savage, P.A., Vosseller, K., Kang, C., Larimore, K., Riedel, E., Wojnoonski, K., Jungbluth, A.A. and Allison, J.P. (2008) Recognition of a ubiquitous self antigen by prostate cancer-infiltrating CD8+ T lymphocytes. *Science*, **319**, 215.

50 Yee, C., Thompson, J.A., Byrd, D., Riddell, S.R., Roche, P., Celis, E. and Greenberg, P.D. (2002) Adoptive T cell therapy using antigen-specific CD8+ T cell clones for the treatment of patients with metastatic melanoma: in vivo persistence, migration, and antitumor effect of transferred T cells. *Proceedings of the National Academy of Sciences of the United States of America*, **99**, 16168.

51 Kolb, H.J., Schmid, C., Barrett, A.J. and Schendel, D.J. (2004) Graft-versus-leukemia reactions in allogeneic chimeras. *Blood*, **103**, 767.

52 Lundqvist, A. and Childs, R. (2005) Allogeneic hematopoietic cell transplantation as immunotherapy for solid tumors: current status and future directions. *Journal of Immunotherapy (1997)*, **28**, 281.

53 Blaise, D.P., Michel Boiron, J., Faucher, C., Mohty, M., Bay, J.O., Bardoux, V.J., Perreau, V., Coso, D., Pigneux, A. and Vey, N. (2005) Reduced intensity conditioning prior to allogeneic stem cell transplantation for patients with acute myeloblastic leukemia as a first-line treatment. *Cancer*, **104**, 1931.

54 Goulmy, E., Gratama, J.W., Blokland, E., Zwaan, F.E. and van Rood, J.J. (1983) A minor transplantation antigen detected by MHC-restricted cytotoxic T lymphocytes during graft-versus-host disease. *Nature*, **302**, 159.

55 Marijt, W.A., Heemskerk, M.H., Kloosterboer, F.M., Goulmy, E., Kester, M.G., van der Hoorn, M.A., van Luxemburg-Heys, S.A., Hoogeboom, M., Mutis, T., Drijfhout, J.W., van Rood, J.J., Willemze, R. and Falkenburg, J.H. (2003) Hematopoiesis-restricted minor histocompatibility antigens HA-1- or HA-2-specific T cells can induce complete remissions of relapsed leukemia. *Proceedings of the National Academy of*

Sciences of the United States of America, **100**, 2742.

56 Sadovnikova, E. and Stauss, H.J. (1996) Peptide-specific cytotoxic T lymphocytes restricted by nonself major histocompatibility complex class I molecules: reagents for tumor immunotherapy. *Proceedings of the National Academy of Sciences of the United States of America*, **93**, 13114.

57 Stanislawski, T., Voss, R.H., Lotz, C., Sadovnikova, E., Willemsen, R.A., Kuball, J., Ruppert, T., Bolhuis, R.L., Melief, C.J., Huber, C., Stauss, H.J. and Theobald, M. (2001) Circumventing tolerance to a human MDM2-derived tumor antigen by TCR gene transfer. *Nature Immunology*, **2**, 962.

58 Schuster, I.G., Busch, D.H., Eppinger, E., Kremmer, E., Milosevic, S., Hennard, C., Kuttler, C., Ellwart, J.W., Frankenberger, B., Nossner, E., Salat, C., Bogner, C., Borkhardt, A., Kolb, H.J. and Krackhardt, A.M. (2007) Allorestricted T cells with specificity for the FMNL1-derived peptide PP2 have potent antitumor activity against hematologic and other malignancies. *Blood*, **110**, 2931.

59 Engelhard, V.H., Bullock, T.N., Colella, T.A., Sheasley, S.L. and Mullins, D.W. (2002) Antigens derived from melanocyte differentiation proteins: self-tolerance, autoimmunity, and use for cancer immunotherapy. *Immunological Reviews*, **188**, 136.

60 Holler, P.D., Chlewicki, L.K. and Kranz, D.M. (2003) TCRs with high affinity for foreign pMHC show self-reactivity. *Nature Immunology*, **4**, 55.

61 Dembic, Z., Haas, W., Weiss, S., McCubrey, J., Kiefer, H., von Boehmer, H. and Steinmetz, M. (1986) Transfer of specificity by murine alpha and beta T-cell receptor genes. *Nature*, **320**, 232.

62 Clay, T.M., Custer, M.C., Sachs, J., Hwu, P., Rosenberg, S.A. and Nishimura, M.I. (1999) Efficient transfer of a tumor antigen-reactive TCR to human peripheral blood lymphocytes confers anti-tumor reactivity. *Journal of Immunology (Baltimore, Md: 1950)*, **163**, 507.

63 Cohen, C.J., Zheng, Z., Bray, R., Zhao, Y., Sherman, L.A., Rosenberg, S.A. and Morgan, R.A. (2005) Recognition of fresh human tumor by human peripheral blood lymphocytes transduced with a bicistronic retroviral vector encoding a murine anti-p53 TCR. *Journal of Immunology (Baltimore, Md: 1950)*, **175**, 5799.

64 Heemskerk, M.H., Hoogeboom, M., de Paus, R.A., Kester, M.G., van der Hoorn, M.A., Goulmy, E., Willemze, R. and Falkenburg, J.H. (2003) Redirection of antileukemic reactivity of peripheral T lymphocytes using gene transfer of minor histocompatibility antigen HA-2-specific T-cell receptor complexes expressing a conserved alpha joining region. *Blood*, **102**, 3530.

65 Morgan, R.A., Dudley, M.E., Yu, Y.Y., Zheng, Z., Robbins, P.F., Theoret, M.R., Wunderlich, J.R., Hughes, M.S., Restifo, N.P. and Rosenberg, S.A. (2003) High efficiency TCR gene transfer into primary human lymphocytes affords avid recognition of melanoma tumor antigen glycoprotein 100 and does not alter the recognition of autologous melanoma antigens. *Journal of Immunology (Baltimore, Md: 1950)*, **171**, 3287.

66 Schaft, N., Willemsen, R.A., de Vries, J., Lankiewicz, B., Essers, B.W., Gratama, J.W., Figdor, C.G., Bolhuis, R.L., Debets, R. and Adema, G.J. (2003) Peptide fine specificity of anti-glycoprotein 100 CTL is preserved following transfer of engineered TCR alpha beta genes into primary human T lymphocytes. *Journal of Immunology (Baltimore, Md: 1950)*, **170**, 2186.

67 Xue, S.A., Gao, L., Hart, D., Gillmore, R., Qasim, W., Thrasher, A., Apperley, J., Engels, B., Uckert, W., Morris, E. and Stauss, H. (2005) Elimination of human

leukemia cells in NOD/SCID mice by WT1-TCR gene-transduced human T cells. *Blood*, **106**, 3062.

68 Zhao, Y., Zheng, Z., Robbins, P.F., Khong, H.T., Rosenberg, S.A. and Morgan, R.A. (2005) Primary human lymphocytes transduced with NY-ESO-1 antigen-specific TCR genes recognize and kill diverse human tumor cell lines. *Journal of Immunology (Baltimore, Md: 1950)*, **174**, 4415.

69 Morgan, R.A., Dudley, M.E., Wunderlich, J.R., Hughes, M.S., Yang, J.C., Sherry, R.M., Royal, R.E., Topalian, S.L., Kammula, U.S., Restifo, N.P., Zheng, Z., Nahvi, A., de Vries, C.R., Rogers-Freezer, L.J., Mavroukakis, S.A. and Rosenberg, S.A. (2006) Cancer regression in patients after transfer of genetically engineered lymphocytes. *Science*, **314**, 126.

70 Rosenberg, S.A. and Dudley, M.E. (2004) Cancer regression in patients with metastatic melanoma after the transfer of autologous antitumor lymphocytes. *Proceedings of the National Academy of Sciences of the United States of America*, **101** (Suppl. 2), 14639.

71 Heemskerk, M.H., Hagedoorn, R.S., van der Hoorn, M.A., van der Veken, L.T., Hoogeboom, M., Kester, M.G., Willemze, R. and Falkenburg, J.H. (2007) Efficiency of T-cell receptor expression in dual-specific T cells is controlled by the intrinsic qualities of the TCR chains within the TCR-CD3 complex. *Blood*, **109**, 235.

72 Sommermeyer, D., Neudorfer, J., Weinhold, M., Leisegang, M., Engels, B., Noessner, E., Heemskerk, M.H., Charo, J., Schendel, D.J., Blankenstein, T., Bernhard, H. and Uckert, W. (2006) Designer T cells by T cell receptor replacement. *European Journal of Immunology*, **36**, 3052.

73 Jorritsma, A., Gomez-Eerland, R., Dokter, M., van de Kasteele, W., Zoet, Y.M., Doxiadis, I.I., Rufer, N., Romero, P., Morgan, R.A., Schumacher, T.N. and Haanen, J.B. (2007) Selecting highly affine and well-expressed TCRs for gene therapy of melanoma. *Blood*, **110**, 3564.

74 Scholten, K.B., Kramer, D., Kueter, E.W., Graf, M., Schoedl, T., Meijer, C.J., Schreurs, M.W. and Hooijberg, E. (2006) Codon modification of T cell receptors allows enhanced functional expression in transgenic human T cells. *Clinical Immunology (Orlando, Fla)*, **119**, 135.

75 Cohen, C.J., Li, Y.F., El-Gamil, M., Robbins, P.F., Rosenberg, S.A. and Morgan, R.A. (2007) Enhanced antitumor activity of T cells engineered to express T-cell receptors with a second disulfide bond. *Cancer Research*, **67**, 3898.

76 Kuball, J., Dossett, M.L., Wolfl, M., Ho, W.Y., Voss, R.H., Fowler, C. and Greenberg, P.D. (2007) Facilitating matched pairing and expression of TCR chains introduced into human T cells. *Blood*, **109**, 2331.

77 Voss, R.H., Willemsen, R.A., Kuball, J., Grabowski, M., Engel, R., Intan, R.S., Guillaume, P., Romero, P., Huber, C. and Theobald, M. (2008) Molecular design of the Calpha interface favors specific pairing of introduced TCRalpha in human T cells. *Journal of Immunology (Baltimore, Md: 1950)*, **180**, 391.

78 Cohen, C.J., Zhao, Y., Zheng, Z., Rosenberg, S.A. and Morgan, R.A. (2006) Enhanced antitumor activity of murine-human hybrid T-cell receptor (TCR) in human lymphocytes is associated with improved pairing and TCR/CD3 stability. *Cancer Research*, **66**, 8878.

79 Thomas, S., Xue, S.A., Cesco-Gaspere, M., San Jose, E., Hart, D.P., Wong, V., Debets, R., Alarcon, B., Morris, E. and Stauss, H.J. (2007) Targeting the Wilms tumor antigen 1 by TCR gene transfer: TCR variants improve tetramer binding but not the function of gene modified human T cells. *Journal of Immunology (Baltimore, Md: 1950)*, **179**, 5803.

80 Li, Y., Moysey, R., Molloy, P.E., Vuidepot, A.L., Mahon, T., Baston, E., Dunn, S., Liddy, N., Jacob, J., Jakobsen, B.K. and Boulter, J.M. (2005) Directed evolution of

human T-cell receptors with picomolar affinities by phage display. *Nature Biotechnology*, **23**, 349.

81 Zhao, Y., Bennett, A.D., Zheng, Z., Wang, Q.J., Robbins, P.F., Yu, L.Y., Li, Y., Molloy, P.E., Dunn, S.M., Jakobsen, B.K., Rosenberg, S.A. and Morgan, R.A. (2007) High-affinity TCRs generated by phage display provide CD4+ T cells with the ability to recognize and kill tumor cell lines. *Journal of Immunology (Baltimore, Md: 1950)*, **179**, 5845.

82 Gattinoni, L., Klebanoff, C.A., Palmer, D.C., Wrzesinski, C., Kerstann, K., Yu, Z., Finkelstein, S.E., Theoret, M.R., Rosenberg, S.A. and Restifo, N.P. (2005) Acquisition of full effector function *in vitro* paradoxically impairs the *in vivo* antitumor efficacy of adoptively transferred CD8+ T cells. *The Journal of Clinical Investigation*, **115**, 1616.

83 Serrano, L.M., Pfeiffer, T., Olivares, S., Numbenjapon, T., Bennitt, J., Kim, D., Smith, D., McNamara, G., Al-Kadhimi, Z., Rosenthal, J., Forman, S.J., Jensen, M.C. and Cooper, L.J. (2006) Differentiation of naive cord-blood T cells into CD19-specific cytolytic effectors for posttransplantation adoptive immunotherapy. *Blood*, **107**, 2643.

84 van der Veken, L.T., Hagedoorn, R.S., van Loenen, M.M., Willemze, R., Falkenburg, J.H. and Heemskerk, M.H. (2006) Alphabeta T-cell receptor engineered gammadelta T cells mediate effective antileukemic reactivity. *Cancer Research*, **66**, 3331.

85 van Lent, A.U., Nagasawa, M., van Loenen, M.M., Schotte, R., Schumacher, T.N., Heemskerk, M.H., Spits, H. and Legrand, N. (2007) Functional human antigen-specific T cells produced *in vitro* using retroviral T cell receptor transfer into hematopoietic progenitors. *Journal of Immunology (Baltimore, Md: 1950)*, **179**, 4959.

86 Zhao, Y., Parkhurst, M.R., Zheng, Z., Cohen, C.J., Riley, J.P., Gattinoni, L., Restifo, N.P., Rosenberg, S.A. and Morgan, R.A. (2007) Extrathymic generation of tumor-specific T cells from genetically engineered human hematopoietic stem cells via Notch signaling. *Cancer Research*, **67**, 2425.

87 Maher, J., Brentjens, R.J., Gunset, G., Riviere, I. and Sadelain, M. (2002) Human T-lymphocyte cytotoxicity and proliferation directed by a single chimeric TCRzeta/CD28 receptor. *Nature Biotechnology*, **20**, 70.

88 Brentjens, R.J., Latouche, J.B., Santos, E., Marti, F., Gong, M.C., Lyddane, C., King, P.D., Larson, S., Weiss, M., Riviere, I. and Sadelain, M. (2003) Eradication of systemic B-cell tumors by genetically targeted human T lymphocytes co-stimulated by CD80 and interleukin-15. *Nature Medicine*, **9**, 279.

89 Wang, H.Y., Lee, D.A., Peng, G., Guo, Z., Li, Y., Kiniwa, Y., Shevach, E.M. and Wang, R.F. (2004) Tumor-specific human CD4+ regulatory T cells and their ligands: implications for immunotherapy. *Immunity*, **20**, 107.

90 Hombach, A.A., Schildgen, V., Heuser, C., Finnern, R., Gilham, D.E. and Abken, H. (2007) T cell activation by antibody-like immunoreceptors: the position of the binding epitope within the target molecule determines the efficiency of activation of redirected T cells. *Journal of Immunology (Baltimore, Md: 1950)*, **178**, 4650.

91 Wang, J., Jensen, M., Lin, Y., Sui, X., Chen, E., Lindgren, C.G., Till, B., Raubitschek, A., Forman, S.J., Qian, X., James, S., Greenberg, P., Riddell, S. and Press, O.W. (2007) Optimizing adoptive polyclonal T cell immunotherapy of lymphomas, using a chimeric T cell receptor possessing CD28 and CD137 costimulatory domains. *Human Gene Therapy*, **18**, 712.

92 Engels, B., Cam, H., Schuler, T., Indraccolo, S., Gladow, M., Baum, C., Blankenstein, T. and Uckert, W. (2003) Retroviral vectors for high-level transgene

expression in T lymphocytes. *Human Gene Therapy*, **14**, 1155.

93 Sasaki, T., Ikeda, H., Sato, M., Ohkuri, T., Abe, H., Kuroki, M., Onodera, M., Miyamoto, M., Kondo, S. and Nishimura, T. (2006) Antitumor activity of chimeric immunoreceptor gene-modified Tc1 and Th1 cells against autologous carcinoembryonic antigen-expressing colon cancer cells. *Cancer Sci*, **97**, 920.

94 Hacein-Bey-Abina, S., Von Kalle, C., Schmidt, M., McCormack, M.P., Wulffraat, N., Leboulch, P., Lim, A., Osborne, C.S., Pawliuk, R., Morillon, E., Sorensen, R., Forster, A., Fraser, P., Cohen, J.I., de Saint Basile, G., Alexander, I., Wintergerst, U., Frebourg, T., Aurias, A., Stoppa-Lyonnet, D., Romana, S., Radford-Weiss, I., Gross, F., Valensi, F., Delabesse, E., Macintyre, E., Sigaux, F., Soulier, J., Leiva, L.E., Wissler, M., Prinz, C., Rabbitts, T.H., Le Deist, F., Fischer, A. and Cavazzana-Calvo, M. (2003) LMO2-associated clonal T cell proliferation in two patients after gene therapy for SCID-X1. *Science*, **302**, 415.

95 Cattoglio, C., Facchini, G., Sartori, D., Antonelli, A., Miccio, A., Cassani, B., Schmidt, M., von Kalle, C., Howe, S., Thrasher, A.J., Aiuti, A., Ferrari, G., Recchia, A. and Mavilio, F. (2007) Hot spots of retroviral integration in human CD34+ hematopoietic cells. *Blood*, **110**, 1770.

96 Gaspar, H.B., Parsley, K.L., Howe, S., King, D., Gilmour, K.C., Sinclair, J., Brouns, G., Schmidt, M., Von Kalle, C., Barington, T., Jakobsen, M.A., Christensen, H.O., Al Ghonaium, A., White, H.N., Smith, J.L., Levinsky, R.J., Ali, R.R., Kinnon, C. and Thrasher, A.J. (2004) Gene therapy of X-linked severe combined immunodeficiency by use of a pseudotyped gammaretroviral vector. *Lancet*, **364** 2181.

97 Recchia, A., Bonini, C., Magnani, Z., Urbinati, F., Sartori, D., Muraro, S., Tagliafico, E., Bondanza, A., Stanghellini, M.T., Bernardi, M., Pescarollo, A., Ciceri, F., Bordignon, C. and Mavilio, F. (2006) Retroviral vector integration deregulates gene expression but has no consequence on the biology and function of transplanted T cells. *Proceedings of the National Academy of Sciences of the United States of America*, **103**, 1457.

98 Bonini, C., Ciceri, F., Marktel, S. and Bordignon, C. (1998) Suicide-gene-transduced T-cells for the regulation of the graft-versus-leukemia effect. *Vox Sanguinis*, **74** (Suppl. 2), 341.

99 Kieback, E., Charo, J., Sommermeyer, D., Blankenstein, T. and Uckert, W. (2008) A safeguard eliminates T cell receptor gene-modified autoreactive T cells after adoptive transfer. *Proceedings of the National Academy of Sciences of the United States of America*, **105**, 623.

100 Quintarelli, C., Vera, J.F., Savoldo, B., Giordano Attianese, G.M., Pule, M., Foster, A.E., Heslop, H.E., Rooney, C.M., Brenner, M.K. and Dotti, G. (2007) Co-expression of cytokine and suicide genes to enhance the activity and safety of tumor-specific cytotoxic T lymphocytes. *Blood*, **110**, 2793.

Index

a

accessory proteins 195
– role 195
acute myeloid leukemia 150
adaptive immune system 29, 64
adoptive T-cell therapy 10, 221
adult cancer patients 115
affinity immune receptors 209
AFP peptide epitopes 24
AFP-specific CTL 24
agonistic antibodies 127
allogeneic hematopietic cell transplant 27
allografted organs 237
AMIDA technology 68, 71
– advantages 71
– disadvantages 71
amino acid
– composition 303
– sequences 263, 303
amniotic fluid 268
amplified genes 52
antibody dependent cellular cytotoxicity 197
– activity 197
anti apoptotic molecule 54
antibody-antigen complexes 69
antibody/interleukin 2 (IL-2) fusion proteins 209
antibody/pseudomonas exotoxin fusion protein 192
antibody therapies 24, 180, 195
anticancer peptide vaccine 303
antigen 3, 4, 5, 6, 7, 8
– epithelial cells 205
– repertoire 6, 8
– specific vaccination 10
– T cells 7, 9, 11, 24, 31
– tumor 53
antigen epitopes 4, 11
– generation 11
antigen presenting cells (APC) 64
antigen processing presentation machinery (APM) 67, 225
antigenic determinants 67
antigenic peptides 4
antigenic protein 48, 57
anti-idiotypic antibodies 143, 210
antisense oligonucleotides 127
anti-tumor immune response, *see* malignant cells
antitumor immunity 210
– immune responses 211, 222
APM-related components 67
apoptosis gene family 121
– inhibitor 121
arginine residues 263
autoaggressive disorders 64
autoimmune toxicity 7
autologous tumor cells 9
avian poxvirus vectors 211

b

B cell lymphoma 121
β-microglobulin subunit 4
bacterial artificial chromosome (BAC) 207
bacterial expression system 55
bacterial pathogens 204
BCR/ABL protein 63, 119
BCR/ABL-specific peptide vaccines 119
bio conductor array 89
bioinformatics systems 89
bioinformatics tools 89
biological contaminants 303
biomarkers 63, 103
BiTE antibodies 197
breast cancer cells 181
Burkitt's lymphoma cells 231, 232, 240

Tumor-Associated Antigens. Edited by Olivier Gires and Barbara Seliger
Copyright © 2009 WILEY-VCH Verlag GmbH & Co. KGaA, Weinheim
ISBN: 978-3-527-32084-4

c

calcium-independent homotypic cell adhesion 180
cancer-bearing host 80
cancer-derived CSC 184
cancer immunotherapies 219
cancer-related autoantigens 146
cancer-specific serum antibodies 64
cancer stem cells (CSC) 184
cancer testis antigens (CTA) 6, 26, 51, 121, 127, 143, 161, 162, 163, 164, 166, 167, 170, 172
– based immunotherapy 148
– based vaccination 148, 149
– definitions 162
– directed vaccines 151
– expression 148, 161
– GAGE 163
– MAGE 163
– positive lesions 148, 149
– promoters 149
– specific TCR repertoire 7
– SSX 163
cancer testis antigens families 124
– BAGE 124
– GAGE 124
– MAGE 124
– SAGE 124
– XAGE 124
cancer vaccines 23
– development 23
carbonic anhydrase expression 92
carboxy-terminal 235
– domain 265
carcino embryonic antigen 26, 203
– based immunotherapy 205
– based vaccines 26
– expression 203, 204
– family 201, 204, 207
– immune therapies 207
– measurements 206
– peptide 205
– poxvirus vaccines 211
– protein vaccines 207
– serum levels 206
– specific cytotoxic lymphocytes 210
– targeted vaccination 204
– transgenic mice 207, 209, 210
– transgenic transplanted tumor models 210
– vaccine-induced effector cells 212
carcinogen-induced tumor cells 30
carcinoma cells 65
CCAAT enhancer binding protein 52
$CD4^+$ T cells 4
– analysis 53
– epitope 5
$CD8^+$ T cells 25
– analysis 53
– epitope 4
cDNA amplification 83
cDNA expression libraries 9, 48, 49, 66
cDNA microarrays 82
– analysis 82
– data 88
– utilization 82
cell lung carcinomas 203
cell lysates 23
cell lysis buffer 83
cell-pulsed dendritic cell vaccines 125
cell signal transduction 181
cell surface mucins 262
cellular/humoral immune response 144, 171
cellular immune responses 126, 209
cellular proteins, see antigen
cellular signaling components 235
central nervous system 29
cervical cancer 249
chemo/antibody combination 187
chemotherapy/cytokine therapy 304
childhood malignancy 115
chromatographic surface 105
chromosomal translocation 63
chronic myelogenous leukemia (CML) 63, 117
circulating tumor cells (CTC) 184
clonal theory 63
– techniques 47
co-stimulatory molecules 211
colorectal cancer tissues 184
complement-dependent cytotoxicity (CDC) 197, 183
– activity 190
Coomassie blue/silver nitrate 70
coxsackie/adenovirus receptor 209
colorectal cancer, 186, 192
– bevacizumab 192
– cetuximab 192
– panitumumab 192
CRC tumors 184
cryptic promoters 6
CTL-mediated lysis 120
CTL precursor frequencies 309
cysteine residues 265
cytokine treatments 211
cytoplasmic proteins 5
cytotoxic T lymphocytes (CTL) 143
cytotoxic Tcells 47

d

DC-based vaccines 32
DC growth factor 170
delayed type hypersensitivity (DTH) 292
dendritic cell 3, 9, 167
dependent gene expression 68
diagnostic imaging 125
diagnostic markers 47
disease-specific multimodal treatment protocols 115
DNA-binding proteins 83, 233
– RBP-J$_k$ 233
– utilization 83
DNA cancer vaccines 287
DNA-dependent RNA polymerase 85, 88
DNA hypomethylating drug 144
DNA methylation 144, 148, 152, 153, 162
– analysis 147
DNA methyltransferase enzymes 144
DNA microarray gene expression data 89
DNA plasmid vectors 223
DNA sequencing 48
DNA vaccines 287
dose-dependent immunosuppressants 304
double-stranded DNA viruses 249
drug-induced demethylation 152

e

EFT, see Ewing family tumors
EGF/PDGF-receptors 251
EGFR, see epidermal growth factor receptor (EGFR)
endocytosed exogenous antigens 5
endoplasmic reticulum 5
EpCAM 190, 194
– antibody 193
– epitope 189
– expression 181, 186, 191, 196
– immunotherapies 183, 192, 194, 196
– positive tumor cells 188
– promoter 193
– protein 189, 194
– protein complexes 195
– specific antibodies 192
– specific therapies 193, 197
– staining 183
epidermal growth factor receptor (EGFR) 30, 128, 220
epigenetic drugs 149, 151
epigenetic mechanisms, see DNA methylation
epigenetic regulators 144
epithelial cancer cells 147, 188, 253, 267
epithelial cell surface 26
epithelial ovarian cancers 261
epithelial tissues 193
Epstein Barr Virus (EBV) 6, 231
– associated tumors 122
– infected cells 231, 237, 239
– nuclear antigen 233
– peptides 240
– proteins 240
– sero-negative recipients 239
EBV-associated post-transplantation lymphoproliferative disease 122, 233
ER-localized peptidases 11
Ewing family tumors (EFT) 126
– pathogenesis 126
EWS gene 126
exogenous proteins 5
extracellular adhesive protein domain 205
extracellular domain 202, 221
ezrin/radixin/moesin (ERM) 267

f

Fc-binding motif 208
fetal acetylcholine receptor (fAChR) 124
fluorescence-activated cell sorting (FACS) 56
fluorescence-labeled cDNA 88
fluorogenic substrates 268
Food and Drug Administration (FDA) 152, 221, 225, 284
fusion genes 119
fusion proteins 6, 146

g

GAGE families 121
gametogenic cells 163
gastrointestinal tract 192
gene expression 81
– data 89
– mutations 6
– ontology 89
– quantification 81
gene profiling arrays 80
gene-specific hypomethylation 164
genomic DNA 233, 263
Glioblastoma multiforme 29
glial fibrillary acidic protein (GFAP) 51
glioma-associated antigens (GAAs) 30
glioma cell lines 30
good manufacturing practice (GMP) 287
GPI-linked membrane proteins 262
graft-versus-host disease (GVHD) 118
graft-versus-leukemia (GVL) 27, 117
growth factor receptors 124
– EGF-R 124

– HER2 124
GVAX vaccines 290
gynecologic cancer intergroup (GCIG) 273

h

HCC-associated antigens 68
HDAC enzymes 145
HDAC inhibitor 147
head and neck squamous cell carcinomas (HNSCC) 106
heat shock proteins 30, 31
heavy chain alleles 4
hemagglutinin protein chimera 208
Hematopoietic malignancies 27, 152
hematopoietic stem cell transplantation (HSCT) 116, 118, 121
heparin-binding growth factor 126
hepatocellular carcinoma (HCC) 67, 163
– autoantibodies 67
HER receptors 220
HER signal cascade 220
HER2 kinase domain 224
HER2-positive tumors 225
heterodimerization signal transduction 225
heterogeneous expression 148
heterotypic interaction 205
hierarchical clustering 89
high-dose chemotherapy 126
high affinity binders 9
high throughput technology 92
histone acetyltransferases (HAT) 145
histone methyltransferase (HMT) 145
Hodgkin lymphoma 240
hormonal therapies 294
hormone-resistant prostate cancer (HRPC) 292
house keeping genes 80
HSP-based vaccines 31
human cancer immunome 54
human chorionic gonadotropin (HCG) 103
human epidermal growth factor receptor 179, 219
human glioma antigens 29
human immune effector cells 185
human immunoglobulin repertoires 190
human leukocyte antigen (HLA) 26, 150, 231
– allele 29
– antigens 118
– class I antigen processing 25
– class I-positive tumors 125
– loading machinery 236
– negative RMS cell lines 124
– target cells 9

human melanoma antigens 28–29
human papilloma virus (HPV) 6, 249
– antigens 254, 255
– basal epithelial cells 249
– encoded proteins 253
– gene expression 251
– induced cancer 249
– infection 249, 250, 255, 256
– malignancies 256
human tumor antigen 5, 49, 54, 220
– identification 49
humoral/cellular immune response 26, 47, 64, 148
– spontaneous 148
hypopigmentary skin disorder 7

i

immature dendritic cells 290
immune cells 194, 195
immune deviation 33
immune effector cells 196
– effector mechanisms 23
immune electron microscopy 203
immune inhibitory factors 10
– IL-10 10
– PGE-2 10
– TGF-b 10
immune-modulators 80
– interleukin-2 80
immune molecules 144
– HLA antigens 144
immune response 255, 284, 286, 287
immune system 3, 23, 24, 55, 167, 204, 231, 238, 283
immunohistochemistry 148, 161, 266
immunopreventive vaccines 107
immunoproteasome 4
immunoreactive spots 67
immunoreceptor-tyrosine-based-activation-motif (ITAM) 235
– motifs 203
immunosuppressive regulatory T cells 255, 256
immunotherapeutic approaches 47, 143
Incomplete Freund's Adjuvant (IFA) 29
inflammatory cells 33
innate/adaptive immune cells 3
insulin-like growth factor (IGF) 127
interferon stimulated genes (ISGs) 92
intestinal intraepithelial lymphocytes 206
intracellular signal cascade 146
intraepithelial lesions 249
intraspinal tumors 115
ionizing radiation 115

l

laser capture micro-dissection (LCM)   82
lectin-like molecules   267
leukemia cells   118
leukocytes   69
lineage-associated expression   51
linear amplification   85
LN-associated chemokines   32
long control region (LCR)   250, 251
low-risk tumors   127
Ludwig Cancer Research Institutes   54
lung carcinoma   68
lymph nodes   3
lymphoblastoid cell lines   235
lymphoid system   253
lysate-pulsed dendritic cells   127

m

macrophage migration inhibiting factor   91
major histocompatibility complex (MHC)   3, 9, 27, 56, 57, 72, 211
– antigen processing   67
– class I molecules   3
– class II molecules   4
– ligands   9, 11, 212
– multimer binding technology   211
– peptides   4, 5
male germline cells   26
malignant bone tumor   125
– identification   126
– osteosarcoma   125
malignant cells   10
malignant tissues   146
– CEBPg   146
mammary epithelial cells   220
MDA-specific T cells   7
Meig's syndrome   270
melanocytic lineage   28
– antigen-processing machinery   28
– gp100/pmel17   28
– tyrosinase   28
melanoma-associated antigens (MAA)   28
– gene transcription   28
– peptide epitopes   29
melanoma antigen (MAGE)   161
– antigens   7
– expression   165
– family   126
– homology-domain   166
– proteins   166
melanoma cells   28, 150
melanoma differentiation antigens   7, 79
melanoma metastasis   8
melanoma mouse models   28
metachronous hepatic metastases   183
metastasizing cells   195
metastatic disease   126, 169
microarray-based technologies   79
– introduction   79
– technical aspects   80
microbial ligands   202
microbial receptors   202
microdissected specimen   105
– lysates   105
microneedle-based injection systems   287
microphtalmia transcription factor   91
minimal residual disease (MRD)   116, 122
– diagnosis   117
– markers   119
– PRAME   119
– WT1   119
missense mutations   25
mitochondrial protein   52
mitotic abnormalities   63
model systems   225
moloney murine leukemia virus   84
monoclonal antibodies (mAb)   71, 128, 143, 180, 221
– based therapies   179
– Herceptin® (Trastuzumab)   71
– technology   47, 65
mouse fibroblasts   72
mucin genes   147, 264
– diagrammatic representation   264
multiple benign diseases   270
– gynecological   270
– non-gynecological   270
multiple sclerosis   64
multiprotein complexes   145
murine antibody   197
murine dendritic cells   24
murine immune system   191
murine monoclonal antibody   267
MYCN-derived peptides   123
MYCN oncogene   122, 123
myelodysplastic syndrome (MDS)   150
myelogenous leukemia (AML)   117

n

N domain of murine   202
N-glycosylation sites   264, 265
N-linked glycan chains   264
N-terminal domain   263
N-terminal-extended precursors   5
N-terminal tails   145
natural killer (NK) cells   31, 64

– inhibitory receptor 205
naturally-occurring autoantibodies (NAAs) 69, 70
NB-associated antigens 122, 123
– identification 123
Neisseria gonorrhoeae 204
Neisseria meningitides 204
neoplastic cells 143, 146
– phenotypic characterization 143
neoplastic diseases 117
neoplastic/malignant melanocytes 28
neurological diseases 119
neuron-specific enolase (NSE) 103
nipple aspirate fluid (NAF) 106
non-cancer-related autoantigens 52
non-gynecological benign diseases 270
non-Hodgkin lymphomas (NHL) 121
non-lymphoid tissues 6
non-myeloablative chemotherapy 11
non-proliferating cells 256
non-small cell lung (NSCLC) 220
non-transformed autologous cells\leukocytes 69
northern blot analysis 52
nuclear signaling 181

o

O-glycosylated residues 264
O-linked glycoprotein 263
oligo-based microarray systems 81
oncofetal antigen 24
oncogenic transformation 221

p

p53 tumor suppressor gene 25, 65
pancreatic ductal adenocarcinoma 165
papillomavirus binding factor 126
pattern recognition receptors 283
PAX/FKHR fusion products 124
PAX3/FKHR peptides 124
PBMC donors 191
pediatric cancer 116, 117
pediatric malignancies 128
pelvic inflammatory disease 270
peptides 285
– based vaccination 126, 293
– binding motifs 4
– epitopes 8, 255
– fragments 57
– ligands 4
– MHC complexes 11
– reactive T cells 9
– specific immune responses 118
– vaccination 285

peripheral blood mononuclear cells (PBMC) 152, 190
peripheral lymphoid organs 3
peripheral tolerance 209
personalized peptide vaccines (PPVs) 309
plasma membrane-associated complex 72
poly-adenosyl-ribosyl transferase 53
polyepitope vaccines 170
polymerase chain reaction (PCR) 82
– amplification 86, 87
– combination 87
– cycles 87
– limitations 86
polymorphic epitopes 118
polymorphic heavy chain 4
polymorphonuclear leukocytes 33
polyvalent vaccines 57
positron emission tomography (PET) 208
post-menopausal women 269, 270
post-translational modifications 26, 48, 64, 67, 145
– glycosylation 26, 48
– identification 67
– phosphorylation 26
pregnancy-specific glycoprotein (PSG) 201
pre-menopausal women 271
pro-inflammatory mediators 254
– interferons 254
professional antigen presenting cells (pAPCs) 5
– DCs 5
– macrophages 5
prognostic factor 191
prognostic markers 125, 146
programmed cell death 166
proinflammatory cytokines 205
proliferation signal 180
prophylactic vaccines 23, 253
prostate-specific antigen (PSA) 71, 103
protein 9, 30
– derived epitopes 236
– endogenous 5
– lysates 105
– periplasmatic 48
protein–protein interaction 204
ProteinChip system 106
ProteinChip technology 106
PROTEOMEX 66
– experiments 67
– schematic view 66
– technology 67
proteomic analysis 275
proteomics-based methods 64

– PROTEOMEX   64
post-transplantation lymphoproliferative
    disease (PTLD)   237, 240
– specific tumor   238
putative tyrosine phosphorylation site   265
PVDF membranes   66

q
quantitative real-time PCR   90

r
radioisotope-conjugates   208
radio-resistant tumors   295
– melanoma   295
RAS genes   25
– mutations   25
RAS oncogenes   25
recombinant pox virus vectors   211
recombinant proteins   48, 143, 286, 287
– implementation   287
renal cell cancers (RCC)   54, 66, 79
reporter molecules   268
respiratory papillomatosis   255
reverse T Cell Immunology   53
RNA amplification methods   82, 84
RNA amplification technology   82, 85
– modifications   82
– optimizations   82
– validations   82
RNasin® Plus RNase inhibitor   83
RNA degradation   82
RNA polymerase   85
RNA protection reagent   83
– RNAlater™   83
reverse transcription polymerase chain
    reaction (RT-PCR)   180
RMS-specific adoptive T-cell transfer   124

s
SART3-positive tumors   126
SART3-specific T cells   126
screening technology   71
screening tumor-derived cDNA expression
    libraries (SEREX)   64
SEA domain sequences   264
SELDI-MS technology   105
– schematic representation   105
SELDI-ProteinChip technology   106
SEREX   146
– approach   48, 117
– database   54
– method   119, 167
– screening   128
– technology   64, 166

serologically-defined antigens   54
sex hormone levels   106
signaling pathway   181
signaling proteins   8
– phosphorylation   8
significance analysis of microarrays (SAM)
    89
single-chain bispecific antibody   193
small-cell lung cancers   208
small interfering RNA   166
solid organ transplantation   122
squamous cell carcinoma express   180
SSX antigen family   7, 51
SSX gene   51, 148
Streptococcus pyogenes   283
synthetic oligonucleotide probes   80
synthetic peptides   167, 224
– epitopes   10
SYT gene   51

t
T-cell antigens   8, 9, 11
– epitopes   8
– identification   8
T cell-based therapy   10
T-cell epitopes   4, 9, 223
– generation   4–5
T-cell receptor-mediated cytotoxicity   204
– genes   203
T-cell responses   193
T-helper lymphocyte   48
T cell-mediated immunotherapy   123
– mechanisms   119
target labeling   88
– direct   88
– indirect   88
TAS-PCR primer   87
T cell receptors   3
tetramer-positive lymphocyte   292
therapeutic agents   32
therapeutic interventions   206
therapeutic markers   72
therapeutic vaccines   253
– development   253
therapeutic vaccination   31
therapeutic vaccines target antigens   253
time-of-flight mass spectrometry
    (MALDI-TOF)   104
TMA staining   182
TMA technology   182
toll-like receptor agonists   92, 283
transcription factor   250
transcriptional profiling   90
– limitations   90

transgenic murine models   207
transitional cell carcinoma   107
translational repressor   53
transmembrane domain   265
transmembrane proteases   233
tumor-associated antigens (TAA)   5, 6, 10, 23, 25, 30, 31, 47, 48, 49, 53, 63, 64, 79, 91, 103, 116, 143, 146, 171, 172, 179, 206, 249, 251, 283, 303, 304
– based active immunization   79
– based immunotherapy   143
– based vaccination   144
– cancer-testis (CT)   26
– classes   24
– epitopes   8
– human melanoma antigens   28
– identification   10, 47
– MAGE-1   49
– minor histocompatibility antigens   27
– molecular characterization   47, 79
– oncofetal antigens   24
– oncogenes   25
– repertoire   10
– specific vaccines   118
– subclasses   5
– targeted immunotherapies   211
tumor-bearing host   34, 47
tumor cells   6, 23, 167
– cell lines   9
– vaccines   289
tumor immunotherapy   116
tumor-derived antibodies   70, 107
– libraries   47
– purification   70
tumor-feeding vessels   82
tumor-host interactions   90
– usefulness   90
tumor-infiltrating lymphocytes   33, 51, 256
– characterization   256
tumor-localized antibody   208
tumor-reactive T cells   11

tumor-specific antigens   55, 283
tumor markers   206, 213
tumor necrosis factor   126
tumor suppressor gene   128
tumor therapies   207
tumor vaccine   119
tyrosine kinase   120
– domain   223
– inhibitors   120, 221, 222
– peptides   286
– receptor   30
– residues   235

u

ubiquitin carboxyl terminal hydrolase (UCHL)   67

v

vaccine development   56
– perspectives   56
vaccine-induced immunity   33
viral genes   233
– epitopes   233
viral oncoproteins   6
– E6   6
– E7   6
viral vectors   172
– pox viruses   172
virus-derived antigenic epitopes   238
virus-encoded antigen   52
virus-induced cancer   311

w

Wilms' tumor protein 1 (WT1)   118
wnt pathway   181, 185
women menstrual cycle   270
World Health Organization   231
WT xenografts   128

x

X-ray crystallography   202